WITHDRAWN

Wildlife Forensic Investigation

Principles and Practice

John E. Cooper
Margaret E. Cooper

CRC Press
Taylor & Francis Group
Boca Raton   London   New York

CRC Press is an imprint of the
Taylor & Francis Group, an **informa** business

CRC Press
Taylor & Francis Group
6000 Broken Sound Parkway NW, Suite 300
Boca Raton, FL 33487-2742

© 2013 by Taylor & Francis Group, LLC
CRC Press is an imprint of Taylor & Francis Group, an Informa business

No claim to original U.S. Government works

Printed on acid-free paper
Version Date: 20130125

Printed and bound in India by Replika Press Pvt. Ltd.

International Standard Book Number-13: 978-1-4398-1374-4 (Hardback)

This book contains information obtained from authentic and highly regarded sources. Reasonable efforts have been made to publish reliable data and information, but the author and publisher cannot assume responsibility for the validity of all materials or the consequences of their use. The authors and publishers have attempted to trace the copyright holders of all material reproduced in this publication and apologize to copyright holders if permission to publish in this form has not been obtained. If any copyright material has not been acknowledged please write and let us know so we may rectify in any future reprint.

Except as permitted under U.S. Copyright Law, no part of this book may be reprinted, reproduced, transmitted, or utilized in any form by any electronic, mechanical, or other means, now known or hereafter invented, including photocopying, microfilming, and recording, or in any information storage or retrieval system, without written permission from the publishers.

For permission to photocopy or use material electronically from this work, please access www.copyright.com (http://www.copyright.com/) or contact the Copyright Clearance Center, Inc. (CCC), 222 Rosewood Drive, Danvers, MA 01923, 978-750-8400. CCC is a not-for-profit organization that provides licenses and registration for a variety of users. For organizations that have been granted a photocopy license by the CCC, a separate system of payment has been arranged.

Trademark Notice: Product or corporate names may be trademarks or registered trademarks, and are used only for identification and explanation without intent to infringe.

Library of Congress Cataloging-in-Publication Data

Cooper, J. E. (John Eric), 1944-
  Wildlife forensic investigation : principles and practice / John E. Cooper, Margaret E. Cooper.
       pages cm
  Includes bibliographical references and index.
  ISBN 978-1-4398-1374-4 (hardback)
  1. Veterinary forensic medicine. 2. Veterinary pathology. 3. Environmental monitoring--International cooperation. 4. Wildlife conservation--International cooperation. 5. Wildlife conservation--Law and legislation. I. Cooper, Margaret E. II. Title.

SF769.47.C664 2013
333.95'416--dc23   2012036534

Visit the Taylor & Francis Web site at
http://www.taylorandfrancis.com

and the CRC Press Web site at
http://www.crcpress.com

To our Families

Three generations who have supported us and cheered us on our way around the world, through thick and thin, and who have helped us to recognise the true values of life

Francis and Elizabeth Vowles

and

Eric and Dorothy Cooper

and

Vanessa

Maxwell and Sarah

and

Moses, Pippa and Hilda

who represent the future in an exciting but uncertain world

In addition, we pay tribute to

Hugh and Margaret Vowles

MEC's grandparents, who also co-authored a book

Contents

Foreword	xi
Preface	xiii
Acknowledgements	xvii
Authors	xxi
Contributors	xxiii
Quotations	xxvii

1 What Is Wildlife Forensics? — 1
JOHN E. COOPER AND MARGARET E. COOPER

2 Types of Wildlife Investigation — 11
JOHN E. COOPER AND MARGARET E. COOPER

3 Legislation — 91
MARGARET E. COOPER

4 Application of Forensic Science to Wildlife Investigations — 113
JOHN E. COOPER AND MARGARET E. COOPER

5 The Wildlife Crime Scene: An Introduction for First Responders — 127
EDGARD O. ESPINOZA, MICHAEL D. SCANLAN, ANDREW D. REINHOLZ, AND BARRY W. BAKER

6 Forensic Entomology — 149
KATE M. BARNES

7 Field Techniques: At Home and Abroad — 161
JOHN E. COOPER AND MARGARET E. COOPER

8 Working with Live Animals — 181
JOHN E. COOPER

9 Working with Dead Animals — 237
JOHN E. COOPER

10	**Dealing with Samples**	**325**
	JOHN E. COOPER	
11	**Genetic Methodologies in Wildlife Crime Investigations**	**367**
	LOUISE ANNE ROBINSON	
12	**Some Aspects of Laboratory Work**	**381**
	JOHN E. COOPER, SALLYANN HARBISON, AND JILL WEBB	
13	**Special Considerations and Scenarios**	**395**
	JOHN E. COOPER, MARGARET E. COOPER, NORMA G. CHAPMAN, ALEXANDRIA YOUNG, REGINA CAMPBELL-MALONE, ANDREA BOGOMOLNI, REBECCA N. JOHNSON, STUART WILLIAMSON, JAIME SAMOUR, MADHULAL VALLIYATTE, JÁNOS GÁL, MÍRA MÁNDOKI, MIKLÓS MAROSÁN, MAURICE ALLEY, GLADYS KALEMA-ZIKUSOKA, AND CHLOE V. LONG	
14	**Collection and Submission of Evidence**	**465**
	JOHN E. COOPER AND MARGARET E. COOPER	
15	**Writing Reports and Appearing in Court**	**481**
	MARGARET E. COOPER AND CHARLES FOSTER	
16	**Conclusions and the Way Forward**	**501**
	JOHN E. COOPER AND MARGARET E. COOPER	

Appendix A: Glossary — 507
JOHN E. COOPER AND MARGARET E. COOPER

Appendix B: Facilities and Equipment Lists — 521
JOHN E. COOPER

Appendix C: Standard Witness Statement (United Kingdom) — 529
JOHN E. COOPER AND MARGARET E. COOPER

Appendix D: Specimen Forms – Wildlife Forensic Cases — 539
JOHN E. COOPER

Appendix E: Sources of Information — 563
JOHN E. COOPER AND MARGARET E. COOPER

Appendix F: Health and Safety: Zoonoses and Other Hazards 569
JOHN E. COOPER AND MARGARET E. COOPER

Appendix G: Preparation and Investigation of Material 573
MARTYN COOKE, ANDREW C. KITCHENER, JOHN E. COOPER, AND JILL WEBB

Appendix H: Scientific Names of Species and Taxa of Animals Mentioned in Text, with Notes on Taxonomy 591
JOHN E. COOPER AND MARGARET E. COOPER

Appendix I: Legal Aspects of Sample Movement in Wildlife Forensic Cases 599
MARGARET E. COOPER

Appendix J: Information and Intelligence Gathering in Wildlife Crime Investigation 603
NEVIN HUNTER

Appendix K: Examination of a Dead Javan Rhinoceros, Cat Tien National Park, Vietnam 615

References and Further Reading 655

Index 723

Foreword

The field of wildlife forensics is new, timely and critical. I am delighted to be asked to introduce this wonderful book written by my long-time friends and colleagues, John and Margaret Cooper.

Like all new fields, wildlife forensics builds on many disciplines. These include the time-honoured discipline of 'natural history', considered now old-fashioned by some. But do you remember that phrase? It conjures up sun-dappled rambles in the woods and breezy strolls on the seashore when we were young. The thrill of discovery started with the simple exercise of picking up a feather, a bone, a shell. Each of these natural objects became a treasure.

Sadly, wildlife forensics has little to do with the innocence of childhood. Today it is no less than the front-line weapon in the fight to protect all living beings from the criminal or irresponsible activities of our own species.

Wildlife Forensic Investigation: Principles and Practice provides guidance on the scientific and practical field techniques used to investigate 'wildlife crime', such as smuggling and poaching, as well as the effects of environmental pollution, climate change and other causes of biodiversity loss. It is aimed not at obtaining convictions but at ensuring that information is collected in a scientifically sound and thorough manner, acceptable as evidence for investigations and prosecutions. Other recently published books have emphasised the detection of crime rather than investigative techniques, which is why this work is so timely.

As well as dealing with incidents involving free-living wildlife, the Coopers address the particular situation of captive wildlife, be it animals in long-term captivity (in circuses, zoos or private breeding centres) or animals in transit – in the iniquitous illegal wildlife trade. They also mention plants and wild habitats, as both are relevant in the wide scope given to wildlife forensics in this book.

The Coopers' varied skills and experiences allow them a fresh and exceptional perspective. John is a veterinarian specialising in exotic animals, including invertebrates, and Margaret has substantial knowledge and experience of conservation and other aspects of animal law. They have worked together for many years in countries in the Caribbean, Africa and the Indian Ocean on projects involving wildlife conservation, training and education, and they energetically continue to do so. They bring to the book a depth of understanding of the principles and practice of wildlife forensics unmatched by others in the field.

John, in particular, knew my late husband well. When Gerald Durrell first took up the cause for the threatened avifauna of Mauritius, John became his advisory veterinary pathologist and worked closely with colleagues at what is now known as the Durrell Wildlife Conservation Trust. John and Margaret both visited the Trust's chelonian breeding facility in Madagascar to train staff in health care, routine sampling and record-keeping for the critically endangered ploughshare tortoise. For many years, John was a member of the Trust's Scientific Advisory Committee. He was a director and Margaret a lecturer on courses in endangered species management offered by the Trust's International Training Centre.

In his formative years, John was very much influenced by Gerald Durrell. He embraced Gerry's passion for conservation and the protection of wild animals and wild places and his conviction that zoos and captive-breeding could truly aid the survival of species. But, perhaps more importantly, John learned from Gerry what it meant to study natural history properly – to attune all one's senses to the rich and diverse inputs of the natural world in order to understand it and, in turn, protect it. He and Margaret refer to Gerry's publication *The Amateur Naturalist* (1982), which describes in detail how to 'be explorers in a complex and fascinating world'.

Thus there is a strong emphasis in this book on the need to have a naturalist's eyes and ears to do wildlife forensic work. The Coopers argue that the curiosity and knowledge of a naturalist often proves invaluable when investigating a potentially criminal incident which might turn out to be nothing more than normal interaction between wild animals. For example, a deer with facial injuries could be the result of a territorial dispute between two stags and have nothing to do with illegal deer-hunting.

Wildlife Forensic Investigation is not only a good read, but it includes a strong plea for supporting good zoos, i.e. those that manage their animals as rescue and assurance populations of endangered species and train people who return to their own countries to work with such species. I was delighted to discover this aspect of the book, for both endeavours were pioneered by the institution created by my late husband, the Durrell Wildlife Conservation Trust.

The Coopers' background in veterinary medicine, animal care, international law and developing countries combined with their knowledge of biology and natural history results in a powerful approach to compiling evidence to fight wildlife crime and related misdeeds. Their interdisciplinary and unique perspective on wildlife forensic investigation is what makes this book essential reading for those who are dedicated to righting the wrongs done to our natural world.

Lee Durrell, MBE, BA, PhD
Honorary Director
Durrell Wildlife Conservation Trust
Trinity, Jersey, Channel Islands

Preface

'Mum and Dad are in a pickup somewhere in Central Africa'.

Letter sent by Vanessa Cooper in response to enquiries about her parents following the 1994 genocide in Rwanda

The writing of this book was prompted by two things. First, following the publication of our earlier book *Introduction to Veterinary and Forensic Medicine* (2007), a broad overview of forensic science and medicine relating to animals, a number of colleagues urged us to produce a book on forensics more specifically focussed on free-living and captive wildlife. At the same time, the need for a rigorous, scientific approach to the investigation of 'wildlife crime' and to involve people from different backgrounds and disciplines has become increasingly urgent. There is growing recognition of the value of special units to combat 'wildlife crime', and scientists have begun to consider how their research might provide the services with assistance with matters such as the identification of animal remains and ascertaining the circumstances of death of wild species that are protected by law. In our book, with the help of contributors, we have endeavoured to explain the principles of such wildlife forensic investigations and how this relatively new and challenging, but exciting, field of endeavour is evolving. The practice of wildlife forensics is a constantly developing subject, moulded by changes in patterns of 'wildlife crime' and by other legal, political and social influences.

We are grateful to many people for their assistance and support. They are listed in the Acknowledgements. To match its international orientation, we have written and edited parts of this book in locations as far apart as Nottingham and New Zealand and as diverse as Barbados, Toronto and Ho Chi Minh City. We thank friends and colleagues in these and many other parts of the world for providing us with the opportunity to do so.

By its very nature, 'wildlife forensics' is a very large subject. It spans topics ranging from anatomy and aviculture to zoo-archaeology and zoonoses, and it involves taxa as varied as aardvarks and ant-lions to zebras and zebu cattle. Strictly, as we explain in Chapter 1, the term 'wildlife' encompasses plants and the habitat on which they depend as well as animals, although we have concentrated on the latter.

We have sought to make our text as wide-ranging as possible and, in particular, to try to fill gaps left by other publications. We have applied our experience of not just travelling but actually living abroad (see quotations at the beginning and at the end of the Preface) in order to give this book an international rather than solely a 'Western' orientation. We have tried to reach out to friends and colleagues, known and unknown, in the poorer or less-well-administered parts of the world where facilities for wildlife forensic work are far from adequate but, nevertheless, investigations still must be performed.

Although we have made every effort to check the statements made and advice given in this book, particularly relating to legal and medical matters, readers must verify this information for themselves.

In keeping with the title of this book, we have – with the help of contributors – introduced considerable detail into explaining some methods of investigation. The reader will, for example, find more information about the examination of feathers and bird pellets in these pages than is included in most other texts. Our intention in so doing is to make our book complementary to others, not competitive with them. We also frequently draw attention to techniques and equipment that, while not specifically designed for use in forensic work, might usefully be adopted (and, where necessary, adapted) to meet the challenges and exigencies that characterise wildlife investigations.

We have likewise broadened the scope of this work, in terms of both species and locations, by inviting a number of colleagues to submit case studies, or what we have called 'scenarios', concerning their own work or describing cases in which they have been involved (see Chapter 13). This approach proved popular in our previous book, and we hope that it will do so again. These contributions, like our own chapters, are presented in a practical way and intentionally orientated not only to forensic work in the westernised world but also in areas of, for example, Africa and Asia, where wildlife is threatened but facilities and infrastructure are often limited.

In all that has been written, we have gone to pains to ensure that in these pages the word 'wildlife' covers the animal kingdom, not only mammals. We have intentionally included information on captive non-domesticated animals because 'wildlife' incidents that require forensic investigation and species that are used in illegal activities are not confined to free-living vertebrates and invertebrates. So the reader will find that our examples often do not focus solely on the large, well-known and charismatic species such as rhinoceroses, polar bears, eagles and sharks but also include ants, butterflies and marine plankton – both in the wild and when kept or studied in captivity.

We have included references in the book to circumstances where wild animals are not the *victims* of human action but the *cause* of incidents. This is an area of increasing importance, both in practical and legal terms, and warrants more scientific study. We also cover the part played by many species as 'sentinels' – that is, circumstances where wild animals provide information that is relevant to an incident or evidence pertaining to a 'non-animal' legal case.

Throughout our text, there is discussion of animal welfare; it is an integral part of our definition of 'wildlife crime'. Welfare is, of course, very relevant when wild animals are kept in captivity, but in addition, ill-treatment of free-living creatures very often occurs during the course of poaching, smuggling and other illegal uses.

In forensic work, it has always been important for the expert to be able to explain words and expressions in lay terms, especially in reports and statements and (verbally) in the courtroom. Wildlife forensics is no exception. It is, however, not only 'lay' people who may need help in this respect. One of the *disadvantages* of the interdisciplinary nature of forensic work, perhaps particularly in the field of wildlife, is that while the majority of those involved are likely to be authorities in their own field of, say, botany or geology, they may not be at all familiar with the technical language of other disciplines. There may even be differences in interpreting the meaning of a term, as in osteology where the 'expert' may either come from a medical (including dental/veterinary) background or be trained primarily in archaeology, anthropology or palaeontology. The need to overcome or to reconcile such differences has never been more important, and it is encouraging to note the appearance of books that help to explain methodologies and terminology to others in a bid to facilitate understanding before and during a court hearing (see, e.g. Prahlow, 2010).

To accommodate this, our Glossary includes a variety of biological, medical and legal terms to reflect the fact that our readers will come from different disciplines.

We have tried to provide a comprehensive Index and, to help colleagues from the other side of the Atlantic to locate words or terms easily, we have included therein some American spellings and word usage where these differ significantly from those used in British/European/Commonwealth English. Despite the influence effect of the Internet, the words of George Bernard Shaw remain true: 'England and America are two countries separated by the same language.' If a reader does not find the information that he or she wants in the text – or coverage is judged insufficient – there may be relevant References and Further Reading (some in languages other than English) to it at the end of the book.

Comments and feedback on what we have written will, as always, be both appreciated and acknowledged!

Our primary hope is that our book will help those involved in legal proceedings and other actions concerning wildlife, especially animals, to gain access to sound evidence that is based on good science and obtained using best available practice. However, conscious that wildlife forensics is a rapidly growing subject, we have explored its background and sought to uncover areas in which such expertise may be needed in the future. In Chapter 2, we have discussed subjects that are either contentious or are likely to be so in years to come. Some readers (and no doubt a few reviewers!) will be surprised at the choice of topics, but we believe that our book should be playing a part in alerting colleagues, especially young people in developing parts of the world, to the challenges that are emerging. Many of these will prove litigious, and sound evidence will then be essential.

Like so many people, we are concerned about the effects of wildlife crime and anxious to see it curbed. However, we also believe strongly in human rights and the underlying concept that guilt must be proved, not assumed. Therefore, the message of our book, as in its forerunner, is that evidence must be sound if justice is to be served. Not surprisingly, therefore, we also feel strongly about the importance of ethics in forensic work. Pressure for an ethical approach to biology is not new (Botzier and Armstrong-Buck, 1985), and the need for biologists to be able to make ethical judgement has subsequently been stressed by authors from a range of backgrounds (O'Riordan, 1995). It was, of course, ethical concerns about welfare and the perceived obligations of humans towards sentient beings that led first to the great debates of the eighteenth and nineteenth centuries in Europe and then to legislation designed to protect animals. In our view, there is a need for those of us who work with wildlife, whether our activities are invasive, minimally invasive or non-invasive (see Chapter 8), to adhere to codes of practice as well as to any legal controls. This approach is, we argue, as important to those who investigate wildlife crime as it is to those who perform scientific research.

Much of the thinking and planning of this book took place in East Africa, where we have spent several years of our life together and where our children, now adults, were both born. In particular, we have gained inspiration and ideas for our writings when we have been at Shamba Musa, the area of land near the coast, in Kenya, where in recent years we have run training workshops. Shamba Musa comprises bush country, on which local zebu cattle graze, but it also overlooks the Mwaluganje Elephant Sanctuary (Figure P.1). Observing some of the problems of human–wildlife conflicts in this area of subsistence agriculture has added a dimension to our views on conservation issues and wildlife law enforcement. Elephants and baboons regularly encroach upon Shamba Musa, and there is interaction between them, domesticated livestock and landowners. The location reflects on

Figure P.1 Shamba Musa.

a small scale the pressures that confront so much of the earth's surface in terms of potential conflict between *Homo sapiens* and wildlife, made all the more challenging because of the rapid growth in the human population (as we penned these sentences, in October 2011, it was announced that the seventh billion person had just been born). Yet, at the same time, the ecological balance that exists at present at Shamba Musa gives us some cause for optimism.

> *'Living abroad is a very different thing from travelling abroad'.*
>
> **Owen Rutter (1936), 'A Traveller in the West Indies'**

John and Margaret Cooper
The Mombasa Club, Kenya

Acknowledgements

'The bird a nest, the spider a web, man friendship'.

William Blake

First, we should like to express our gratitude to Lee Durrell, Honarary Director of the Durrell Wildlife Conservation Trust, on Jersey, Channel Isles for agreeing to write the Foreword to this book and for her kind and generous comments.

We are grateful to those who have contributed chapters or sections to this book. They all appear in the list of contributors and are named individually in the relevant parts of the text.

Becky Masterman, our commissioning editor at Taylor & Francis Group, has throughout guided us with (as she promised at the outset) alacrity and good humour. Mr Ed Curtis and Mr Alfred Samson directed the final stages of this book with unfailing courtesy and diligence.

Our family has continued to give us encouragement. We thank our daughter, Vanessa, and our son, Maxwell, and daughter-in-law, Sarah (Hutton) Cooper, along with Moses and Pippa, for their interest in, and resigned acceptance of, our various adventures. We have been delighted to welcome the arrival of Hilda Grace, during the final days of proof-reading. They and our parents are the subject of the dedication at the beginning of the book.

We sincerely thank Sally Dowsett who is a long-standing friend of our family and a supporter of so many of our activities. We greatly appreciate her invaluable help, especially many months of typing, her sound advice on design and her regular good cheer and convivial company.

Emily Neep, on train visits to Cambridge and at home in Norfolk, has helped to check chapters and collate references. We wish her well for her new life at university.

A number of people either reviewed chapters or appendices or advised us on content. In this respect, we are most grateful to Hany Elsheikha, David Bailey, Zoltan and Agnes Bakos, Bernard Baverstock, Martin Böttcher, Anne-Maria Brennan, Ivan Broadhead, Paul Budgen, Jason Byrd, Charles Cadwallader, Lorna Cleverley, Penny Cooper, Chris Daborn, Chris Davis, Corinne Duhig, Charles Foster, Alan Gunn, Martin Hall, Rebecca Johnson, James Kirkwood, Andrew C. Kitchener, Chris Laurence, Martin Lawton, Adrian Linacre, Pat and Mary Morris, David Owino Ojigo, Donald Peach, Henrietta Price, Kristel-Marie Ramnath, Jack Reece, Tony Sainsbury, Dick Shepherd, Benjamin Swift, Helen Skelton, Alan Stewart, Ulrike Streicher, Faris al-Timimi, Jill Webb, Sean Wensley, Amoret Whittaker, David Williams, Bernard Baverstock and Alisdair Wood. The task of those who have reviewed sections or made helpful comments has not been easy; as William Cowper put it, 'After all, reviewing is a detested sport, that owes its pleasures to another's pain'.

We also thank those who responded to our original email messages in 2009–2010, asking for comments on our proposal to write this book: Tom Adams, Mike Appleby, Phil Arkow, Tom Bailey, Barry Baker, Regina Campbell-Malone, Norma Chapman,

Jim Groombridge, Carlos Hasbun, Thijs Kuiken, Andrew Lawrence, Ian McDowall, Howard and Eleanor Nelson, Rob Ogden, Kristel-Marie Ramnath, Tony Sainsbury, Jaime Samour, Brian Stuart, Tiawanna Taylor, Steve and Anna Tolan, Vic Watkins, Sean Wensley and Peer Zwart.

We thank Ben Garrod for providing us with information about geometric morphometrics and Pamela Athene Smith for producing the text and drawings of gorilla bones that appear in Appendix D. We are grateful to staff of the Medical School of the University of the Witwatersrand for their support and assistance with our mountain gorilla study in South Africa in 1998, especially the late Professor Phillip Tobias who served as our host, Professor Beverley Kramer who gave permission for the examination of the gorilla material and publication of our findings and Anna Biscardi who performed the bone density studies. Peter Dawson, Mary-Ann Costello, Llewellyn Sinclair and Peter Montja were responsible for the provision and preparation of the specimens and for taking radiographs; we are indebted to them all.

Marc Driscoll, Renee Lezama and Adana Mahase, former students of ours at the University of the West Indies and now qualified veterinarians, kindly gave us permission to modify and reproduce in this book protocols and forms developed by them for their studies on the pellets of black vultures in Trinidad.

We acknowledge the British Trust for Ornithology (BTO), Natural England and the Disease Risk Analysis and Health Surveillance Project at the Institute of Zoology, London, for giving permission for us to reproduce their form (checklist) entitled (*sic*) 'Post-mortem of radio tagged birds – additional information required'.

The report on the death of a Javan rhinoceros is reproduced with the permission of Worldwide Fund for Nature (WWF) which is the copyright holder and publisher. It is the work of the WWF but the authors, Ulrike Streicher, Ed Newcomer, Douglas Mc Carty, Sarah Brook and Bach Thanh Hai, are credited here.

We acknowledge with thanks the Libraries of the Royal College of Veterinary Surgeons, the Department of Veterinary Medicine, University of Cambridge and the National Museums of Kenya for their help with references and continuing support in other ways.

This book provides an opportunity to thank Richard (Dick) Stroud for his invitation to visit the United States Fish & Wildlife Forensics Laboratory in 2000, his hospitality during our stay and his generous review of our earlier volume in the *Journal of Wildlife Diseases* (2007).

Unless otherwise stated, the majority of the photographs in this book were taken by, and remain the property of, Margaret E. Cooper or the authors of individual contributions. Some were the work of Richard Spence, our friend and medical photographer at the University of the West Indies (UWI) in Trinidad. Our veterinary colleague at UWI, Ravi Seebaransingh, advised on a number of forensic necropsies and took photographs during the course of those investigations. Donald Peach very kindly allowed us to use some of his pictures of hair, and Richard Jakob-Hoff provided the group picture taken at the New Zealand Centre for Conservation Medicine during the First Australasian Workshop and Seminar on Wildlife Forensics. The pictures of the late Kenneth Carr, left to those who tended him in his last few years, were kindly passed to us by Edward Le Conte with agreement that we should use them. Maureen Harper had the foresight to take the image that we have used with our biographical notes. Chris Daborn sent us a picture of vultures, appropriately labelled 'lunch by the roadside', that we immediately, with his consent, added to the book. We are most grateful to them all.

Acknowledgements

Our involvement in wildlife forensics reflects our academic background. We are a husband-and-wife team, veterinary surgeon and lawyer, who share an interest in natural history and global issues. Over the years, many people have helped and encouraged us in our studies on wildlife forensics and thereby, directly or indirectly, played a part in the genesis of this book. We mention several individually later in these Acknowledgements or in the text. There are other people, though, who over the years have inspired or encouraged us in terms of our interests and pursuits, particularly those relating to wildlife, veterinary medicine and animal law. We take this opportunity to mention some of those who are no longer with us, but whose influence was strong and whom we should not want to forget. They are the naturalist, broadcaster and former MI5 agent Maxwell Knight; the father of lower vertebrate pathology Edward Elkan; entomologist Brian Baker; bird artist Rena Fenessey; ornithologist and author Leslie Brown; naturalist and filmmaker Joan Root; palaeontologist Louis Leakey; veterinary surgeon and truly 'best man' Roy Clutterbuck; deer biologist Donald Chapman; our dear friend who was part of the history of UWI and the Imperial College of Tropical Agriculture (ICTA) in Trinidad, Joyce Gibson Inniss; and those two confidantes to Dian Fossey—and friends to many of us who worked with gorillas in Rwanda—Rosamund Carr and Alyette de Munck. Finally, we remember Steve Bennett who died while we were in the final stages of writing this book. He was the doyen of the Caribbean veterinary profession for many years, a true, swashbuckling Trinidadian who was ever a good friend and colleague to us ('the limeys') when we lived in Trinidad. *Valete et ite in pace!*

In our earlier incursions into veterinary forensic studies during the 1970s and 1980s, the Cromwell Veterinary Group, in Huntingdon, kindly provided facilities for clinical and radiographical examination of animals that were submitted to us by the Royal Society for the Protection of Birds (RSPB) and others. The Cromwell Veterinary Group recently celebrated the centenary of the establishment of their practice in 1911. We congratulate them on their commitment to animals and humans alike and express our gratitude for the welcome and assistance that they have always afforded us. We also thank Henry Berman and Joan Berman who, for so many decades, have been involved in educational and charitable activities in the Huntingdon area. These have included the St Ivo School Entomology and Natural History Society, established in 1957, that influenced and inspired many young naturalists, not only one of us (JEC) but also our two children, who were both taught biology by Henry Berman.

We may have left our Caribbean life behind, but we have been sustained as we completed this book by Karen Yorke, who has spurred us on with cheerful messages and news, liberally backed up with supplies of Trinidad music together with traditional (and notoriously well-laced) 'black cake'.

This book has been written and edited in various parts of the world; we are grateful to friends and colleagues in those locations for providing us with inspiration and support.

John and Margaret Cooper
The Mombasa Club, Kenya

Authors

John and Margaret Cooper are a husband-and-wife-team from the United Kingdom.

John E. Cooper, who has had a lifelong interest in natural history, trained as a veterinarian and is now a specialist pathologist with particular interests in wildlife and exotic species, tropical diseases and comparative medicine. Margaret E. Cooper is a lawyer who qualified originally as a British solicitor and has made the study of animal and conservation law her special interest.

The Coopers have lived and worked in East and Central Africa, including a period in Rwanda with the mountain gorillas, in the Arabian desert, in the Caribbean and in France and Germany. In 2009, they returned from Trinidad and Tobago. There, during almost seven years at the University of the West Indies, they combined their medical and legal backgrounds in the promotion of an interdisciplinary approach to veterinary and biological education, wildlife conservation and forensic science.

Now based once again in the UK, the Coopers are involved in teaching at various universities and pursuing their professional interests, especially in ecosystem health, diagnostic pathology, animal law and forensic science. They continue overseas travel and lecturing, including their work with wildlife, domesticated animals and local communities in East Africa.

The Coopers believe strongly that justice is best served when veterinarians, biologists and others involved in legal cases concerning animals are appropriately trained in forensic techniques so that evidence is correctly presented and reports are of maximum value to litigants and courts of law. Greater competence in these respects will promote the conservation of wildlife, the welfare of animals, the health of humans, the promotion of professional standards and the reputation of the judicial process.

Contributors

Maurice Alley, BVSc, PhD
Associate Professor of Veterinary Pathology
New Zealand Wildlife Health Centre
Massey University
Palmerston North, New Zealand
e-mail: M.R.Alley@massey.ac.nz
Contribution: *Predation of Endangered New Zealand Weka by Dogs*

Barry W. Baker, BA, MA
Supervisory Forensic Scientist
US Fish & Wildlife Service
US National Fish & Wildlife Forensics Laboratory
Morphology Section
1490 East Main Street
Ashland, Oregon 97520-1310, USA
e-mail: barry_baker@fws.gov
Contribution: *The Wildlife Crime Scene: An Introduction for First Responders*

Kate M. Barnes, BSc, MSc, PhD
Lecturer in Forensic Biology
Department of Biological and Forensic Sciences
University of Derby
Kedleston Road
Derby DE22 1GB, UK
e-mail: kbarnes1@derby.ac.uk
Contribution: *Forensic Entomology*

Andrea Bogomolni, BA, BS, MA
Research Associate II
Woods Hole Oceanographic Institution
266 Woods Hole Road
Woods Hole, Massachusetts 02543, USA
e-mail: andreab@whoi.edu
Contribution: *Marine Mammals: A Special Case*

Regina Campbell-Malone, BSc, PhD
Guest Investigator
Woods Hole Oceanographic Institution
MS#50, Woods Hole, Massachusetts 02543, USA
e-mail: regina@whoi.edu
Contribution: *Marine Mammals: A Special Case*

Norma G. Chapman, BSc, CBiol
29 The Street, Barton Mills
Suffolk, IP28 6AA, UK
e-mail: ngchapman@btopenworld.com
Contribution: *Mammalian Teeth*

Martyn Cooke
Head of Conservation
The Royal College of Surgeons of England
35-43 Lincoln's Inn Fields
London WC2A 3PE, UK
e-mail: mcooke@rcseng.ac.uk
Contribution: *The Preparation and Fixation of Specimens for Research and Museum Display*

John E. Cooper, DTVM, FRCPath, FSB, CBiol, FRCVS
RCVS Specialist in Veterinary Pathology
and
Diplomate European College of Veterinary Pathologists
and
European Veterinary Specialist Zoological Medicine
Visiting Professor
University of Nairobi
Nairobi, Kenya

and

Department of Veterinary Medicine
University of Cambridge
Cambridge CB3 0ES, UK
e-mail: ngagi2@gmail.com
Contribution: Principal Author

Margaret E. Cooper, LLB, FLS
Solicitor (not in private practice)
Visiting Lecturer
University of Nairobi
Nairobi, Kenya

and

Honorary Research Fellow
Durrell Institute of Conservation and Ecology
University of Kent
Kent CT2 7NR, UK
e-mail: ngagi2@gmail.com
Contribution: Principal Author

Edgard O. Espinoza, BS, MPH, DrPH
US National Fish & Wildlife Forensics Laboratory
Office of Law Enforcement
US Fish & Wildlife Service
1490 East Main Street
Ashland, Oregon 97520-1310, USA
e-mail: ed_espinoza@fws.gov
Contribution: *The Wildlife Crime Scene: An Introduction for First Responders*

Charles Foster, MA, Vet MB, PhD, MRCVS
Barrister
Fellow of Green Templeton College
University of Oxford
Outer Temple Chambers
London WC2, UK
e-mail: Charles.Foster@gtc.ox.ac.uk
Contribution: *Writing Reports and Appearing in Court* (with Margaret E. Cooper)

János Gál, PhD, Dipl. ECZM
Department of Pathology and Forensic Veterinary Medicine
Faculty of Veterinary Science
Szent Istvan University
PO Box 2
Budapest, H-1400, Hungary
e-mail: Gal.Janos@aotk.szie.hu
Contributions: *Forensic Expert Work in Wildlife Medicine in Hungary*
and
Areas of Wildlife Forensics in Hungary, with Practical Examples

SallyAnn Harbison, BSc, PhD
Senior Science/Technical Leader
Forensic Biology
ESR Mt Albert Science Centre
Private Bag 92021
Auckland, New Zealand
e-mail: SallyAnn.Harbison@esr.cri.nz
Contribution: *Quality Management and Wildlife Forensic Science*

Nevin Hunter, BSc
Detective Inspector
Head of Unit
National Wildlife Crime Unit
Old Livingston Police Station, Almondvale Road, Livingston
West Lothian, EH54 6PX, Scotland, UK
e-mail: nevin.hunter@nwcu.pnn.police.uk
Contribution: *Information and Intelligence Gathering in Wildlife Crime Investigation*

Rebecca N. Johnson, BSc, PhD
Australian Centre for Wildlife Genomics
Australian Museum
6 College Street
Sydney, New South Wales 2010, Australia
e-mail: Rebecca.Johnson@austmus.gov.au
Contribution: *Investigation of 'Aircraft Wildlife Strike' Using Forensic Techniques*

Gladys Kalema-Zikusoka, B Vet Med, MRCVS, MSPVM
Founder and Chief Executive Officer
Conservation Through Public Health
Plot 1 Katalima Crescent, Naguru, PO Box 10950
Kampala, Uganda
e-mail: gladys@ctph.org
Contribution: *Mountain Gorilla Disease: Implications for Conservation*

Andrew C. Kitchener, PhD, BSc, FRZSS
IUCN Cat Specialist Group
Principal Curator of Vertebrates
Department of Natural Sciences
National Museums Scotland
Chambers Street
Edinburgh EH1 1JF, UK
e-mail: a.kitchener@nms.ac.uk
Contributions: *Mammal Measurements*
Cosmetic Post-Mortem Examination of Mammals
Cosmetic Post-Mortem Examination of Birds

Chloe V. Long, BSc, PhD
Loughborough University
Loughborough, Leicestershire, LE11 3TU, UK
e-mail: C.V.Long@lboro.ac.uk or degutopia@btinternet.com
Contribution: *Bats and Wind Turbines*

Míra Mándoki, DVM, PhD
Senior Lecturer
Department of Pathology and Forensic Veterinary Medicine
Faculty of Veterinary Science
Szent Istvan University
PO Box 2
Budapest, H-1400, Hungary
email: Mandoki.Mira@g-mail.com
Contribution: *Forensic Expert Work in Wildlife Medicine in Hungary*

Contributors

Miklós Marosán, PhD
Department of Pathology and Forensic Veterinary Medicine
Faculty of Veterinary Science
Szent Istvan University
PO Box 2
Budapest, H-1400, Hungary
e-mail: Marosan.Miklos@aotk.szie.hu
Contribution: *Areas of Wildlife Forensics in Hungary, with Practical Examples*

Andrew D. Reinholz, BSc
Forensic Scientist – Criminalistics
US National Fish & Wildlife Forensics Laboratory
Office of Law Enforcement
US Fish & Wildlife Service
1490 East Main Street
Ashland, Oregon 97520-1310, USA
e-mail: andy_reinholz@fws.gov
Contribution: *The Wildlife Crime Scene: An Introduction for First Responders*

Louise Anne Robinson, BSc, PhD
Lecturer in Forensic Biology
Department of Biological and Forensic Sciences
University of Derby
Kedleston Road
Derby, DE22 1GB, UK
email: l.robinson@derby.ac.uk or robinsonlouiseanne@yahoo.co.uk
Contribution: *Genetic Methodologies in Wildlife Crime Investigations*

Jaime Samour, MVZ, PhD, Dip ECZM (Avian)
Director Wildlife Division
Wrsan
PO Box 77338
Abu Dhabi, UAE
e-mail: jaimesamour@hotmail.com
Contribution: *Identification Systems in Hunting Falcons in the Middle East*

Michael D. Scanlan, BS
Senior Forensic Scientist – Criminalistics
US National Fish & Wildlife Forensics Laboratory
Office of Law Enforcement
US Fish & Wildlife Service
1490 East Main Street
Ashland, Oregon 97520-1310, USA
e-mail: mike_scanlan@fws.gov
Contribution: *The Wildlife Crime Scene: An Introduction for First Responders*

Madhulal Valliyatte, BVSc & AH PGDWLM
Sree Nandanam
PO Feroke Calicut
Kerala, India
e-mail: drvalliyatte@gmail.com
Contribution: *Welfare Issues of Captive Asian Elephants*

Jill Webb, CBiol, MSB, MIET, FRMS
Science Director, Life Sciences
Cairngorm Scientific Services Ltd
Southern Office, Maplehurst, Green End Road
Radnage, High Wycombe HP14 4BY, UK
e-mail: jillwebb.css@gmail.com
Contribution: *The Value of Microscopy*

Stuart Williamson
General Investigator
Compliance and Response
Ministry for Primary Industries
608 Rosebank Road, Avondale, PO Box 1026
Auckland, New Zealand
e-mail: Stuart.Williamson@mpi.govt.nz
Contribution: *Dealing with Illegal Wildlife Trade: A Case Study of Coral Imports into New Zealand*

Alexandria Young, MA, MSc
School of Applied Sciences (Forensic & Biological Sciences Group)
Bournemouth University
Talbot Campus, Fern Barrow
Poole, Dorset BH12 5BB, UK
e-mail: younga@bournemouth.ac.uk
Contribution: *Advances in Scavenging Research*

Quotations

'Science is that lamp which man has himself kindled. It has built him lighthouses on the dark shores of the unknown; but his dreams, his quests for truth, lead him beyond the waters which his little lamp of knowledge illumines'.

Sidney Royse Lysaght (1860–1941)

'No man is an island, entire of itself every man is a piece of the continent, a part of the main if a clod be washed away by the sea, Europe is the less, as well as if a promontory were, as well as if a manor of thy friends or of thine own were any man's death diminishes me, because I am involved in mankind and therefore never send to know for whom the bell tolls; it tolls for thee'.

John Donne (1572–1631)

What Is Wildlife Forensics? 1

JOHN E. COOPER
MARGARET E. COOPER

Contents

Introduction	1
Theme of This Book	4
Scope of Wildlife Forensics	5
Animals as Sentinels	8
Methods in Wildlife Forensics	9

'Let us make man in our image, after our likeness: and let them have dominion over the fish of the sea, and over the fowl of the air, and over the cattle, and over all the earth, and over every creeping thing that creepeth upon the earth'.

Genesis 1:26

Introduction

What is 'wildlife'? In most dictionaries and reference texts it is taken to imply 'animals and plants that are not domesticated, tame or cultivated', but definitions vary. In the *Little Oxford Dictionary* (1986), for example, the definition given was 'wild animals collectively'. Insofar as this book is concerned, the word 'wildlife' is taken to comprise those species of animals or plants that are normally to be found occurring naturally in the wild, as opposed to 'domesticated' or 'cultivated' species that have, over many years, been brought into captivity and within which selection for certain characteristics has been carried out.

However, within these pages the definition of wildlife will be narrowed to animals – both vertebrate (Figure 1.1) and invertebrate (Figure 1.2). Plants will be mentioned but not in detail. Many species of plant are threatened (Figure 1.3), sometimes by the very same pressures that affect animal populations. In certain instances theft and illegal importation or exportation are the salient features of wildlife crime cases involving plants, but often the situation is more complex involving, for example, medicinal plants (Hamilton and Erhart, 2012). The writings of Gunn (2009) and Hardey and Martin (2012) provide a useful account of the role of plants in wildlife forensic work. Some knowledge of botany and sources of information about plants are often valuable when investigating wildlife crime and performing the necessary forensics. This book places the emphasis on wildlife forensic investigation as it relates to animals, but there is also reference within its pages to plants and habitats because of their relevance to the conservation of animals (PAW, 2005).

A wild animal is, by definition, of a species that has not been domesticated; thus a wolf is a wild animal whereas a pet dog is domesticated. However, the wolf may be in captivity or be 'free-living' (often termed 'free-ranging' in the United States of America (USA)).

Figure 1.1 An African elephant, a species that has suffered greatly from poaching.

Figure 1.2 East African butterflies following emergence. Invertebrates comprise 90% of all living animals.

The definition of 'domesticated' in this book is that given by Mason (1984) – 'an animal that breeds under human control, provides a product or service useful to humans, is tame and has been selected away from the wild type' (see Glossary). Some animals, while not domesticated, have long been established in captivity as well as continuing to be found in the wild. An example is the Asian elephant, which is used by humans for a variety of purposes (see 'Welfare Issues of Captive Asian Elephants' section in Chapter 13). A few species

What Is Wildlife Forensics?

Figure 1.3 Many species of flowering plant are endangered. This Jersey pink (*Dianthus gallicus*) is found only on the island of Jersey in the British Channel Isles and is on the International Union for Conservation of Nature (IUCN) List of Threatened Plants.

of animal that might be considered 'wild' are in fact 'feral' – for example, dogs, cats or ferrets that have returned to a free-living or a semi-wild situation of their own volition or because they have been rejected by owners.

An understanding of the aforementioned points requires some knowledge of how living things are classified. We have included in Appendix H information about the International Commission on Zoological Nomenclature (ICZN), as well as the scientific name of the species of animal referred to in the text.

And what do we mean by 'forensics'? As explained by Cooper and Cooper (2007), the word 'forensic' essentially means relating to, used in or connected with a court of law. It is derived from the Latin word *forensis*, which meant 'public', and was an adjectival form of *forum*, originally meaning a market, later a place of debate. The history and evolution of forensic science, with particular reference to medicine, is outlined in Chapter 5.

As far as this book is concerned, 'wildlife forensics' is considered to be the application of forensic science to the conservation and protection of non-domesticated animals, both in the wild and in captivity, together with, where relevant, that of plants and the environment. Offences that relate to the foregoing matters are now commonly termed 'wildlife crime' – that is, activities that threaten wild animals, plants or their habitats and which constitute a breach of national, regional or international law. Some indications of the practicalities of investigating wildlife crime are given in a number of books – for example, Stewart (2009) in the United Kingdom and Neme (2009) in the United States.

Definitions, however, differ. In an editorial to a Special Issue of *Forensic Science, Medicine and Pathology*, Wilson-Wilde (2010) defined wildlife crime as follows: 'It involves the illegal trade in animals, plants and their derivatives and can result in the depletion of natural resources, invasion of pest species and the transmission of diseases'. In this book our definition is far wider than that and includes not only other activities that adversely affect wild animals but also insults that wildlife inflicts on humans and their property. Wilson-Wilde categorised wildlife crime offenders into three main groups: minor offenders, organised illegal trading and serious major criminal activity (McDowell, 1997). Minor offences generally relate to such matters as permits and constitute opportunistic types of crime. Organised illegal trading refers to deliberate clandestine poaching with intent to make a gain and to meet the needs of a certain market or to the order of an individual. In marked contrast to both of the foregoing, serious major criminal activity involves groups which are professional law breakers. They are financially backed and specifically market products; such offenders may also be involved in other illegal acts, including the shipping of drugs.

It follows that combating wildlife crime usually requires, *inter alia*, a well-equipped forensic facility that can provide cutting-edge technology. Establishing such a facility is not easy (see Appendix B).

Wilson-Wilde also discussed what drives demand in wildlife trade. It is not possible to be certain, but factors such as fashion, rarity of the species, trends in alternative remedies and medicine and criminal elements all play a part. Products from particularly endangered species sell for large sums and therefore are more sought-after by collectors who stand to make significant profits despite the dangers and penalties that they risk. Placing a species on the CITES list, Appendix I, Wilson-Wilde argued, may serve to make it more appealing.

Wilson-Wilde went on to say 'Wildlife crime also includes offences involving domesticated species, such as animal cruelty cases and where an animal may be used to link an individual to the commission of an offence (e.g. dog hairs on a suspect)'. We would not place animal cruelty, *per se*, under the umbrella of 'wildlife crime' because, in the majority of such cases, domesticated animals are the victims. Further, depending upon the jurisdiction, the animal welfare legislation may not apply to free-living non-domesticated species. Nevertheless, it can be said the investigation of a wildlife crime, particularly the smuggling of endangered species, will often reveal evidence of suffering during the time that the 'wildlife' is in captivity or under a person's control. This quite often leads to charges under the animal welfare law (and, for that matter, a variety of other offences, such as those listed in Appendix J, Table J.1), being laid as part of a wildlife crime prosecution.

Theme of This Book

This text provides an introduction to the rapidly evolving field of 'wildlife forensics'. Within its pages, the crucial role of science in the investigation of contraventions of national, regional and international wildlife legislation is explained and discussed. In order fully to understand the importance of wildlife crime and wildlife forensics, one must know something of the history of wildlife conservation and environmental protection and their global significance in the twenty-first century; these are therefore discussed briefly, with references, in Chapter 2.

Although the primary concern of wildlife investigations may be free-living animals, a knowledge of how and why such species are kept in captivity is also necessary because some wildlife forensic cases concern the taking of species for sale (as in the pet trade), for display (in public zoos and private collections) or for other purposes, such as rehabilitation.

What Is Wildlife Forensics?

Scope of Wildlife Forensics

Wild animals, both free-living and captive, are increasingly the subject of litigation in Europe, North and South America, Australasia, Africa and the Far East. They thereby may become the subjects of forensic investigation.

There are three main ways in which wild animals, free-living or captive, are involved in litigation. They can

1. Be the *cause* of an incident – for example, if they injure a person, damage property or spread an infectious disease (Figure 1.4). This aspect is discussed in more detail in Chapter 2.
2. Be the *victim* – if they are killed, injured, poached, exported illegally or treated inhumanely.
3. As mentioned by Wilson-Wilde earlier, provide information that is relevant to an incident – for example, because they were present when a crime was committed and their hair or other material remains as evidence ("see 'Animals as Sentinels' later").

Animals of different species, singly or in groups, can cause injuries, death and financial loss.

Wildlife forensic cases can involve domesticated animals, as well as humans; for instance, a pet dog may be responsible for killing or maiming a protected species (Figure 1.5).

Figure 1.4 The Nile crocodile is still widespread in sub-Saharan Africa and remains an important cause of injury and death to humans, domesticated livestock and wild animals.

Figure 1.5 An example of the role that domesticated animals may play in wildlife crime incidents. It was not clear whether this dog killed the iguana or merely retrieved the body.

Insofar as aforementioned category 2 is concerned, where the animal is the victim, legal cases generally fall into five categories:

1. *The animal is found dead under unusual, unexpected or suspicious circumstances.*
   In this category a dead animal or a group of animals is investigated with a view to determining the circumstances – that is, the cause, mechanism and manner – of death. See Chapter 9.
2. *The animal is alive but exhibits unusual, unexpected or suspicious clinical signs.*
   Here, investigation relates to live animals that are found in a sick, injured or incapacitated condition and the circumstances appear unusual or suspicious. Such cases can range from a single animal, such as a deer caught in a trap in Wisconsin, the United States, or a stranded whale on the coast of England, to an outbreak of infectious disease or poisoning that is affecting animals at a waterhole in East Africa (Figure 1.6, buffalo at water-hole). Not all of these cases may be the result of illegal activity but if the facts are to be established, they warrant a full, painstaking, forensic investigation.
3. *The animal's welfare apparently is, or has been, compromised.*
   In these cases there is a need to determine whether an animal is being (or has been) subjected to unnecessary pain, discomfort or distress (Figure 1.6, buffalo with scar). Cruelty may be alleged (see Chapter 8). In some, but not all, instances the ill-treatment of a wild animal may be linked to abuse of livestock or domestic pets or violence to humans (Cooper and Cooper, 2007); this too is discussed in Chapter 8.
4. *The animal appears to have been taken, killed or kept in captivity unlawfully.*
   Cases that fall in this category are those that are most often termed 'wildlife crime' – that is, as stated earlier, activities directed at wild animals, plants or their habitats which constitute an offence under national, regional or international law (Figure 1.7).

What Is Wildlife Forensics?

Figure 1.6 In Africa buffalo congregate at waterholes, especially in droughts, and are there susceptible to predation and the spread of infectious disease. The buffalo on the right has a white scar, indicative of depigmentation following a previous injury.

Figure 1.7 In East Africa local people with knowledge of the area are sometimes employed to search for and remove snares that have been placed to catch small mammals. Here one such employee demonstrates how this particular type of snare is set.

The motives for carrying out such activities vary (see Wilson-Wilde before and Chapter 2). It should not be forgotten that wildlife crime is sometimes a response to poverty and a need to obtain food, not just driven by greed and criminality.

Animals as Sentinels

As stated earlier, wild animals can form part of a forensic investigation because they provide important data that are relevant to humans or other species. For example, birds or rodents may be found dead or dying in a farm building following a fire or explosion. Clinical examination, necropsy or laboratory testing of such victims may yield information as to why and when the accident occurred and evidence as to what might have been responsible for the incident – such as traces of chemicals or explosives. European starlings are an example of a species that often serves as a sentinel in this and other respects (Figure 1.8).

On a broader front, as stressed in Chapter 2, wild animals have provided valuable information about the hazards of certain poisons, such as the catastrophic poisoning of birds of prey by chlorinated hydrocarbons in the 1960s and the precipitous decline over the past two decades of Asian vultures due to diclofenac (see Chapter 2). Captive wild animals are occasionally used for toxicological studies that are as exacting and methodical as the best forensic investigations in order to learn more about the safety of new agricultural, veterinary and medical products.

Figure 1.8 The European starling is widespread and has also been introduced into many other countries, where it is usually considered to be a pest. Because of its ubiquity and habits, the starling is often a good indicator of environmental change.

Methods in Wildlife Forensics

Methods used in wildlife forensics follow closely those used in other criminal investigations. Many, but not all, are covered in this book, with a particular emphasis on the scientific aspects.

Most forensic cases involve a combination of the following:

- A visit to the scene of the alleged crime and an assessment of what is seen or found (see Chapter 5)
- Information gathering, primarily by interviewing people who are, or are believed to be, involved in the incident or may have relevant information (see Appendix J)
- Examination of live animals (see Chapter 8)
- Examination of dead animals (see Chapter 9)
- Collection and identification of evidence, including derivatives and samples for laboratory testing (see Chapter 10). This may necessitate a certain amount of field work (see Chapter 7)
- Correct transportation of specimens to the laboratory for testing (see Chapters 10 and 12)
- Laboratory tests, sometimes in the field (see Chapters 7 and 10)
- Correct storage and presentation of evidence (see Chapters 10 and 12)
- Production of report(s) (see Chapter 14)
- Appearance in court (see Chapter 15)
- Retention of material for further court proceedings or for reference purposes (see Chapter 10)

Crime scene investigation is a very important component of wildlife forensic work and is discussed in detail in Chapter 5. Scenes of crime that involve wildlife may be isolated areas – for example, in desert, forest, mountains or marine environments. The investigator operating in such locations must be able to exercise ingenuity and adaptability and be prepared to use portable field equipment (see Chapter 7) or a mobile laboratory (Maas, 2006). It must be remembered, however, that nowadays wildlife crime scene investigations may not be restricted to one, well-circumscribed, location; as Cooper and Cooper (2007) pointed out, the advent of computer-generated criminal activity combined with easy electronic communications can introduce a new and very different dimension.

Throughout this book, as in Cooper and Cooper (2007), the international dimension of wildlife crime – and, as a corollary, the practice of cross-border wildlife forensics – is stated clearly. Investigators with particular expertise may find themselves working in an unfamiliar country, where the language, culture and legislation are very different from home. The particular challenges of such duties are outlined in Chapters 3 and 7.

Many standard forensic methods, whether for live animals, dead animals or samples, can be applied directly to wild animals. However, the variety of species involved in wildlife crime means that extra, different, methods of investigation and interpretation may be needed.

In Chapters 8 to 10, it is stressed that many observational, clinical and *post-mortem* diagnostic techniques, used routinely in veterinary medicine, have an important part to play in wildlife forensic investigations. As with work involving domesticated animals, these usually have to be reinforced or supplemented with special tests such as routine laboratory analysis, radiography and other imaging, analysis of stable isotopes and molecular (mainly DNA) technology. An essential prerequisite to any wildlife legal case is the correct identification of an animal's species; however, in countries where funding is limited and resources are scarce, this is likely to be based, initially at least, on traditional techniques such as gross, morphological examination of body parts rather than more sophisticated and expensive DNA-based molecular methods. Forensic entomology is an important aspect of wildlife forensic work and can provide valuable information about the circumstances of death and the movements of both the perpetrator and the victim. Its potential application is examined in Chapter 6.

As stressed at the beginning of this chapter and repeated throughout the book, wildlife forensics requires an interdisciplinary approach. Many cases are likely to include an input by scientists from various specialities, such as ecologists, zoologists, botanists and veterinarians and both professional and lay representatives of conservation or animal welfare organisations.

As in all fields of forensic science, wildlife crime investigation necessitates the correct collection, processing and storage of evidence for possible production in court (see Chapters 14 and 15). Essential requirements are the use of standard techniques following established protocols, meticulous investigation, proper selection, labelling and transfer of samples, a reliable chain of custody and accurate record-keeping. The importance of quality management, at all stages of forensic investigation, is stressed in Chapter 12.

In Chapter 2 there is discussion of the many, diverse, areas in which litigation and lawsuits seem likely to evolve in years to come. Factors contributing to this include the lucrative global trade in animals, alluded to earlier in this chapter, and the growing concern shown by many international bodies, sovereign states, non-governmental organisations (NGOs) and members of the public about environmental issues and these are likely to lead to an increase in the need for forensic investigations in the future.

Wildlife forensics is an exciting new discipline but presents many challenges. Those involved in investigating wildlife crime may face threats and physical dangers, especially when working in the field. Likewise, those accused of contravening conservation or animal welfare legislation may be exposed to unpleasant publicity and contentious allegations. If the case gets to court, the expert witness – like others – is likely to be subjected to rigorous, sometimes apparently hostile, cross-examination. However, meticulous forensic examinations may be wasted if the experts who are to present the results as evidence are not properly prepared and skilled as witnesses. Chapter 15 examines the legal framework of expert evidence and provides advice on writing reports and appearing in court.

It was stressed by Cooper and Cooper (2007) that the continued conservation and protection of wildlife and the environment – as well as the freedom and rights of individuals – depend in no small part on the integrity of those people who investigate a case and produce reliable evidence and sound reports. This book on wildlife forensic investigation is intended to assist in that respect.

Types of Wildlife Investigation 2

JOHN E. COOPER
MARGARET E. COOPER

Contents

Introduction	12
Wild Animals as the *Cause* of an Incident	13
Effects of Wildlife on Agriculture and Food Production	15
Attacks and Predation by Wild Animals	17
Carnivore–Livestock Conflicts	20
Deer and Road-Traffic (Vehicular) Accidents	21
Scavenging	22
Infectious Hazards	23
Bites and Other Wounds	25
Poisoning and Envenomation	26
Injuries from Reptiles	27
Wild Animals as *Victims*	27
Evolution of Wildlife Crime and Wildlife Conservation	28
Range of Human-Induced Injury in Wildlife	31
Poaching and the Trade in Animals and Animal Parts	32
Ivory and Education	36
Poisoning	36
Investigating Human-Induced Injury in Wildlife	39
War and Civil Disturbance	39
Disasters	40
Bushmeat	40
Traditional Medicines	42
Hunting, Shooting and Fishing	43
Wildlife in Sport	46
Oil and Gas Releases	46
Accidents	47
Captive Wildlife	48
Zoological Collections	48
Aquaria	50
Invertebrates	50
Animals in Entertainment	51
Circuses	52
Keeping of Wildlife as Pets	52
Use of Wildlife in Research	53
Effects of Research	56
Forensic Testing and Research	56
Movement of Animals	57

Challenges of Welfare	57
Abuse	58
Wild Animals as *Sentinels*	59
Examples of Issues Involving Wildlife	59
Conservation Biology and the Maintenance of Biodiversity	60
Species Protection	60
Environmental Forensics	63
Mounted (Taxidermy) Specimens and Collections of Dead Animals	64
Identification of Animals or Their Products	65
Conservation and Cultural Attitudes	66
Releases, Introductions/Re-Introductions and Translocations	66
Biological Control	69
Pest Species	70
Pests and Conservation Programmes	71
Interventions	72
Management of Wildlife Habitat	72
Climate Change	73
Use of Wildlife Resources	75
Wildlife Species for Sustainable Food Production	75
Game-Ranching	76
Breeding Programmes and Conservation Genetics	77
Animal Sanctuaries	78
Habituation of Wild Animals	78
Rehabilitation of Wildlife	78
Confiscations	81
Disease	82
Significance in Wildlife Forensics of 'Global Health' and 'One Health'	83
Effects of Leisure Activities	85
Tourism	85
Disposal of Waste, Including Litter	86
Conclusions	88

'Wildlife conservation has become one of the visible and contentious areas of contact between Africa and the West'.

Adams and McShane

Introduction

In this chapter the types of wildlife forensic case are covered. The text is divided into four main parts. The first three parts follow the categories outlined in the previous chapter – that is, where an animal is

1. The *cause* or *perpetrator* of an incident
2. The *victim* of an incident
3. A '*sentinel*', providing information that is relevant to an incident or a field of enquiry

These three categories are relatively clear-cut. Any one of them can lead to a criminal action, a civil action or both (see Chapter 13). They may, alternatively or in addition, prompt an enquiry or other investigation in relation to the incident.

The fourth category that is covered in this chapter is, however, less well defined and, to a certain extent, includes some forensic 'star-gazing'. It comprises a number of topics and situations relating to animals and conservation, some of them rapidly evolving, which may lead to litigation and with it the need for forensic investigations in the future. It takes the reader into such areas as conservation biology, sustainable utilisation and global health, all of which can be contentious and subject to argument and debate. Some of the areas covered are, perhaps, speculative, but they should be in the back of the minds of anyone who is involved in wildlife forensics and anxious to be part of new developments. This aspect, the evolution and ramifications of the subject, is referred to again in the concluding chapter.

As elsewhere in this book, in this chapter we use examples about different species and from disparate parts of the world to illustrate relevant points.

It is, of course, difficult to quantify and categorise animal forensic cases. What is illegal in one country may not be so in another; cockfighting, for example. In poorer countries only certain cases may reach court, for financial reasons, because of politics, social pressures or on account of corruption or inadequate forensic facilities. The types of wildlife cases that may need involvement of an investigator also differ from country to country. Again, there are many reasons for this. As an example, commonly encountered wildlife forensic cases in the United States concern gunshot/shot gun/rifle/archery incidents, pesticide poisoning, other environmental contaminants, trauma (including road kills), predator kills or electrocution. In the United Kingdom, on the other hand, where firearms are strictly controlled and where hunting of animals by shooting is less prevalent (see later), a comparable list starts with poisoning and neither predator kills nor electrocution features to any significant extent – because Britain has no large mammalian predators and power lines that might be the cause of electrocution are on a smaller scale. In contrast, welfare cases (allegations of cruelty or causing suffering) – primarily, but not exclusively, relating to captive wild animals but also those that are trapped or snared – are amongst the issues that routinely involve British biologists and veterinarians who provide forensic advice.

The four categories will now be considered.

Wild Animals as the *Cause* of an Incident

Wild animals, both vertebrate and invertebrate, cause damage to humans, property, livestock, crops and the environment. In so doing, they prejudice health and safety, animal health and welfare, food production and conservation. Such effects are particularly likely to be of relevance in terms of the law, or in respect of insurance or malpractice claims, if they result in physical injury to humans, livestock or property; toxic damage to humans and other animals; psychological damage to humans and possibly to other animals; or infection, with or without clinical disease, in humans and other animals. Animals may do damage to one another as part of natural behaviour or in response to environmental or human pressure.

At present, information about the sort of injuries that are sustained by humans and how best to interpret these in a forensic context is largely confined to medical texts. It would be advantageous if members of the veterinary and medical professions were to collaborate in comparative studies and produce authoritative publications on the injuries inflicted on humans and domesticated livestock by wildlife.

Animals may either affect humans, livestock or the environment directly – for example, by damaging or injuring them – or indirectly by transmitting, or serving as reservoirs of, disease. Whatever the effect, these incidents can be broadly classified as 'human–wildlife conflicts'.

Sometimes it is indigenous wildlife that is the cause of human–wildlife conflicts, sometimes introduced animals (see also later). An example of the latter is the maintenance of spread of bovine tuberculosis in New Zealand by the introduced brush-tail possum (Coleman, 1988; Coleman and Cooke, 2001; Montague, 2000). Domesticated animals that become feral can cause damage to the environment as well as have an effect on local native species – for instance, feral horses (Levin et al., 2002), feral goats (Maas, 1998) and feral cats (Molsher, 1999).

Human–wildlife conflicts have been analysed by various authors in recent years, notably by Conover (2001, 2002). In his 2001 work, Conover covered a wide range of factors in the interaction between animals and humans, including the economic impact, hazards to humans such as environmental damage and exotic species, zoonoses, lethal control, fertility control, wildlife translocation, fear-provocation stimulation, chemical and sonic repellents and habitat manipulation. Specific papers on wildlife damage to crops include those by Conover and Decker (1991), and human injuries, illnesses and economic losses caused by wildlife by Conover et al. (1995).

Various species of wild animals may cause injury or death to humans, intentionally or accidentally, in a variety of ways. Animals can be the source of bites, trauma, electrocution, stings and hypersensitivity reactions (Strickland, 1991). Cooper and Cooper (2007) tabulated examples of these 'non-infectious insults' and stressed that both vertebrates and invertebrates could be responsible. Some effects of wild animals are unusual: For example, in Canada, quills from porcupines may penetrate deep into the body (Banfield, 1974), and Daoust (1991) describes the finding of a quill in the brain of a dog.

Personnel who work with, or in close contact with, free-living or captive wild animals are particularly vulnerable to injuries – for example game rangers (Figure 2.1).

Figure 2.1 A ranger in central Africa who was gored by a buffalo has his healing wounds checked.

Types of Wildlife Investigation

The various effects that wildlife can have on humans mean that a variety of disciplines may be involved in investigations. For instance, members of the medical profession will take a leading part in the examination of people injured or killed by wild animals. However, a veterinarian or biologist may need to collaborate with them in the examination of the animal(s) alleged to be the perpetrator(s).

Effects of Wildlife on Agriculture and Food Production

Wildlife has long been recognised to have a significant effect on agriculture – in the case of crops, for example, the damage that is caused by geese to oilseed rape (McKay et al., 1993) and by badgers to pastures (Moore et al., 1999). Rodents are an important cause of loss to farmers in many parts of the world: Sometimes it is possible to forecast when outbreaks will occur, as shown (for example) by applying ecological principles to studies on *Mastomys* in Tanzania (Leirs et al., 1996).

Competition can also have a direct effect on the acquisition of food. Fishermen, for instance, may lose a significant part of their catch to piscivorous birds (Figure 2.2).

Measures used to counter wildlife damage can themselves have unexpected deleterious effects. For example, various methods can be used to discourage deer, including chemical deterrents, scaring devices and appropriate fences; all can have an adverse effect on non-target species. Trees can be protected from debarking by wrapping their trunks in wire mesh; but when this method was used in Japan to deter sika deer, it reduced the diversity of bryophytes (Oishi, 2011). This was found to be due to zinc toxicity caused by galvanised iron mesh.

The use of poisonous substances to control 'pest' species can result in disease and death of non-target wild animals. This is a particular problem in poorer countries where

Figure 2.2 Fishermen in small boats on the coast of Mexico compete with gannets for the fish that they have just caught.

Figure 2.3 Wildlife in a privately owned sanctuary in Kenya is subject to rigorous management.

arthropods and other invertebrates transmit a range of diseases, usually necessitating effective ways of killing them (Cooper and Cooper, 2001).

The need to *manage* rather than to protect wildlife was pioneered by Leopold (1933) and by many others since (Figure 2.3). Nevertheless, this is often not fully understood, especially by people in the more 'developed' countries where urbanisation has reduced the opportunities for direct contact with nature. This is supported by Miller and Jones (2005, 2006) who studied perceptions of wildlife management objectives and priorities in Australasia, and by Conover (1997a,b) who carried out broadly similar research in the United States. Reducing contact between wild animals and humans clearly plays an important part in minimising 'human–wildlife conflicts', and much has been published on this – for example, the booklet by Hocking and Humle (2009), which provides guidelines for the prevention or reduction of conflicts between humans and great apes.

In all wildlife control strategies – and in any analysis of damage caused by wildlife – it is essential to have an understanding of the behaviour and physiology of the species. These are relevant to stress and disturbance and to reproductive success. There is increasing interest in behavioural rhythms of wild animals: these influence management, welfare and conservation.

Wildlife damage control necessitates a sound understanding of ecology (Hone, 2007) as much in a European town where urban foxes and badgers are, or are perceived as, causing problems as it does in a rural community in India where jackals and monkeys raid homesteads. Interventions may also have to be tailored to comply with legislation on the protection of species, the control of pests, human safety and the intricacies of private and public ownership of land.

The conflict between urban growth and conservation is complex and will continue to attract interest – and, probably, litigation – in the coming years. An interesting concept, that of 'sustainability assessment', was put forward by Devuyet et al. (2001) in their book *How Green is the City?* The authors addressed the environmental problems associated with increased urbanisation and suggested ways of attaining a more sustainable condition in urban areas, so that people can maintain or improve their health, productivity and quality of life in harmony with nature.

Attacks and Predation by Wild Animals

There is much historical evidence of attacks by wild animals, both vertebrate and invertebrate, on *Homo sapiens* or his hominid ancestors. Such attacks may date back a long time in human evolutionary history. A particularly interesting example, first proposed by Berger and Clarke (1995), and subsequently the subject of considerable debate (see, e.g. McKee [2001]), relates to the famous Taung Skull (an infant *Australopithecus africanus*), a fossil found in South Africa in 1924. The Taung prehistoric site is unusual in that it has yielded only the one hominid specimen and in that the associated fauna consists of various small animals, many of which exhibit unusual damage to the bones and skulls. Berger and Clarke presented data that demonstrated similarities between the Taung assemblage and those characteristic of an extant, large bird of prey. They suggested that a bird of prey may have been responsible for the accumulation of most of the Taung fossils, including the australopithecine skull. The Berger and Clarke hypothesis stimulated much-needed research on the hunting behaviour of large raptors, especially that relating to the taking of primates. This in turn led to further defence of the 'Taung raptor' theory, in particular, the demonstration of damage to the orbital floors of the Taung child that corresponds to that considered characteristic of (non-human) primates preyed upon by contemporary crowned hawk eagles (Berger, 2006).

The occurrence of injuries inflicted on humans by wild animals can be seasonal (e.g. during the breeding season when young are protected) or can represent a change in behaviour either in the condition of the animal itself (such as elephants in musth) or be caused by a change in group behaviour or environmental conditions (e.g. attacks by gulls, which in some locations in Europe is reaching epidemic proportions).

In any attack or interaction, however mild, wild animals may simultaneously infect humans with sundry pathogenic organisms, some of which might be zoonotic – see later.

Both free-living and captive wild animals can inflict physical harm to, sometimes kill, humans. An accident in a zoo, such as when a keeper is killed (sometimes accidentally) by an elephant, is an example of the former (see, e.g. Benirschke and Roocroft [1992]).

Attacks or interactions generally result from closer contact between humans and wildlife and the ensuing competition for land and resources, especially where the human population is growing rapidly (Figure 2.4). Such competition can cause serious enough practical and legal problems in 'developed' countries, but for millions in rural communities in poorer parts of the world they are an everyday threat to life, health and livelihood. In Tanzania, for example, with a population of only 32 million people, lions killed at least 563 and injured many more over a 15 year period (Packer et al., 2005).

Tourists, as well as residents (both rich and poor), in developing countries are attacked by wild carnivores and other potentially dangerous species. There is, it appears, an

Figure 2.4 A view of small farms (*shambas*) on a hillside in Rwanda where there is competition for land and resources.

increasing tendency for tourists to claim compensation, which has led to a defensive position by tour organisers in terms of insurance and exclusion of liability. Such incidents can range from a scratch or bite from a monkey to severe injury or death as a result of an encounter with a buffalo or other game species. Tour operators and insurance companies are becoming increasingly aware of these and other dangers, and as a result often go to great lengths to protect their clients from close contact with wild animals – a measure that is not always appreciated by those who visit Africa or other parts of the world in order to see wildlife and experience adventure.

There is long history relating to attacks on humans by carnivores. One of the best known relates to the 'Man-eaters of Tsavo', a pair of lions that, in 1898, attacked those working on the railway line being built in order to reach the shores of Lake Victoria in what was then called British East Africa and is now the Republic of Kenya (Miller, 1971). Despite much public interest in those events, leading to the production of books and films, it was not until the beginning of the twenty-first century that information about the 'man-eating' lions themselves became available. Kerbis Peterhans and Groske (2001) examined the skulls and skins of the Tsavo lions that had been kept at the Field Museum of Natural

History in Chicago, the United States, for over 75 years. They also studied the regional history of the time. Their work on the skull of the lion that was the primary culprit revealed a traumatic injury that may have limited his ability to take normal prey. There had also been debate about possible dental lesions in the man-eaters (Neiburger and Patterson, 2000).

The Tsavo episode closely followed the pandemic of rinderpest in Africa, which decimated cattle and buffalo, both important prey items for lions. Kerbis Peterhans and Groske also ascertained, from a review of the literature, that attacks on humans by lions occurred before the man-eating episode and continued well afterwards; they suggested that this may have evolved into a local behavioural tradition. The Tsavo story is not only of historical interest but also relevant to contemporaneous investigations into attacks by animals and illustrates the importance, even many decades after the event, of linking meticulous examination of material (in this case skulls, skins and hairs) with historical review.

Similar 'man-eating' incidents have been recorded in other parts of the world, especially Asia (Corbett, 1944).

Competition so often revolves around access to land. In many parts of Africa, local farmers face a regular threat from wild animals such as baboons which steal their crops, and elephants that may trample and damage their property. Action often needs to be taken over such animals. Sometimes it is necessary to kill them; often it is possible to deter them or even to capture and to translocate them elsewhere. Each case has to be judged on its merits.

Injuries may be closely linked with religious or cultural traditions. For example, Orthodox Hindus consider monkeys to be sacred animals, to be revered and protected. Rhesus monkeys and people have co-existed for many years in Vrindaban in Mathura District, Uttar Pradesh, India. The monkeys are highly valued both by locals and pilgrims who visit the area, in part because of their status; but during the past two decades increasing human and monkey populations have led to human–monkey conflict and a decline in popular respect for the monkeys. To ease this situation, one of the largest reported translocations of monkeys was undertaken (Imam et al., 2002). Twelve groups, a total of 600 monkeys, were translocated in January 1997 to eight sites in semi-natural forested areas. A post-translocation study indicated that they had settled and appeared to be exhibiting normal behaviour. This study suggests that translocation of commensal macaque monkeys to forested areas can be a successful technique for their rehabilitation.

Climatic conditions can contribute to unexpected interactions between wild animals and humans. On Mombasa Island, for example, in Kenya, buffalo periodically swim across from the mainland in search of grazing and water (Roland Minor, personal communication). In the absence of available 'natural' water on the island, they make for swimming pools in the homes of wealthy residents. Elephants will do the same and will cause considerable damage in the process, especially trampling on the crops of local people.

Conflict between humans and great apes has attracted particular attention because of the high profile and scarcity of the latter. In South East Asia, the survival of orang-utans is threatened by loss of habitat to logging activities, but in Africa the threats include not only encroachment but also the health risks that such animals and the local population may present to each other due to the proximity of villages, livestock and fields to the gorillas' rainforest habitat. The need for management and good health practices has been recognised. Circumstances vary, but useful guidelines were produced by International Union for Conservation of Nature (IUCN) (Hocking and Humle, 2009). Various projects incorporate into their work the education of villagers whose homes and *shambas* (small-holdings) are in the proximity of protected areas where gorillas and other primates live. An example of how

Figure 2.5 A welcome to the laboratory of CTPH in Uganda, which serves humans and livestock as well as wildlife (see 'Mountain Gorilla Disease: Implications for Conservation' in Chapter 13).

measures instigated to protect mountain gorillas can be integrated with measures to conserve other wildlife and help local communities is given by Kalema-Zikusoka (see 'Mountain Gorilla Disease: Implications for Conservation' section in Chapter 13) (Figure 2.5).

Habitat destruction is mentioned again later.

Carnivore–Livestock Conflicts

Killing and injuring of domesticated livestock by wild animals is well recognised in many parts of the world. It can cause grave losses, devastating to poorer communities and a limitation on rural development. It is often a cause of conflict between farmers, conservationists and animal welfare groups, sometimes leading to legal action.

There have been many published studies on the issue, for example Meriggi and Lovari (1996) reviewed wolf predation in southern Europe and posed the question 'does the wolf prefer wild prey to livestock'; Conner et al. (1998) studied the effect of coyote removal on sheep depredation in northern California; and Linnell et al. (1991) explored the idea that, amongst large carnivores that kill livestock, 'problem individuals' may exist. Treves and Naughton-Treves (1999) discussed 'risk and opportunity' for humans co-existing with large carnivores.

In Africa, species such as African wild dogs, cheetahs, lions, leopards and spotted hyaenas may kill livestock and are therefore themselves hunted and killed by local pastoralists. Such conflict has led to the extirpation of such species from many areas. Woodroffe et al. (2007), using a case–control approach, investigated the possible value of co-existence of people, livestock and large predators in community rangelands, by measuring the effectiveness of traditional livestock husbandry in reducing depredation by wild carnivores. Different measures were effective against different predator species but, overall, the risk of predator attack by day was lowest for small herds, accompanied by dogs as well as human herders, grazing in open habitat. By night, the risk of attack was lowest for herds held in enclosures (bomas) with dense walls, few gates, where both men and domestic dogs were present. Unexpectedly, the presence of scarecrows increased the risks of attack on bomas. These authors suggested that improvements to livestock husbandry can contribute to the conservation and recovery of large carnivores in community rangelands, although other measures such as prey conservation and control of domestic dog diseases are also likely to be necessary for some species.

Differentiation of predated and scavenged carcasses is not always easy and certainly not always appreciated by aggrieved stock owners. The Raptor Conservation Group of the Endangered Wildlife Trust in South Africa has produced educational literature, 'Innocent until proven guilty', in both English and Afrikaans. This emphasised that domesticated dogs, jackals and caracals are more likely to predate lambs in southern Africa than are birds of prey. The literature depicts morphological features of predation and provides guidelines on how to distinguish lambs that are stillborn or which died soon after birth from those that have survived for a few hours or longer; greater detail is provided in papers by Roberts (1986) and (earlier work in Australia) by Rowley (1970).

Kaartinen et al. (2009) looked at wolf depredation on sheep farms in Finland where wolves had recently expanded their distribution range, raising concerns and public debate over wolf–livestock conflicts. Between 1998 and 2004, there were 45 wolf attacks on sheep on 34 farms. Sheep farms with the highest risk of wolf depredation were those located in regions where wolves were abundant. These were usually located close to the Russian border. Here the landscape is a mosaic of forest, wetlands and open areas, sparsely populated by humans and with farms far from each other. They produced probability maps to help predict the risk of wolf predation on livestock on farms in the region.

Allegations of predation by wild animals on stock, such as lambs, always need very careful investigation. Many species have traditionally been considered predators on flimsy evidence. Often there are other causes of stock death. This point was graphically illustrated in a paper on Arctic foxes in Iceland by Hersteinsson and Macdonald (1996). They pointed out that, in Iceland, legislation to exterminate Arctic foxes dates back to about AD 1295. Hersteinsson and Macdonald suggested that, as with red foxes (Macdonald, 1987; Macdonald and Johnson, 1996), a proportion of carcasses found and death attributed to Arctic foxes had been scavenged, not killed. Many lambs succumb as a result of neglect or natural causes. As long ago as 1783, an Icelandic writer expressed disgust at the numbers of lambs that died 'due to poor condition and husbandry' (cited by Hersteinsson and Macdonald, 1996).

Deer and Road-Traffic (Vehicular) Accidents

Deer appear to the uninitiated to be innocuous herbivores, but the danger of injuries from some species has long been recognised. The annual toll of deer involved in animal–vehicle collisions in the United Kingdom was estimated 7 years ago to be at least 40,000

(Langbein et al., 2004), and estimates in 2011 by the Deer Initiative suggested that the number is rising. Such incidents are generally higher during the rutting season and at dawn and dusk, when deer usually travel (the 'rush hour' for motorists as the daylight period shortens). More scientific data are needed on accidents involving deer and more reliable figures as far as numbers of deer and species of deer are concerned. Chapman (2006) asked veterinary practitioners and others to send her details of any such incidents. She drew attention to the frequency of encounters between (introduced) muntjac and dogs; while, as a result, some muntjac are maimed or killed, in other incidents the dogs themselves suffer wounds, particularly inflicted by the canine teeth of muntjac bucks. The increase of deer populations in the United Kingdom has resulted in substantial damage to crops (see later).

'Collision investigation' is a recognised forensic speciality and while it usually implies conventional traffic accidents involving humans and vehicles, wild animals are often affected, killing or damaging themselves or humans (see Chapter 9). Sometimes threatened species are affected – see, for example, the survey of koala road kills in New South Wales, Australia, by Canfield (1991).

While wild animals of many species are injured or killed by traffic, it should be noted that reptiles and amphibians are amongst those that are particularly vulnerable (Ashley and Robinson, 1996). Indeed, such road kills already often provide useful information about reptile abundance and may be used to survey the presence of infectious disease (Wright et al., 2004). Forensic investigations could perhaps benefit from a similar enterprising approach. There is current interest in factors that may affect trends in road kills, as illustrated by the work of Barthelmess and Brooks (2010) looking at the influence of body-size and diet on such incidents in mammals.

Some injured wild animals harm those who try to help them.

Scavenging

Both vertebrates and invertebrates may mutilate and devour human and animal bodies after death, and an investigator may be called to advise on the species of animal responsible.

Animal-inflicted injuries inflicted *ante mortem* that prove fatal must be differentiated from those that were caused *post mortem*. The latter are the result of predation, in Northern Europe for example, by dogs and badgers on land (see 'Advances in Scavenging Research' section in Chapter 13) or by fish and crabs in water (Shepherd, 2003; Williams et al., 1998). Much has been published on *post-mortem* scavenging as an aid to the assessment of dead stranded marine mammals (Read and Murray, 2000). Animals of diverse species may mutilate and devour both animal and human carcases (see Figure 2.6).

Such activity may involve carcases found by the scavengers in, for example, woodland or the sea. In the case of some captive (usually carnivorous) wild animals, the body of the owner or a keeper may be scavenged if he or she dies, the body is not immediately found and animals have access to it. Scavenging of human bodies can also take place on a large scale in uncontrollable situations such as the 1994 genocide in Rwanda. Stray dogs roamed and they and wild corvids fed on the corpses of those who had been killed. As a general rule, *post-mortem* scavenging can be differentiated from *ante-mortem* predation by the lack of haemorrhage or inflammation; in humans it tends to be concentrated on exposed, unclothed, areas where scavengers can gain access.

Figure 2.6 East African vultures and marabou storks feed on the carcase of an animal. (Courtesy of Chris Daborn.)

Infectious Hazards

Zoonoses can conveniently be defined as 'those diseases and infections that are naturally transmitted between vertebrate animals and humans' (based on World Health Organization [WHO] definition) and are covered in more detail in Appendix F. Zoonoses are of considerable legal significance, increasingly so in highly safety-conscious countries (Palmer et al., 2011). The emergence or spread of a zoonosis, such as an *Escherichia coli* infection, might, under certain circumstances, constitute a criminal act, especially if appropriate risk assessments have not been carried out (Walsh and Morgan, 2005). A case or cases of such an infection – for example, in children following a visit to a zoo or at an outside restaurant frequented by free-ranging peafowl (Figure 2.7) in – may, in addition, be grounds for civil law actions, insurance claims or allegations of professional (e.g. veterinary) negligence.

Zoonoses associated with wild animals can occur in urban areas as well as in rural locations – perhaps more so. A simple, but topical, example concerns salmonellosis in wild birds. In many westernised countries there is great interest in putting out food for wild birds, attracting them to bird tables or feeding stations. However, there is also increasing evidence that such practices can play a part in spreading infectious diseases of birds, including salmonellosis which can prove zoonotic (Lawson et al., 2001). Salmonellosis has caused the death of passerine birds in a number of countries, as well as the United Kingdom (Kirkwood and Macgregor, 1998; Prescott et al., 1998). In most cases, the type of *Salmonella* involved is *Salmonella typhimurium* phage type 40. In addition to their effects on birds, *E. coli* O86 and *S. typhimurium* phage types 40 and 160 can cause disease in humans.

Concern about salmonellosis and other possible human health risks from feeding wild birds has prompted the production of guidelines about hygiene (including the choice of an appropriate disinfectant and how to use it properly) and methods of dealing with the carcases of birds that are found in the garden. Manufacturers of foodstuffs for wild birds are

Figure 2.7 Two peafowl compete for food on the table in an outside dining area. Such birds may be a source of zoonotic infection but, possibly, not such a threat to human health as the cigarettes that they are helping to advertise!

generally conscious of the need to educate purchasers and, where appropriate, to carry out quality control checks on the diets that they are providing (see Chapter 10).

Molina-Lopez et al. (2011) identified wild (i.e. free-living) raptors in Spain as carriers of antimicrobial-resistant *Salmonella* and *Campylobacter* strains. They argued that the existence of *Salmonella* and *Campylobacter* reservoirs in wildlife is a potential hazard to animal and human health, but that the prevalence of these species in wildlife is largely unknown. In their study, multidrug-resistant *Salmonella* and quinolone-resistant *Campylobacter* strains were present in several species of birds of prey that had no record of antimicrobial treatment.

Both the medical and veterinary professions are likely to be involved in legal cases concerning zoonoses. However, it cannot be assumed that such a background automatically equips a medically or veterinary trained person to be called as an expert witness. A recent study in India (Kakkar et al., 2011) showed that a relatively small number of medical students in that country were familiar with zoonoses, their diagnosis and treatment.

In any legal dispute about zoonoses, tracing the source of the infection and determining when transmission occurred are of prime importance. While the former is often not difficult to ascertain, the latter can be very problematic. Remarkably little work has been carried out on the epidemiology of some of the most important infections and diseases. Some data do exist and may be helpful. For example, the U.S. Centers for Disease Control and Prevention define the time from infection of a person with *Salmonella* to shedding of the organism in the faeces as 1–3 days, sometimes longer. A recent study by

Aceto et al. (2011) calculated the time from *Salmonella* infection in animals to the first detection of the organism in the faeces, and the onset of clinical signs, to see whether the U.S. Centers for Disease Control and Prevention three-day criterion could be used to identify nosocomial infection in animals. The literature was searched for journal articles, theses and book chapters that contained studies in which cattle, horses and sheep were experimentally infected orally with *Salmonella enterica* sub-species *enterica*. The time to detection of *Salmonella* in faeces and the onset of diarrhoea and pyrexia were recorded and the variables were analysed. Forty-three relevant studies were found; 2 were published in the 1920s, 15 in the 1970s and there was a median of 4 studies from each of the other decades between 1940 and 2010. In these papers the time that elapsed before *Salmonella* could be detected in faeces ranged from 0.5 to 4 days. Times to onset of diarrhoea and pyrexia ranged from 0.33 to 11 days and from 0.27 to 5 days, respectively. Time to onset of diarrhoea was related to the host age and *Salmonella* serovar; no other associations were identified. As a result, the authors conclude that the time to detection of *Salmonella* in faeces is not a reliable measure for identifying the source of an infection; adhering to a three-day criterion is unreliable. The authors added that relying on clinical indices, such as time to onset of diarrhoea or pyrexia, before sampling for *Salmonella* would increase the risk of environmental contamination and nosocomial spread, as animals may begin shedding *Salmonella* in the faeces several days before the onset of clinical signs.

The aforementioned example relates to one type of zoonosis, salmonellosis, in the cases referred to earlier derived from domesticated mammals. Similar research is needed on infectious diseases that involve wildlife.

A few specific studies have been carried out on the zoonotic risks presented by certain wildlife species when they come into contact with humans – for instance, marine mammals (Mazet et al., 2004). There is also an extensive volume of literature on diseases and organisms transmissible from fish, much of it in the public health area but, overall, a dearth of reliable data that might be cited in a legal case.

Bites and Other Wounds

Bite wounds that are inflicted by animals *ante mortem* are of considerable importance in forensic work (see 'Predation of Endangered New Zealand Weka by Dogs' section in Chapter 13). Animals bite people and people sometimes bite animals. Animals also inflict bite wounds on others of their own or different species. Bites not only damage tissues but may also introduce infectious organisms, especially bacteria; different species harbour different organisms and the oral flora can easily be influenced by the feeding habits and diet of the perpetrator (Cooper and Cooper, 2007). In all cases involving animals, regardless of species, a suitably experienced veterinary clinician or knowledgeable pathologist can usually assist with information on

- The species inflicting the bite
- Identification of the individual animal
- The health status of the animal inflicting the bite, including identification of oral flora and the presence or absence of dental disease
- Type of bite (incisors, cheek teeth) and the mechanisms involved (e.g. shearing, tearing)

A comparative odontologist, usually with a dental, medical or veterinary background, may also be able to advise. Bitemark analysis is a specialised aspect of human forensic odontology (see Chapters 9 and 10) and is becoming increasingly important in veterinary and comparative forensic studies.

In any investigation involving a bite where legal action is threatened, it is important to assess the wounds that the bite has caused. Such assessment is often not an easy task in practical terms because the victim will require a detailed examination. In addition, however, the way in which the victim's tissues respond to a bite and how rapidly changes take place in those tissues will depend upon the species of victim. In particular, there will be differences in endothermic ('warm-blooded') animals depending upon their age and size and in ectothermic ('cold-blooded') animals depending upon the ambient temperature to which they are exposed. These differences are explained in more detail in Chapter 9.

Poisoning and Envenomation

Cooper and Cooper (2007) discussed at some length the part played by animals, both domesticated and wild, in poisoning and envenomation. The latter is a feature of some species of snakes, lizard, scorpion and marine species (Figure 2.8). In consequence, these will not be covered in detail here. Cooper and Cooper (2007) also provided information about bufotoxicosis ('toad poisoning'), mainly as it affects domesticated mammals.

Until recently, little scientific information was available on the local and systemic effects of bufotoxins in wildlife. The need for such data has, however, never been more important, particularly throughout tropical Australia, where the introduced cane toad has recently been implicated in mass mortality of freshwater crocodiles (Letnic et al., 2008). The situation regarding birds is interesting and poorly studied. Many raptors eat

Figure 2.8 Monitor lizards (*Varanus* spp.) can inflict painful bites and their saliva contains toxic substances.

These are important considerations in any discussions about the evolution of wildlife crime and how it might be tackled in the long term.

In particular, the value of 'protected areas' has to be re-assessed, especially where these are expressly for wildlife and local people are excluded. The effectiveness of protected areas is never easy to calculate and the criteria used may vary – as will the political and legal environment in which the national park, game reserve or sanctuary functions. A questionnaire developed by the World Bank and the World Wide Fund for Nature (WWF) was used by Quan et al. (2011) to survey 535 nature reserves in China.

Du Toit (2002) discussed community-based wildlife management in southern Africa and concluded that exponential human population growth and the adoption of non-indigenous and unsustainable land-use practices have resulted in a land crisis in that part of the world. This in turn is causing the emigration of people from regions in which natural resources have been depleted, and the immigration of people into regions where natural ecosystems are still relatively intact. The latter, which had not previously been intensively settled due to the land tenure policies of previous governments, their marginal potential for dryland cropping and/or the presence of tsetse flies, may still support important wildlife resources. Associated with the land crisis are inevitable and intractable socio-political tensions, which heighten the urgency for conservation agencies to assist rural Africans to recognise wildlife as an asset rather than a liability. Community-based natural resource management systems hold more promise for the conservation of African wildlife than does a dependence on largely unenforceable state-legislated controls.

Range of Human-Induced Injury in Wildlife

Humans may injure or kill wildlife in a variety of ways – sometimes intentionally, sometimes by neglect, sometimes by accident. The task of the investigator who is presented with an animal that has been found injured or dead under unusual circumstances is not to apportion blame (this is the responsibility of the court) but – with the assistance of a veterinary pathologist – to decide whether or not the injuries or death were likely to have been human-induced (anthropogenic), caused by other species or the result of such a factor as hypothermia, hyperthermia or starvation (see later). Thus, for example, if large numbers of dead lizards and snakes are found on moorland following a public holiday and human action is suspected, the investigator will need to initiate

- A crime scene visit and assessment
- Specialist examination of the dead reptiles and their tissues
- Involvement of other experts, for example herpetologists, as necessary

Assaults on wild animals can be *malicious* or *accidental* and can be further divided into three main categories:

- *Physical*: trauma, excess heat or cold, immersion in water or other insults
- *Sexual*: attempted copulation, damage to reproductive organs
- *Psychological*: teasing, taunting or threatening, deprivation of company, unsuitable social grouping

A general review of the effects of non-infectious insults on wildlife, including trauma, electrocution, burning, drowning, hypothermia and hyperthermia, was provided by Cooper (1996).

Much has been published on the infliction of traumatic injuries in certain taxa – cetaceans, for example (see 'Marine Mammals: A Special Case' section in Chapter 13). As an example, Read and Murray (2000) documented anthropogenic trauma in small cetaceans (dolphins, porpoises) and listed entanglement in gill nets, purse seines, ropes and lines, gunshot, vessel collisions and blast-induced trauma. They provided valuable information on how to differentiate and diagnose such incidents. They re-stated that entanglement in fishing gear is the most common human-induced cause of mortality in small cetaceans in Europe and emphasised that 'the physical evidence associated with entanglement is specific to each combination of cetacean and fishing gear'.

A whole spectrum of investigations may be needed in the study of human-induced injury in wildlife. The veterinarian is often a key part, but he or she is likely to be working as part of a team, alongside zoologists, botanists, ecologists and representatives of conservation bodies (Stroud, 1998). Investigations may include clinical or *post-mortem* examination of wild animals, analysis of samples for poisons or evidence of firearm damage and investigation of vegetation, soil or water.

Each of these is discussed in more detail elsewhere in this book.

Poaching and the Trade in Animals and Animal Parts

As was pointed out earlier in this chapter, the term 'poaching' usually refers to unlawful hunting. At the local level, it is usually carried out in order to provide food for the poacher or his family. However, large-scale poaching fuels the illegal trade in wild animals and their derivatives and has serious effects such as decimation, even the disappearance, of target species from that area, pressure on other species as the demand for meat or other products grows, socio-economic disturbance and political controversy. The last of these may be very profound if, as has often been the case in recent years, senior figures in government are involved in the trade or, because they or their relatives are beneficiaries, choose not to take appropriate action.

An example of the effects of large-scale poaching can be taken from the history of the Tsavo National Park. This has been well documented elsewhere and will only be summarised here. Tsavo has a rich and diverse ecosystem (Figure 2.11). In the nineteenth century, it provided the location for the construction of the Kenya–Uganda railway and then, in the First World War, it was the centre of fierce battles between British and German forces who were fighting for supremacy over East Africa. In 1948, Tsavo was established as a National Park and served as a breeding and viewing ground for many large African mammals, especially elephants, rhinoceroses and gazelle. Poaching was always a threat because Tsavo was large (443,000 km^2, 3% of Kenya's land surface, and one of the largest conservation areas in Africa) and very inaccessible. Up until Kenya's Independence from Great Britain in 1963, poaching was either on a small scale or effectively contained. However, in the 1970s and 1980s the situation changed. In the early 1980s, for example, poachers from Somalia were responsible for killing about 5000 elephants per year in order to harvest their ivory. Kenyan Rangers were killed in their efforts to try to control the situation. The 1970s and 1980s also saw a severe, prolonged drought and the direct and indirect effects of that on the elephants and other species have been debated and published. Elephant populations numbered 40,000 in the early 1970s, but by 1988 only 5,000 elephants remained. Rhinoceroses (black) suffered even more: 99% were killed in order to harvest their 'horn', which was then sent to the Middle East and Asia where it fetched – and continues to fetch – a high price.

Types of Wildlife Investigation   33

Figure 2.11 Tsavo National Park, Kenya, photographed in 2011.

The response of the Kenyan government was to establish the Kenya Wildlife Service (KWS) and to equip it with modern vehicles and appropriate weapons. It took 3 years for KWS to regain control of Tsavo National Park. Much of the story is recounted by Leakey and Morrel (2001). In 1989, an international ban on trade in ivory rhino horn helped to reduce poaching. In the 1990s, Tsavo acquired new vehicles and heavy duty equipment, including earthmoving machines. As a result, anti-poaching measures could be carried out with more ease on a larger scale. Twenty years later (2011), Tsavo still faces many threats but its history indicates how strong, often draconian, methods can help to reverse what appears to be a desperate situation.

Well-equipped, sophisticated, organised crime syndicates present a growing threat to endangered wildlife – and often a danger to those who seek to confront them. As an example, in March 2011 the IUCN SSC Species e-Bulletin reported that such syndicates had killed 800 African rhinoceroses over the previous 3 years.

Illegal wildlife trade can be defined as

> an environmental crime that directly affects the environment and includes live pets, hunting trophies, fashion accessories, cultural artefacts, traditional medicine ingredients and wild meat (bushmeat) for human consumption.

In 2003, it was calculated that such illegal wildlife trade was second only to the drugs trade, being worth approximately £10 million. In 2008, it was estimated to be worth at least $5 billion, though this figure may actually have been closer to $20 billion per annum. In Europe, the wildlife trade is approximately worth between €18 and €26 million. Some details are provided on websites (see, e.g. http://www.eurocbc.org/wildlife_guidelines.pdf http://en.wikipedia.org/wiki/Wildlife_trade). However, as pointed out in Chapter 10, caution needs to be taken in interpreting the figures that are given in discussion and, particularly, lobbying about wildlife trade.

Available data include figures as follows. Between 1994 and 2006, 783 skins and body parts of tigers, 2766 leopards and 777 otters were seized in India alone. In Spring 2008, two shipments containing approximately 23 tonnes(t) of dead pangolins and scales were discovered in Vietnam, and in July 2008 there were 14 t of frozen pangolins and 50 kg of scales seized in Sumatra. In 2003–2004, there were more than 7000 seizures (including 3.5 million wildlife specimens) in Europe. From 2002 to 2006, nearly 1000 illegally traded Egyptian tortoises were seized in the European Union (EU) – a number that represents approximately 13% of the believed entire population. In the year from April 2006 to March 2007, Customs officials intercepted more than 165,000 illegal wildlife products crossing UK borders.

Reviews of the illegal trade in wildlife have been published from various individual countries – for example Australia (Alacs and Georges, 2008).

Scientific surveys are often the most reliable sources of information. For example, a study by Tapley et al. (2011) compared the trade in reptiles and amphibians in the United Kingdom between 1992–1993 and 2004–2005. In particular, the impacts of captive-breeding and colour and pattern morphs on price structures were examined. The number of amphibian and reptile species in the trade more than doubled over this period, and less than a third of the species traded were common to both trading periods. More traded species were listed by CITES in 1992–1993 than were in 2004–2005. Taking inflation into account, the study showed that the price of reptiles and amphibians recorded increased over the 10 year period, and that some snake species had done so dramatically when colour and pattern morphs were considered. The price change of chelonians was probably the result of responses to changes in various trade regulations. Price increases for amphibians seemed to represent their increased popularity, coupled with the overhead costs of captive-breeding on a commercial scale being transferred to the hobbyist. The increased popularity of captive-bred colour and pattern morphs could alleviate pressure on wild stocks. On the other hand, because such animals are predominantly being bred outside their countries of origin, no benefits accrue to local people; trade could thereby undermine sustainable use programmes.

The trade in live animals, dead animals and their derivatives is clearly a major challenge to those combating wildlife crime. Its investigation necessitates specialist forensic investigation, including the identification of species from trace samples and mixtures (Tobe and Linacre, 2007). Cooper and Rosser (2002) outlined the relevant legislation and organisations.

The species traded illegally are not just the large mammals that attract much public attention. For example, seahorses are globally threatened by overexploitation and habitat destruction. They command a high monetary value and marketability. Seahorses are listed on Appendix II of CITES (Convention on International Trade in Endangered Species of Wild Fauna and Flora – see Chapter 3), but a study by Rosa et al. (2011) showed that in Brazil the dried trade was unregulated, without formal records, and primarily domestic, it was primarily sustained by incidental captures in trawl nets. The live seahorse trade was mainly destined for exports, and regulated through national quotas.

Another example from Brazil concerns molluscs and the marine curio and souvenir trade. Dias et al. (2011) pointed out that molluscs are sold all over the world as curios and souvenirs and that the international market encompasses about 5000 species of bivalves and gastropods. However, the species involved are undocumented in all but a very small number of countries. In a study in northeastern Brazil, a total of 126 species (41 bivalves and 85 gastropods) was found to be sold individually as decorative pieces or incorporated into utilitarian objects. Large gastropod shells were observed in all of the localities surveyed – used mainly as part of table lamps. The majority of the species involved are harvested from the Atlantic Ocean (68%), but many are imported from Indo-Pacific countries. Among the species harvested in the Atlantic Ocean, 11.2% are endemic to Brazil. The majority of the species sold tend to inhabit shallow habitats, which facilitates their capture. As there are no official statistics available concerning the marine curio and souvenir trade in Brazil, some species may be endangered. The authors advocate the introduction of harvesting regulations, which should include minimum capture sizes, capture quotas, specific periods for harvesting and the use of non-destructive techniques.

Poaching occurs in much of the world. It can be extremely difficult to investigate and prosecute due to the nature of the evidence available. Previous studies have focussed on the identification of endangered species in such cases. Difficulties arise if the poached animal is not endangered. Species such as deer have hunting seasons whereby they can legally be hunted; however, poaching is the illegal take of deer, irrespective of season. Therefore, identification of deer alone has little probative value as samples could have originated from legal hunting activities in season. After a deer is hunted, it is usual to remove the viscera, head and hind limbs. The legs that are removed are a potential source of human touch DNA. Tobe et al. (2011a) investigated the potential to recover and profile human autosomal DNA from poached deer remains. DNA from the legs of 10 culled deer was extracted, quantified and amplified to determine if it would be possible to recover human STR profiles. Their study demonstrated the potential of recovering human touch DNA from poached animal remains and they suggested that there is potential for this test to be used in relation to other species of poached remains or other types of wildlife. It should be mentioned that, in their Summary, Tobe et al. stated that

> If a species is protected by international legislation such as the Convention on International Trade in Endangered Species of Wild Fauna and Flora then simply possessing any part of that species is illegal.

This is an inaccurate statement as possession of such material is not, *per se*, unlawful (see Chapter 3).

The legal trade in wildlife brings into play a variety of legislation from several distinct areas of law (Cooper and Rosser, 2002) – see Chapter 3. Many species are subject to restrictions on international movement with the aim of protecting wild populations from over-exploitation. Animal health legislation is strictly applied to the movement to most animals to prevent the spread of infectious diseases between importing and exporting countries. The welfare of animals in the course of trade requires consideration and relevant legislation has been put into place, particularly in respect of transportation.

Ivory and Education

Described by John Donne as 'Nature's great masterpiece, an elephant; the only harmless great thing', elephants have borne the brunt of much of the world's illegal trading (see earlier, regarding Tsavo in Kenya). There is, however, some encouraging international collaboration in tackling the elephant ivory problem. In July 2011, for example, the Lusaka Agreement Task Force (LATF) (see Chapter 3) was reported to have burnt 335 tusks and 41,553 signature seal blanks, weighing 4,967 kg (Stiles, 2011a). The KWS had been storing the ivory on behalf of the LATF since March 2004 when it was received from Singapore. The ivory was part of a stated 6.5 t seizure made in Singapore in 2002. DNA analysis indicated that the ivory originated from elephants from Zambia, Malawi and Tanzania, consistent with the export from Lilongwe, Malawi.

The burning of ivory, exemplified graphically by the action taken in Kenya some years ago, is sometimes put forward as a way of discouraging poaching and sale. However, Stiles (2011b) questioned the rationale behind the burning of ivory. He accepted that most commentators strongly believe that burning large quantities of ivory serves to deter elephant poaching by 'sending a message'. But what, Stiles said, is this message? Instead, he suggested, seeing a pile of burning tusks may just stimulate hunters to find more ivory before it has all gone.

The laboratory examination of elephant ivory is considered in Chapter 10.

Poisoning

A poison (otherwise known as a toxicant or, usually if originating from an organism, a toxin) is a substance, a chemical, that can harm the body if it is taken in sufficient quantity. Most substances can be poisonous if taken in sufficiently large amounts. This was pointed out 400 years ago by Theophrastus Philippus Aureolus Bombastus von Hohenheim (1493–1541), usually known as Paracelsus, to whom the quotation '*Dosis sola facit venenum*' ('It is the dose that makes (determines) whether or not a substance is poisonous') is attributed.

It can be useful, especially in legal and forensic investigations, to divide poisoning incidents into three categories – natural, accidental and malicious. Each will be mentioned briefly later.

It is sometimes forgotten by investigators that poisoning can be due to toxins that are present in nature, not just toxic substances that are produced by humans. Some animals produce or contain toxicants (see earlier). Many plants harbour toxic substances; a widespread example is bracken (*Pteridium* spp.), which can cause ill-health and death, in different ways, in ruminants and equids. Other natural toxins that are relevant to wildlife health include those that are released when toxin-producing algae rapidly grow and multiply (harmful algal blooms or HABs); these can, for example, kill marine mammals (see 'Marine Mammals: A Special Case' section in Chapter 13). In addition, various inorganic poisons may be present in soil or water.

Poisons of many types are used, legally and illegally, to control (kill) wild and feral animals. The situation differs from country to country; what is legal, so long as it is used in accordance with guidelines, in one country may be totally banned in another. Poisoning can be *accidental* – for example, when a substance that is being used lawfully inadvertently enters a river or contaminates pasture, or *maliciously*, as when poison baits are put out to kill birds of prey and other protected species.

Wild animals (including invertebrates) are susceptible to a wide range of poisoning and most incidents require at least an initial forensic investigation (Elliott et al., 2011; Roscoe and Stansley, 2012; Stroud, 1998). Legal cases relating to alleged poisoning necessitate information on the type of poison present and when or how it is used. There may also be the question of who is permitted to possess, provide or supply the poison. In many countries – most of those within the EU, for example – farmers and landowners who need to control pest species (foxes and crows, in particular, in the case of Europe) are either required by law or encouraged by education to use safe alternatives to poisons such as shooting, trapping or deterrents.

Another important consideration is that some poisons cause a slow and painful death; as a result, charges of cruelty may be made. This may, however, not be applicable to free-living wild animals (see Chapter 3).

Acute poisoning cases are usually more likely to be detected by the public or others because animals are found dead or dying. There is therefore often a bias in reports. Toxic substances may be ingested, inhaled (e.g. gases) or passed through the skin. Susceptibility can depend upon the species involved, for example there may be ready absorption of certain chemicals through the mucous skin of frog but less so through the heavily keratinised integument of an alligator. Insects, such as bees, are susceptible to many toxic substances and, by virtue of their social behaviour, may spread the poison to others of the same or different species.

Chronic toxicosis may have profound secondary effects on ecosystems. If poisons enter the food chains, they can have particularly adverse effects on certain species, such as predators. The classic, much-quoted, story of chlorinated hydrocarbons in North America and Europe in the 1950s and 1960s is an excellent example (for background, see Cooper [2002a] and Mellanby [1967]). In Britain, the story started with reports of deaths in foxes, some of which showed neurological signs before death. Initially, distemper or some other infectious disease was suspected. In the early 1960s, unexplained deaths in birds began to be reported; in 1961, for example, 142 dead birds were collected over a period of 12 hours(h) on the Royal Estate, Sandringham, in Norfolk. From 1962, deaths in birds of prey and a marked decline in their numbers alerted scientists and others to the fact that toxic chemicals might be involved. Years of research, involving *post-mortem* and toxicological examination of dead animals and retrospective studies on the eggs of birds of prey in museum collections, confirmed that chlorinated hydrocarbon insecticides, such as DDT, aldrin and dieldrin, were for the most part responsible. The spectre that widespread use of pesticides and other chemicals presented was graphically portrayed in the book by Carson (1962).

A useful summary of the direct and indirect effects of pesticides and pollutants on birds, graphically illustrated with a drawing of a peregrine on its nest with crushed thin-shelled eggs, is given by Newton (1998). Ratcliffe (1967, 1980) conducted seminal studies – truly forensic in their detail and investigative quality – on this and other aspects of chlorinated hydrocarbon pesticide poisoning in birds of prey, with particular reference to the peregrine (see Chapters 9 and 10).

Some toxic substances can be immunosuppressive and thus predispose wildlife to infectious disease; polychlorinated biphenyls (PCBs) in porpoises are an example (see 'Marine Mammals: A Special Case' section in Chapter 13). Petroleum-based chemicals can kill animals by toxicity as well as by having an adverse effect on pelage and thermoregulation (see later).

Wildlife may be poisoned as a result of ingesting toxic substances from domesticated livestock – for instance, farm animals that have been poisoned accidentally, euthanased, or contain veterinary drugs. Poisoned wildlife, especially if chronically affected, may itself attract predators, which may then be secondarily affected. A topical example of the ingestion of toxic substances from domesticated livestock concerns the analgesic diclofenac, which for some years was used in India and elsewhere for cattle, especially cows which in several Indian states are considered sacred and may not therefore be euthanased. Vultures that ingest diclofenac die of renal failure (Oaks et al., 2004). Diclofenac is now recognised as the major cause of decline of *Gyps* vultures in South Asia: Three species have been brought to the brink of extinction (http://www.vulturedeclines.org/). The results of the decline of vultures illustrate the profound effects that poisoning can have on humans and on both wild and domesticated animals.

The decline in vultures in South Asia has also prompted establishment of captive-breeding centres – supported, interestingly, by such organisations as the (British) Royal Society for the Protection of Birds (RSPB), which has not traditionally been supportive of captive-breeding or other ventures relating to the keeping of birds in captivity. At the time of writing (2011), the Bombay Natural History Society, with support from the RSPB and the newly formed Consortium Saving Asia's Vultures from Extinction (SAVE), manages three conservation breeding centres in India.

The effects of the vulture decline extend into the local communities. 'Carcase dumps' in Asia formerly attracted thousands of vultures. Now, however, vultures are scarce and in their place, packs of dogs, many feral, visit the dumps and can, in turn, be responsible for dog bite incidents and spread of diseases, including rabies. One study attributed an estimated 47,300 human deaths to rabies over a period (Markandya et al., 2008). The Parsi community, which had traditionally placed the bodies of their dead on Towers of Silence so that vultures could consume them had to seek alternative methods of disposal. Without the vultures to dispose of carrion, there has been a concomitant rise in stray dog populations.

The manufacture and sale of diclofenac for veterinary use is now illegal in India, Nepal and Pakistan, but there is growing evidence that in some places farmers and livestock owners are purchasing the human preparation and using it for their cattle. There are new dangers on the horizon, in addition, including the use (misuse) of replacements for diclofenac, such as ketoprofen and meloxicam (Naidoo et al., 2009; Ng et al., 2006).

Another example is poisoning due to carbofuran. This carbamate systemic (and contact) pesticide is sold under various trade names including Furadan (FMC Corporation). It is used to control insects in field crops, such as potatoes, maize and soya beans. Its deployment has increased in recent years because it is one of the few insecticides that will kill soya bean aphids. Incidences of intentional and unintentional wildlife poisonings involving carbofuran are increasing, especially in Africa and India. A multi-authored book edited by Richards (2011) is concerned entirely with the issue of carbofuran and wildlife mortality. It provides a summary of the history and mode of action of carbofuran and its current global use. It covers wildlife deaths stemming from legal and illegal uses, outlines wildlife rehabilitation, forensic and conservation approaches, and discusses global trends in responding to the wildlife mortality.

Although carbofuran has been banned for use in the United Kingdom and, at the time of writing, the United States is in the process of removing it from the market, the product is widely available elsewhere. For example, it is legally available in most agro-vet shops in Kenya and is there misused to kill a wide range of animal species that are deemed a 'nuisance' by farmers and pastoralists (Odino et al., 2008). Confirmed mortalities due to carbofuran in Kenya include birds, especially vultures and other raptors, lions, hyaenas and hippos.

There is still a paucity of information about the effects of many familiar chemicals on wildlife. For example, Cooper (2008b) discussed possible health hazards posed to animals as a result of the establishment and use of aluminium smelters. Although much is known about fluoride poisoning (fluoride is a by-product of aluminium smelting), the direct and indirect effects of aluminium remain largely unknown. Cooper advocated routine monitoring of wild animals in the vicinity of aluminium smelters and other industrial plants from which effluent may be entering the environment and there having an effect.

A growing problem in many parts of the world is the question of agricultural and other chemicals that remain on farms and elsewhere but which are redundant, often because they are no longer approved for use. In some countries there is a black market in these products and so poisoning incidents associated with their use continue to occur in humans, domesticated animals and wildlife. Even in westernised countries, the cost of collection and disposal is usually prohibitive and so many farmers, gamekeepers and pest-controllers find themselves in possession of unwanted toxic substances that might still threaten the environment. In the United Kingdom there is now a scheme, 'Project soe (security in the operational environment)', to assist in the safe, cost-effective removal of such chemicals.

The forensic investigation of poisoning in live animals is discussed in more detail in Chapter 8, in dead animals in Chapter 9 and in the laboratory in Chapter 10. In all instances where poisoning is suspected, intentional or accidental, it is important to ask key questions from the outset:

- Have there been previous incidents in the location attributed to, or, confirmed as, poisoning?
- What agricultural or forestry practices are carried out in the area?
- What pest control measures are in use locally?

Many other factors help to pinpoint poisons as the cause of death, even before live or dead animals or samples are examined, including careful analysis of the location where suspect poisoning has taken place, its prevalence and distribution, species affected and the clinical signs/*post-mortem* lesions (or absence of them) that have been reported.

Investigating Human-Induced Injury in Wildlife

This is likely to include clinical examination of live animals with injuries or possible signs of physiological shock, stress or metabolic derangement (see Chapter 8). Alternatively, or in addition, it may be necessary to investigate a wild animal's carcase with a view to determining the cause, mechanism and manner of its death (see Chapter 9). The processing of samples originating from such situations is dealt with in Chapter 10.

War and Civil Disturbance

The adverse effects that war and civil strife can have on the environment and wildlife are well recognised. The authors of this book were in Rwanda from 1993 to 1995 and were able to see for themselves the results of the 1994 genocide – not only in terms of human deaths and suffering but also the destruction of habitat, such as areas of the Virunga National Park.

Draulans and Van Krunkelsven (2002) drew attention to the paucity of data on the impact of war on biodiversity and provided an overview of the consequences of prolonged war on forest areas in the Democratic Republic of the Congo. They stated that only a few of the effects of that conflict have been beneficial, the most important one being the collapse of the wood industry. However, the war increased the number of people who relied on wood for fuel and on bushmeat for protein. The presence of soldiers and refugees exacerbated the situation. Hunting continued and illegal products, including ivory, were merely stocked in order to be sold when the situation improved. Draulans and Van Krunkelsven considered that war only benefits the environment when it keeps people out of large areas of forest.

Disasters

Disaster management is an increasingly important subject. Management of both natural and human-induced disasters is necessary if human and animal lives are to be saved and casualties minimised. The book by Arora and Arora (2012) discusses what is described as best practice for the practice of disaster medicine in both developed and developing countries. Other texts cover animals, usually only domesticated species. Post-disaster relief, rehabilitation and recovery can all have an effect, direct and indirect, on wildlife.

Bushmeat

Bushmeat (termed 'wild meat' in some areas, such as the Caribbean) can be defined as 'wild animals killed for food' and the species involved can range from grasshoppers to guineafowl and gorillas (Cooper, 1995d; Cooper and Cooper, 2007). Millions of people in South America, Africa and Asia continue to depend upon bushmeat, as either an essential source of animal protein or as an important cultural addition at weddings, baptisms or funerals (Koppert et al., 1998).

In recent years, bushmeat has become big business and instead of relatively small numbers of animals being taken, sold and eaten at a local level, millions of tonnes of meat and other products are being exported all over the world.

The trade in bushmeat is important, or potentially so, for the following reasons:

- Contravention of legislation, both national and international (e.g. CITES) (see Chapter 3)
- Depletion of populations of wild animals, sometimes with resulting increases in other species and changes in vegetation (Rapport et al., 1998)
- Spread of diseases of humans and animals and probable role in the appearance of emerging infections (see Global Health, later)
- Socio-economic effects, especially on local communities

The relevance of the bushmeat trade and conservation is well recognised. The extinction of at least one sub-species, Miss Waldron's red colobus monkey, has been attributed to bushmeat hunting. However, in a thought-provoking article, Milner-Gulland (2002) posed the question 'Is bushmeat just another conservation bandwagon?' She went on to discuss why interest in the bushmeat issue had grown (and continues to do so).

She argued that bushmeat hunting is 'the toughest challenge yet for human-centred conservation', and pointed out that the practice is often deeply embedded in the general economy, widely distributed geographically, frequently in countries with few legal controls, involves a range of people (hunters, dealers, vendors and consumers) and supplies both subsistence needs and commercial markets with complex commodity chains leading to big cities. It extends across national borders. Bushmeat can, Milner-Gulland said, be a key part of people's livelihoods. However, she asked, is the level of attention being given to bushmeat hunting warranted? It could be argued that habitat loss and degradation are the major dangers to wildlife. However, it was important to remember, Milner-Gulland stressed, that threats to wildlife do not act independently but synergistically; this is particularly true of bushmeat and logging. Attention to the social and economic environment within which bushmeat harvesting takes place can help in tackling the problem.

The detection, prevention and control of the bushmeat trade are important in countering wildlife crime. However, in many countries law enforcement is weak, making policies of selective hunting unreliable or impossible to enforce (Rowcliffe et al., 2004). Some argue that relatively wealthy people in developed countries want to stop those in poorer parts of the world from benefiting from utilisation of their own wildlife (Cooper, 1995d).

As Cooper and Cooper (2007) pointed out, investigation of bushmeat cases can be dangerous, especially when it is associated with other interests such as logging, drug-dealing, guerrilla activity or political rivalry.

Forensic examination of bushmeat may reveal important information such as the species of animal, its age, sex, provenance and health status. A pathologist may be needed. He or she might be asked to 'match' carcases (or part thereof) where more than one animal has been butchered (Stroud and Adrian, 1996).

Cooper and Cooper (2007) suggested that an approach to strengthening existing conservation legislation relating to animals taken for bushmeat might be to use public health legislation. In Kenya, the Meat Control Act requires that all meat be inspected and certified safe for consumption before being offered to the public and in a statement on 14 September 2005, the Kenya Veterinary Association warned that those eating gamemeat, including animals killed by vehicles, risked contracting zoonotic diseases such as anthrax. This has been strongly reinforced by a recent report on zoonotic viruses associated with illegally imported wildlife products (Smith et al., 2012) in which the authors claimed that the global trade in wildlife has historically contributed to the emergence and spread of infectious diseases. The report details the findings of a pilot project to establish surveillance methodology for zoonotic agents in confiscated wildlife products. Initial findings from samples collected at several international airports identified parts originating from non-human primate (NHP) and rodent species, including baboon, chimpanzee, mangabey, guenon, green monkey, cane rat and rat. Pathogen screening identified retroviruses and/or herpesviruses in the NHP samples. These results are the first to demonstrate that illegal bushmeat importation into the United States could act as a conduit for pathogen spread, and suggest that implementation of disease surveillance of the wildlife trade could help facilitate the prevention of disease emergence.

In a study from the United Kingdom, Snary et al. (2012) performed a qualitative release assessment to estimate the likelihood of henipavirus entering the United Kingdom.

The genus *Henipavirus* includes Hendra virus (HeV) and Nipah virus (NiV), for both of which fruit bats (particularly those of the genus *Pteropus*) are considered to be the wildlife reservoir. The recognition that henipaviruses occur across a wide geographic and host range suggests the possibility of the virus entering the United Kingdom. From their assessment, the authors concluded that the importation of fruit bats, bushmeat, horses and companion animals from certain specified areas carries a low annual probability of release of henipaviruses into the UK and people entering the United Kingdom were estimated to pose a very low risk. The annual probability of release for all other release routes was assessed to be negligible. Nevertheless, continued research on bushmeat, bats and viruses remains important.

Traditional Medicines

The human race has for millennia relied on medicinal products derived from natural sources, and animals have always formed an important component. Thousands of uses have been claimed, sometimes substantiated, for medicines made from leaves, herbs, roots, bark, mineral substances and animals (Figure 2.12).

Wildlife is commonly poached for use in traditional medicines. Although the use of rhino horn is probably the best known, many other species are killed, or in some cases

Figure 2.12 Throughout Africa, plants and other substances are sold for medicinal purposes in local stores as well as in towns and cities.

kept alive (e.g. bears for bile production), in order to harvest tissues or secretions that are considered to be therapeutic. The effects on wild animal populations vary. Guo et al. (1997) looked at the sustainability of wildlife use in traditional Chinese medicine.

Much detective work and various tests may be needed to determine the origins of some medicines (Linacre and Tobe, 2008). Species primers have been successfully developed to identify endangered species used in traditional medicines (Tobe and Linacre, 2011) (see Chapter 11).

Interestingly, according to Alves and Souter (2010), canids are among the mammal species most frequently used in traditional folk medicine around the world. In their paper, Alves et al. assessed the global use of canine material and its implications. Their review indicated that 19 species are used in traditional medicine, representing 54.2% of described species of the family Canidae. Of the species in medicinal use, two are listed on the IUCN Red List (see later) as 'Endangered' and three as 'Near Threatened'.

Alves et al. (2010) published a 'world overview' of the use of primates in traditional folk medicine, and da Nóbrega Alves et al. (2008) an overview of the global use of reptiles in traditional folk medicine and the implications of this for conservation. They ascertained that at least 165 reptile species, belonging to 104 genera and 30 families, are used in traditional folk medicine around the world. Some species are used as sources of drugs for modern medical science. Of the reptiles recorded, 53% are included on lists of endangered species.

Hunting, Shooting and Fishing

The hunting of animals has played a significant part in the evolution and development of humans. Originally used solely as a means of obtaining food and other animal products, it is also practised as 'sport' (see later). Hunting is also inextricably associated with the study – and, in recent years, the management and conservation – of wildlife. Many of those who contributed to our knowledge of the natural world hunted animals. For example, the celebrated Franco-American ornithologist, artist and naturalist, John James Audubon (1785–1851), famously stated:

> Hunting, fishing, drawing, and music occupied my every moment. Cares I have not, and care naught about them.

Strictly, 'hunting' is the pursuit of any living thing, usually wild animals, for food, recreation or trade. In North America, the word 'hunting' usually means the shooting of animals, whereas in Great Britain it has become associated with the use of hounds to chase (hunt) foxes, deer and other mammals. The British also tend to use the term 'field sports', or 'country sports', for outside activities that involve catching or killing animals; as a result, it can refer to such diverse and different pursuits as shooting, hunting, fishing, ferreting and the use of birds of prey (falconry) to take quarry. Opponents of such activities tend to prefer the rather pejorative expression 'blood-sports'.

Attitudes towards hunting vary in different parts of the world. Sometimes marked differences exist between neighbouring countries; for example, at the time of writing (2011), all forms of hunting are banned in Kenya, whereas in Tanzania, its East African neighbour, the legal shooting and trapping of wildlife persist and represent an important source of income for this poor country. In many countries there is a continuum between legal hunting and illegal poaching, and both can contribute to the bushmeat trade (see later). In Trinidad and Tobago in the West Indies, for example, where hunting of many species is legal during the hunting season (in 2005 there were 9000 legal hunters), there is also much illegal killing of wildlife; but since this is difficult to control, the country has approximately 200 honorary game wardens to help to protect habitats and wildlife. These wardens consist of members of law enforcement groups, hunters with knowledge of the forest wildlife species and members of non-governmental organisations.

Many studies have dealt with the effects of unregulated subsistence hunting on game animals, and most have documented significant declines of at least some of the target species. Studies on Guianan forest game birds by Thiollay (2005) showed that increased hunting pressure was often associated with road building and logging operations, which greatly facilitated access to large, previously inaccessible forest areas. They concluded, in general, that birds become more important hunting targets in areas where large mammals are depleted and in countries where sport hunting is widespread or where cartridges are cheap and accessible. Regardless of whether or not they are preferred game or only killed opportunistically, most large tropical forest birds have demographic (high survival, low productivity) traits that make them particularly sensitive to increased adult mortality rate.

Hunting (shooting) is generally more acceptable in North America and continental Europe than it is in Britain. Such is the opposition to the killing of wildlife in Britain that its use as part of wildlife management (to maintain a population of healthy animals within the carrying capacity of an environment) and even pest control are often the subject of heated debate. This is referred to in more detail in the following.

'Fishing' can be defined as the capture and release, or capture for food, of fish and sometimes other (usually invertebrate) species, such as molluscs. Fishing is widely practised throughout the world and is important in terms of providing food, both for industrialised societies and for local communities (Figure 2.13). However, it is not without its legal complications. Laws relating to fishing are often complex and may date back a considerable period. Contravention of fishing legislation can lead to fines, imprisonment and confiscation of fishing tackle (equipment). Often control of fishing is covered by a combination of legal requirements and codes of practice. On the island of Jersey, for example in the British Channel Isles, recreational fishing by members of the public is governed by a number of regulations. Any boat used for commercial fishing (where the catch is sold) must hold a valid licence issued by the States of Jersey Fisheries and Marine Resources Department. Shellfish, other than crabs (!), may not be taken from the sea while totally or partially submerged, and using under-water breathing apparatus, or wearing a face visor, mask or goggles. The Code of Practice, which is not legally binding, includes such instructions as to replace in all fish to the sea that are not to be used and 'Turn it and Return it' – an exhortation to replace in to their original position any rocks that have been turned over.

In many countries there are voluntary codes of conduct for those who fish. In addition, there is increasing concern about the welfare of fish – particularly the extent to which they may perceive pain (and thus be adversely affected by the use of hooks) and the fact that some, not all, fish taken for food die as a result of hypoxia, a method of killing that would not be acceptable in the case of pet or laboratory fish in the United Kingdom and certain other countries.

For over 70 years, concern has been expressed about overfishing and the effect it might have – and is now having – on fish numbers on the marine environment (Clark, 2007).

The debate about the welfare of wild animals was fuelled in the United Kingdom by the decision of Parliament to ban the hunting of foxes and other species with dogs – pastimes

Types of Wildlife Investigation

Figure 2.13 A fisherman with his net in the river in a sanctuary in East Africa. Fishing is not sanctioned here but little can be done to prevent it because the watercourse belongs to the local community to which the man belongs.

('sports' – see later) that had been practised for centuries and considered by many to be part of British rural tradition. Before and after abolition, there was public and scientific discussion as to the extent to which animals that were chased and killed 'suffered' and how this rated in comparison with, say, their being shot and injured (Fox et al., 2005) or dying from inanition or infectious disease. It has also introduced into forensic work the need for a wildlife pathologist to be able to recognise bites and other injuries that have been inflicted by dogs rather than by other species (PAW, 2005).

Although, as explained earlier, hunting with hounds has now been banned in Britain, dogs (e.g. terriers) can still be used to control certain species, for example rabbits. However, those working such animals have to be aware that they might be accused of targeting protected species, especially the badger. In an attempt to avoid this, the British Field Sports Society (BFSS) (see Appendix E) has issued information on how to recognise a badger sett, including signs of bedding near a sett entrance, regular paths to dung pits, scratches on trees, coarse grey hairs with black near the tip on barbed wire or low trees and signs of excavation linked by well-defined paths.

The information provided by the BFSS can also be of assistance to those investigating alleged baiting, gassing or killing of badgers. Although this example relates specifically to the United Kingdom, it illustrates the value of having access to literature and expertise from different quarters, including those who hunt, shoot and fish, when investigating possible wildlife crimes.

Concerns over the effect of the banning of hunting by hounds on the welfare of both the wildlife (e.g. foxes dying as a result of shotgun or rifle wounds) and the environment prompted the Veterinary Association for Wildlife Management (VAWM) (see Appendix E) to produce a series of 25 questions and responses entitled 'Answers to misconceptions about hunting'. In another publication the Association seeks to rebut the claim that killing of animals by hounds is painful; they argue that the kill is almost instantaneous because the neck is broken or the thorax is crushed as a result of the considerable power weight ratio and height advantage of the hound over a fox.

In any legal case relating to such issues, the investigator should be prepared to read appropriate literature and, of course, as part of his or her objectivity, material produced by organisations that hold different views. Likewise, when being questioned in court about the welfare of wildlife, it is important that the expert witness is objective (see Chapter 13) and concentrates on the animal(s) in question, rather than straying into politics or being swayed by personal points of view. He or she should be cognisant of relevant research and publications on the welfare of wildlife, which is often widely scattered in the literature.

Wildlife in Sport

Wild animals have been used in 'sport' for centuries, as evidenced by the large numbers and variety that were taken to Imperial Rome for combat 2000 years ago and the subsequent persistence in Europe, until relatively recently, of such activities as bear-baiting.

Some sports involving animals still exist and are legal, but this depends upon the country and its legislation. Activities that come under the broad heading of 'sport' range from those that are non-invasive, such as racing, to others that involve killing, such as the shooting of deer, other mammals or galliform birds.

Some 'sports' may involve more than one species of animal – for example, badger-baiting, which is still practised illegally in some parts of Britain. Dogs, especially terriers, are used in the baiting (see also earlier), and both they and the badger can sustain severe injuries.

Oil and Gas Releases

The release of oil (petroleum products) into the sea, sometimes on to the land, has attracted attention and concern for many decades because of the effect that such spillages can have on wildlife and the environment. There are many examples and it is outside the scope of this book to go into great detail. Nevertheless, such incidents have resulted in legal action on a number of occasions and litigation seems likely to be more frequent in future. The *Deeepwater Horizon* offshore oil drilling rig explosion in the Gulf of Mexico in April 2010 was the largest accidental oil spill in history. Eleven crewmen were killed. The fireball that resulted from the explosion could not be extinguished and, on 22 April 2010, *Deepwater Horizon* sank, leaving the well discharging oil (4 million barrels in all) at the seabed. Large numbers of animals, including birds and fish, died in the immediate aftermath and ongoing ecological studies suggest that the major environmental effects will not be apparent for a decade. There is particular concern about organisms near the seafloor (the benthos), that are important in the food chain and the decline or death of which will affect other species, including many that are relevant to fisheries. At the time of completing this book (March 2012), the BBC has reported that BP has reached a GB£4.9 billion ($7.8 billion) deal with the largest group of plaintiffs. That sum will be used to benefit some 100,000 fishermen, local residents and clean-up workers whose livelihoods or health have suffered. The company has not admitted liability and still faces claims from the United States and state governments and drilling firms.

Lessons learned from the *Deeepwater Horizon* episode may help to avoid similar spills in future. However, many argue that the oil industry remains complacent about the risks that its activities pose. If this is true, the implications for the environment and the continuing relevance of carefully collected and accurately interpreted forensic evidence are clear.

The pathogenesis of oiling in birds is summarised in Chapter 9. Much has been said and published about the effects of oiling incidents on the environment. A useful publication is the book by Nel and Whittington (2003), which recounts the work done to rehabilitate oiled South African penguins. Oil spills do not only affect birds and mammals: Reptiles, fish and invertebrates may be killed or incapacitated. Nor is oil only of significance to marine species: Penetration of the pollutant into coastal areas can affect many other animals.

A major problem in evaluating the effects of oil on animals and the environment is that there are substantial differences in the chemical composition of oil from different sources, as well as between the various refined fractions. Fuel oils tend to exert their toxic effect because of the presence of volatile aromatic fractions (e.g. benzene). Ingestion by birds of emulsifying agents used in the dispersion of oil slicks has been recognised as an additional hazard, but these are now less toxic than those compounds used three decades ago.

Concern over the effects of successive oil spills on the Atlantic coast has prompted calls for international or European regulations to speed the timetable for making double-hulled tankers a statutory requirement. There is also pressure to make changes to the current system whereby oil companies opt for the lowest transport costs and, as a result, compromise marine safety. Oil and gas development in the World Heritage and wider protected area network in sub-Saharan Africa and its environmental implications was the subject of a study by Osti et al. (2011). Increasingly, there is pressure to seek compensation from those who spill oil, whether or not this is intentional.

Not all effects of oil on wildlife relate to spillage in the sea. Oil production operations generate various waste fluids that may be stored in pits, open tanks and other sites accessible to wild animals. Birds in particular visit these fluid-filled pits and tanks ('oil pits') because they often resemble water sources; many birds become trapped and die. Trail (2006) reviewed the problem in the United States and suggested how it might be solved, or at least ameliorated. Forensics laboratory and USFWS field data suggest that oil pits currently cause the deaths of 500,000–1,000,000 birds per year. Although the efforts of law enforcement and industry have produced genuine progress on this issue, oil pits remain a significant source of mortality for birds in the United States.

Accidents

Numerous 'accidents' can befall wildlife, of all taxa. Some such accidents are covered under different headings in this chapter, earlier and later, and in Chapters 8 and 9 as well. An example of how the amateur naturalist who can assist in compiling data about – and possibly take action over – such incidents is well illustrated by the article in a herpetological magazine by England (1999), who was only 14 years old at the time. He surveyed mammals, reptiles and amphibians that fell into storm drains near his home. Although it might be argued that such accidents are inevitable, part of trying to survive on a day-to-day basis, they can be significant in terms of reducing numbers of animals, particularly small localised populations, and there are welfare considerations as well. Perhaps an advantage of the current emphasis (pre-occupation, some would say) on health and safety and risk assessment in Europe is that better precautions are now taken to prevent accidents. The legislation is intended to protect humans, but in some cases will be of benefit to wildlife. On the other hand, pre-occupation within Britain and elsewhere with (e.g.) blocking holes and drains in case a child comes to harm may restrict access to animals that are seeking shelter. Another example of human–animal competition!

Some accidents are very much a result of human action – often neglect. Kirkwood et al. (1997) described cases of entanglement in fishing gear and other causes of death in cetaceans stranded on the coasts of England and Wales. Between August 1990 and September 1995, the carcases of 422 cetaceans of 12 species that had died around the coasts of England and Wales were examined. The cause of death was diagnosed in 320 (76%) of them. The most frequent cause of death in harbour porpoises and common dolphins was entanglement in fishing gear ('bycatch') (see also 'Marine Mammals: A Special Case' section in Chapter 13).

New technology needs careful appraisal before it is adopted on too wide a scale because of the effects it may have on wildlife. Sometimes such an assessment is required by law, but it may be limited in scope. The spread of wind-farms in Europe and further afield may be an example on this topic and is covered in Chapters 8 and 13. Concern has been expressed over the effects of wind-farms on large eagles in south-eastern Spain (Martínez et al., 2010) and diverse species elsewhere (Douglas et al., 2012).

Captive Wildlife

Cooper and Cooper (2007) stressed that captive wildlife presents its own legal challenges, relating to both species conservation and welfare. They pointed out that there is often not a clear-cut distinction between captive and free-living – for instance, because of movement of casualty animals from the wild into captivity and then release back to the wild. Ranched wild animals are also an example – neither completely free-living nor confined in captivity. Although wild animals of some species have for long been kept in captivity, in recent years a whole range of vertebrate and invertebrate species has become popular with the public at large. Their welfare can be compromised and this has led in some countries to calls for their confinement in captivity to be banned or strictly controlled.

Assessment of the welfare of captive wildlife presents many difficulties. Often little is known of the natural history and biological requirements of these animals. *The Five Freedoms* (see Chapter 8) are generally applicable – to captive birds, for instance (Cooper, 2002a). Ectothermic animals should either be kept within their known 'preferred optimum temperature zone' (POTZ) – see also PBT (Chapter 8) – or be offered a choice by the provision of a gradient (Cooper and Williams, 1995).

Some examples of 'captive' situations are as follows.

Zoological Collections

Many vertebrate and some invertebrate animals have been popular in captivity for display for centuries. The emphasis in such cases was usually on entertainment, but slowly changed to encompass scientific study. More recently, conservation and education have begun to play a far greater part in zoo management.

In many countries there is no specific legislation covering zoos but the situation is changing, in part due to the concern of the public and because of lobbying by pressure to be a particular risk because species are mixed, hygiene is poor and conditions of holding and transportation facilitate microbial transfer (Karesh et al., 2005). Concern over salmonellosis, in particular, has prompted the production of guidelines outlining methods of reducing the risks from captive reptiles. Liability for human case of reptile-associated salmonellosis could well rest, in whole or in part, on the owner of a pet shop that provided the reptile, a veterinary surgeon who gave it a health check or the manufacturer of the disinfectant that, despite its claims, proved ineffective when used.

A much-quoted example of the hazards of keeping reptiles in private hands concerns the question, in the United States, of the keeping of hatchling turtles. This was historically popular until epidemiological studies revealed that 14%–18% (approximately 280,000) cases of reptile-related human salmonellosis (RRS) annually were turtle-associated. In 1975 the trade in these animals was banned. In the year following, there was 77% reduction in RRS infections (Mermin et al., 2004). Reptile-keeping in the United States began to change to other species, mainly snakes and lizards; 3%–4% of households held approximately 9–11 million amphibians and reptiles (APPA, 2005). Reptile-related salmonellosis prevalence rose to around 5%–7% of all 'related infections' to involve more than 70,000 cases annually (Mermin et al., 2004).

At the same time as this debate continues, there is a growing awareness of the benefits that 'companion animals' (pets) may bestow on humans. As an example, a meeting at the Royal Society of Medicine in London in May 2011 highlighted the benefits of keeping domesticated animals (even though these, too, can present dangers to humans) and attention was drawn by speakers to the many contributions made by non-domesticated species, including rodents, to the health, safety and well-being of humans (Bonner, 2011).

In some countries, such as the United Kingdom, the keeping of certain species may need a licence. Elsewhere, the holding of any wildlife by members of the public may or may not be prohibited.

A topical example of how the pet trade can adversely affect the conservation of an endangered species concerns the cheetah (Ogada, 2011). Cheetahs have been trained to hunt for about 5000 years. Akbar the great, the Mogul Emperor, who ruled Hindustan (present day India) from 1556 to 1605, was reputed to have kept 1000 of these animals for hunting. Cheetahs feature in Arabic literature – for instance, in the writing of Usamah ibn Munqidh (1095–1188) – and, as well as falcons, were used for sport hunting. The appeal of the cheetah has, however, in no way declined. According to Ogada, 'There is a vast and growing nouveau riche in the Middle East, and cheetahs have unfortunately become acquisitions and objects of desire'. Few, if any, of these animals reproduce, and so the demand for more cheetahs remains – and grows. Sport-hunting using cheetahs is enjoying a renaissance and, according to Ogada, in the United Arab Emirates (UAE) a young cheetah can retail for $ 9,000–11,000.

Use of Wildlife in Research

Research on free-living wild animals very often constitutes a form of intervention. Such interventions can, in theory, be described as non-invasive, minimally invasive or invasive, but the distinctions are often blurred. Some research projects may appear to be non-invasive – for example, the observation of animals from a distance using binoculars – but even these may have a subtle, but significant, effect on the animals in question. Other forms of intervention are very clearly invasive, for instance, the capture of wild animals in order to apply radiotelemetry equipment, to take blood samples or even to give treatment or vaccination.

Certain types of research on wildlife attract adverse criticism – mainly those that are clearly invasive, where the animal may be pursued, captured, restrained and subjected to procedures that are at least stressful and at worst cause pain or compromise survival (Mellor and Monamy, 1999).

The role of stressors on wildlife cannot be overestimated. While all wild animals are exposed to a range of stressors, a healthy animal should be able to respond to these satisfactorily. However, if there are too many stressors, or if the effect of the stressors is excessive, the animal may not be able to react adequately (its 'general adaptation syndrome' cannot cope) and it will become stressed. A stressed wild animal may show non-specific signs and is also likely to exhibit an increased susceptibility to disease. It may even die (see below).

Research using animals is a contentious subject and has led to legislation in many countries, the regulation of research institutions, veterinary supervision, institutional animal care and use committees, the formulation of guidelines and codes of practice and high standards of training of staff.

Research on wild animals may or may not follow similar lines. If scientific investigative work that is invasive is carried out in the field, the legislation may well apply, even though the work is being done far from the laboratory – as is the case with the UK's Animals (Scientific Procedures) Act (see Chapter 3). Macaldowie (2011), in an editorial entitled 'Working on the Wildside', explained that, from a (British) regulatory viewpoint, the major concerns are that the studies can be scientifically justified and are compliant with the 3Rs (Replacement, Refinement and Reduction), the reasons for doing them outside a designated facility are clear, all procedures are being performed by competent licensees and sufficient independent support is made available to ensure that the animals' interests are as fully safeguarded as possible at all times. Complications are that such work often takes place at remote sites and frequently with unrestricted public access. The research may involve work on wild animal species or environmental issues of particular public concern and special considerations usually need to be taken into account before licensing such work. It is recognised that with wildlife it is often difficult to stay completely aware, let alone in full control, of what may be going on outside an establishment. Also, there may be additional legal responsibilities concerning the health, safety and welfare of staff working at remote sites and with species that could be dangerous on account of their size and lack of previous handling, or exposure to zoonotic diseases. Appropriate training of new or inexperienced researchers is particularly important. Other valuable sources of support and advice from outside the immediate scientific community are also available and should be sought as necessary, including help from experienced agricultural workers, wildlife rangers and local veterinary surgeons. More information about the British situation is to be found at

http://www.homeoffice.gov.uk/science-research/research-statistics/science/
http://www.fera.defra.gov.uk/wildlife/ecologyManagement/wildMammalsTraining.cfm

The effect of stressors on wild animals as a result of interventions has been studied by a number of authors. In the African context, the scenario that has attracted particular interest and controversy concerns the African wild dogs (Figure 2.16) that disappeared from the Serengeti National Park in Tanzania two decades ago. It was postulated (Burrows et al., 1994, 1995) that this local extinction might have been a result of various interventions that were carried out, including the vaccination of the dogs against rabies. That hypothesis

Types of Wildlife Investigation

Amongst the threats that wild (free-living) animal populations face are competition for food and other resources, predation, parasitism, toxic chemicals and habitat destruction. Most of these may be exacerbated by such influences as inbreeding depression. Often the decline of a species or a population is multi-factorial.

Habitat destruction is one of the most significant causes of decline of species and reduction of biodiversity. Paradoxically, such activities, even if on a small scale, may be more likely to result in legal action in westernised countries than the extensive damage caused by logging and burning of bush in certain other parts of the world. This reflects the extent to which the law is adequately enforced and the willingness and confidence with which members of the public report environmental damage to the authorities. In the United Kingdom damage to, or destruction of, a pond that harbours protected amphibians will very often attract attention and complaints from naturalists, ramblers and local residents; prosecution and conviction of those responsible may follow. Sometimes an activity may or may not be unlawful – for instance, the uprooting of hedges – but might lead to investigation and legal action if accompanied by (for example) pollution of waterways (Figure 2.17).

Figure 2.17 A hedgerow and copse are cleared in France, polluting a stream.

Some species are more likely to be subject of a legal case – and thus of a forensic investigation – than are others. This reflects public concern about certain groups of animals, mainly those that have appeal because they are attractive, spectacular or appealing in other respects. For example, there was a far greater reaction to news that the last mainland Javan rhinoceros, the sub-species formally found in Vietnam, had been found dead, poached (WWF/International Rhino Foundation). WWF Report confirms Javan rhino extinct in Vietnam http://www.wwf.org.uk/what_we_do/press_centre/?unewsid=5367) (see Appendix K), than when, some years previously, scientists reported that, as a result of the intentional introduction of the predatory snail *Euglandina rosea*, all the species of Moorean *Partula* snail became extinct in the wild (Murray et al., 1988).

Some species tend to feature particularly frequently in legal cases and thereby in forensic investigations in many parts of the world. An example is birds of prey (raptors) – see later. Other taxa that attract particular attention in the United Kingdom include bats (Chiroptera) and badgers – the latter in part because of the controversy that rages in the United Kingdom regarding their role in the maintenance and spread of bovine tuberculosis.

Birds of prey (raptors) are frequently the subject of legal cases. The reasons for this are complex but relate to such issues as the decline of many species in the past – and to some extent at present – as a result of pesticides, historical persecution by gamekeepers in some countries and areas and by farmers in others, the position of raptors at the top of many food chains and their perceived status in society.

As a result of these and other factors, forensic investigation of raptorial species is relatively advanced. Heindenreich (1997) provided a specific chapter on forensics of birds of prey, including data on age determination and identification, while Cooper (2002a) covered many aspects of health, welfare, disease and legislation. In view of the particular vulnerability to theft and smuggling of some raptors, there is a specific section in this book about micro-chipping of falcons (see 'Identification Systems in Hunting Falcons in the Middle East' section in Chapter 13).

In the eyes of many, the illegal killing of birds of prey is a classic example of the heinous nature of wildlife crime. In the United Kingdom, there is a political aspect to this because some of those organisations that are committed to species' protection tend to associate the persecution or decline of free-living birds of prey with the management and use of land for the recreational shooting of red grouse. There is some evidence for this – Amar et al. (2012) reported lower productivity and 'growth rates' of peregrine populations on grouse moors than on non-grouse moor sites and said that analysis of wildlife crime data confirmed the persecution of the species was more frequent on grouse moors. They discussed other possible causes of what appeared to be poor breeding performance of programmes on grouse moors, but concluded that 'the most logical explanation for these differences is that illegal persecution associated with grouse moor management is responsible'. Others, however, particularly but not only the owners of grouse moors where conservation is given high priority, challenge these claims. It was also pointed out by Blackmore (2011) that more recent figures, some of them from the RSPB's own *Birdcrime* reports, show that only 21% of confirmed incidents of illegal persecution of peregrines occurred in areas that could be associated with grouse moors.

Differences of approach and opinion on such a sensitive subject as the conservation of birds of prey seem likely to continue to lead to dissent and the use of legislation to resolve issues rather than genuine discussion and co-operation. There is need for far greater collaboration between ecologists, ornithologists, field naturalists, falconers and rehabilitators.

spent on controlling three invasive species that were threatening waterways, exacerbating flooding and – by reducing oxygen – harming fish. Wang et al. (2011) surveyed (mainly terrestrial) plants in rapidly urbanising areas of Beijing, China. They drew up a comprehensive list of naturalised and invasive flora and identified 112 naturalised (including 48 invasive) plants within the Beijing Municipality. They ascertained that North and South America are the main contributors. The Beijing districts with the highest number of naturalised and invasive species of plants are also those that have experienced the highest human population growth, urban expansion and economic growth in recent years. Urban expansion in Beijing is predicted to continue in the near term; additional invasions seem likely. An integrated management strategy for the whole municipality is urgently needed.

Elimination is the most difficult management goal for exotic species of animals and plants; successes are rare. The dearth of publications on the methodology and outcomes of eradication programmes is disturbing. Reports are needed on the lines of Hoffman (2011), who detailed the successes and failures of eradication efforts on six populations of African big-headed ant in northern Australia.

Concern over invasive species has grown (Hilton and Cuthbert, 2010; Pimentel, 2011) and has prompted IUCN to establish an Invasive Species Specialist Group, which produces its own Newsletter, entitled *Aliens: The Invasive Species Bulletin*. The IUCN SSC Invasive Species Specialist Group mission is to reduce threats to natural ecosystems and the native species they contain by increasing awareness of the impacts of invasive species, and of ways to prevent their spread and control or eradicate them. In November 2011, an Action Plan to harmonise data needed to tackle threats to biodiversity from invasive species was announced. Government experts meeting under the Convention on Biological Diversity (CBD) endorsed a wide-ranging programme to strengthen information available on the spread of invasive alien species (IAS). The Joint Work Programme (http://www.cbd.int/doc/?meeting=sbstta-15) is aimed at harmonising data from a wide range of different databases and networks, with a view to eradicating priority invasive alien species by 2020.

Recently, however, authors of a number of articles have called upon the conservation community to approach invasive species in a different way. In a News Focus story Embracing Invasives (18 March, page 1383), it was suggested that the Galápagos embrace the aliens. An article in *The New York Times* accused environmentalists, conservationists and gardeners who were targeting invasive species of being unreasoningly dogmatic and xenophobic. These articles suggested that concern over invasive species derives from an unreasonable desire to maintain pristine ecosystems and to exclude all alien species. In fact, the protagonists argue, we should recognise that distributions of animals and plants are constantly changing, that community structure is dynamic, that alien species enter and are introduced into natural communities and that modified (and even degraded) ecosystems have conservation value. The scene is set for some legal challenges.

Biological Control

Biological control has long attracted attention and many species have been used, often introduced, in order to try to eliminate or reduce numbers of an unwanted animal or plant. Many have had disastrous effects (mongooses in the West Indies and the predatory snail *Euglandina rosea* on Moorea, e.g.) but some have had the desired effect. Arthropods are commonly used in the biological control of weeds, which are a major cause of crop loss in many parts of the world (Muniappan et al., 2009). The problem has increased substantially on account of the burgeoning world trade in plants and foodstuffs. Chemical control is becoming less acceptable and has limited applicability in the tropics. A recent topical example of possible biological control relates to the fungus *Batrachochytrium dendrobatidis*, the cause of the disease chytridiomycosis, which is believed to be responsible for unprecedented population declines and extinction of amphibians. Buck et al. (2011) described consumption of zoospores of *Batrachochytrium* by the zooplankton *Daphnia magna*. This inhabits amphibian breeding sites where *Batrachochytrium* transmission occurs. Buck et al. suggested that consumption of zoospores may lead to effective biological control of *Batrachochytrium*.

Pest Species

Some species are considered in law to be 'pests' and their control is often controversial and thus of some forensic relevance. Pests are often either excluded from protected lists or are in a category whereby they can be killed or taken under permit by an authorised person. In the United Kingdom, for example, cormorants, Egyptian geese and other avian species may be killed under licence in order to prevent damage to fisheries. In these cases, there may be debate as to what constitutes 'damage' and whether alternative methods might have been used.

Trapping is often selective in terms of species that are caught, as illustrated many years ago by work on rodents in Africa by Neal and Cock (1969).

Some methods of pest control are carried out on a very large scale. For example, the red-billed quelea, a small passerine bird, is a major pest of grain crops in many parts of Africa (Figure 2.19). In Zimbabwe, its pest status in summer has caused many subsistence farmers in semi-arid areas to plant maize instead of the better-adapted small grains (Mundy, 2000). As a result, harvests are small and this has contributed to the need for the country to be supported by food aid programmes. The preferred control measure is spraying with the chemical fenthion. This, however, can have adverse effects. Operators can be poisoned through careless application, children may collect the poisoned *Quelea* and become exposed to the chemical and numerous non-target species are killed by the spray. In addition, some of the sprayed birds leave the roost and take the poison elsewhere, where the breakdown products may prove even more toxic. This example illustrates the various potential legal implications of using pesticides, some of which may lead to wildlife forensic investigation.

There is a growing tendency to question the humaneness of methods used to control pests. This has for long been an interest of UFAW (the Universities Federation for Animal Welfare) (see Appendix E), with particular reference to rodenticides. UFAW has published, on its website, *Guiding principles in the humane control of rats and mice.* (http://www.ufaw.org.uk/rdents.php), which was developed by the UFAW working group concerned with this subject.

Research on more humane methods is not a new concept. For instance, the effect of three configurations of the jaw of kill-traps on time to loss of palpebral reflex of brushtail possums, a widespread pest in New Zealand, was assessed by Warburton et al. (2000). The use of leg-hold traps had received persistent opposition from animal welfare proponents. The authors concluded that the killing effectiveness of kill-traps designed for capturing possums can be improved by offsetting the jaws without the need to increase the power of

Figure 2.19 Part of a large flock of red-billed quelea fly over a waterhole in Africa.

the trap. Therefore, for some target species, such modifications might satisfy the demands of animal welfare proponents for traps that kill rapidly, without compromising trapper safety.

The live-trapping and release elsewhere of rodents is becoming an increasingly popular method of rodent control by the public and is seen as being a more humane option than alternatives. However, it appears that little is known about the fate of live-trapped and released mice. It seems very likely that the welfare of translocated and released mice would depend on many factors, including the place and time of release.

Anticoagulants are the most widely used rodenticides at present, but there are growing concerns about (a) their declining efficacy due to the evolution of resistant strains, (b) the risks to non-target species and (c) their adverse welfare consequences. UFAW is encouraging the development of new and more humane methods.

Research on pest control methods can also attract criticism – for instance, studies carried out to evaluate poisons such as cyanide (Morgan et al., 2001).

Pests and Conservation Programmes

The science of game management, pioneered by Leopold (1933), continues to play an important part in promoting a balance between the maintenance and welfare of species that are considered desirable for (e.g.) hunting/shooting and the control of other 'pest' species. This, however, can be a sensitive and divisive issue that may, on occasion, lead to litigation.

The black or ship rat has become the dominant rat species throughout the subtropical and tropical regions of the Western Hemisphere. The species was taken to the West Indies from Europe around the mid-seventeenth century, and there in its new range it caused great economic and ecological havoc. This was compounded by further deleterious effects on native wildlife brought about by an explosion in numbers of the Indian mongoose, itself introduced from the late nineteenth century in an attempt to control the rats.

Johnston et al. (1994) described rat eradication in order to protect the St Lucia whiptail lizard population, which was confined to Maria Island Major and Maria Island Minor in the Caribbean, numbering about 1000 individuals. Part of the conservation programme for this species was to establish a colony on a suitable island; Praslin Island appeared to be the most appropriate. However, black rats had colonised Praslin and established a population there, so it was necessary to eliminate the rats from the island. Johnston et al. described the successful eradication.

Interventions

There appears to be an increasing desire by, and need for, humans to intervene in issues and circumstance regarding wildlife. Some practical interventions are relatively minor – for example, the translocation of amphibians from a wet area that is threatened by drainage to another swamp or pond where they can live and breed – a commonplace activity in Britain and other 'westernised' countries. In the poorer parts of the world, however, interventions may be more complicated and, often, controversial – see later. The reception, treatment and care of wild animal casualties are examples of intervention and are discussed later in this chapter.

The official policy regarding sick and injured wildlife in some countries may be strictly non-interventional. In many African national parks, for example, an animal that is seen to be injured may be left to its fate and no action permitted in terms of either treating or killing it. This can create difficulties, not only for members of the public (especially tourists) but also sometimes for national and expatriate veterinarians and others concerned about welfare. This illustrates once more the dilemma of how to reconcile species protection with individual welfare (see earlier).

Management of Wildlife Habitat

The management of land can be contentious and legal cases may ensue, often needing a forensic opinion. For example, the moorlands of Britain, so important in terms of biodiversity, have been damaged by human activity in the past and probably face new challenges, such as climate change, in the future. In the Peak District of England, for example, where peat formed over a period of 8000 years, erosion has taken place and new peat is not forming (Campen, 2010). Factors that have contributed to this include uncontrolled fires, excessive grazing by sheep and recreational trampling by walkers and tourists.

A major project, funded by the European Commission, commenced in July 2010 in order to conserve Peak District moorlands. Such programmes are not without their divisions however; in the Peak District, as in many other locations, there are landowners who manage this land with a view to shooting while others strongly oppose it (see also birds of prey, earlier). Such contention can easily fuel accusations, allegations and legal proceedings.

Another example of where there can be such a conflict of interests concerns heather. On traditionally managed grouse moors there is usually an internal burning rota, in order to encourage low heather which favours red grouse. Landowners who burn less tend to have longer heather which ground-nesting birds of prey, such as merlins, may prefer.

Livestock grazing is an important management tool of agri-environment schemes initiated within the EU to maintain and reserve biodiversity of grassland birds. However, grazing can affect bird populations negatively by depressing reproduction through nest trampling and increasing nest predation (Pakanen et al., 2011).

Climate Change

Interest in climate change goes back over a century, when many papers on the subject were published in Europe, especially Germany. The subject even formed the basis of a lecture to the Linnean Society of London during the first year of the Second World War (Simpson, 1940). At the present time, the diverse effects of a rise in global temperature are the subject of much concern and some concerted international action (Dessler, 2011). The significance is reflected by the recent statement of Ban Ki-Moon, the United Nations Secretary-General, who said 'For my generation, coming-of-age at the height of the Cold War, fear of nuclear winter seemed the leading existential threat on the horizon. But the danger posed by war to all humanity – and to our planet – is at least matched by climate change'.

Climate change is considered by many to be the world's most pressing environmental/economic issue (FitzRoy and Papyrakis, 2009). It is very relevant to human health, as explained by Philp (2001).

A recent book (Beever and Belant, 2011) provides a much-needed synthesis of climate-change information from different taxa, ecological processes and ecosystem contacts and presents trends expected in wildlife and ecological responses to climate change. Shifts in temperature and rainfall can prompt the spread, emergence or re-emergence of some infectious diseases while increases in exposure to ultraviolet radiation may contribute directly and indirectly to the decline of amphibians (Cooper, 2002b). The adverse effects of increased ultraviolet light on skin are increasingly recognised. Most of the studies have been carried out on humans and on domesticated animals. Climate change even threatens marine mammals (see 'Marine Mammals: A Special Case' section in Chapter 13). Newton (1998) discussed global warming in the context of birds and stressed that the various projections of the effect of global warming must be interpreted with caution.

Some evidence of past climate variation is dramatic – for example, the discovery of bones of hippopotamus in England. Allen (1996) discussed how plants cope with changes in temperature, pointing out that, throughout the Quaternary period, the climate has been characterised by alternating glacial and inter-glacial conditions. Plants have had to respond to these changes; evidence showing how this has happened comes from studies of past climates and past vegetation distribution patterns. Palaeoecology provides an opportunity to assess how plant distributions and vegetation patterns have changed in response to climate changes, especially over the past 18,000 years, since the maximum of the last glacial.

Allen argued that an understanding of the mechanism involved, and the rates of past changes, is important for the prediction of the effects of future climate change (be it anthropogenic global warming or global cooling at the start of another glacial stage) on vegetation. Schulman and Lehvävirta (2011) edited a special volume on the role of botanic gardens in this age of climate change.

Changes in the distribution of certain animals ('habitat associations') have been attributed to changes in climate. For example, the silver-spotted skipper butterfly was previously restricted to south-facing slopes in the United Kingdom where it reaches its northern range limit (Thomas et al., 1986). In recent years butterflies of this species have colonised on east-, west- and south-facing hillsides, as an increase in ambient temperature has meant micro-climatic conditions for egg-laying are now suitable on these aspects.

International action to control or to mitigate the effects of climate change has included economic incentives (such as carbon taxes) and international agreements such as Kyoto.

Changes in climate may increase the vulnerability of wildlife to the effects of adverse weather. Avian ecologists, in particular, have investigated the role of such change on birds. Climatic extremes may comprise prolonged periods, such as hard winters or droughts (Figure 2.20), or brief episodic events such as rain or hailstorms. Migrating birds in particular can be very susceptible to bad weather and Newton (1998) gave a number of examples of heavy losses. Oceanic extremes, such as changes in sea currents, can have major effects on seabird populations, mainly by affecting their food supplies and birds may be washed ashore or blown inland by gales – so-called wrecks.

Figure 2.20 Drought in Kenya, resulting in a shortage of food and water for both domesticated animals and livestock.

Catry et al. (2012) described landscape and weather determinants of prey availability (mole crickets) and the implications of these for the lesser kestrel. They showed that an understanding of the ecological requirements of preferred prey can help in the implementation of management measures to improve food supply for target species.

Use of Wildlife Resources

Humans have been using wildlife for food, clothing and medicine for millions of years, with dramatic effects on population viability in a number of cases – the ratites in New Zealand and passenger pigeons in the United States, for example. There is pressure on wildlife species, not only directly through over-exploitation but indirectly because they occupy space which could be used for other purposes. Bolton (1997) argued that wildlife utilisation must be part and parcel of wildlife conservation. However, there are those who strongly oppose this view. For example, the supply of feral (introduced) macaque monkeys from Mauritius for medical research in the West has helped to provide funds for the conservation of endangered endemic species on the island. However, the political power of activist groups in the West opposed to the use of monkeys in biomedical research, and the establishment of breeding colonies in the United States and Europe, could lead to reductions in the export of monkeys from Mauritius, further depredation of endangered species by the macaques and the loss of funds for conservation.

Wildlife Species for Sustainable Food Production

An increasing challenge is how to feed the growing human population (see Preface) and at the same time conserve the world's natural resources and biodiversity.

The task is great. Norton (1994) predicted that the size and yield of irrigated land, cropland, pasture and rangeland would not be able to keep pace with the 30% rise in the world's population, from 5290 million in 1990 to 7000 million in 2010. Already malnutrition, in varying degrees, affects two-thirds of the human population.

In recent years, some scientists have turned their attention to animal protein and looked critically at whether 'new' (different) species might make a useful contribution. At its simplest level this has consisted of encouraging the keeping of animals such as domestic rabbits that might be better suited, in terms of size, constitution and husbandry requirements, to particular circumstances. Another step has been to promote the harvesting and/or farming of species that are widespread, or even agricultural pests, but which have potential as a food source (Cooper, 1992, 1994, 1995c; Jori, 2001). For example, the collecting and rearing of land snails is an excellent way of producing much-needed protein in many parts of the world (Cooper and Knowler, 1991). An inevitable progression from this has been to look at other wild (free-living) species and to question whether they, too, might be harvested – or in some cases farmed and domesticated – in order to produce more food, especially in areas where domestic animals fail to flourish or are having a damaging effect on the environment. In a thought-provoking paper entitled *Wildlife utilization: use it or lose it*, Kock (1995) argued that such an approach might be essential if certain species were to survive. There will be legal and ethical considerations, of course, but these are not insurmountable (Cooper, 1995a).

The human race has hunted and eaten animals, ranging from invertebrates to mammals, since time immemorial. Vegetarianism is increasing in popularity in the 'developed' countries and is traditional in certain societies elsewhere. It seems unlikely, however, in the foreseeable future, to have any effect on the greater part of the world where the main source of protein for the family remains the ox or the village chicken, supplemented where appropriate with bushmeat taken from the wild (see earlier).

The question of whether wildlife should be further promoted as a food source is a controversial one. Proponents argue that many wild species will not survive unless steps are taken to make wild animals 'pay their way'; for example, Eltringham (1994) stated:

> If wildlife is to survive, some means must be found to reconcile the needs of the animals with the legitimate aspirations of the human population. One solution is to give the wildlife a value so that local people will want to conserve it.

An example of the present and potential value of wildlife as a food source concerns the Class Aves. The domestic fowl, turkey, duck and pigeon are important sources of food in most parts of the world. However, although there are approximately 9000 species of birds, fewer than a dozen of these have been successfully domesticated. Many other avian species are, however, taken (harvested) from the wild and used for food, including pheasants, partridges, guineafowl, seabirds, pigeons and diverse small passerines such as weavers.

It is argued that greater use could be made of wild birds to provide food, particularly protein, in localities where people are malnourished or disadvantaged. For a start, it should be possible to bring certain new species of birds into domestication, such as chachalacas in South and Central America and Hartlaub's duck from West and Central Africa.

In addition to domesticating new species, more extensive and better organised harvesting of free-living birds has been advocated. The killing of wild birds for food is a traditional and essential part of the lifestyle of many people in poorer countries. Harvesting must, however, be carried out efficiently, humanely and sustainably, and local people must be involved in decision-making. It seems that greater use of wild birds as a food source could benefit humans in terms of nutrition and also, perhaps, encourage the sustainable conservation of wild species and their habitats. Nevertheless, utilisation of wildlife brings with it dangers of abuse, accusations of illegal activity and ample scope for litigation.

The argument extends to more than just birds, however. Concerns about the maintenance of biodiversity and the relevance of this to human health and well-being (Grifo and Lovejoy, 1997), as well as to wildlife, are integral to any debate. Increased food production using conventional sources is leading to the destruction of ecosystems and of indigenous animals and plants, as anticipated over 40 years ago by Fraser Darling (1960) who wrote

> ... to exchange the wide spectrum of 20–30 hoofed animals, living in delicate adjustment to their habitat, for the narrowed spectrum of three ungulates exotic to Africa – cattle, sheep and goats is to throw away a bountiful resource and a marvellous ordering of nature.

Game-Ranching

The principles of game ranch management were explained by Bothma (1996), in his popular manual. Important considerations include careful planning, habitat evaluation and the provision of fencing, water, roads, vehicles, buildings, *bomas* and holding pens. It is vital that the correct species are chosen for the available habitat. Successful harvesting, capture, transportation and meat processing need experience. Other considerations include plant succession and carrying capacity, grazing systems, control of bush and the correct use of burning. Game-ranching can prove effective in conservation as well as in the generation of income but, as Bothma emphasises, it will only be viable in the long term if the local population benefits; the game rancher must become involved in rural development.

Breeding Programmes and Conservation Genetics

Attempts to propagate wildlife have resulted in the establishment of captive-breeding establishments for many different species, some on a large scale (Figure 2.21). Others are more modest. Some such enterprises have prompted the voicing of concerns and complaints about welfare and a few – for example, purporting to be breeding birds of prey–have been the subject of legal cases relating to conservation legislation.

In some cases, captive-breeding has become an important tool in species conservation programmes. Examples are legion and include, for example, the sterling work on Mauritius, supported by Jersey Zoo and other bodies, which has resulted in saving from extinction a number of species of birds and reptiles. The establishment of breeding centres for vultures in India, Nepal and Pakistan was alluded to earlier in this chapter. Despite controversy in the past, few conservationists now doubt the value, in certain instances, of breeding species in captivity in subsequent release.

Captive-breeding brings with it many challenges, particularly relating to animal welfare, to health and disease (particularly the dangers of spread of pathogens to free-living animals) and to the genetic status of the animals that are being bred. Witzenberger and Hochkirch (2011) reviewed molecular studies on the genetic consequences of captive-breeding programmes for endangered animal species. They pointed out that current management strategies for *ex situ* populations are based on theoretical models, which have mainly been tested in model species or assessed using studbook data. In recent years a

Figure 2.21 A large aviary in the Middle East, designed for the captive-breeding of bustards.

number of molecular genetic studies have been published on captive populations of various endangered species, but a comprehensive analysis of these studies is still needed.

Legal cases relating to *bona fide* captive-breeding establishments are rare, but some individuals and organisations are tempted to use the label 'captive-breeding' for facilities and methods that may contravene conservation or animal welfare legislation. Investigators may be involved in such cases – an example, as pointed out by Cooper and Cooper (2007), of where the practice of wildlife forensics can involve the investigator in studies on both captive and free-living animals.

Conservation genetics is not only a feature of properly planned captive management programmes but also an increasingly important part of many aspects of wildlife forensics. For instance, Johnson (2012) discussed its relevance in the context of birds.

Animal Sanctuaries

Well-meaning individuals and organizations often establish 'sanctuaries' for the care of wild animals (Cooper and Cooper, 2007). These can be the subject of enquiry or legal action on various grounds – conservation, welfare or nuisance.

Habituation of Wild Animals

Animals may be habituated intentionally or inadvertently. The best example of the former is probably the mountain gorilla in Central Africa (see 'Mountain Gorilla Disease – Implications for Conservation' section in Chapter 13). Inadvertent habituation occurs when, for example, troops of baboons become so accustomed to humans that they raid *shambas* (small-holdings).

Habituated animals provide opportunities to study species at close quarters. This facilitates health-monitoring and, especially relevant to the detection of wildlife crime, the recognition of injuries associated with snaring or shooting.

On the other hand, habituated animals are more vulnerable to infectious disease of humans (see 'Mountain Gorilla Disease – Implications for Conservation' section in Chapter 13) and, because they tolerate close proximity of *Homo sapiens*, to persecution. They are also usually more likely to encroach on agricultural land or to venture near human habitation where they may be subjected to stressful episodes, injury or death.

Rehabilitation of Wildlife

The care of sick, injured, orphaned and displaced wildlife is practised in many parts of the world, especially the richer, usually industrialised, countries. It presents many challenges – both logistical and in terms of animal welfare (Cooper and Cooper, 2006; Mullineaux et al., 2003). There are ethical and legal considerations (BWRC, 1989; Cooper and Sinclair, 1990). Wildlife rehabilitation often presents special challenges in developing countries (Karesh, 1995). There is a need for training of rehabilitators and veterinarians (Penzhorn, 1997), especially when non-mammalian species such as reptiles are involved (Gould, 1998).

Studies on the role and efficacy of wildlife rehabilitation have been carried out in some, but not many, countries. For example, Wimberger et al. (2010) surveyed wildlife rehabilitation centres in South Africa with a view to seeing if there was a need for improved management, and Wimberger and Downs (2010) analysed annual intake trends of one particular urban rehabilitation centre, also in South Africa.

The care and rehabilitation of many species remains specialised and poorly understood – marine mammals, for instance (Maniere et al., 2004). If carried out properly, rehabilitation can help to boost numbers when individuals of a small threatened population are concerned. It can also be valuable in terms of education, and of collecting scientific data. However, rehabilitation may also serve as a cover for illegal acts or, if not properly carried out, elicit charges of cruelty (see later).

In the *Oxford English Dictionary* 'rehabilitation' is defined as 'restoration to a normal life or good condition'. In this book, in the context of wildlife, it is taken to mean 'The care in captivity (or temporary restraint/confinement) and attempted return to the wild of sick, injured or displaced wild animals'. Such sick, injured or displaced animals can be termed 'casualties'.

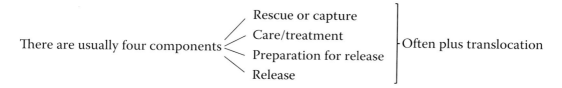

Wild animals may be presented for rehabilitation on account of

1. Injuries, for example road-traffic accidents (RTA), shooting, snares
2. Infectious diseases, for example rabbits with myxomatosis, foxes with mange
3. Non-infectious diseases, for example oiled seabirds, reptiles caught in grass fires
4. Other circumstances, for example 'orphaned' young animals, individuals separated from others of the same species for some other reason

In the majority of countries there is no legislation that relates specifically to wildlife rehabilitation; however, it is usually covered in part by laws relating to welfare and species protection. In most 'westernised' countries there are laws relevant to the rescue, rehabilitation and release of wildlife. In the Republic of Ireland, for example, there is the Wildlife Act 1976 and the Republic, like all EU countries, is subject to EU Directives and Regulations.

In Britain, under the Wildlife and Countryside Act, 1981, it is legal to take a sick, injured or disabled wild animal of a protected species and to kill it if it is beyond recovery; alternatively it may be kept and given first aid and care (or treatment by a veterinary surgeon) until it is fit to be returned to the wild. A few raptors need to be ringed (banded) and registered if kept in captivity – even as casualties. General licences permit a few 'authorised' people to keep these species temporarily (15 days), and a veterinary surgeon (veterinarian) to keep them for treatment (for 6 weeks) without ringing and registration.

Ethical considerations are different from, but may be linked to, legal provisions. Increasingly, questions are being asked about the welfare considerations in rehabilitation. In Britain, for example, the question of treatment of injured bats, all species of which are protected by the law, is controversial because these animals cannot readily be returned to the wild and the quality of their life, if they are retained in captivity, is debatable. Similar arguments apply to other species in many parts of the world. If legal actions are to be avoided, rehabilitation projects should either adhere to existing guidelines or those of comparable organisations elsewhere. The British Wildlife Rehabilitation Council (BWRC) has published ethical guidelines for the care of wildlife and the International Union for the Conservation of Nature and Natural Resources (IUCN) has produced guidelines concerning re-introductions and translocations (see earlier).

Rehabilitation is a complex subject which involves not only the animals but also the humans concerned, many of whom are very committed but do not necessarily have an understanding of population dynamics and the role of disease and death in the wild.

The problems presented by wildlife casualties in poorer countries were outlined in an earlier paper on rehabilitation in East Africa (Cooper, 1995b). Shortage of funding and often unavailability of equipment and expertise mean that the care of sick and injured wildlife cannot always be undertaken satisfactorily. Some projects, especially involving large charismatic species such as elephants, attract substantial sums of money and, as a result, a considerable amount of work can be done. Nevertheless, it is often argued that the funds that are spent on such efforts in order to save, and possibly in some cases to return to the wild, only a small number of animals, might be better used in broader-based conservation exercises, either relating to a range of species or to improving the well-being of local people who work with or live adjacent to them.

A particular dilemma is posed in the case of the great apes, such as the chimpanzees of Africa (Figure 2.22) and the orang-utans of South-East Asia. There may be successful

Figure 2.22 A view of Ngamba Island on Lake Victoria in Uganda, where confiscated and orphaned chimpanzees are rehabilitated.

rescue and rehabilitation but the animals often cannot be released, for a variety of reasons. For example, the chimpanzees that have been confiscated or otherwise taken into captivity are often not suitable psychologically for return to the wild, while the shortage of habitat restricts the release of orang-utans. An added difficulty at this level is the prevailing philosophy that certain species are entitled to their full life-span and that alternative provision for them must therefore be found. Suitable accommodation for apes is very difficult and expensive to provide on a long-term basis that is sustainable in a poor or unstable country.

The cost of rescuing great apes can be considerable. On 19 April 2010, the United Nations Peacekeeping Mission used United Nations helicopters to airlift a group of orphaned gorillas rescued from poachers in Rwanda and Congo. This exercise involved a number of organisations including the Gorilla Rehabilitation and Conservation Education (GRACE) Centre, initiated by the Dian Fossey Gorilla Fund International, in partnership with Disney's Animal Programs, the Congolese Wildlife Authority (ICCN) and the Pan African Sanctuary Alliance (PASA). PASA was formed in 2000 to unite the sanctuaries across Africa that rescue and rehabilitate chimpanzees, gorillas, bonobos and thousands of other endangered primates.

Cooper and Cooper (2007) pointed out the problems that can arise when an animal has been 'rescued' (perhaps a natural casualty, perhaps because it has been caught and injured in a trap) or confiscated pending or following a court case, and as a result has to be found a home or returned to the wild. The subject has been debated in many quarters and the arguments will not be repeated in detail here.

It is ironic that an animal that has been brought into captivity or removed from an individual on account of circumstances that might be classified as part of the spectrum of wildlife crime can then become the subject of another, separate, enquiry or even a legal action. Such can be the case under various circumstances, including the following, where the animal

- Is still in captivity but its welfare is inadequate
- Has been returned to the wild, where it is either unable to cope in terms of (e.g.) feeding itself, or is bullied or killed by others of the same or different species

In both such scenarios allegations of cruelty or neglect can be alleged. Concern by animal welfare organisations or enforcement agencies as to how casualties are looked after, or the circumstances in which they returned to the wild, may prompt investigations that can lead to legal action. In any enquiries, investigations or legal actions relating to such species, expert advice must be sought.

In most countries, it will be conservation laws that are of the greatest relevance to casualties (see Chapter 3). However, as indicated earlier, the fact that sick or injured wild species are held in captivity may mean that animal welfare legislation is also applicable. In some countries, such as the United Kingdom, the United States and Canada, the introduction of voluntary codes of practice for rehabilitators has played an important role in encouraging higher standards of care.

Confiscations

Both confiscated wild animals and those being held as evidence often present problems. They may spend many months, even years, away from their owner or keeper before the case is resolved. Sometimes the animals are in the care of agencies that are concerned primarily with conservation and have little or no experience of keeping wildlife in captivity.

Often, especially in poorer countries, facilities for holding seizures either do not exist or are inadequate (Cooper and Cooper, 2007). Even in the United Kingdom there have been many allegations that confiscated animals have suffered or died needlessly while being held.

Disposal of live animals following a case can present even more challenges. In some instances, if the accused is found not guilty or the case is not proved, the animals can be returned to their keeper. Under different circumstances other arrangements have to be made and the main options are then

- Return to the wild
- Retention in captivity in a zoo or private hands
- Euthanasia

An option that is being used increasingly in the United Kingdom, in instances where there is little risk that the live animal will be removed, is to leave it with the suspect, pending the outcome of the case. A legally binding undertaking is signed, declaring that the animal will not be disposed of in any way.

A similar dilemma faces those who have to deal with other 'unwanted' animals, such as stock that remains when a zoological collection has to be closed or the owner of a private collection dies (Graham, 1996).

Disease

In some forensic cases, wildlife disease is the issue – for example, when an infectious condition kills animals and there is legal action against persons who either encouraged its spread or failed to take appropriate action to minimise its effects. An example is myxomatosis. The virus causing this disease is sometimes spread or introduced in order to kill rabbits; this may be unlawful. Definitive diagnosis of the disease, necessary as evidence, is likely to require sophisticated techniques, including transmission electron-microscopy (TEM) (Figure 2.23).

When dealing with such cases, the investigator needs to have knowledge and understanding of how diseases (both infectious and non-infectious) affect wildlife and the ecological factors that can encourage or limit their effect. The book edited by Hudson et al. (2002) provides valuable information in this respect. Another useful reference is the *IUCN DRA Tool Needs Survey* (Jakob-Hoff, 2011), available in English, French and Portuguese. This, while concentrating on disease risk assessment (DRA), attempted, with a questionnaire, to answer questions such as 'Who cares about wildlife disease risk?' and 'What do they care about?' The first is answered by listing those disciplines (27 in all) that responded and the various organisations to which they are affiliated. Answers to the second question reveal that 'wildlife disease scenarios of concern' are as follows:

- Wildlife management – *ex situ*
- Wildlife management – *in situ*
- Domestic animal/wildlife interactions
- Human/wildlife interactions
- Domestic animal translocations
- Wildlife translocations
- Other, for example emerging new diseases, mass mortality investigations, poachers and traps, trade

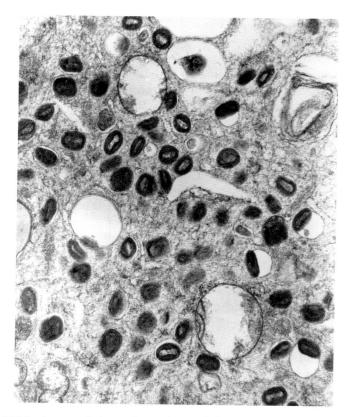

Figure 2.23 A TEM picture of tissue from the eyelid of a rabbit, showing large numbers of virus particles consistent with myxomatosis. A definitive and important diagnosis.

Perhaps the most telling results of the survey relate to the 'wildlife taxa of main concern'. Nearly 35% of respondents listed mammals, just over 25% birds, 15% reptiles and just over 10% amphibians. Fish, numbering over 25,000 species and key components of many ecosystems, were of concern to only 6% and invertebrates (89% of the Animal Kingdom and key components of biodiversity – see earlier) a mere 5%. This 'speciesism' is seen in many other areas of wildlife research and conservation and is reflected in forensic work, where the majority of cases relate to charismatic mammals and birds and relatively few to the so-called lower vertebrates and invertebrates. The report specifically states that 'Currently they (DRA tools) are poorly applicable to insects (*sic*), particularly non-pest and non-domesticated insects. They need to lose their vertebrate-centricity …'

Many authors in recent years have discussed the part played by infectious diseases in the health status and survival of threatened species; for instance, Cooper, (1989c) and Pedersen et al. (2007).

Significance in Wildlife Forensics of 'Global Health' and 'One Health'

The need in wildlife forensic investigations to tackle challenges in a broad-based, 'holistic' way is particularly true when the health of animals is of prime concern. In recent years there has been an upsurge of interest in 'One Health' – that is, the concept of promoting stronger links (possibly even unifying) the veterinary and medical professions and using their knowledge and experience in a synergistic way. It is argued that such collaboration will produce great advances in the study and control of diseases, including those brought about by environmental changes such as global warming and disasters. The 'One Health' paradigm encompasses the human population, domesticated animals and wildlife. It follows that forensic investigations into infectious and non-infectious factors relating to wildlife morbidity and mortality may also benefit from involving those active in both human and domesticated animal medicine – and probably other disciplines as well.

Okella et al. (2011) stated that, while 'One Health' is becoming a twenty-first century exhortation, the intricate relationship and linkages between animal health and the well-being of humans have been noted over many centuries. Examples of the relationship between zoonotic disease and human tragedy are seen throughout history, the most notorious perhaps being the 'Black Death', spread by rats, that killed millions of people in the Middle Ages.

The term 'ecosystem health' is usually taken to refer to a similar interdisciplinary approach, but one largely driven by biologists. The editors of a standard work on the subject (Rapport et al., 1998) define ecosystem health in terms of 'the interrelations between human activity, social organisation, natural systems, and human health'.

The thinking summarised earlier suggests that health, whether it be that of animals, humans or plants, can and should no longer be looked at in isolation. The study and promotion of the health of forests offers one example: The subject has taken on a new perspective in recent years, with many 'sub-disciplines' involved, ranging from entomology to oceanography (Castello and Teale, 2011).

The importance of an ecological approach to the study of wildlife diseases was demonstrated in the book by Hudson et al. (2002), referred to earlier. A similar ecosystem-based orientation is increasingly the norm in fisheries management (Christensen and Maclean 2011; Link, 2010). This approach clearly needs to be applied to wildlife forensic investigations.

Bossart (2011) was practising 'One Health' when he explored the role of marine mammals as sentinel species (see earlier) for oceans and for human health, linking his studies with current concern about climate change and environmental degradation.

A study from Brazil further illustrates the concept. Quaresme et al. (2011) investigated possible hosts of *Leishmania* species in an area endemic for American tegumentary leishmaniasis (ATL) in northern Minas Gerais State. Domestic dogs and small forest mammals are known reservoir hosts for *Leishmania infantum*, but their role in the transmission cycle of the *Leishmania* species that cause cutaneous leishmaniasis is not established. The results of this research were that *L. (V.) braziliensis, L. (L.) infantum* and *L. (V.) guyanensis* are circulating among wild and synanthropic mammals present in the Xakriabá Reserve, thus highlighting the epidemiological diversity of ATL in this region.

Recently there has been increased concern that the movement of animal and plant material from one country to another could be leading to the spread of infectious diseases, including those transmissible to humans from animals. Coupled with the conversion on a large scale of forest and of 'wilderness' for human use (to grow crops, to graze cattle and to produce timber), the risk of disease spread from wildlife is increasing. This needs to be tackled from both ends, ranging from the education of hunters and those who handle bushmeat to negotiation with governments and food distributors in the Western world.

Wildlife ecology is complex and there is a need for veterinarians, ecologists and others to understand it better. Mathematical modelling (computer simulations) can help research workers predict the spread of pathogens from one species to another and the effect that climate and environmental change can have on these transmission pathways.

Legislation 3

MARGARET E. COOPER

Contents

Introduction and Background Information	91
General Aspects of Law	93
European Institutions and Law	94
UK Law	94
Wildlife Law	95
International	96
Regional	97
National (Domestic or Internal)	98
Local Laws	101
Wildlife Trade	101
International Trade	101
Wildlife in Captivity	105
Keeping Animals	105
Welfare	106
Liability for Animals	106
Other Relevant Legislation	107
Animal Health	107
Disposal of Animal Waste	107
Occupational Health and Safety	108
Employment Law	109
Veterinary Laws	109
Law Enforcement and International Collaboration	109
Conclusion	111

'At his best man is the noblest of all animals; separated from law and justice he is the worst'.

Aristotle

Introduction and Background Information

This chapter discusses the fields of law that are relevant to wildlife forensic investigation.
    Much of it will relate to legislation that provides for the protection and conservation of 'wildlife', but it will also refer to other legislation that has to be considered in the course of wildlife forensic investigation.
    The term 'wildlife' is defined in Chapter 1 as 'species of animals or plants that are normally to be found occurring naturally in the wild'. For the purpose of this chapter, the term will refer primarily to animals although it is well recognised that the survival of all species depends upon the conservation of their habitat.

As mentioned in Chapter 15, legislation and litigation involving wildlife can apply to both free-living and captive species. Those in captivity are subject not only to conservation laws but also to legislation relating to animal health, animal welfare, theft, damage to property, and authorisation to keep animals for certain purposes such as education, display, sport, trade, rehabilitation or captive-breeding. Forensic investigation and evidence may be required not solely in crime but in insurance disputes, planning inquiries, civil law claims in negligence, nuisance, trespass or liability for injury or damage caused by animals and in professional negligence (malpractice). Other issues that may need the support of forensic evidence include veterinary professional discipline, medicines law, health and safety, the movement and transportation of animals and biosecurity.

In forensic investigations and expert evidence, it is important to state clearly the terminology that is being used in written reports and oral evidence. As discussed in Chapter 15, the term 'wildlife' is used very loosely, both in common parlance and in scientific usage. It is applied, often without any particular thought, to species that are commonly found 'in the wild' when they should be more accurately described as 'free-living', that is not living in captivity. For example, the title of the Convention on International Trade in Endangered Species of Wild Fauna and Flora (CITES) refers to wild flora and fauna but, although intended to protect free-living populations of species, the controls that it imposes can only apply to captive specimens. The term 'wildlife' is often used for

- Species that are actually living freely in the wild
- Specimens of the same species even when they are actually living in captivity
- Dead specimens, parts and derivatives of such species

Although it is useful at times to use the word generically, it is more precise to refer to 'wildlife' as 'non-domesticated species' to distinguish them from 'domesticated' species (as defined in Chapter 1) because these terms apply whether an animal is free-living or captive (Figure 3.1).

Figure 3.1 Orang-utan are often referred to as 'wildlife' yet this one lives in captivity, in a zoo.

Legislation

season, is often included. Depending on the precise legal provisions, in most cases, for the law to be broken, these actions would have to be committed recklessly or intentionally.
- Some species may warrant special protection, such as stricter provisions, higher penalties or special legislation.
- Protection for some species only during the breeding seasons.
- Habitat protection that controls access to, and use of, land or is essential for species conservation.
- Trade controls on protected species (with some authorised trade in captive-bred specimens).
- Controls on hunting: the seasons, weapons, vehicle use and other indiscriminate methods are regulated. Alternatively, hunting in some or all forms may be prohibited.
- Regulation of taking game.
- Prohibition of indiscriminate forms of capture such as traps or snares although there may be authorisation of certain methods of capture for the purposes of taking small game, pest control and scientific research (see Chapters 2 and 9).
- Authorised control of protected species that are causing damage – for example, to property, crops, public health or air safety.
- Enforcement powers, offences and penalties.

Species protection laws are often not absolute because there are many situations in which it is necessary to manage conflicts of interest between species conservation or habitat protection and other forms of land use. Resolution of such issues may require the control of wildlife if it interferes unduly with other legitimate use of land; alternatively, as a matter of policy, the legislation may permit the sustainable use of natural resources. Usually, such exceptions are allowed under licence or permit, either issued to individual also by way of a general authorisation. Most permits are subject to conditions that require strict compliance.

Wildlife legislation may allow the rescue and rehabilitation of sick and injured protected species. The legal position varies; in the United Kingdom anyone may do so provided that they release the animal once it has recovered or can justify long-term captivity if it will never be fit to return to the wild. They must also comply with any other relevant legislation. On the other hand, many countries only allow people who are trained and registered as wildlife rehabilitators to care for wildlife.

Sometimes there may be legislation for a particular species that is considered to be in special need of protection – for example, badgers in the UK laws, the Philippine eagle in the Philippines and the bald eagle in the United States.

Information on wildlife legislation for the United Kingdom is to be found in Cooper (1987a), Parkes and Thornley (1997, 2000), Rees (2002) and Palmer (2001); for New Zealand in Wells (2011); for Australasia in Sankoff and White (2009); and for the United States in Bean and Rowland (1997). A number of institutions and organisations, many with specialised expertise, that provide information on wildlife law are listed in Table 3.2, and sources of some national wildlife legislation are given in Table 3.3. For an examination of the state of wildlife crime in the UK (see HCEAC, 2012a; HCEAC, 2012b).

Table 3.2 Sources of UK Wildlife Legislation

Species	Organisation	Website/Publication
Bats	Bat Conservation Trust	Bats and the Law http://www.bats.org.uk/pages/bats_and_the_law.html
Badgers	Badger Trust	Badgers and the Law http://www.wcbg.org.uk/about-badgers/badgers-and-the-law/
Birds	Royal Society for the Protection of Birds (RSPB)	Wild Birds and the Law: England and Wales http://www.rspb.org.uk/Images/WBATL_tcm9-132998.pdf Wild Birds and the Law: Scotland http://www.rspb.org.uk/Images/WbatlScotland_tcm9-202599.pdf
Birds of Prey Keeping (UK)	Animal Health (q.v.)	Birds of prey – What you need to know http://animalhealth.defra.gov.uk/about/publications/cites/CITES%20Birds%20of%20Prey.pdf
Invertebrates	Buglife – The Invertebrate Conservation Trust	Legislation and Policy http://www.buglife.org.uk/conservation/policy
Taxidermy	K. McDonald	UK Taxidermy Law http://www.taxidermylaw.co.uk/index.cfm
Wildlife Scotland	Scottish Natural Heritage	Scotland's Wildlife: The Law and You http://www.snh.org.uk/pdfs/publications/wildlife/wildlifelaw.pdf
Northern Ireland (NI)	A Stewart Environment and Heritage Service, NI	Wildlife and the Law (Stewart, 2012) Wildlife: The Law and You in Northern Ireland http://www.doeni.gov.uk/niea/wildlifeandthelawscreen.pdf
Wildlife law and species status (UK)	Joint Nature Conservation Committee (JNCC)	Conventions and Legislation http://jncc.defra.gov.uk/page-1359 UK Legislation http://jncc.defra.gov.uk/page-1376 Conservation Designations for UK Taxa http://jncc.defra.gov.uk/page-3408
England wildlife law and licensing; animal law	Natural England	Wildlife Management and Licensing http://www.naturalengland.org.uk/ourwork/regulation/wildlife/default.aspx Legislation on conservation and links to other animal legislation http://www.naturalengland.org.uk/ourwork/regulation/wildlife/policyandlegislation/legislation.aspx
Sporting shooting	The National Gamekeepers' Organisation	Frost (2011)

Table 3.2 (continued) Sources of UK Wildlife Legislation

Species	Organisation	Website/Publication
Gamekeepers	The National Gamekeepers' Organisation	Anon (2010b)
Wildlife offences	Crown Prosecution Service	Legal Guidance to Crown Prosecutors
		http://www.cps.gov.uk/legal/v_to_z/wildlife_offences/
CITES offences	TRAFFIC	TRAFFIC Bulletin
		http://www.traffic.org/bulletin/
CITES Controls and law for United Kingdom	Department for Environment Food and Rural Affairs Animal Health	CITES Management controls for United Kingdom
		http://animalhealth.defra.gov.uk/cites/index.htm
		Guidance Notes
		http://animalhealth.defra.gov.uk/cites/applications/guidance.html

Local Laws

City, town and rural authorities may make bye-laws to protect specific areas and landscapes. These may also protect the species to be found there and control access by the public and the methods of management.

In some countries customary practices or law may serve to protect wildlife

- For example, in Uganda, termite mounds are owned by people, which may help to conserve them (Cooper, 1995a).
- In the Cook Islands, traditional methods are still used to prevent over-exploitation of coastal natural resources (http://www2008.spc.int/DigitalLibrary/Doc/FAME/InfoBull/TRAD/19/TRAD19_11_Tiraa.pdf)
- In New Zealand, the government Department of Conservation collaborates extensively with Maori to ensure that conservation methods accord with Maori customs and their traditional attitudes towardxs their sacred species (http://www.doc.govt.nz/about-doc/role/maori/).
- In the Solomon islands, a dichotomy exists between one section of the population that reveres the dolphin and protects them whereas another sector actively uses traditional rights of exploitation for food (Anita, 2004). Certain countries claim traditional rights to whaling (some approved by the International Whaling Commission) and sea mammal hunting.

Wildlife Trade

International Trade

The Convention on International Trade in Endangered Species of Wild Fauna and Flora lays down the rules for the movement (commercial and non-commercial) between countries of the species listed on its three appendices. The CITES website (http://www.cites.org/)

Table 3.3 Sources of Other National Wildlife Legislation

Country	Source	Legislation/Link
United Kingdom	UK Legislation	Wildlife and Countryside Act 1981 (and see Table 2) http://www.legislation.gov.uk/ukpga/1981/69
Australia	Department of Sustainability, Environment, Water, Population and Communities	http://www.environment.gov.au/biodiversity/wildlife-trade/index.html
Ireland	Scott Cawley Ltd	http://www.scottcawley.com/newsletter/Species%20Protection.pdf A Quick Guide to Protected Species in Ireland
New Zealand	Department of Conservation	http://www.doc.govt.nz/about-doc/role/legislation/wildlife-act/ Wildlife Act 1953 (text) http://www.legislation.govt.nz/act/public/1953/0031/latest/DLM276814.htm
United States	US Fish & Wildlife	Facts about Federal Wildlife Laws
United States and other countries	Michigan State University	College of Law Animal Legal and Historical Centre http://animallaw.info Animal law including wildlife, primarily in the United States, but other countries too
Canada	Department of Justice	Canada Wildlife Act http://laws.justice.gc.ca/PDF/W-9.pdf Species at Risk act http://laws.justice.gc.ca/PDF/S-15.3.pdf http://www.ec.gc.ca/default.asp?lang=En&n=E826924C-1
Kenya		Wildlife (Conservation and Management) Act (Rev 2009) http://www.law.co.ke/klr/fileadmin/pdfdownloads/Acts/Wildlife__Conservation_and_Management_Act___Cap_376_.pdf
India	Ministry of Environment and Forests	The Indian Wildlife (Protection) Act, 1972 (amended) http://www.law.co.ke/klr/fileadmin/pdfdownloads/Acts/Wildlife__Conservation_and_Management_Act___Cap_376_.pdf
Malaysia		Wildlife Conservation Act 2010 http://www.wildlife.gov.my/pengumuman/Wildlife%20Conservation%20Act%202010%20Act716.pdf
China		Law of the People's Republic of China on the Protection of Wildlife Citation: Order No. 9 (1989) http://www.animallaw.info/nonus/statutes/stcnwildlaw.htm See also Zhang (1996), Li and Li (1998), Wang (1998) and Hill (2011)
European Union		Conservation http://www.eu-wildlifetrade.org/html/en/wildlife_trade_regulations.asp http://www.eu-wildlifetrade.org/pdf/en/1_international_legislation_en.pdf

and Stewart (2008, 2009, 2011) illustrate the complexities of wildlife law enforcement and the skills, time and costs involved. There are occasionally formal reports of cases, but these are available only for the few wildlife actions that get to the higher courts. Useful accounts of enforcement activities are provided in

- The regular publications of the RSPB, *Legal Eagle* and *Bird Crime* (http://www.rspb.org.uk/ourwork/policy/wildbirdslaw/legalpublications.aspx)
- The *TRAFFIC Bulletin* (http://www.traffic.org/bulletin/)
- The news webpage of PAW (http://www.defra.gov.uk/paw/news/)
- The reports and websites of the multinational bodies mentioned earlier and elsewhere in this book

Conclusion

This chapter has been written in a broad and general manner with the hope that it will provide sources, models, concepts and information that are either of direct use, or can be adapted and applied, in different parts of the world, in the service of wildlife conservation.

The laws that provide the backbone of wildlife conservation and associated activities depend, for their enforcement, upon the forensic investigation techniques and skills provided in the rest of this book.

The applicability of many standard forensic science methods to work with wildlife, especially the procedures for dealing with evidence, was emphasised more recently in a chapter by Wallace and Ross (2012).

Special Features of Wildlife Forensics

If one looks at wildlife forensics in the most simplistic way, one can ask 'What is needed of the forensic scientist that may help in the investigation and possible prosecution of wildlife crime?' The answer can be given under the categories listed in Table 4.1.

It appears, therefore, that investigation of alleged wildlife crime can be carried out using many of the standard techniques and approaches followed in other, more traditional, areas of forensic science.

There are, nevertheless, some important differences and peculiarities that do, to a certain extent, mark out 'wildlife forensics'. Some of these relate to the fact that in such work, any one of thousands (millions, if one includes invertebrates) of species of animal

Table 4.1 Techniques in Wildlife Crime Investigation

Category	Examples	Comments
Clinical work	Examination, diagnosis, sampling and treatment of live animals.	In some countries, for example United Kingdom, diagnosis and treatment require, by law, the services of a registered veterinary surgeon. Can include radiology and other imaging techniques and laboratory investigations.
Identification and morphological studies	Species' determination using bones, skins, hair, feathers, scales and derivatives.	Involves gross, microscopic and DNA-based techniques.
Pathology	*Post-mortem* examination of whole carcases, organs and tissues from dead animals – and samples from live animals.	See above regarding diagnosis. Can include various supporting tests, for example radiology and other scanning techniques and laboratory investigations.
DNA technology	Species, sex and parentage determination; geographical origin; linking an animal to a human, other animal or object; other investigations as mentioned earlier.	A very important component of many investigations. See Chapter 11.
Stable isotopes	Geographical origin.	See Chapter 10.
Toxicology	Natural, accidental or malicious poisoning.	See Chapters 2, 9 and 10.
Crime investigation	Studies on ballistics, chemical assay, soil analysis, fingerprints, tyre marks, etc.	'Standard' investigations, usually involving police and specialists.
Crime scene studies	As mentioned earlier in the text, plus other investigations.	As mentioned earlier. May require experience of fieldwork and knowledge of natural history. See Chapter 6.

Source: Adapted from Cooper, J.E. and Cooper, M.E., *Introduction to Veterinary and Comparative Forensic Medicine*, Blackwell, Oxford, UK, 2007.

might be involved. Others are attributable to the varied, often different, questions that may be asked in a wildlife forensic case – for example, regarding an animal or its derivatives (see Box 4.2).

The provenance (place of origin, derivation) of material is very often a key consideration in wildlife forensics. As Box 4.3 shows, this can relate not only to a whole live or dead animal but also to diverse derivatives – some of which may be only microscopical in size.

There are various ways of answering this question about provenance, and the point made by Linacre (2009) in the following text illustrates how wide a range of tests, many

> **BOX 4.2   SOME QUESTIONS THAT MAY BE ASKED ABOUT ANIMALS**
>
> - Species/sub-species/hybrid?
> - Age?
> - Sex/entire/neutered?
> - Sexually mature/fertile/in oestrus?
> - Pregnant/gravid?
> - Lactating/incubating?
> - Geographical origin?
> - Migrant or resident?
> - Origin – captive or free-living?
> - Captive or free-living – for how long?
> - Health status?
> - Age of skin wound, bruise, fracture or other lesion?
> - (For carcases or tissues – how long dead?)

> **BOX 4.3   THE ORIGIN (PROVENANCE) OF ANIMALS AND THEIR DERIVATIVES**
>
> - Live animals
> - Dead animals
> - Hairs, feathers, scales, including shed skins (sloughs) of reptiles
> - Horns, antlers, hooves
> - Blood and tissues
> - Eggs, embryos, fetuses and placentae
> - Ingesta, faeces, urine and saliva
> - Bones and teeth, including elephant ivory
> - Macro- and micro-parasites

but not all of them modern molecular techniques, may be needed in order to do so adequately:

A question that is increasingly asked is "From what part of the world did this sample originate?" In part, DNA can answer this question, provided that there is extensive research

Aspects of Wildlife Forensic Investigation

Scene of Crime Visit and Preparation of Evidence

Scene of crime visits are particularly important in wildlife forensic investigations as many incidents take place in the field. Sometimes the issue is an environmental one (e.g. alleged misuse of pesticides) in which event that location must be visited. In cases of (say) suspected badger baiting or poaching of an endangered species, animals may need to be examined, alive or dead, *in situ*.

The correct preparation and presentation of evidence is as essential in wildlife cases as it is in human and domesticated animal forensic work. Particularly important aspects are as follows:

- Record-keeping, including use of photographs and/or video footage
- Maintenance of chain of custody
- Taking appropriate samples
- Correct handling, packaging, labelling and storage
- Retention of material
- Compilation of report(s)

Laboratory Investigations

As in other fields of forensic work, these are a key aspect of most wildlife cases. The quality of the results depends upon the sampling methods used, transportation and storage, the techniques employed and the skills of those processing the specimens. Animal health regulations may influence how and to where pathological material may be sent and the Convention on the International Trade in Endangered Species of Wild Fauna and Flora (CITES) will dictate the movement from one country to another of derivatives from endangered species of animal, even when this is for legal purposes (see Chapter 3).

A secure chain of custody remains essential in laboratory investigations. So too does quality control (see Chapter 10 and 'Why Is Quality Management Important?' section in Chapter 12).

DNA technology is an important part of laboratory work in animal forensic investigations work (PAW, 2005) and is discussed in detail by Robinson (Chapter 11). DNA-profiling techniques can provide information about such matters as parentage and kinship, sex (gender), species and populations of animals. Although DNA technology plays an essential part in many wildlife cases (Linacre, 2009) (Chapter 11), it cannot answer all the questions that may be asked – see earlier.

The investigation of wildlife crime is not, essentially, very different from any other type of forensic work. Its special, but by no means unique, features are that it can involve a whole range of species, an equally wide range of techniques (see Table 4.1), and is very often performed under field conditions (see Chapter 7 and Appendix B). Methods that are used in more traditional forensic studies still have a part to play, however – for instance, fingerprinting, casting of footprints and tyre marks and blood spatter analysis.

Shortcomings

There are aspects of wildlife forensics that have slowed its metamorphosis into a truly global discipline. One relates to the paucity of laboratories currently devoted to such investigations. The US Fish & Wildlife Service (USFWS) laboratory in Oregon, the United States (see earlier), describes itself on its website (see Appendix www.fws.gov/oregonfwo) as 'the world's only ...' There are, in fact, other wildlife forensic laboratories elsewhere, in Europe, Africa and Asia, for example, but these are generally small and staffed by only one or two people.

In many jurisdictions, wildlife forensic material is, or can be, examined by existing government or private laboratories. Sometimes, however, such institutions are limited in their scope, for example, in the range of toxicological tests that they can offer. In the United Kingdom, where there is not as yet (2011) a dedicated wildlife forensic facility, the establishment of the England Wildlife Health Strategy (EWHS) – which, *inter alia*, is concerned with wildlife poisoning – may help to rectify this (Hartley and Lysons, 2011).

Conclusions

The situation and developments outlined earlier amply illustrate that, while in some respects a new subject, 'wildlife forensics' is not a novel discipline. Its tenets and its practice are essentially the same as those in other fields of forensic science. As the foregoing and succeeding chapters in this book indicate, such activities as intelligence and information gathering; the investigation of the crime scene; dealing with samples, collection and submission of evidence; report writing; and appearing in court are based in a wildlife case on exactly the same principles as those relating to homicide, malicious damage or theft. Most of the laboratory techniques employed in wildlife work are identical to those used in human and veterinary forensic work. Specialised tasks such as clinical examination and necropsy differ only in minor respects when dealing with wild animals from those used for domesticated animals and humans.

Wildlife forensic work is rapidly evolving and progressing, and is developing into a well-defined area of study. It presents unusual, but not unique, challenges to those involved in its practice. However, much extrapolation is possible, and necessary, from other, longer established, fields of forensic investigation.

which we all are susceptible (see Appendix F). The potential hazards may or may not be present or severe enough to cause symptoms until later in the processing, hours, weeks, months and even years later. Keep in mind that wildlife crime scenes may contain toxic chemicals, pathogens (Figure 5.1) and pesticides/poisons not expected in a human crime. Appropriate personal protective equipment and safety considerations are critical in such instances.

Investigators need to exercise caution when approaching an area of interest, whether it ultimately proves to be a crime scene or not. Heightened awareness and acting on clues help assure that the first responders enter the area safely. The extent of the area that encompasses a crime scene is usually unknown. Furthermore, the safety boundary assessment is an ongoing responsibility beginning with the first person on the scene and extending to additional personnel who arrive later. Safety boundary assessment occurs throughout the entire crime scene investigation process, including clean-up.

The personnel responding to the area must keep in mind their own physical limitations. Individual differences in physical conditioning will play a part in how the scene can be processed. Pre-existing injuries and conditions such as fear of heights or claustrophobia should be shared with other members of the team to facilitate different assignment of tasks.

Hazard(s) at a crime scene can arise from physical, biological or chemical factors. For examples and reviews, see Gómez and Aguirre (2008), Karesh et al. (2005), Laurent (1977), Lee and Ladd (2001), Sampson and Sampson (2003), Sheffield et al. (2005). It should also be noted that one hazard can have the elements of more than one of these categories involved, for example, a broken bottle (physical) that has blood on the surface (biological). Each hazard would be handled separately in a different way, but when combined may require novel solutions.

These lists are not exhaustive, but reflect some of the possible hazards personnel need to keep in mind when working in the field. Investigators should be especially mindful of suspected pesticide/poisoning cases, and exercise extreme caution when handling any animal carcase (see Appendix G).

Safety concerns do not stop at the scene. If hazardous items are collected, care in packaging and labelling needs to be practised to assure safe conditions for laboratory personnel. Postal and delivery companies also have very strict regulations regarding the shipment of potentially hazardous materials (see Chapter 3). Examples of physical, biological and chemical hazards are listed in Box 5.1.

BOX 5.1   HAZARDS

Physical/environmental hazards: hazardous weather/physical conditions: extreme weather, flashfloods, lightning, fog, getting lost, fire, booby-traps, etc.
Hazardous locations/terrain: landslide, avalanche, bodies of water, cliffs, vehicle traffic (train/car), hunting area, etc.
Biological hazards: blood-borne pathogens, fungus/mould, body fluids, viruses, bacteria, etc.
Hazardous animals: bear, cougar, snakes, etc. (depends on location – see Chapter 7).
Hazardous invertebrates: spiders, bees/hornets/wasps, ticks, leading in some cases to arthropod-borne illnesses, etc.
Hazardous plants: poison oak/ivy, cacti, etc. (depends on location – see Chapter 7).
Chemical hazards: various pesticides/herbicides, poisonous bait areas, clandestine laboratories, dump sites, chemicals used by the investigator to process the scene, etc.

Securing and Protecting the Crime Scene

The initial boundary of the crime scene is established based on a preliminary, and thus imperfect, assessment of the potential crime and its setting (Figure 5.2). The information gathered as the investigation proceeds will allow the initial responder to narrow the protected scene. The marking or securing of the initial scene should accommodate adjustments as facts are learned and observations are made. The purpose of the boundary is a visual indicator to all that care should be taken to minimise the loss of potential evidence and to protect investigators from hazards.

Command posts or staging areas may be established either inside or outside the secured crime scene boundary. It is often useful to designate additional areas within the outer boundary to bring attention to specific area(s) and to have a central secured location when multiple scenes are discovered. Having multiple secured areas aids in the division of labour for personnel assisting in the protection of the scene, especially if the areas are large. An essential task is to evaluate which evidence to protect first. For example, small pieces of paper that can be lost in the wind or fragile evidence such as footwear and tyre (tire) tracks may require immediate attention depending on location, weather conditions or traffic.

If the scene is located in a relatively public area, personnel need to keep in mind the method(s) used to protect and secure the area because of the volume of onlookers and vehicular traffic. It may be necessary to put up additional barriers, to block the view of the

Figure 5.2 Crime scene processing. Note the crime scene boundary, the evidence markers and the process of photo-documentation.

onlookers and to document important elements or evidence that need to stay confidential for purposes of the investigation. In the past few years, the prevalence of recording devices in the public domain has skyrocketed. The technology that provides the ability to capture still images, video and audio is affordable, and such media are easy to send or post on the Internet (see Chapter 7). This adds a novel element to protecting and securing crime scenes.

In more remote areas, the crime scene investigator also needs to plan how evidence may be protected and secured from animals that will enter the scene, leaving tracks and scat and possibly displacing trace evidence. Scavenging of the victim's remains may already have taken place (see Chapter 2) and personnel will have to stop or at least minimise it to preserve those remains.

Caution should be used when allowing the entry of responding personnel into the crime scene. Many individuals do not need to enter the scene. A crime scene is no place for a large group of people, which only adds to the possibility of damaged, missed or contaminated evidence. A recording of the details of each person wanting to enter the scene needs to be made by the officer in charge. Simply wanting to take a look is no reason to jeopardise or complicate the case. This includes supervisors and management personnel, along with individuals from other agencies. The documentation of all persons entering the scene is essential. This includes the initial investigator, all laboratory personnel and anyone else who enters the scene. One should not hesitate to ask why entry is requested. If there is more than one scene, care should be used to limit crossing-over to other scenes. This will reduce the potential for crime scene contamination and again limit the total number of persons involved per scene.

Searching for and Locating Evidence

Crime scenes are generally discovered by someone other than a law enforcement officer. Someone walking their dog, hiking, hunting or checking their farmland may find a dead animal. Rural land owners report hearing shots during the night and find a dead elk lying in one of their fields. If the circumstances appear to be suspicious, they report the finding to their local game enforcement agency (in the United States) for further investigation. The reporting party is usually able to point out the location of the victim and the general area of the crime scene to the responding officers.

Initial Walk-Through

One or two investigators conduct the initial walk-through of the scene. Typically, they take a direct route towards and away from the carcass. However, care should be taken not to take the most likely route taken by the perpetrator(s). If necessary, take a circular route to the scene and avoid rushing to the carcass. The purpose is to further assess the crime scene and to evaluate what equipment may be needed for further processing. At least one person should take photographs as they approach the victim. These photographs will depict the scene as it is first seen by the investigators. People should both walk in and out using the same path, being aware of small items of potential evidence such as cartridge cases, cigarette butts, blood drops, foot or tyre prints. Carry evidence flags and tape with you to mark small items as you see them (Figure 5.3). These may be hard to spot later when the light changes. Do not collect evidence at this time.

Figure 5.3 Evidence flags are used to mark possible evidence items during the initial assessment and walk-through of a potential wildlife crime scene. With further evaluation, formal evidence items are then identified with numbered placards.

Re-evaluate the scene after the initial walk-through to see if you need additional resources for processing, such as a metal detector, additional personnel or specialised safety equipment for pesticides. You may be able to decrease the size of the original protected perimeter.

As you work, remember to monitor weather conditions. Inclement weather can affect your processing priorities. Tyre tracks and shoe prints are rapidly destroyed by rain. Snow will blanket everything at the scene, or the sun or rain may melt prints in the snow. These and other physical factors may be even more significant in some parts of the world (see Chapter 7).

Scene Search

This is the formal, systematic, organised search for evidence. Don't be in a hurry. A thorough search takes time. Evidence can be destroyed, lost or overlooked if people rush to the victim. Be aware of your responsibilities to avoid contaminating any human crime scenes that may also be present. Avoid allowing other animals to suffer, even if it means destroying evidence (see Chapter 8). Always ask if the scene is declared safe before entering it.

One person should be in charge of the scene. That person is responsible for assigning tasks to other searchers. Optimally, the search team should consist of the person in charge, one or two photographers, searchers and two people to take and record measurements (one to make a scene drawing, and someone to record, bag and tag the evidence). All information is funnelled to the investigator in charge, who will also be the liaison to the news media or legal prosecutor, if necessary.

Do not contaminate the scene. Searchers should not eat, drink or smoke in a crime scene. Cigarette butts, candy wrappers (sweet papers) and drink containers discarded by searchers can be confused with actual evidence items. This can lead to false assumptions and hypotheses at the scene, along with unnecessary laboratory work, and embarrassing questions in court. Items such as evidence bag peel-off strips, casting materials and other debris introduced by the search team should be removed at the end of the examination.

Natural light is best for the search (see also Appendix B). It is often better to postpone the search until daylight or suspend a partially completed search until there is better light. If the search is interrupted, someone should stay at the scene for security until the search is conducted or continued. Weather conditions may force a search that is not optimal. Collect as much evidence and document as much as you can. A roll of plastic sheeting or tarps can be used to cover fragile evidence if weather is a problem.

A good search is slow and methodical. Different search styles may be used depending upon the scene. Indoor scenes are usually limited in size and scope and may consist only of a search of a garage, freezer or outbuilding. Outdoor scenes can be relatively small or extensive, covering large areas or multiple, related areas. Terrain can also be a factor. Some scenes include bodies of water along with shoreline, steep mountainous areas or desert.

A standard search pattern is the line search, in which searchers start in a straight line side-by-side and walk through the area. The searchers should be sufficiently close enough together that there is no unsearched area between them. The distance may be arms-length or closer in brushy areas, or wider where visibility is greater.

A grid search pattern is another common method used. Here, the crime scene is divided into a grid and each section of the grid is thoroughly examined. The grid boundary and divider intersections can then be used as further controls for documenting and mapping the scene.

Another method is a spiral pattern, either starting from the victim and working outwards or starting from the perimeter and searching in towards the victim. You may be able to limit your search to the width of a road or trail plus a little extra width to allow for the distance that something could be thrown from a vehicle. Combinations of search patterns can be used depending upon the scene and weather conditions.

Don't forget to look up and around you. In cases of suspected poaching, check the area for permanent tree stands or evidence that a portable tree stand had been attached to a tree. Check tree branches for evidence of bullet strikes. This information will help establish the path of a bullet, and by backtracking, may lead you to the shooter's position. Once you find where the shooter was shooting from, you can search for spent cartridge cases, cigarette butts or other items of evidence. Keep in mind that the animal may have travelled some distance from where it was originally struck. Backtracking the blood trail may lead to the area where the animal was originally hit.

Re-search the area after the sun changes direction. Searches in the morning, noon and afternoon may highlight cartridge cases or other small items that were overlooked initially. The key to remember is that the search should be thorough. In many cases, a good search will produce more items than are needed to prosecute the case.

Only items relevant to the crime need to be collected, but it may be difficult to know what evidence items are important to the case while processing the scene. The scene investigators have to decide which items are relevant to the crime. For example, a beer bottle lying in the ditch with a label so weathered that it is faded and falling off is probably not relevant to a scene that is a few days old. Cartridge cases that are rusty or heavily tarnished are not likely to have been used in the past few days either. These items do not need to be collected. The scene processors should document any items that are not collected and the reason for not collecting them. This will refresh your memory if the case goes to court and the defence or prosecutor asks why you didn't pick up those items that are shown in the scene photographs. If there is any doubt about the relevance, collect it.

Evidence not collected is evidence lost. Once a scene has been searched, the evidence is collected and stored. At the end of the crime scene search, the scene is typically released by the officer in charge. In many jurisdictions, once a scene is released, evidence cannot be legally collected from it at a later time. Thus, if in doubt, collect suspected evidence prior to release of the scene. If later you determine that the item was not connected to the scene, it does not need to be sent to the crime laboratory for further analysis.

What Is Physical Evidence?

Physical evidence is anything large or small that ties the victim, suspect and crime scene together. Edmond Locard's Exchange Principle (see earlier) mentions that even minute particles of material will be exchanged between the suspect and the victim or suspect and crime scene. The material left behind may be DNA left on an object touched by a suspect, or from cells distributed by a sneeze or cough. The challenge in crime scene processing is identifying evidence, collecting and preserving it so that it is available for further examination in a laboratory.

Evidence includes the victim. The carcase is usually collected and submitted to a forensic laboratory veterinary pathologist to determine the cause of death (see Chapter 9).

The pathologist may collect bullets or shot pellets from the animal, which will serve as evidence. Other examinations may focus on evidence related to electrocution, poisoning or accidental death (some natural deaths may initially be treated as suspicious).

Once an evidence item is chosen to be collected, a separately numbered placard is placed next to the item (Figure 5.3). Evidence is then photographed and mapped prior to collection.

Photographing Evidence

Photography plays a role in all wildlife forensic work, but is of prime importance in the field. If possible, two cameras should be available, in case one fails. Spare batteries must be packed, making allowance for the effects of temperature on battery usage.

When in the field, cameras should not be exposed to extremes of temperature and other environmental effects. Cameras should be protected from dust and accidental damage or interference. Guarding against excessive moisture and, particularly, immersion in water is vital; seawater and alkaline (soda) water are especially harmful to photographic equipment. It is essential that every effort is taken to protect both exposed and unexposed photographic film (if used), disks and other essential kit. A padded, water- and dust-proof container offers protection for photographic equipment, provided that the container itself is looked after carefully. When in the tropics, shade should be used but an alternative is to wrap the containers in banana or taro leaves, which provide excellent insulation. In cold situations body-warmth and appropriate insulation will assist (Cooper, 2013).

Smith (2011a) described the use of a medium-cost camera and lenses to obtain high-quality close-up images (in this case of invertebrates), which could prove useful if photographic evidence is to be collected in areas where it would be inappropriate or risky to use expensive or bulky equipment.

The purpose of photographing a crime scene and the associated evidence is to produce documentation for future reassessment, to provide transparency to the scene investigation, and to preserve images that may be used in court to explain the investigators' hypotheses and reconstruction of events. If there is a limit on time or number of investigators at a crime scene (due to safety fears, adverse weather, etc.) and full documentation is not possible, then photos are the next best type of documentation. Images are also useful to an investigator or an expert as an aide-memoire to help him/her recall the scene, the animal or other material or objects.

It is important to note that, with limited exceptions (e.g. latent print photographs), the photographs taken at a scene are generally not considered evidence in themselves. Rather, they are documentation of the observations of the investigators. Therefore, these photographs, when accepted in court, are typically accompanied by the sworn testimony of the investigator that these images are an accurate reflection of what was observed or discovered at the scene of the crime.

An investigator may be fortunate to have the services of a professional forensic photographer, but in many circumstances involving wildlife forensics the investigator may have to make his or her own digital image record. Several useful texts have been published on the subject of forensic photography, such as Redsicker (2000); Robinson (2010) and Weiss (2009). Other publications that might be considered are Moorehead (2011) and Duncan (2010). Additional relevant sources are mentioned in the text elsewhere. The techniques discussed here can produce high-quality photographs of evidence such as bloodstains, gunshots or tracks. However, practical advice for those who may have to work with limited resources is provided by Munro (1998).

If a specialist forensic photographer is not available, the following recommendations for crime scene photography are based on the assumption that the reader is a competent photographer and aware of photographic nomenclature and photographic techniques (e.g. depth of field, ISO, automatic versus manual settings, etc.). Additionally, this section will presume that the photographer is using a digital camera and that the photographs will be printed through some batch process or the photos will be processed using computer software (e.g. Adobe Photoshop, GIMP, etc.). Technological advances have made crime scene photography much simpler than it was in past decades. Although the equipment, style and personal photographic preferences vary between individuals, there are some basic recommendations that are rooted in good practice, legal (court) concerns and experience.

Photographic Considerations before Arriving at the Scene

It is recommended that

1. The investigator is familiar with the operation of the digital camera, beyond the 'automatic' mode.
2. The camera has a flash system that can be activated or deactivated on demand.

3. The date and time on the camera are accurate. This allows the investigator to rely on associated metadata.
4. The CSI kit contains extra recording media (e.g. compact flash, etc.).
5. The CSI kit contains extra batteries (all charged and tested).
6. The CSI kit contains an assortment of rulers of different lengths and different colours including 18% grey ruler (sometimes called neutral grey) for post-processing colour correction.
7. The CSI kit contains flashlights (torches), location placards, evidence flags, to assist the photographic process, etc.
8. Provision should be made for extremes of temperature or other difficulties, such as exposure to difficult terrain or water.

Photography of the Scene and the Evidence Items

Photographs are probably the easiest way to collect and archive crime scene documentation. A careful, methodical approach is essential, resisting the natural impulse to rush to the victim or object.

1. Before photographing a crime scene, it is good practice to write on a notepad the date, time, name of the officer, relevant weather observations (e.g. overcast, 28°C, with wind from the north, etc.), and then take a photograph of this plaque. The plaque records the logistical and ambient conditions at the scene.
2. The next series of photographs should document the scene as first perceived (and without any evidence placards, flags or rulers). These are overall panoramic photographs that give the general overview of the scene. In cases of very large scenes, the investigator can take sequential photographs.
3. At this point, the photographer should pause and wait for the lead investigator to process the scene as described in crime scene processing section. This process results in the crime scene being littered with evidence flags and location placards, each numbered in ascending order (Figure 5.3). These are the locations where evidence may be present. The photography, documentation, collection and preservation will help the investigator with leads, and with reconstructing the suspected crime.
4. The next series of photographs documents the actual evidence items as indicated by the placards (Figure 5.4). Each placard will indicate an evidence item, and the photographer must take three photographs of each. The first photograph shows the evidence item next to the numbered placard. The second photograph is a close-up of the evidence item with just a portion of the placard, and the last photograph is a close-up of the evidence item (in which just the item and the adjacent ruler are visible).

    The technical conditions required to photograph a cartridge case are very different from those required to photograph a carcase, and the investigator is expected to be proficient enough in the operation of the camera so that each photo is of sufficient quality for future needs. Examples of photographs that require a high level of expertise include
    a. Latent prints
    b. Tyre tracks

The Wildlife Crime Scene

Figure 5.4 Example of an evidence photograph. Note the inclusion of an evidence marker and a ruler for scale.

  c. Evidence that is partially obscured by shadows, while the remaining portion is in bright sunlight
  d. Photography in dark conditions or at night
  e. Animal tracks and shoe impressions left in soil, snow, etc. (see Chapter 14)
5. In order to be thorough, the last image of the crime scene should be followed by a photograph of the placard (including time of departure) used to initiate the photo shoot.

A whole range of photographs should be taken, even if some appear to be of little or no relevance at the time. An apparently minor lesion or a dead animal, for example, may prove later to be of importance (see Chapter 9). An image of it may prove critical. This can be particularly relevant in such instances as when a second necropsy is performed. It may be difficult to distinguish wounds (e.g. skin incisions) that were present when the animal was first presented from those inflicted during the first *post-mortem* examination.

If more than one person is involved in the photography, record who took each picture. This is an important precaution because the photographer may need to be identified during court proceedings. As a general rule, avoid asking other people to take photographs at the crime scene because they may be asked to testify.

Photographic Considerations after the Crime Scene

The investigator is now in possession of a series of digital photographs that are essential documentation of the crime scene. These photographic files must be transferred from the camera and archived. A prudent procedure for archiving digital photographs is to copy the entire content of the compact flash card (or SD) to a CD (or DVD) and close the CD so the files cannot be over-written. An easy method to determine if the photographic files are transferred successfully is to view the newly created CD with the thumbnail viewer. Corrupted files will not display an image. It is important that the investigator does not delete any photographs from the crime scene series because the numerical continuity of the digital files is more important than concerns over the quality of the photos.

By archiving the native files, the original images will be preserved intact. Any subsequent processing (e.g. contrast adjustment, grey-scale correction, clarifying a shadow area, etc.) should be carried out on copies or clones of the original native files. All such adjustments should be recorded.

These recommendations are a counsel of perfection and, particularly in the case of wildlife crime investigation, it may not always be possible to meet such standards. For example, field conditions are hazardous, good-quality photographic equipment and skilled photographers are not necessarily available. In addition, in many jurisdictions, crime scene photography is not a task carried out by the first responder. Further, in countries with limited resources, it is not unusual for a single investigator to be responsible for all aspects of wildlife crime scene preservation and documentation (see Chapter 7). This may also be the situation at the *post-mortem* examination of an animal or in the laboratory examination of plants or other objects, in which case the best photographic record of the evidence should be made as is possible in the circumstances.

Each country and legal system has its own rules of evidence. Digital evidence is widely used by courts, but it should be noted that some jurisdictions may not accept such images. If a witness is giving evidence in an unfamiliar country, he or she should ascertain whether the current rules of court procedure allow digital photographic material.

The standards for the use of digital images by the British police forces were laid down in ACPO (2007), and Cohen and MacLennan-Brown (2007) and IVS (2009) reviewed the then current advice from the UK Home Office Scientific Development Branch and the U.S. Scientific Working Group Imaging Technology on digital imaging.

Documenting and Mapping the Crime Scene

Note-taking, record-keeping and crime scene documentation are thoroughly discussed in the forensic literature (Cooper and Cooper, 2007; Hart and Budgen, 2008; Swobodzinski, 2004; Wertheim, 1992). Rather than repeat these reviews, here we focus on the importance of crime scene mapping. There are two reasons to create a map of the crime scene: (1) to accurately document the precise location where each evidence item was collected and the relationship of these to one other, and (2) to illustrate the crime scene and the important features relevant to the investigators' hypothesis. Producing a map will also prevent data loss in instances of camera or camera card failures.

Accurate Documentation

The widespread use of geographic positioning systems (GPS) has unfortunately led many crime scene investigators to neglect mapping the crime scene. This is a serious omission: such maps are essential for a thorough investigation. Although there are very sophisticated

The Wildlife Crime Scene 141

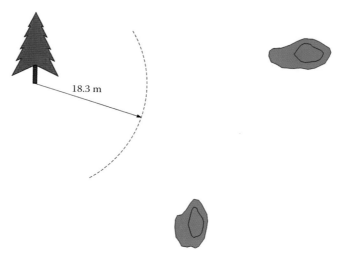

Figure 5.5 Example of an inaccurate map using single-point measurement.

instruments that create digital vector maps of every object encountered in a crime scene, the cost of these systems can exceed normal operating budgets. While GPS devices are excellent for providing navigation assistance, their use at a crime scene is limited to providing a general location where a crime was committed, and not the exact place were a carcase or firearm was found. Their accuracy is less than can be achieved through detailed mapping. In order to be precise and accurate in documenting and recording evidence location, a map must be created that uses triangulation.

A single measurement from a fixed point, such as the tree in Figure 5.5, can be precise; but if the investigator needs to return to the scene and find the exact spot of the evidence item, then the measurement could represent any location within the arc illustrated in Figure 5.5. This is an example of a precise measurement, but inaccurate location.

Two measurements from two distinct fixed points, such as the tree and rock in Figure 5.6, can also be precise; but if the investigator needs to return to the scene and find

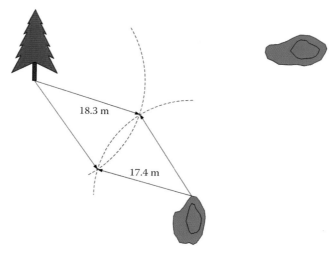

Figure 5.6 Example of an inaccurate map using dual-point measurements.

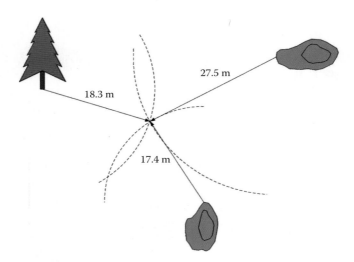

Figure 5.7 Example of an accurate map using triangulation measurements.

Table 5.1 Example of a Data Table Used for Documenting and Triangulating Evidence Locations

Evidence Item	Fixed Point 1	Fixed Point 2	Fixed Point 3
1 (gun)	9.4 ft	24.6	29.5
2 (carcase)	18.3	17.4	27.5
3 (cartridge case)	1.7	27.2	34.2

the exact spot of the evidence item, then the measurements provide two locations where the arcs intersect as shown in Figure 5.6. This is another example of precise measurements, but ambiguous location of the evidence item.

Three measurements from three distinct fixed points, such as the tree and two rocks in Figure 5.7, are very precise and accurate; if the investigator needs to return to the scene and find the exact spot of the evidence item, there is only one location that would match the measurements as shown in Figure 5.7. Where all three arcs intersect is the exact location where the evidence was collected.

It is clear from these illustrations that proper documentation and mapping of a crime scene requires measurements for each item from three fixed points. In a notebook sketch, this looks something like the data in Table 5.1. The fixed points (permanent markers) used for triangulation should also be documented in the notes.

Illustrating the Crime Scene

Large crime scenes, such as many encountered by wildlife crime investigators, are complex and views are often obscured by trees, shrubs, rocks, etc. Therefore, the most efficient way of illustrating a crime scene is to create a sketch map or diagram. This map illustrates where each evidence item was located, shows the fixed points used for precise measurements and displays the complexity of the scene. Photographs are unable to capture the overhead

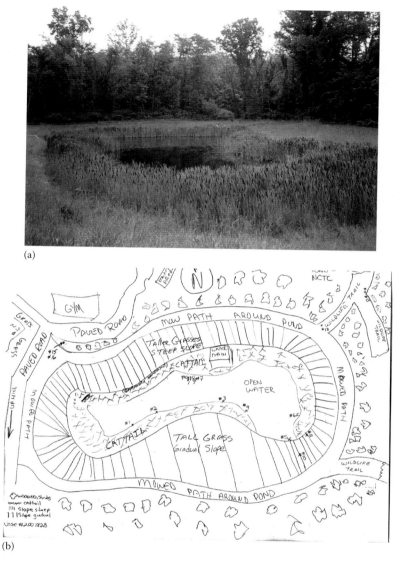

Figure 5.8 (a) Photograph of crime scene lacking detail. (b) Sketch map of the same crime scene illustrated in Figure 5.8a. Maps show details not evident in photographs.

spatial perspective that can be documented in a map of large crime scene. Figure 5.8a and b are both from the same crime scene. A casual look at both demonstrates that the photograph (Figure 5.8a) does not convey the detail offered by the map (Figure 5.8b).

Collecting and Preserving Evidence

Once evidence has been found, there are two parallel tasks that guide the investigator at the scene: (1) the accurate collection and (2) the accurate preservation of the evidence (Gorea, 2005). The following example illustrates the different criteria for each. On a foggy morning, a brass cartridge case is found at a crime scene. Pursuant to his investigation, the officer would like to obtain latent prints from the fired cartridge. His first concern is to *collect* the cartridge case while protecting it from extraneous latent prints and avoiding smudging any latent prints that may be present. Then the officer must *preserve* the evidence by removing any moisture encasing the cartridge that could oxidise breach mark striae (spiral lines that reflect the interior of the barrel of the gun from which it was fired). It is obvious from this example that a breach in protocol of either the *collection* or the *preservation* would compromise the value of the evidence and may make the hypothesis untestable.

Detailed discussion of how to deal with particular evidence types is beyond the scope of this chapter, but can be found in numerous reference works (Cooper and Cooper, 2007; Federal Bureau of Investigation, 2008; Fisher, 2003; Gunn, 2009; Lee and Ladd, 2001; Waggoner, 2007; Walker and Adrian, 2003). Here, we provide a concise table that lists the preferred collection and preservation methods of common evidence types seen in wildlife crime scenes (Table 5.2). All evidence should be accurately documented and sealed (Figure 5.9) at the crime scene. Some evidence may need to be repackaged or resealed once it has been transported to the office or laboratory.

Note that collection of some of these types of evidence (e.g. latent prints, poisons, and firearms) requires specialised training to avoid compromising the evidence and to protect personnel. Basic supplies that should be included in a wildlife crime scene investigation kit are listed in Appendix A. This list was prepared on the assumption that the evidence will be submitted to a laboratory for analysis, rather than analysed in the field (e.g. in the case

Table 5.2 Preferred Collection and Preservation Methods for Common Evidence Types Found at Wildlife Crime Scenes

Evidence Type	Preferred Collection	Preferred Preservation
Carcass	Collect in plastic bag.	Freeze
Tissue samples	Cut about ½ in. (1 cm) cube and place in plastic bag.	Freeze
Blood spot – dried	Collect sample and place in paper envelope.	Freeze
Blood spot – wet	Collect blood with a clean, Q-tip.	Dry and freeze
Latent prints on bottles, eggshells, etc.	Collect without touching and immobilise.	Store dry
Latent prints on documents, tape, etc.	Collect without touching and immobilise.	Store dry
Suspected poison	Collect less than 1 mL in plastic tubes.	Freeze
Firearm	Allow a firearms officer to collect and make safe. This is typically done by removing the magazine and clearing the chamber, taking care to preserve any latent prints.	Store dry
Spent cartridge	Place in paper envelope.	Store dry
Spent cartridge with latent prints	Collect with the aid of a pencil and immobilise.	Store dry

Figure 5.9 Example of properly sealed evidence showing the initials of the person who sealed the evidence, along with the date on which the sealing took place.

of a field necropsy). Other wildlife crime scene supply lists were published by Cooper and Cooper (2007) and Cooper et al. (2009) (see also Appendix B of this book).

Initiating a Chain of Custody

A chain of custody (see Chapter 14) is a written record of everyone who has had the evidence in their possession. This extends from the time the evidence is picked-up at the crime scene or the scene of a search warrant, until the evidence item is released by the court or returned to the submitting agency for permanent storage or disposal.

Every time the evidence is transferred, the transaction is recorded with the date, time and name (typically signature or initials) of the persons transferring and receiving the evidence item. This ensures that the evidence item is under the control of a specified individual and in a secured location at all times to prevent unauthorised alteration, tampering or loss.

Some commercial evidence bags are pre-marked to record the chain of custody. Alternatively, a tag with the chain of custody record can be attached to each piece of evidence. Once the evidence has been transferred to a property room or laboratory, a master chain of custody is often computerised to keep track of all of the evidence items in the same case.

A typical chain of custody starts with the person who picks up the evidence at the scene. If the evidence is transferred to a person responsible for storing the evidence (often called the 'evidence custodian'), that transaction is recorded. The item is then placed in a locked evidence storage area. If the evidence needs to be sent to a crime laboratory for analysis, it is then checked-out of the storage room and transferred to the crime laboratory by the evidence custodian where it is signed-for and placed in the laboratory's secure storage room. When a forensic scientist is ready to analyse the item, he or she checks the evidence out of the storage room. When the examination is finished, the item is transferred again to the evidence storage room. The item is returned to the submitting agency where it is again placed in a secure evidence storage room until it is needed for court. If the case goes to trial, relevant evidence items are signed-out of the agency evidence storage area and transferred to the custody of the court. After the trial, the judge may order the release of some or all of the evidence to the defendant – or the evidence may be transferred back to the original agency for storage, use as laboratory standards or disposal. Each transfer of evidence requires a record of the date, time, persons transferring the evidence (typically signature or initials) and the storage location for each item.

Summary and Conclusions

The basic steps involved in wildlife crime scene investigation are the same as those that apply to cases in which the victim is a human (see Box 5.2). These procedures include (1) securing and protecting the scene, (2) searching for and locating evidence, (3) photographing evidence, (4) documenting and mapping the scene, (5) collecting and preserving evidence, (6) initiating a chain of custody. The crime scene investigator follows this process in an attempt to link a victim with a suspect and a crime.

This chapter addresses each of these steps from the perspective of the first responders to the crime scene, regardless of whether or not those responders are law enforcement officers,

---

BOX 5.2   BASIC STEPS OF CRIME SCENE INVESTIGATION

- Obtain search warrant or relevant permission to access scene (if required)
- Arrive with CSI kit
- Designate team leader
- Assign team tasks
- Take photo of placard, noting date, time and location (before entering scene)
- Evaluate for safety (required throughout process)
- Define and secure crime scene boundary (may change throughout investigation)
- Re-evaluate supply needs
- Conduct initial walk-through (flag items of potential interest)
- Identify evidence using systematic searches (mark location with numbered placards)
- Photograph evidence
- Map and document evidence (use triangulation)
- Collect, seal and preserve evidence
- Initiate a chain of custody
- Photograph crime scene when exiting
- Take final photo of placard, noting time of exit
- Develop crime scene hypothesis (reconstruct events)

wildlife forensic scientists, veterinarians, biologists or other individuals called to a crime scene as part of a legal investigation. The ultimate goal of the crime scene investigation is to determine whether or not a crime has been committed – and if so, to identify suspects and to preserve forensic evidence for presentation in a court of law. This may seem a daunting task given the variety of scenarios and animal species that might be encountered at wildlife crime scenes. While the logistics of such investigations may vary greatly in diverse environmental and urban settings, the basic steps of locating, documenting and preserving evidence remain the same.

Acknowledgements

We thank John and Margaret Cooper for inviting us to write this chapter. Editorial improvements were provided by Pepper W. Trail, two anonymous reviewers and the editors. Ken Goddard provided Figures 5.5, 5.6 and 5.7. The opinions, findings and conclusions in this article are those of the authors and do not necessarily represent the views of the US Fish & Wildlife Service.

Appendix: A Suggested List of Basic Crime Scene Supplies

Supplies will vary depending upon the crime scene. For additional wildlife crime scene supply lists, see Appendix B, Cooper and Cooper (2007) and Cooper et al. (2009).

Camera (with data cards, extra batteries, lenses, flash, etc.)
Chalk
Compass/GPS
Duffle bag/backpack (rucksack) (container to hold supplies)
Envelopes (legal size to coin)
Evidence boundary tape
Evidence flags and or evidence markers (tents)
Evidence tape and packing sealing tape
First aid kit
Flashlight (torch) and batteries
Forensic scales, 'L'-shape
Forms (evidence, photo, chain of custody)
Gauze pads (sterile)
Graph paper
Index cards (4 × 5)
Large L-shape plastic rulers for tyre track or footwear photos
Magnetic powder – black, white, applicator, lift tape (for latent prints)
Magnetic powder applicator (for latent prints)
Magnifier
Mask – from dust to full respirator, depending on need
Multi-tool/knife
Notepad(s) of different sizes
Paper bags, lunch bags/grocery size
Paper clips

Pencils
Pens
Permanent markers, large/small
Plastic bags, various sizes
Plastic garbage (rubbish) bags
Plastic sheet (large)
Razor blades
Reference material (identification guides, maps, etc.)
Rope – 100 ft (30 m)/nylon cord
Rubber (elastic) bands
Rubber gloves
Safety glasses
Sealable plastic containers/tubes
Swabs (Q-tips) (sterile)
Tape-measure
Tweezers (forceps)
Water/food, energy bars, etc.
Wooden rulers
Zip-ties

Forensic Entomology

KATE M. BARNES

Contents

Introduction	149
Calculating Minimum *Post-Mortem* Interval	152
By Age of Blowfly Larvae	152
Insect Succession	153
Seasonality	154
Cases of Neglect and Abuse	155
Entomotoxicology	155
Wound Sites	156
Trace Evidence	156
Collection and Packaging of Entomological Samples	156
Collection from Live Animals	157
Collection from Dead Animals	158
Collection from the Surrounding Area	158
Collection Techniques	158
Packaging Entomological Material	160
Conclusion and Summary	160

Introduction

Forensic entomology is an area of forensic science that uses the insects or invertebrates found at a crime scene to provide information about the crime committed. Identification and analysis of insect specimens collected from a human corpse or an animal carcase can allow the estimation of the minimum *post-mortem* interval (PMI) (see Chapter 9) as well as supply information on toxicological aspects of the body, wound-site location, interval of abuse or neglect or identification of the victim, and act as geographical trace evidence.

Insects have been used in crime investigations for several centuries in China (Gennard, 2007; Greenberg and Kunich, 2002; McKnight, 1981). However, forensic entomology did not reach Europe until the early 1800s, with insect evidence being presented in court for the first time in Britain in 1935. From these roots, forensic entomology has grown into a vibrant and fast-paced field of research, with insect evidence currently being used and accepted in courts in most parts of the world.

Insect evidence is traditionally associated with crimes against humans but can be equally applicable in cases concerning animal abuse, neglect, poaching and illegal killings (Anderson, 1999; Anderson and Huitson, 2004). Most research is conducted using domestic pig (*Sus scrofa*) carcases on account of the ethical restrictions associated with utilising human remains (Table 6.1). See also 'Advances in scavenging research' section

Table 6.1 Summary of the Types of Animal Carcases Used to Investigate the Temporal Pattern of Insects during the Decomposition Sequence (For Scientific Names, see Appendix H)

Authors	Location	Animal Used	Year
Davies	England	Mouse/Sheep	1999
Smith and Wall	England	Mouse/Quail	1997
Isiche et al.	England	Mouse	1992
Dekeirsschieter et al.	Belgium	Pig	2011
Bourel et al.	France	Rabbit	1999
Leclercq	France	Wild boar	1996
Leclercq and Verstraeten	France	Llama	1992
Anton et al.	Germany	Pig	2011
Reibe and Madea	Germany	Pig	2010
Kentner and Streit	Germany	Rat	1990
Matuszewski et al.	Poland	Pig	2008, 2010b
Nabaglo	Poland	Bank vole	1973
Arnaldos et al.	Spain	Chicken	2004
Garcia-Rojo	Spain	Pig	2004
Castro et al.	Portugal	Pig	2011
Grassberger and Frank	Austria	Pig	2004
Kocarek	Czech Republic	Rat	2003
Özdemir and Sert	Turkey	Pig	2009
Abd El-bar and Sawaby	Egypt	Rabbit	2011
Kelly et al.	South Africa	Pig	2009
Braack	South Africa	Impala	1981
Bharti and Singh	Punjab, India	Rabbit	2003
Shi et al.	China	Rabbit	2008
Wang et al.	South China	Pig	2008
Chin et al.	Malaysia	Pig	2008
Chin et al.	Malaysia	Pig	2007
Vitta et al.	North Thailand	Pig	2007
Voss et al.	Western Australia	Pig	2011
Voss et al.	Western Australia	Guinea pig	2009
Bornemissza	Western Australia	Guinea pig	1957
Lang et al.	Tasmania, Australia	Possum	2006
Archer and Elgar	Victoria, Australia	Pig	2003
Eberhardt and Elliot	New Zealand	Pig	2008
Valdes-Perezgasga et al.	Mexico	Pig	2010
Jirón and Cartin	Costa Rica	Dog	1981
Barrios and Wolff	Colombia	Pig	2011
Segura et al.	Colombia	Pig	2009
Martínez et al.	Colombia	Pig	2007
Perez et al.	Colombia	Pig	2005
Wolff et al.	Colombia	Pig	2001

Table 6.1 (continued) Summary of the Types of Animal Carcases Used to Investigate the Temporal Pattern of Insects during the Decomposition Sequence (For Scientific Names, see Appendix H)

Authors	Location	Animal Used	Year
Gomes et al.	Southeastern Brazil	Pig	2009
Carvalho and Linhares	Southeastern Brazil	Pig	2001
Horenstein et al.	Argentina	Pig	2010
Horenstein et al.	Argentina	Pig	2007
Centeno et al.	Argentina	Pig	2002
Davis and Goff	Hawaii	Pig	2000
Richards and Goff	Hawaii	Pig	1997
Early and Goff	Hawaii	Pig	1986
Watson and Carlton	Louisiana, USA	Bear/deer/alligator/pig	2005
Tabor et al.	Virginia, USA	Pig	2005
De Jong and Chadwick	Colorado, USA	Rabbit	1999
Payne	South Carolina, USA	Pig	1965
Reed	Tennessee, USA	Dog	1958
Michaud and Moreau	New Brunswick, Canada	Pig	2009
Sharanowski et al.	Saskatchewan, Canada	Pig	2008
VanLaerhoven and Anderson	British Columbia, Canada	Pig	1999
Anderson and VanLaerhoven	British Columbia, Canada	Pig	1996

by Young in Chapter 13. The adult pig is a good model for human decomposition as it has similar hair coverage, the size and layout of organs are similar, and the intestinal bacteria are similar due to the animal's omnivorous diet. The pig also releases similar volatile organic compounds during decomposition and therefore attracts the same insects as does a decomposing human corpse (Dekeirsschieter et al., 2009). The use of different animal carcases in forensic research has provided a large amount of information on how animals decompose and demonstrated that insects are attracted to a dead animal in the same way as they are attracted to a dead human (Rodriguez and Bass, 1983). This information is useful to veterinarians dealing with both wildlife crimes and cases of abuse and neglect in domestic animals (see Chapters 9 and 10).

This chapter provides an overview of how insect evidence can supply information in crime investigations. First, the three different methods for calculating the minimum PMI are outlined. Cases of how insects can be used as indicators of intervals of abuse and neglect are then examined, followed by an outline of the field of entomotoxicology (using insects to provide toxicological data from a dead body). In addition, descriptions are given of how insects can be utilised to indicate the presence of *pre-mortem* wound sites and provide geographical trace evidence, such as whether a body was moved after death. The section ends with detailed instructions on how to collect and preserve insect evidence and how to find a forensic entomologist should one be required.

Calculating Minimum *Post-Mortem* Interval

Determining the time of death is important in focussing the occurrence of a crime into a specific time frame. A pathologist can give a fairly reliable indication of PMI – the length of time that has passed between death and discovery of a body – for the first 72 h following death (see Chapter 9). However, beyond this point, changes in the corpse, such as rigor mortis, algor mortis and livor mortis (see Chapter 9) have completed their cycles and so cease to provide any relevant information on PMI (Dix and Graham, 2000). From this time onwards, insect evidence is most useful and can provide estimates from days to weeks and even months after death has occurred.

Insects do not indicate the actual time of death but the minimum time since death by determining when the carcases were first colonised by insects. There are three main methods that a forensic entomologist can use to determine minimum PMI from insect evidence: by ageing blowfly larvae from the corpse, through insect succession or by seasonality of insect activity. The method used will depend on the type of insect evidence available which will in turn be determined by the period of insect colonisation. An overview of each method is given in the following text.

By Age of Blowfly Larvae

Blowflies (Diptera: Calliphoridae) are the initial colonisers of a dead body (Greenberg, 1991; Rodriguez and Bass, 1983) and are attracted to human or animal remains very rapidly after death. Therefore, these insects are the most reliable indicators of minimum PMI.

Adult blowflies detect the first odours of decomposition using sensilla located on the third segment of the antenna (Erzinçlioğlu, 1996). Once at the corpse, blowfly activity is determined by size, smell, condition and position of the carcase. Female flies deposit eggs in the dark, moist areas of the dead body, such as the natural orifices (eyes, nose, mouth, anus and genital regions), the underside of the body and any wound sites (Figure 6.1).

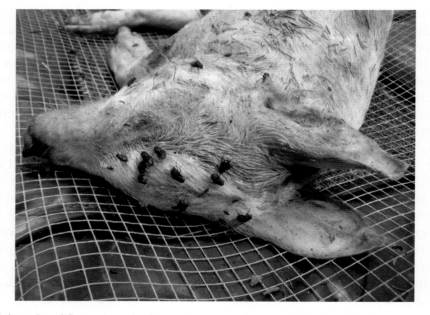

Figure 6.1 Rafts of fly eggs are laid on a pig carcase very rapidly after death.

Forensic Entomology

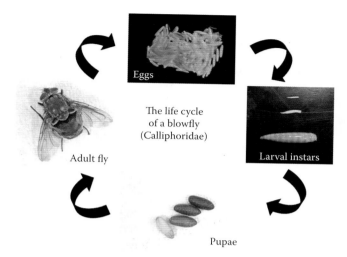

Figure 6.2 Life cycle of the blowfly – egg stage, three larval instars, puparial stage and adult fly.

The rate at which eggs hatch into larvae is dependent on weather conditions. Larvae pass through three growth stages (instars): first, second and third, moulting between each to enable growth. During the second and third instar, they form larval masses and feed voraciously on the soft tissues of the body. Larvae then enter a prepupal period in which feeding ceases (Figure 6.2). During this post-feeding stage, the larvae of most blowfly species disperse from the corpse in search of a dry place in which to pupate. Once there, the larval cuticle contracts, hardens and darkens to form a puparium inside which the larval organs break down during metamorphosis via a pupa into an adult.

Blowflies are ectothermic/poikilothermic (completely reliant on external sources of heat) and develop through their life cycle at a predictable rate that is dependent on the ambient environmental temperature (Anderson, 2000; Byrd and Allen, 2001; Donovan et al., 2006; Grassberger and Reiter, 2001, 2002; Richards et al., 2009). Therefore, the minimum PMI can be estimated by determining the time since oviposition (egg-laying). This can be calculated by obtaining local meteorological data, together with the identification of the species and stage of development. Minimum time since death may range from hours to weeks and can be given in hours or days depending on the availability of meteorological data for the time period in question.

Insect Succession

A dead body typically undergoes five main stages of decomposition: fresh, bloated, active decay, advanced decay, and skeletonisation (Bornemissza, 1957). Different species of insects are attracted to a corpse at disparate stages of decomposition, responding to the different odours released as decay progresses. In general, fly species dominate the initial stages of decomposition, whereas Coleoptera (beetles) dominate the later stages (Mégnin, 1894).

Although insect species colonising the body vary according to season, geographical region, microclimate and environmental factors such as temperature, humidity, time of day and rainfall, the faunal succession of insect families follows a distinct pattern (Archer and

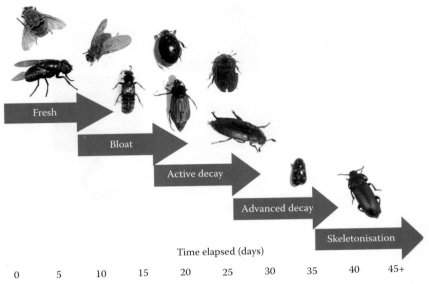

Figure 6.3 General outline of the successional colonisation of a pig carcase by different insect families through the decomposition sequence: Calliphoridae, Muscidae, Staphylinidae, Histeridae, Silphidae, Dermestidae, Nitidulidae and Tenebrionidae.

Elgar, 2003; Arnaldos et al., 2004; Campobasso et al., 2001; Greenberg, 1990; Kocarek, 2002; Walker, 1957). This predictable pattern of insect succession can be used to link the insects found on a body to a specific stage of decay, thus giving an indication of minimum PMI (Figure 6.3).

Determining minimum PMI by insect succession is not as accurate as ageing blowfly larvae found on a body. However, it can be useful in determining the time frame of less recent crimes when blowfly larvae are no longer present. It is best used in tropical regions with relatively constant climates, such as Hawaii (Goff, 2000), but can be useful in temperate regions when no other evidence is available. Seasonality of temperate climates has a profound effect on succession; so, for this method to be successful, all insects colonising a body at the time of discovery must be collected and sent to a forensic entomologist who has knowledge of the local area and habitat. However, seasonality itself can be a useful tool (see the following text).

Seasonality

Forensic entomology is usually of most value during the first few weeks after death, depending on the season. However, it may be possible to determine the season of death, based on the species of insect remains (e.g. empty puparial cases) present several months or years later (Nuorteva, 1987; Staerkeby, 2001; Teskey and Turnbull, 1979).

Insect activity and development is dependent on environmental conditions and therefore varies with season (Archer and Elgar, 2003; Arnaldos et al., 2004; Carvalho and Linhares, 2001; Centeno et al., 2002; Voss et al., 2009). During periods of harsh weather conditions, such as cold winters and very hot summers, most insects go through a period of suspended development termed 'diapause'. Knowledge of different insect species' seasonal activity patterns means that crimes can be tied into specific times of the year, thus indicating when a death occurred. In animal cases, where there is a legal hunting season (see Chapter 2), knowing the minimum time since death could conclusively indicate whether an animal was killed legally in season or illegally out of season (Anderson, 1999).

Cases of Neglect and Abuse

Myiasis is the colonisation of a living vertebrate host by dipteran larvae which feed on the host's dead or living tissue, body fluids or ingested food (James, 1947). Adult flies are attracted to an animal on account of an injury or the presence of contaminating excretory material on the hair, fur or fleece and lay eggs which develop into larvae (Anderson and Huitson, 2004). As on a corpse, the larvae pass through three developmental stages (instars) whilst feeding on animal tissue, then finally cease feeding and drop off the animal to pupate.

Forensic entomology can be used to determine the interval of abuse or neglect in the same way that flies can be used to determine the minimum PMI by ageing of blowfly larvae (Benecke, 2004). That is, with knowledge of the body temperature, fly species and stage of insect development, the minimum length of time of neglect or abuse can be determined. However, this could be greatly shortened in living hosts due to a more rapid progression through the fly life cycle as a result of the host body temperature (Anderson and Huitson, 2004). In dead hosts, abuse and neglect can be proved to have occurred before death due to the stage of development of specific indicator species compared with the more recent stages of *post-mortem* colonisers present (Benecke and Lessig, 2001).

Myiasis often occurs not only in wild species but also in domesticated animals such as sheep, where it causes considerable economic damage each year (Wall et al., 1995). In pet animals, myiasis can occur when a wound is left untreated or when there is an accumulation of faeces or urine due to neglect, poor housing conditions or poor health. In a survey conducted by Anderson and Huitson (2004), dogs were the species most frequently presented with myiasis in British Columbia, Canada, followed by rabbits, cats and then horses. Most cases of myiasis appeared to be the result of an untreated wound or matted hair coupled with the presence of excrement, indicating that owner neglect (see Chapter 8) has a major role in the occurrence of myiasis.

Entomotoxicology

Necrophagous insects colonising a dead body ingest the decomposing tissues along with any toxins they may contain. Therefore, in cases where animal or human remains that no longer contain sufficient amounts of tissue or body fluids for testing are found, these insects can be used as an alternative source for toxicological analyses. In fact, some studies have shown that insect analyses are more successful in providing a result than are highly decomposed body tissues (Kintz et al., 1990; Nolte et al., 1992). Toxicological analyses have successfully been performed on a wide array of insect material from blowfly larvae and puparial cases to beetles and beetle frass (Beyer et al., 1980; Bourel et al., 2001; Gosselin et al., 2011a; Miller et al., 1994; Nolte et al., 1992). However, to date, the amount of drug found in the insect is only indicative of its presence or absence and cannot be reliably correlated with the amount ingested prior to death (Gosselin et al., 2011b). In addition, the presence of toxins in a dead body can disrupt the temporal pattern of the insect life cycle; therefore, any known toxins should be declared to the forensic entomologist who is analysing the material.

Wound Sites

Blowflies generally lay their eggs in the natural orifices of the body. These areas are dark and moist (see earlier text) and provide an ideal environment for egg incubation and for first instar larvae that hatch from the eggs. Blowflies will also lay eggs in recent wounds, and therefore masses of larvae in areas other than the natural orifices may indicate a *pre-mortem* wound site.

Trace Evidence

Insect succession on a dead body is strongly influenced by habitat type (Hwang and Turner, 2005; Walker, 1957). In some cases, distinct insect species are only found in very localised environments and can therefore be used to link bodies or suspects to specific habitats, thus providing geographical trace evidence. To date, specific marker species have been used to show that a corpse was moved after death (Benecke, 1998) and arthropod bites have provided a link between suspects and a crime scene (Prichard et al., 1986; Webb et al., 1983).

Collection and Packaging of Entomological Samples

As with any forensic evidence, entomology is only of value if the evidence is collected correctly and reaches the forensic entomologist in a usable state (see Chapter 12). Both live and dead insect specimens and their remains (e.g. empty puparial cases) should be collected because all document insect presence and activity and even dead insects can provide useful information to the investigation (Staerkeby, 2001).

There are two areas of a crime scene that should be examined for insect evidence: the body and the surrounding area. The carcase should be searched in an orderly sequence so that all insect evidence is collected. The head area should be examined before moving on to the underside of the animal and the ground below and then finally into the surrounding area (at least 2 m away from the body). This technique should ensure that a representative number and species of entomological samples are collected from the crime scene. Samples should be stored in vials labelled with an item code, crime scene details, collector's name, time and date of collection and location from which the sample was collected (head, wound, soil, etc.). Samples collected from different areas of the body should be kept in separate vials. In forensic cases, as with all evidence, packaging should be sealed and signed to show that it has not been tampered with (see Chapter 14).

The equipment needed to collect insect specimens is widely available (see also Appendix B) and should be packed in a holdall ready to take to a crime scene when needed (Figure 6.4). The kit should include forceps, a variety of different-sized collecting containers with lids, flask of boiled water for killing larvae, food source for feeding larvae (such as porcine liver), a killing jar (containing ethyl acetate), paper towelling, labels, small paint

Figure 6.4 A typical entomological kit containing items needed to collect insect evidence from a living or dead animal.

brushes (for collecting eggs or first instar larvae), small spoons (for collecting larvae), a hand-held insect net, thermometer, small trowel, small sieve and a preservative solution (commonly 70%–95% ethanol).

Collection protocols are similar for living and dead animals, whereas collection techniques differ according to the species and developmental stage of the insect.

Collection from Live Animals

Fly eggs and feeding larval stages may be found within a wound and should be carefully removed without damaging them. Small numbers of larvae can be removed by hand or flushed with water. Larvae should only be removed manually rather than by killing with a chemical treatment whilst still in the wound because the latter can cause anaphylaxis, shock and even death of an animal due to the larvae dying and releasing foreign proteins into the bloodstream (Anderson and Huitson, 2004).

Collection from Dead Animals

Eggs and fly larvae may be found in and around the natural orifices of a body and on the underside when the carcase is lifted. They may also be collected from folds of skin behind the ears or in the crevices of joints. The soil underneath the carcase should also be explored for any dispersing fly larvae, fly pupae or beetles – to a depth of 3–5 cm.

Collection from the Surrounding Area

At the end of the feeding stage, blowfly larvae move into a post-feeding stage where they cease feeding and, in most cases, leave the body to find a dark, dry place to pupate. If more advanced stages of the fly life cycle, such as post-feeding larvae, pupae or puparial cases are overlooked, then an underestimate of time since death will be made from the entomological evidence.

At outdoor crime scenes, post-feeding larvae may be found crawling away from the body or burrowing into nearby soil (commonly to a depth of 3–5 cm). Indoors, pupae can be located in many remote places including under skirting boards, furniture, floorboards, rugs and carpets. Adjacent rooms should be searched for insect evidence as larvae may have dispersed to some distance from the body. Outdoors, an area of at least 2 m from the body should be grid-searched for post-feeding and puparial stages (Amendt et al., 2007). Soil may have to be sieved over a tray or hand-searched in order to collect pupae. Leaf litter within the search area should also be investigated for crawling insects, such as beetles, which may prove to have forensic relevance.

In cases of abuse or neglect in living animals, post-feeding stages and pupae may be collected from areas which the animal has frequented, such as a basket, blanket, bedding or rug. However, in all cases where more advanced life stages of fly are found, they must be checked to see whether it is the same species of fly as that found feeding on the animal before any conclusions on time frames can be made.

The crime scene environment should also be recorded (see Chapter 5), including information about the condition and location of the body, any surrounding vegetation or furnishings and the current weather conditions including temperature (ambient, body and larval masses). Photographs of the crime scene, animal and/or wound site before and after larval removal are also essential and should be taken with a metric scale in the shot (see Chapter 5).

Collection Techniques

All life stages should be collected from the crime scene and each has a specific procedure for collection (Figure 6.5). Living samples should be split into two batches: half should be killed and preserved, so that an accurate record of each life stage at the time when it was collected from the crime scene is recorded; and half should be kept alive, so that they can be reared to adulthood for reliable species identification. Dead samples can be placed straight into 70%–95% ethanol. If possible, a subsample of each collection should be killed and preserved by placing in a freezer at −20° for at least 1 h in case molecular or toxicological analyses are required.

Insect eggs are small (~1–2 mm), pale and oval-shaped; fly eggs are laid in clumps and resemble grains of rice (Figure 6.2) whilst beetle eggs are laid singularly. All should be collected using a small moist paintbrush and transferred to damp tissue or paper towelling to prevent them from drying-out in transport. Containers should allow air diffusion but

Forensic Entomology

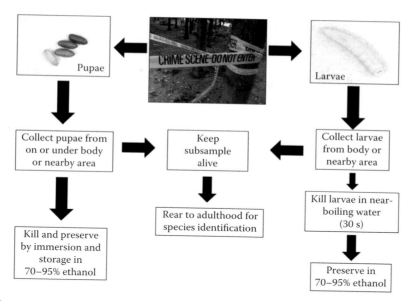

Figure 6.5 Basic outline of insect collection techniques from a crime scene.

prevent escape if the eggs hatch into larvae during transport. A subsample of specimens should also be killed and preserved by immersion in 70%–95% ethanol.

The feeding larval stages should be collected using forceps (or a spoon if there are large numbers!). Half the sample should be placed in very hot water for at least 30 s to destroy the internal enzymes, maintain larval length and improve preservation (Adams and Hall, 2003). They should then be poured through a small sieve and transferred to a vial with 70%–95% ethanol. The other half should be kept alive in an aerated plastic container, with food (such as porcine liver) and crumpled dry paper towelling over the top which soaks up excess fluids and allows air diffusion.

Pupae are small (~12–14 mm), oval-shaped and white to dark brown, depending on age (Figure 6.2). Pupae can be collected with forceps and then placed in a vial with crumpled paper towel to prevent damage. Half of the sample should be preserved by placing them directly in 70%–95% ethanol solution. Empty puparial cases look similar to the pupae but have the anterior end split or broken-off and should also be collected and retained as additional evidence.

Adult flies can be collected from the scene using a hand-held insect net. Ideally, they should be collected prior to the body search as they may disappear when the crime scene is disturbed. The net should be swept in a left-to-right motion to trap flies and then the end flicked-up and over the frame with a flick of the wrist to ensure the flies are secured inside. Flies can be placed in a killing jar (containing ethyl acetate) whilst still inside the net by placing the area in which they are trapped into the jar and securing the lid. After a few minutes, the ethyl acetate should have taken effect and the flies can be transferred to labelled vials with lids.

Adult beetles, beetle larvae and other crawling insects can be collected from the body and surrounding area by gloved hand and placed individually into labelled containers. This is to prevent cannibalism or predation that would probably occur should multiple specimens be placed in the same jar. Adult beetles can be killed by placing them in a freezer.

Puparial cases, insect skins or beetle frass should be stored in dry conditions or placed directly into vials containing 70%–95% ethanol.

Packaging Entomological Material

All the entomological evidence, together with details of the case, should be forwarded to a forensic entomologist for analysis. In Europe, the website of the European Association for Forensic Entomology (http://www.eafe.org/) provides contact details on forensic entomologists from various countries and the North American Forensic Entomology Association provides information for those living in North America (http://www.nafea.net/NAFEA/Home.html).

Living insect samples should, ideally, arrive within 24 h of collection so as to maximise their survival rate and minimise the chance of escape. As stressed earlier, packaging of living samples should allow air diffusion and contain a food source, if the animals are still feeding.

Containers and vials containing insects should be individually labelled and then packed in cardboard boxes, stabilised with foam if sending long distances or in a cool box if being taken a short distance back to the laboratory. Entomological evidence should ideally be transported at low temperatures (2°C–6°C) so as not to advance living stages (Amendt et al., 2007; Gennard, 2007). If possible, the temperature of the journey should be logged so that the complete thermal history of the specimens can be accounted for in entomological analyses.

Conclusion and Summary

Insects collected from a crime scene are a useful and reliable source for admissible information relating to a crime scene. To date, insect evidence has provided detailed information on the minimum *post-mortem* interval (PMI) (see Chapter 9) in both animal and human killings, determined the toxicological content of a corpse, provided trace evidence to link suspects to crime scenes and victims to death scenes, indicated the location of *pre-mortem* or *ante-mortem* wounds and positively determined cases where humans have been abused or neglected. However, this information is only possible through the correct collection of insect evidence and subsequent analyses by a professional entomologist. In the past century, the field of forensic entomology has advanced greatly in both its ability to provide novel evidence and also in its acceptance and use in courts in many parts of the world. Therefore, it can be used as both a stand-alone indicator of criminal activity and to strengthen and corroborate other evidence that is presented in court. As technology advances and research continues, the field of forensic entomology will become more robust and reliable and should be able to provide more precise and meaningful data to crime investigations around the world.

Field Techniques
At Home and Abroad

JOHN E. COOPER
MARGARET E. COOPER

7

Contents

Introduction	161
Practical Considerations	162
Legal Considerations	164
Cultural Considerations	165
Personnel	165
Performing Forensic Work in the Field	166
Planned Fieldwork	166
Opportunistic Forensic Work in the Field	167
Use of Field Equipment	168
Special Equipment	171
Examination of Live Animals in the Field	172
Examination of Dead Animals in the Field	172
Basic *Post-Mortem* Technique for Use in the Field	176
Laboratory Work in the Field	177
Sample-Taking in the Field	179
Recording and Collating Findings in the Field	179
Conclusions	180

'Indeed, a field biologist's greatest danger lies not in encounters with fierce creatures and treacherous terrain, but in being seduced by the comforts of civilization'.

George Schaller

Introduction

There are frequent occasions when wildlife forensic work has to be performed away from the investigator's home base and normal environment and, often, without access to optimal equipment and back-up facilities. This aspect of fieldwork means that such investigations may be carried out anywhere in the world, not just in isolated locations or in poorer, 'developing', countries. A wildlife crime scene in a moorland in Germany or on an island off the Scottish coast may well present as many challenges and need as much planning and care as forensic investigations in the Amazon rainforest.

    Some types of fieldwork involving wildlife are dictated by the species with which one is dealing. A marine mammal case is an example. Whether alive or dead, the animal will be either in the sea or on the shore and both locations present challenges. Campbell-Malone

Figure 7.1 A necropsy on a whale commences. Those involved have only two hours to obtain as much information as they can before the incoming tide makes further work dangerous.

and Bogomolni (see 'Marine Mammals: A Special Case' section in Chapter 13) put this graphically when they say that 'coastal cliffs, remote islands, unstable pack ice and rocky, wave-inundated beaches often prove to be insurmountable obstacles preventing carcase retrieval ... currents and tides can also move buoyant carcases great distances from the original place of death, making it impossible to analyse the scene' (Figure 7.1).

Traditionally, fieldwork involving animals may be carried out for various purposes:

1. Investigation and treatment of disease and/or health–monitoring
2. Application of preventive measures such as vaccinations or vector control
3. *Post-mortem* examination
4. Laboratory investigation of material from live or dead animals or the environment
5. Basic and applied research
6. Forensic examination – the subject of this chapter

Practical Considerations

There are various special, sometimes unique, features of fieldwork. These are often exacerbated when working overseas and particularly so in certain countries where language, religion, traditions and legal framework are different. And yet it is often in these very

Field Techniques

locations that high-quality wildlife forensic investigations are most needed. Some of the world's rarest and endangered animals are to be found in 'developing' countries with the least infrastructure.

Many wildlife forensic investigations have an international dimension and the investigator may need to travel – and possibly, for a period, live – overseas (see Chapter 3). Visas and work permits are likely to be needed, as well as passports. Authorisation may be necessary to take or to use veterinary medicines and chemicals. The Customs of the host country may demand letters of introduction and copies of permits and may wish to charge duty on incoming equipment and goods. In all such dealings, signatures and appropriate rubber stamps add weight to such documentation. Apparently simple tasks can often take a long time, especially when they involve more than one country's officials. Unnecessary difficulties can often be avoided if there is sound preparation beforehand – see, for example, the guidelines produced by the British Veterinary Association (http://www.bva.co.uk/overseas/working_volunteering_overseas.aspx).

Dealing with people may not be easy. All forensic work requires interaction with others but field investigations often present particular challenges because of the need under such circumstances to work with local personnel, sometimes a whole community. This can be especially problematic in countries where one is unfamiliar with local culture and traditions (Figure 7.2).

Figure 7.2 A Rwandese villager dances. Cultural sensitivity is important in forensic work.

In some countries, the fact that one is investigating a suspected illegal act does not mean that all assistance will be given. On the contrary, the authorities (or individuals) may conspire not to do so.

Another difficulty can be gaining access to the area where the offence is believed to have taken place. This may be a protected area such as a national park or game reserve, government land, or private property. Important considerations include obtaining the appropriate permission and avoiding accusations of trespass. Working with or under the umbrella of in-country institutions, such as the Department of Veterinary Services and/or University Veterinary Faculty, can assist in finding a navigable path through the myriad of logistical challenges faced in a foreign environment.

Legal Considerations

There are usually legal requirements in conducting wildlife forensic investigations in the field because one is working with protected species or carrying out what would normally be considered prohibited activities. This means that, even in a wildlife crime investigation, one may need to obtain a permit to kill, capture, handle, keep or ring (band) animals and/or use nets, traps or take samples.

Licences may also be needed for other purposes. In some countries, especially where officials are sensitive, the wildlife forensic investigator neglects this at his or her peril. Considerations in this respect include the question of whether a special licence is needed for radio, firearms or vehicles. Compliance with local legislation is always wise – for instance, relating to motor vehicles (seat belts, overloading) or gun safety (which may require special authorisation if imported – see earlier text).

One may need to use drugs in a forensic investigation – for instance, to immobilise animals – and appropriate authority is likely to be needed for these. Considerations insofar as drugs and medicinal products concerned are availability, importation if necessary (see earlier text), safekeeping and disposal.

In some countries, registration with the authorities may be necessary before any work can be carried out – for example, for a veterinarian to practise or for a biologist to perform what may be seen by local officials to be 'research'. The fact that such registration necessitates the payment of a fee (as, indeed, can gaining access to a protected area – see earlier text) often encourages authorities to insist on compliance with what they view as a statutory requirement.

Legislation relating to health and safety may or may not be in force in a country, or it might not be properly enforced (see Chapter 3). However, this is no excuse for wildlife investigators to ignore the need to protect those involved in forensic work, especially in the field. The prevention of accidents requires appropriate risk assessment and codes of practice, and in this context insurance may be important to cover liability, accident, health, vehicle and equipment. Useful guidance on field safety is to be found in publications by such organisations as the Royal Geographical Society and on numerous university websites.

If one is likely to bring evidence (e.g. samples) back to a home laboratory, all necessary arrangements should be made before departure. Legislative requirements may involve both animal health officials (Ministry/Department of Agriculture) and those concerned with the conservation of endangered species (usually Ministry of Environment or its equivalent) (Cooper, 1987a; Cooper and Cooper, 2007) (see Chapter 3). CITES regulations apply not only to live and dead animals but also to recognisable derivatives such as blood smears and small samples for DNA studies – both of which are commonly used as evidence in forensic cases.

An awareness of legislation is clearly important in any forensic investigation but this does not only relate to the statute(s) which are the subject of the enquiry but others that may be relevant. Both national and international laws may apply. The three main categories that are likely to be pertinent in wildlife forensic investigations overseas are legislation pertaining to wildlife conservation, animal health and animal welfare. Employment-related legislation may be relevant (see Chapter 3).

Cultural Considerations

As mentioned before, when carrying out forensic investigations in the field, it is important to be sensitive to the feelings and knowledge of local people (Figure 7.2). Issues in this respect include the following:

- Adequate consultation
- Respect for customs, religion
- Appropriate dress, conduct
- Caution regarding photography and videotaping
- Awareness of perceptions and attitudes towards wildlife and their environment

Photography and videotaping are both very useful (arguably essential) in fieldwork but, if not discussed beforehand (consultation, see earlier text), may offend local people in some parts of the world. Having an awareness of perceptions and attitudes towards wildlife and their environment can equally prove very important. For instance, certain societies have strong feelings, sometimes backed by the law, about how wild animals and plants or their remains should be handled and whether or not they should be removed. An example of this concerns the Maori of New Zealand, for whom some species of bird have deep cultural significance (Cooper and Cooper, 2007) (see Chapter 9).

Respect for the community is not only a courtesy but can contribute positively to the investigation. If local people are taken into the confidence of investigators they may be able to provide information about incidents and perhaps identify people in the locality who might have knowledge or expertise that could help with the enquiries.

During the course of an investigation, one may be staying in the area, possibly in a local town or village. The points mentioned previously about respecting culture are clearly particularly important in such circumstances. In addition, however, the investigator must be prepared for facilities, including access to communications (Figure 7.3), that are rather different from those that pertain at home.

Personnel

Wildlife crime investigations should involve a team approach. The team members, representing different disciplines and expertise, may themselves come from disparate countries and backgrounds. This can work well but problems of communication and compatibility can arise, especially under field conditions.

Figure 7.3 If one is working in a rural area overseas, communications may not be easy. This is a post office in a village in East Africa. Mail is being collected for a project.

If a team is involved in an investigation, the report may need to be composite, with contributions by multiple authors (see Chapter 14). However, insofar as is possible, specific areas of opinion should be attributed to one person – for example, the pathologist should report on wounds or bones, the herpetologist on live frogs, the palynologist on pollen and the carpologist on fruits and seeds.

Performing Forensic Work in the Field

Field forensic investigations can be conveniently divided into those that are *planned* and those that have to be *opportunistic*.

Planned Fieldwork

This is what takes place on most visits to a wildlife crime scene. On such occasions, the investigator knows that he or she is visiting a specific location and, within limits, can plan the work that is likely to take place and the equipment required. Advanced preparation, including the compilation of SOPs and the preparation of equipment, helps to ensure that such fieldwork is performed proficiently (Cork and Halliwell, 2002); it is then more likely to yield reliable results which will stand up well to scrutiny in cross-examination.

Preplanned fieldwork, using appropriate equipment and techniques, requires not only careful planning but also good teamwork (see the following text). Useful background

Field Techniques

reading includes Cooper (2013), Cooper and Cooper (2007), Cooper and Samour (1997), Cork and Halliwell (2002), Frye et al. (2001) and Keymer et al. (2001).

Opportunistic Forensic Work in the Field

Opportunistic fieldwork may be necessary in wildlife forensic investigations when dealing with the unexpected – for instance, when an injured, sick or dead animal is discovered or an illegal act is suspected and immediate action is needed. The optimal equipment may, however, not be available. In any wildlife investigation, but especially in the field, personnel must be prepared to consider techniques and equipment – possibly what is available locally – that, while not specifically designed for use in forensic work, might prove useful in the prevailing circumstances. Some simple examples of such improvisation in the field are listed in Table 7.1.

A lightweight folding umbrella is always an asset in the field (Figure 7.4). Although intended primarily for protection from the rain and sun (important in fieldwork), it can in addition be used for such diverse purposes as collecting rainwater, catching and containing invertebrates and other small animals, marking important sites and searching for and securing trace evidence when 'beating' undergrowth.

Table 7.1 Field Techniques: Improvisation

Requirement	Ideal Situation	Possible Improvisation	Comments
Protective clothing	Gloves, gowns, aprons, masks, hat and overshoes.	Small plastic bags as gloves, hats, masks and overshoes; large plastic ('garbage') sacks as gowns or aprons.	Elastic bands or string can be used to attach small bags to hands or feet. Holes can be cut in large sacks to accommodate the wearer's neck and limbs.
Handling of contaminated or infected material	Rubber gloves, forceps.	Plastic or paper bags, wooden spatulae.	See Cooper and Samour (1997).
Sample collection	Laboratory containers.	Film pots.	See Frye et al. (2001) and Cooper (1996).
Note-taking	Purpose-made clipboard of hardboard or strong cardboard. Notebook or hand-held computer (see text).	A piece of cardboard and a large (bulldog) clip.	The cardboard can be incinerated after use.
Chilling or freezing	Refrigerator and freezer.	Wrapping specimens in layers of newspaper or leaves.	Using a sheltered area (under trees) will help as will burying in leaf litter or damp soil.
Fixation of tissues	Formalin or laboratory alcohol.	Industrial 'meths' (methylated spirits). Alcoholic drinks, including local brews, rum, wine and strong beer.	See Cyrus (2005), Spratt (1993) and Cooper and Cooper (2007).

Source: Adapted from Cooper, J.E. and Cooper, M.E., *Introduction to Veterinary and Comparative Forensic Medicine*, Blackwell, Oxford, UK, 2007.

Figure 7.4 Examples of portable, lightweight equipment. An umbrella has many uses in the field.

Use of Field Equipment

Portable kits are an integral part of fieldwork, even when motor vehicles are available (Figure 7.5), as the wildlife crime scene may extend into very inaccessible places. The development of field kits owes much to various pioneers in the practice of tropical medicine who recognised the need to be able to perform basic diagnostic tests in isolated areas. Some background to such work is given in the book by Rickman (2006), who developed such valuable instruments as the 'Spindoctor', used to measure whole blood parameters, including packed cell volume (PCV), using only centrifugal force.

The design and composition of field kits depend upon the circumstances, in particular

- The type of case – live animal, dead animal, derivatives, environmental pollution
- The location – whether the kit has to be carried by hand, on a bicycle or motorbike
- Whether assistance is available – for instance rangers, mountain guides or porters
- The distance back to appropriate laboratories, clinics or a police station, where follow-up investigations can be carried out and/or material can be stored

The assembling of equipment for immobilising a wild animal, using a field kit deep in the forest, is shown in Figure 7.6.

Certain field investigations require special equipment. A typical entomological kit containing items needed to collect insect evidence from a living or dead animal is given by Barnes (Chapter 6). During the course of examination of a dead marine mammal it may be helpful to have a metal detector in order to scan the carcase for evidence of shot (Neme and Leakey, 2010) (see 'Marine Mammals: A Special Case' section in Chapter 13).

Field Techniques

3. Photograph or draw the carcase *in situ*. Do not disturb the site unnecessarily as it may constitute a crime scene, in which case it will need to be investigated fully by a forensic team (see also the following text)
4. Pending detailed examination, if the ambient temperature is high, a carcase should ideally be chilled (kept at refrigerator temperature) (see Chapter 9), but this is not usually practicable under field conditions and it may not be permissible to move the dead animal. The best one can do, so long as it is consistent with such forensic needs as not destroying evidence, is to keep the carcase cool by either covering or moving it to a cool location. The situation may be reversed in cold regions or high in the mountains: here there is a danger that the carcase will freeze before necropsy commences; freezing will preserve the specimen but it damages tissues and can make certain investigations (e.g. histology) very difficult. Again, it may need to be covered
5. The carcase may also need to be protected from scavengers, including humans (see 'Advances in Scavenging Research' section in Chapter 13).

If you have to carry out the *post-mortem* examination yourself in the field

1. Follow a 'clean/dirty' system throughout. Initially establish and cordon off an area where the necropsy is to be performed; only those who are 'dirty' should be permitted within that area (Figure 7.9). Alternatively, the back of a pick-up vehicle can be turned temporarily into a necropsy table (see Chapter 9)
2. Adhere to strict hygienic precautions in all the work. Rubber gloves or their equivalent (see Table 7.1) must be worn by those handling the carcase. Avoid contamination of paperwork and the outside of specimen containers by ensuring that only

Figure 7.9 A field necropsy area surrounded by a simple fence is used for the examination of dead flamingos.

'clean' personnel, who are stationed outside the cordoned-off area, handle such items. Boil or disinfect instruments after use and, if possible, between specimens
3. Follow a basic, but standard, protocol when performing the necropsy – see the following text
4. Keep written records of all findings plus (if possible) tape-recordings and/or photographs/videotapes. Preferably, one ('dirty') person should carry out the examination while another ('clean') writes. Report systematically everything you see, even if it has to be in non-technical language. Try to have available a camera with back-up batteries and disks and water-proofing (see Chapter 5) so that specific features can be photographed. Alternatively, or in addition, make sketches – even rough diagrams can be useful
5. Be prepared to take tissue samples and store these fresh (cool) and fixed (in formalin or an alcoholic preparation – see Table 7.1). Ensure that some material (fresh/frozen tissues or blood) is retained for DNA studies. Also, if practicable, take swabs for microbiology from tissues that show abnormalities. Remember that one can never take too many samples. They need not all be processed.

Basic *Post-Mortem* Technique for Use in the Field

1. External
    a. Weigh and measure the animal (use standard techniques)
    b. Identify the species and attempt to determine age and sex
    c. Examine externally (including all orifices) – record any parasites, lesions or abnormalities. Retain samples, fixed or fresh, as necessary – see earlier text
    d. Check important structures, for example, preen gland in birds, teeth in rodents
    e. Photograph or draw as much as possible
2. Internal
    a. Open carcase – examine internal organs
    b. Record any lesions or abnormalities and note whether the alimentary tract contains food or other material
    c. Check important structures, for example, bursa of Fabricius in birds
    d. Retain samples, fixed or fresh, as necessary, including (frozen if possible) lung, liver and kidney for toxicological investigation
    e. Photograph or draw as much as possible
3. Immediate investigations, if feasible
    a. Open portions of intestines and look with naked eye for parasites, such as worms. Count them or estimate total numbers
    b. Make wet preparations of intestinal contents in saline and examine them under a microscope for parasites, food items and other material. Photograph or draw what you see
    c. Do 'touch preparations' (impression smears) of liver and any organs showing abnormalities; stain and examine.
       Photograph or draw as much as possible
4. Longer-term investigations
    a. Preserve lung, liver, kidney, plus any abnormality in formalin, alcohol or other fixative
    b. Retain other samples as appropriate – see earlier text

Field Techniques

5. Retention of carcases
   a. Retain remains of the animal in a fridge, frozen or fixed in formalin (see the following text). If you have access to several carcases, store them in different ways
   b. General guidelines are to retain specimens in a fridge (+4°C) for up to 7 days, frozen/fixed thereafter. Remember that samples for DNA studies should be either frozen or fixed in an alcohol, not in formalin
   c. Record in a diary or on the computer as well as on a *post-mortem* sheet how the carcase and samples have been saved and include a reminder that they may need to be processed/discarded at a later date

Laboratory Work in the Field

Although forensic laboratory investigation is discussed in detail in Chapter 10, some more specific features of fieldwork are discussed here.

Samples collected in the field have the advantage that they are relatively fresh and therefore, in theory at any rate, more likely to be a 'representative specimen' (see Chapter 10). However, obtaining such samples may entail careful searching and possible exposure to dangers from animals, humans and the elements. Whenever possible, two people should be involved and proper protocols followed (Figure 7.10).

Figure 7.10 Collecting faeces and hair samples in the field, in this case from the night nest of a gorilla.

Many field investigations require only a hand-lens, operating loupe or solar/battery-powered dissecting–microscope. In some forensic work, however, access to a microscope that will operate in the field is essential. For this, it will need an external battery-operated or solar (panel or mirror) light source. Good quality microscopes that use a mirror rather than electricity are still available, often second-hand, and, although they are bulky, often prove ideal out-of-doors so long as there is sunlight (Figure 7.11).

Microscopes that have been assembled specifically for fieldwork have been used extensively in the past. They include the Swift FM-31 LWD TM, which is constructed from durable aluminium and brass and the McArthur – no longer in production but sometimes available second-hand – for which there are both aluminium and brass models and a lightweight plastic model. Other portable microscopes are currently being produced – for example, as a sequel to the Millennium Health initiative (Dunning, 2007), the Newton Nm1 (www.cambridgeoptronics.com) (Cooper, 2013).

In Chapter 10, Webb and Cooper discuss the question of processing and staining tissue samples (animal and plant) in the field, pointing out that standard embedding, sectioning and H&E staining are not practicable there. Instead, they recommend the use of toluidine blue, which is simple, quick and allows many different constituents to be distinguished.

Figure 7.11 Microscopy can be very important in field forensic investigations. Here, a binocular instrument is being used that has a mirror attachment and therefore does not rely on a source of electricity. It is bulky in comparison with a purpose-built field microscope but less expensive.

Field Techniques

Environmental monitoring can be an important part of forensic investigations. In the context of investigating ill-health and death in animals, useful information may be gleaned from veterinary publications. One recent paper in particular is worthy of reference here even though some of its contents may not immediately seem relevant to field forensic work; this is by Carr (2010) entitled 'Environmental medicine tool box for pig vets'. The author provides an approach to (domesticated animal) environmental medicine and emphasises the importance initially of observing animals 'in an undisturbed natural state' – so essential in wildlife work (see Chapter 8). Carr gives valuable information on such subjects as the assessment of air and light levels (the emphasis is on farm buildings but applicable to wildlife in the context of a zoo or rehabilitation centre). Items that may be needed in this context include equipment to measure air temperature and humidity, a light metre to measure light levels, anemometers to determine air speed and movement, infra-red thermometers, tachometers to calculate fan speed, remote sensors to monitor changes in temperature and humidity, smoke emitters and 'analytical sticks' to measure concentration of gases (such as CO_2, CO and H_2S) and thereby detect possible pollutants.

Quite apart from its value in terms of monitoring, Carr's paper provides useful information in terms of the sort of equipment that may be of value in field investigations, including those of a forensic nature.

A useful text dealing with (human) laboratory practice in tropical countries is the two-volume work by Cheesebrough (2006).

Sample-Taking in the Field

It is very likely to be necessary to take representative or significant samples in the field for transportation back to the laboratory. The principles of accurate sampling are addressed in Chapter 10.

Recording and Collating Findings in the Field

As indicated in Table 7.2, data in the field can be recorded and processed as handwritten notes and tape-recordings or on a mobile device such as a small battery-powered personal digital assistant (PDA), smartphone, tablet or netbook computer (Cooper, 2013).

PDAs have been largely superseded by smartphones that offer a wider range of functions, notably the ability to be used as a phone (!), messaging, e-mail, inbuilt global positioning system (GPS), maps, camera, bar code reader and – for fieldworkers with a penchant for cricket – that all important function of following the latest test score! There is specific surveillance software available to facilitate data capture by smartphones such as Episurveyor or Epicollect. The parameters of the collected information such as sensitivity, specificity, timeliness and representativeness can be increased by up to 100% by use of IT (Ojigo and Daborn, 2011). One should be wary of poor screen visibility in strong sunlight that can affect certain smartphones. Other considerations include the choice between touch screen and keypad, depending on the ability of the person collecting the information. PDAs may still offer advantages by way of extended battery time, specificity for a particular data collection purpose and cost. The keyboard of a netbook offers significant advantages for capturing more complex data and is the mobile device of preference when data collection can be taken whilst seated in field chair at a table.

There are small portable charging devices, such as the 'minigorilla', that can be used to extend battery time for a mobile device. A solar-powered version enables protracted periods of mobile device use away from generators or mains power.

Other important features of mobile devices are their ability to (1) transfer stored data back and forth between the computer and a notebook or desktop personal computer providing a back-up facility if the original data are lost, (2) make tapeless digital recording of notes or sounds, (3) take and rapidly transmit photographs as may be required and (4) access or receive reference material to assist the field investigation.

In all circumstances successful transmission depends upon a network connection.

Conclusions

Not surprisingly, wildlife forensic work often has to be performed under field conditions, away from the investigator's home base and without access to many of the facilities that this implies. It is not though only those people who live in poorer countries who stand to benefit from the development of better field equipment; even investigators working in less accessible areas in richer parts of the world, such as Europe, North America and Australasia, need to have access to reliable *in situ* technology if work away from the office or laboratory is to yield meaningful and reliable results. The forensic scientist who is adaptable and resourceful should be able to perform investigations and generate data of high evidential value under a range of different, sometimes challenging, conditions.

It is fitting, at the time of writing (February 2012), 100 years since the fateful expedition of Captain Robert Falcon Scott and his team to the Antarctic, resulting in the death of its leader and others, to recall that, although they did not achieve their goal to be the first to reach the South Pole, during that journey they collected, and carried despite their failing strength, quantities of geological and other scientific specimens. These, destined to be retrieved and analysed several years later in Britain, were still with the team, properly labelled and catalogued, when they died.

The words in the leader of the expedition's last letter are apt:

'These rough notes and our dead bodies must tell the tale'.

Captain Scott

Working with Live Animals

JOHN E. COOPER

Contents

Introduction	182
Before Clinical Investigations Commence	183
History and Identity	184
Identification and Recognition of Individual Animals	184
Identification of Individual Animals in the Field	188
Identification of an Animal's Species	188
Morphometrics: Weighing and Measuring	189
Condition	190
Age Determination	192
Methods of Examination of Live Animals	193
Indications for Examination	194
Clinical Examination Techniques	194
Clinical Signs	197
Lesions	197
Describing and Ageing (Age Determination) of Lesions	198
Ageing (Age Determination) of Bruises	198
Special Techniques in Forensic Cases: Imaging	199
Normal Anatomy	201
Emesis and Gastric Lavage	202
Emesis	202
Emetics and Stomach/Crop Flushing (Lavage)	202
Neck Ligatures for Sampling Ingesta	203
Treatment of Infectious and Non-Infectious Diseases	203
Welfare Investigations	204
Euthanasia	207
Retention of Wildlife in Captivity/Hospitalisation	207
Monitoring (Screening) of Wild Animals Prior to Release	207
Screening of Birds of Prey Prior to Release	209
Supporting Investigations – Laboratory Tests	209
Specific Examination	210
Central Nervous System	211
Eyes and Ophthalmoscopy	211
The Importance of the Oral Cavity	213
Examination of the Oral Cavity and Teeth	213
Bitemarks	214
Bite Wounds	215
The Bite	215
The Material/Tissues	216

Dealing with Psychological or Behavioural Disorders	216
Dealing with Infectious Diseases	217
Dealing with Non-Infectious Diseases	218
Traumatic Injuries	218
Investigation of Wounds	219
Assessment of Wounds	220
Traps, Nets and Snares	220
Road-Traffic (Vehicular) Accidents	222
Effects of Aircraft	222
Wind-Farms	222
Abuse	222
Starvation	223
Dehydration	226
Obesity	226
Shock	226
Neglect	227
Pollutants	227
Poisoning	228
Hyperthermia	228
Hypothermia	229
Burning	230
Explosions	231
Electrocution and Drowning	231
Electrocution	231
Drowning	232
Ionising Radiation	232
Transportation of Animals	233
Smuggled Animals	233
Other Environmental Stressors	234
Magnetic Fields	235
Conclusions	235

'Observe, record, tabulate, communicate. Use your five senses. Learn to see, learn to hear, learn to feel, learn to smell, and know that by practice alone you can become expert'.

Sir William Osler

Introduction

Many wildlife forensic investigations involve live animals. These may be free living (in the wild) or in captivity. A wild animal may need to be examined clinically in order to help ascertain whether it (a) has been the victim of an insult (e.g. a knife attack on a swan in a public recreation area) or (b) itself has been responsible for an incident (e.g. a buffalo that has gored and injured a tourist in a game reserve). Live animals may also require examination on account of concerns about infectious disease, such as the spread of zoonoses, or because health checks are required prior to release.

As emphasised elsewhere, it is always important at a crime scene to avoid allowing animals to suffer, even if this means treating or euthanasing those that are sick or injured and thus possibly destroying evidence.

This chapter focusses on the examination of live animals – 'clinical investigation' – particularly in respect of situations when they are victims of an insult, but also, to a lesser extent, when they are alleged to be responsible for an incident (e.g. bites, when a thorough examination of the animal's oral cavity is likely to be necessary).

It is neither within the scope of this book to cover all aspects of clinical work with all species, nor is the aim to teach the reader how to be a veterinarian. Nevertheless, it is important for those involved in wildlife forensic work to be aware of how to approach live animals—for reasons of personal safety as well as to minimise the risk of inadvertent injury or distress to an animal that is being examined.

Ideally, of course, a book of this kind which discusses insults caused by wild animals to people as well as the injuries and other damage done to animals by *Homo sapiens* would include more information about the sort of injuries that are sustained by humans and how best to interpret these in a forensic context. At present, such information is largely confined to medical texts but, in due course, it would be advantageous if members of the veterinary and medical professions were to collaborate in producing publications. Many of the injuries inflicted on humans do, of course, differ little from those inflicted by wild animals on domesticated livestock, but a comparative study appears to be long overdue.

Before Clinical Investigations Commence

Prior to the forensic investigation of a live animal, the first question to be asked is 'Who should be involved?' In many countries the clinical examination of a live animal in order to make a diagnosis is restricted to a registered veterinary surgeon (veterinarian). In other countries, the clinical examination, diagnosis and treatment of wildlife (as opposed to domesticated animals) may be performed by non-veterinarians. Additional personnel may be involved, either in a supporting role (e.g. a veterinary nurse [the United Kingdom] or a veterinary technician [the United States]) or in a consulting role because, in order to answer the questions that are being posed, some aspects of the examination require input from a person from another discipline – for example, an animal behaviourist.

Although a practising veterinarian (a general practitioner) can play a key part in forensic examination, there is an increasing need to involve specialists. The general practitioner may well be needed to carry out an initial examination (usually because he or she is the person immediately available at that time); but if a case is complex or if special information is required for a court case, input from other more specialised disciplines will be necessary.

Clinical expertise is not the only important part of examining a live animal that is the subject of a legal hearing. Wild animals differ greatly in their behaviour and requirements. Involving personnel with a sound knowledge of biology and natural history, who are familiar with the species, is always advisable. The 'biology' of the species encompasses such aspects as its anatomy, physiology, nutritional requirements and behaviour. 'Natural history', on the other hand, a rather less precise term, implies not just formal knowledge of a species' biology but also an understanding of – an empathy with – the animal's lifestyle (see Chapter 4).

The other important question in a forensic examination, before the work commences, is how to ensure continuity of evidence (see Chapter 14). In any work with live animals, especially if samples are taken, a chain of custody is created. It is advisable that the number of persons in that chain is kept to a minimum, but this consideration should not outweigh the benefit of having adequate assistance with the clinical examination and access to professional and support staff who can deal properly both with the animal and with the samples.

History and Identity

Obtaining as full a history as possible, not always easy when dealing with wildlife, is a vital prerequisite (see Box 8.1).

An aside, but of some relevance, concerns birds that are capable of mimicking humans or other animals. When such a bird is part of a legal case, it is worth noting and recording its vocalisations (what it says/whistles or whom it impersonates) because there have been legal cases where a talking bird has provided pertinent information. For example, the (British) *Daily Telegraph* reported, on 4 September 1996, that when a man was accused of handling a stolen cockatoo, the case against him was much strengthened when the bird identified itself with the words 'Hello Primrose' as the owner arrived at the police station. Such instances are rare but evidence based on such observations may be significant. So also, of course, can be an animal's behavioural response to a person or another animal. As always, observation is paramount – see later.

Identification and Recognition of Individual Animals

Correct identification and recognition of the individual animal is vital in any legal case. This applies whether the animal is alive or dead and the topic is therefore covered both here, later and in Chapter 9. Description and recognition are distinct from identifying the animal's species (see later), but some aspects may be common to both – recognition of certain characteristic morphological features, for example. Description should include the animal's sex and age, where these can be determined, weight (body mass) and at least one measurement. It is helpful to use a datasheet that bears an outline of an animal (usually left and right views, sometimes dorsal and ventral) and on this to indicate any specific features or special markings that may be important in recognition or are relevant to the case.

> ### BOX 8.1 IMPORTANT FEATURES OF A GOOD HISTORY
>
> - Accurate dates and times of incidents or findings. It can make a lot of difference to a forensic enquiry to know whether an animal was found sick or dead early in the morning or later in the day.
> - The circumstances regarding the animals, clinical signs noted, how many were involved, the social grouping, accommodation, diet, etc.
> - Other circumstances – for example, other animals and people in the vicinity.
> - Supporting photographs or drawings.

Morphometrics: Weighing and Measuring

The weighing and measuring of dead animals are discussed in some detail in Chapter 9. Some recommended methods of measuring live animals, both vertebrate and invertebrate, are given in Table 8.2 and examples (reptiles and amphibians) in the line drawings (Figure 8.4). It should be noted that the body length of a chelonian—tortoise, terrapin or (sea) turtle—is measured as a straight line, as if the animal is placed between two book-ends, not by the length of curvature of the carapace.

Using such a measurement, often referred to as 'standard length', is recommended for a wide variety of species, including cetaceans and birds. Adhering to conventions established in zoological morphometry is important. If accurate weights and measurements are to be recorded, regardless of the species in question, appropriate equipment will be required – see Box 8.2.

The two parameters that are of greatest potential value and usually amongst the easiest to determine are weight (mass) and one or more body measurements. It is important that both are recorded.

Weighing is usually straightforward but if the animal bears extraneous items such as collars, hoods, bells, splints, plaster casts or dressings, these must either be removed or allowance be made for them in any calculations. It is also important to record if a raptor has a crop full of food, or a snake has a prey item in its stomach, because these will give

Table 8.2 **Measuring of Live Animals**

Group	Method	Comments
Mammals	Either (a) crown – rump (CR) length or (b) snout to base of tail plus (where appropriate) length of tail. Specific measuring techniques apply to some species, e.g. marine mammals and bats.	The tail may be truncated – for example, because of an injury – and therefore a whole body length, including tail, can be unreliable. See Appendix G—'Cosmetic *Post-Mortem* Examination of Mammals'.
Birds	Carpal length.	A specially made metal rule with a raised stop is recommended. If a reliable carpal length proves impossible because of moulting, feather damage or pinioning, a tarsal length should be recorded.
Reptiles	Snout-vent (SV) and vent-to-tail (VT) lengths.	See line-drawings. A whole body length is to be discouraged because of the possibility of tail damage (see earlier). Care must be taken when straightening (putting the animal into a plastic tube will assist).
Amphibians	Tailless species (frogs and toads) – snout – vent length (SVL). Tailed species (newts and salamanders and tadpoles) – as reptiles.	See line-drawings. A rule with a stop is advisable. Care as mentioned earlier.
Fish	Various methods used. If in doubt in legal cases, measure SV and VT.	Different techniques are used by aquaculturists and ichthyologists (see literature).
Invertebrates	Vary, depending upon the species. Molluscs, for example, are usually measured in two planes – height and breadth. Lepidoptera, Odonata, etc. can have their wingspan measured.	Consult specialist texts. Use high-quality sliding callipers.

Source: Adapted from Cooper, J.E. and Cooper, M.E., *Introduction to Veterinary and Comparative Forensic Medicine*, Blackwell, Oxford, UK, 2007.

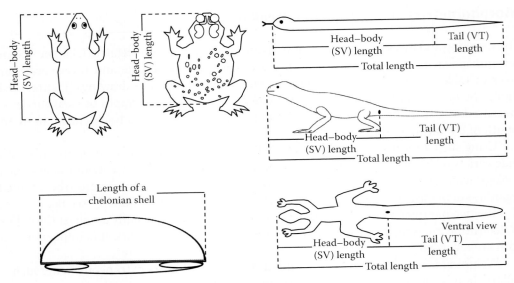

Figure 8.4 Some methods of measuring reptiles and amphibians.

> **BOX 8.2 EQUIPMENT FOR WEIGHING AND MEASURING**
>
> *Weighing*:
> A balance or scales. A cloth bag or similar to restrain small 'exotic' species.
>
> *Measurements*:
> Plastic rules and/or tape and/or vernier (sliding) callipers. Plastic or glass tubes and strong plastic sheets facilitate 'straightening' elongated specimens, such as snakes and certain invertebrates.
>
> - If photographs are to be taken, a non-reflecting (non-glare) metric (centimetre or metre) scale should be available to place by the specimen.
>
> *Source:* After Cooper, J.E. and Cooper, M.E., *Introduction to Veterinary and Comparative Forensic Medicine*, Blackwell, Oxford, UK, 2007.

erroneous figures. While it is acceptable for an aquatic or amphibious species to be wet when weighed, care should be taken to reduce the amount of water to a minimum. When weighing small animals in the field using a spring balance, the wind can give false results – minimised by either folding the bag so as to reduce the surface area or carrying out the weighing in a bucket (without touching the sides).

Different methods of measuring are used for different taxa. Mammalogists, for example, use a number of procedures for species ranging from bats to bottle-nosed dolphins (Lundrigan, 1996). Table 8.2 provides guidelines for those involved in forensic work with vertebrates or invertebrates. If in doubt, a range of measurements should be taken.

Condition

Defining and assessing 'condition' may be important, especially in welfare case. A specific condition score is valuable but in its absence, evidence of increase in weight of an animal following its being put on a normal feeding regime can prove a useful measure in court or in a written statement, as evidence of malnutrition (see later).

The definition and determination of 'condition' in non-domesticated species can present particular difficulties. 'Condition' is an ambiguous term because it is commonly used both as an indicator of an animal as a whole and also parts of the body (e.g. 'condition of the plumage'). In only a few wild animals has the assessment of condition been properly studied but there are some useful reports, such as the work by Adamczewski et al. (1997) describing the effects of season and reproductive states on seasonable changes in body composition of female muskoxen. Hewson (1968) weighed mountain hares during live-trapping and demonstrated changes related to season, time of birth, disease, starvation, adverse weather and injury. Condition in fish was addressed by Weatherley and Gill (1987).

In birds, condition has been defined as 'some qualitative assessment (which may be determined quantitatively) of the bird that has a direct bearing on its fitness' (Gosler, 2004). There have been many relevant studies; for instance, Hargitai et al. (2012) investigated winter body condition in relation to age, sex and plumage ornamentation.

Methods used in birds that are discussed by Gosler are shown in Box 8.3.

The condition of wild animals may fluctuate depending upon the time of year and the species' requirements – for example, in so-called "migratory fattening" when birds lay down reserves to sustain them on long flights (Newton, 1998).

Condition scores have been established for most domesticated species but relatively few have been well-tested in wildlife. Some used for captive animals can be applied to the same species in the wild. Valliyatte (see 'Welfare Issues of Captive Asian Elephants' section in Chapter 13) describes the use of condition scores for various types of elephant in India.

Non-mammalian species present particular difficulties. For example, reptiles vary greatly in morphology and this can present problems in terms of assessing 'condition'. Land tortoises usually have a protective, often hard, carapace and plastron. Methods of measuring 'body mass condition' in tortoises were reviewed and discussed by Hailey (2000) and by Willemser and Hailey (2002). The easiest method was to calculate the mass of the tortoise in relation to its size, a concept that was originally devised by Jackson (1980). Hailey and colleagues

BOX 8.3 ASSESSMENT OF CONDITION IN BIRDS

- Relative mass – the mass (weight) relative to the size of the bird. Often called 'condition' or 'condition index' but this method can be misleading.
- Fat reserves – subcutaneous, internal (*post mortem* or during laparoscopy) or using total body electrical conductivity (TOBEC). The former can be misleading; the latter is expensive, cumbersome and not necessarily better than using non-destructive fat 'scoring' methods.
- Muscle protein – assessment and scoring of pectoral muscles, or recording their shape using fine wire or dental alginate.
- Physiological analysis – usually of blood.
- Moult and plumage – age and condition of feathers, including ptilochronology (see Chapter 10).
- Parasites – numbers and species.
- Biopsy techniques, permitting tissues to be examined or isotopes to be measured.

devised a condition index (CI) and applied this to both free-living and captive (European) *Testudo* species, and as a result produced a programme that reliably calculates body condition of these species so long as the tortoise has a carapace length of 100 mm or more.

Gosler (2004) suggested that fluctuating asymmetry (FA) might be used as an indicator of condition (see Chapter 9), but he warned of bias because most observers are themselves asymmetrical (left- or right-handed).

Age Determination

The age determination of animals can be very important in forensic cases but may prove difficult, even in domesticated species. The dentition can contribute usefully to the assessment of age in most species (see Chapter 9 and 'Dentition of Mammals' section in Chapter 13) because it bears more relevant evidence than does any other organ system. However, wherever possible, the estimation of age is best made by using all the available criteria, dental and non-dental.

Techniques based on dentition, used for centuries in horses and ruminants (Dabas and Saxena, 1994), have come under scrutiny in recent years. Even in humans, while comparison with standard charts provides an accurate method during the period of tooth development, after about puberty reliability decreases because it becomes dependent on the third molar only.

Some techniques that can be used for age determination of live vertebrate taxa are given in Table 8.3.

As will be apparent from Table 8.3, reliable methods of telling the exact age of a live animal are few. There are so many variables relating to the biology and history of the individual animals. Some useful references exist – mainly to domesticated animals (see standard veterinary and animal husbandry texts) but also a few to wild species, for example primates (Erwin and Hof, 2002). Age estimations at death of *post-mortem* remains (see Chapter 9) – as well as species identification and sexing (see below) – have been much refined by the research of zoo-archaeologists (see e.g. Davis, 1987).

Table 8.3 Some Methods of Age Determination in Vertebrate Animals

Methods	Species	Comments
External appearance including secondary sexual characteristics	Many	Much variation
History, behaviour	Many	Evidence of parturition, egg-laying and reproductive activity may distinguish adults
Specific age-related external features, such as rings or horns	Ungulates especially bovids. Reptiles (shields)	See literature
Specific age-related internal features, such as 'growth rings' in cortices of long bones, sometimes the number of corpora lutea on ovaries	Mammals, birds, reptiles (bone), other species (ovaries)	See literature
Dentition	Mammals, especially equids, ruminants and cetaceans	May make use of eruption patterns or (probably more reliable) microscopical examination of teeth
Stage of metamorphosis	Amphibians, some fish	See text
Pelage	Some mammals, some birds	Hair and feathers can be indicative of age group (see text)

Source: Adapted from Cooper, J.E. and Cooper, M.E., *Introduction to Veterinary and Comparative Forensic Medicine*, Blackwell, Oxford, UK, 2007.

It is often less difficult to put an animal into an age group or age band – such as determining whether a mammal (e.g. a fox) is a cub or an adult, on the basis of whether or not it still has deciduous teeth. As always, care must be taken with certain taxa: cetaceans, for instance, may have no teeth or only permanent dentition. Birds can usually be put into three age groups – adult, sub-adult and juveniles – by analysis of the feathers, eyes or feet. Likewise, young birds can usually be roughly classified as nestlings or fledglings, on the basis of their plumage development. Feather deuterium values may also help determine the age group (Yohannes et al., 2012).

Methods of Examination of Live Animals

It is outside the scope of this book to discuss in detail the clinical examination of all types of wild animals. The general rule is that the animal, regardless of species, should initially undergo a full basic clinical examination, similar to that performed during a normal (diagnostic) veterinary investigation. This means that the animal is examined systemically (as well as systematically), even though there may be an obvious presenting wound or lesion of probable importance. The examination should, as a general rule, incorporate the standard stages of (a) visual inspection, (b) auscultation, (c) percussion, (d) palpation and (e) other diagnostic procedures. In order to minimise stress, equipment likely to be needed should be available *before* the examination begins, and the choice may be influenced by the history or allegation. This should include equipment to obtain samples, such as urine or faeces, where they are presented during the examination. Examining the animal on a clean sheet of paper taped to the examination table can be extremely useful in order to obtain such samples – and also, sometimes, to collect parasites – for evaluation by bacteriology or microscopy without undue contamination from the table. As emphasised elsewhere in this book, having a notebook or similar to hand can be invaluable during the examination, or better still a tape-recorder in order to have a note of findings without the need to write at the same time as handling the animal. Again, a camera should, whenever possible, be available as 'a picture paints a thousand words'.

It is important in any clinical investigation to make full use of all one's senses. Touch and hearing are key parts of the aforementioned stages. To them should be added 'use of the sense of smell' because forensic cases, especially those associated with poisoning or wounding, are often associated with the development of a characteristic odour.

The clinical examination of domesticated animals is fully covered in standard veterinary texts. Non-domesticated species are usually more problematic, but there is an increasing volume of literature relating to this insofar as captive individuals are concerned.

Recommended techniques for restraint, handling and sampling wild animals are also to be found in specialised texts for biologists and field workers – for example, for those working with Australian species (Tribe and Spielman, 1996) or for amphibians (Heyer et al., 1994).

Use of a clinical record sheet is essential (see Appendix D) and, like the necropsy sheet (see Appendix D), is often best compiled specially for the forensic case rather than relying upon a standard clinical form, which may or may not permit answers to be given to specific questions of forensic importance.

As might be expected, different species have different normal features. For example, while the mucous membranes of most vertebrates are pink or red in colour, those of crocodiles and certain other species may be yellow. Familiarity with what is normal is, as emphasised throughout this book, essential.

A 'scoring' system is recommended when there is a need to quantify. Colour codes are useful and can be used to help describe faeces, urine, etc. Objectivity is enhanced if reference can be made to colours on a chart or the Internet (such as those used in photography). The *consistency* of, say, faeces can be difficult to describe in an unambiguous way. Use should be made of the grading systems (usually 1–5, liquid–solid) advocated by some of the nutritional companies.

Indications for Examination

In a forensic case live animals may need to be examined for a number of reasons:

- Clinical signs and lesions due to illegal activities
- Naturally occurring infectious or non-infectious disease
- Disease induced by treatment, lack of treatment or poor management

Clinical Examination Techniques

Before handling a wild animal for clinical examination, it is well worth reducing the light intensity. In many species – most birds, for example – this has a calming effect and will thus improve the welfare of an animal that is already likely to be in a state of shock. Subdued lighting also eases clinical examination from the perspective of the handler/veterinarian and (important in forensic work) lessens the likelihood that damage occurring during examination will complicate or obscure relevant clinical signs or lesions.

Published texts have rarely to date dealt specifically with the forensic examination of live wild animals. Guidance on clinical examination of certain 'exotic' animals and wildlife is available in, for example, the papers dealing with the forensic avian case (Cooper and Cooper, 1986, 1991; Forbes, 1998). A very useful recent contribution has been the publication by Newbery and Munro (2011). Their paper, illustrated with colour images of specific cases and equipment, takes the investigator – in their case, the veterinary practitioner – through the important stages of dealing with a live animal that is possibly the subject of a legal case (see Box 8.4).

BOX 8.4   STAGES OF EXAMINATION

- Initial information to be recorded
- Details to collect when an animal arrives
- Protocol for the initial examination of the animal
- Features to observe when forming a 'general impression' of the animal, from a distance
- Features to note in a clinical examination
- Creating a video/photographic record of evidence
- Security of evidence
- Report-writing
- Appearing in court

Source: After Newbery, S. and Munro, R., *In Practice*, 33, 220, 2011.

Examination of the forensic case brings with it extra considerations that may not form part of standard veterinary clinical investigation. Laurence and Newman (2000) discussed this in some detail and the general rules below largely reflect their advice as well as that of Cooper and Cooper (2007) (see Box 8.5).

There are tried and tested methods of carrying out a basic clinical examination of most species of domesticated animal, and a practising veterinarian will be familiar with these.

Clinical examination of wild animals can, however, be challenging, especially (but not exclusively) when they are free-living. This is in part because even handling the animal may present difficulties and dangers, and in part because of the lack of information about values and significance of lesions or ill-health in many species. The veterinarian asked to carry out such work must be sure that he or she is confident to do so; if in doubt, a knowledgeable colleague, preferably one specialised in the taxon involved, should take the case.

BOX 8.5 SOME GENERAL RULES FOR THE FORENSIC EXAMINATION OF A LIVE ANIMAL

- Carry out as full and methodical an examination as possible in the time available and with the equipment that is available. Record all clinical signs and lesions (see later).
- Although the examination should initially be general, pay particular attention to any organ systems or parts of the body that may be particularly relevant to the case.
- Depending upon species, certain structures may be especially pertinent to a forensic investigation. Thus, for example, when examining a bird, the preen gland, important in water-proofing of plumage, should be located and, if present, examined and a comment made on its appearance and whether or not it is functioning. A brood patch may be found in a bird that is incubating eggs. The teeth and scent glands of rodents should always be examined, as should the horns or antlers of ruminants and the pouch of marsupials.
- If feasible (not always the case with wildlife), the animal's body temperature should be recorded in degrees Celsius. It is not acceptable to state that the temperature was 'normal' or 'within normal limits' (Laurence and Newman, 2000). It must be remembered that the body temperature of an ectothermic ('cold-blooded') animal such as a reptile, amphibian, fish or invertebrate will be the same as that of the environment (see later).
- If time, facilities, or the veterinarian's lack of experience of the species in question limit a full examination, this should be stated in the report. It is preferable at an early stage to admit that certain examinations were not carried out (or feasible) rather than for this to come to light during cross-examination in a court case or similar hearing.
- Detailed contemporaneous notes (hard copies and/or computerised) should be taken, together with tape-recordings and photographs.
- Laboratory tests may be carried out which would not form part of a normal standard clinical examination: samples should therefore always be taken (see later and Chapter 10).

> **BOX 8.6 CLINICAL CHECKLIST**
>
> 1. History and management
> 2. Clinical examination, with particular attention to
>   a. Skin and coat
>   b. Nares
>   c. Ears and eyes
>   d. Oral cavity, including teeth
>   e. Anogenital region
> 3. Procedures
>   a. Body temperature
>   b. Palpation
>   c. Auscultation
> 4. Laboratory tests
>   a. Faeces
>   b. Exudates
>   c. (Blood, if practicable)

Protocols do not exist for most exotic species and therefore will have to be formulated by the veterinarian – after checking with colleagues and literature.

A suggested clinical checklist for a wild (free-living or captive) animal is given in Box 8.6.

Invertebrates present particular challenges (Figure 8.5) but are increasingly likely to be the basis of a legal or insurance action (some species are worth considerable sums of money). Texts are available on the clinical examination of these species (see, e.g. Cooper [1990, 1992, 1999]).

Figure 8.5 Postgraduate students learn how to observe and examine live invertebrates.

Clinical Signs

A key part of the clinical examination, whether for veterinary or forensic purposes or both, is the detection and recognition of clinical signs. In order to assist the non-medical investigator, a number of these are defined in the Glossary (Appendix A). Video-taping has much to commend it in order to obtain a permanent record of clinical signs, some of which may be only transient (see the following).

Clinical signs may be best detected before an animal is handled or even made aware of the observer's presence, as veterinary practitioners recognise (Carr, 2010). Dian Fossey drew attention to this in her work with mountain gorillas when she described contact with the free-living animals as either 'obscured', when gorillas did not know she was watching them, or 'open' contacts, when they were aware of her presence. As far as the former were concerned, she stated 'Obscured contacts were especially valuable in revealing behaviour that otherwise would have been inhibited by my presence' (Fossey, 1983). This, as the authors know from their own work in Rwanda with the gorillas, includes hiding and disguising any signs of injury or ill-health. It is a feature of most, if not all, wild animals.

Clinical signs can be influenced by a number of other factors also. In particular, as emphasised earlier, most mammals and birds are endothermic (warm-blooded) meaning that they can, within limits, maintain their body temperature by the internal control of their metabolic rate. Clinical signs are, therefore, relatively predictable in terms of onset and duration. In contrast, most reptiles, amphibians, fish and invertebrates are ectothermic (cold-blooded) and their responses to both infectious and non-infectious insults can be easily influenced by changes in environmental temperature (Cooper and Jackson, 1981; Frye, 1991). There are two ranges of temperature that are important in ectothermic animals:

1. An activity range, within which they move, may feed and outwardly appear healthy.
2. A preferred body temperature (PBT), a narrower range than 1, within which such processes as digestion and immunoglobulin production best function. The PBT is the range that an ectotherm will choose if given the opportunity and to which captive specimens should have access (see also 'preferred optimum temperature zone' (POTZ) – Chapter 2).

If the environment for an ectothermic animal is not optimal, a chain of events can occur, usually culminating in death. This is often still termed the 'maladaptation syndrome' – see Box 8.7.

Lesions

Care must be taken during examination of a wild animal to record and describe accurately any lesions that are present, even if they may appear to be of no immediate relevance to the case. Lesions often provide information the importance of which is not appreciated until later, perhaps during court proceedings (Cooper and Cooper, 2007).

The examination, description and interpretation of lesions are discussed in detail in Chapter 9. Descriptions for forensic purposes need to be as accurate as possible, not least

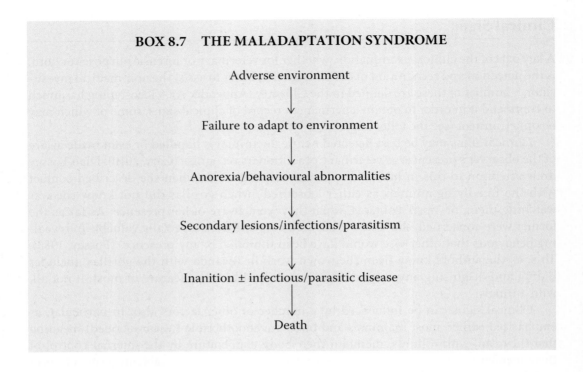

because they may be checked by someone else. In addition, as was stressed by Cooper and Cooper (2007), it is important in legal cases to be able to say whether interpretation of lesions or clinical signs and an opinion based upon this are, or are not, evidence based. Clinical acumen and intuition have an important part to play in routine veterinary and medical work but, if used to substantiate the claim in court, may be challenged. Scientific opinions change, of course. For example, the standard approach to skin lesions was based on what was termed 'pattern analysis'; this is now often passed over in favour of a step-by-step, problem-based, investigation. The expert witness will need to be aware of both, even though he or she will have views as to which is more relevant in the case that is being discussed.

Describing and Ageing (Age Determination) of Lesions

This is discussed in more detail in Chapter 9. Here, however, bruising is mentioned because of its importance in legal cases where the animal is still alive. In all assessments the investigator must remember that healing and other tissue responses are temperature-related.

Ageing (Age Determination) of Bruises

This question has long-challenged forensic investigations, as catalogued by Langlois (2007). A decade earlier, Stephenson (1997) reviewed the ageing of bruising in children and pointed out that, while earlier forensic texts had suggested that a sequence of colour changes could be used to deduce the age of a bruise, it was not that simple. There were many variables, including the amount of blood extravasated, the depth of the bruising below the surface of the skin, the severity of the blunt force that caused the injury, the vascularity of the underlying tissue, connective tissue support at the site of injury (bruising may follow a different sequence if the skin is loose, such as around the eyes or genitalia), the age of the victim (in human work bruises in adults may take longer to resolve) and the colour of the skin.

Traditional opinion in human work has been that bruises are initially red/blue and sometime later turn brown, green and yellow. However, Stephenson argued that the various factors listed earlier mean that it is unlikely that any one particular bruise can be precisely aged and that bruises of the same age could be of different colours.

In animals, ageing (age determination) of wounds has been studied by histochemical and biomechanical methods (Rackallo, 1972). In rats of different ages enzyme changes occur in the same chronological order, but older rats take longer to reach the maximum intensity.

There are also species differences. In poultry, bilirubin can be detected biochemically in bruises by 14 h and disappears by 4 days; the colour change and rate of healing are affected by environmental temperature (Handy et al., 1961). Histological ageing of bruises in lambs and calves was explored by McCausland (1978). In adult cattle, bruising progresses slowly, with bilirubin undetectable until 60 h and disappearing by 8 days. In experimental bruising in cattle, the sequence of colour changes was red from 15 min to 2 days, green from days 3 to 4 and yellow and orange from days 4 to 6; however, the rate of change varied in animals of different ages. In calves, the yellow coloration appeared earlier, at 48 h. Even with detailed histological examination it was only possible to age bruises crudely as 24 h or 24–72 h.

So how might this information be applied to wildlife? The answer has to be 'with great caution'. Some extrapolation is probably reasonable but sound data concerning bruising (and, indeed, other lesions) in wild animals are sadly lacking. The situation is complicated by the fact that 'wildlife' can include mammals, birds, reptiles and amphibians and fish – of different sizes, with disparate biological features – and therefore much variation is to be expected. Some data can be obtained from texts dealing with meat inspection of game animals, for example, but far more work is required and there is a need for those who deal with (e.g.) recently killed wild animals to record observations.

Special Techniques in Forensic Cases: Imaging

Radiography (carrying out x-ray examination) has long played a part in human forensic work and in domesticated animals too it has been employed for many years, mainly as a means of detecting fractures, lead shot or other significant features in cases where illegal wounding or killing is alleged. Radiography can also contribute to other aspects of animal forensics, including *post-mortem* examination (see Chapter 9), in conjunction with appropriate tests, to assess nutritional/metabolic status (e.g. bone density) or to investigate non-animal/non-human material (e.g. the detection of metallic fragments or other radio-dense material in foodstuffs, soil or water samples).

Digital radiography is increasingly used in veterinary practice and permits manipulation and enhancement so as to produce high-resolution images (Hagen, 2012). One potential problem with the use of digitised radiographic images, however, is fraud. It is a simple matter to increase the visibility of lesions by altering (e.g.) brightness and using an 'unsharp mask' filter over the entire image. Such enhancement can be used legitimately for teaching and publication. However, it is also relatively easy to alter specific features of a radiograph, including the ablation or creation of a lesion. Fraud is also possible by conventional photographic techniques, but is generally more difficult.

There are methods to be used when transferring files to check whether the sent and received versions are identical. This may be applied simply to see that data burned to a disk have not been corrupted to ensuring that files, comprising data, including photographic images, that are intended for use as evidence have not been altered or tampered with. It is also of value to the medical, dental, veterinary and other professionals who have to issue certificates based on digital imagery – for example digital radiographs. If the professional cannot verify that the digital material is the same as the original, it may be necessary to modify the certificate (McEvoy and Svlastoga, 2009). Those who certify matters that are not within their own knowledge risk disciplinary measures from their professional bodies.

Other imaging techniques that can play a part in animal forensic work include

- Ultrasonography
- Magnetic resonance imaging (MRI)
- Computerised tomography (CT)
- Doppler
- Endoscopy

The investigator should be aware of their existence and be prepared to consider their use by an appropriately qualified person. This applies particularly to ultrasonography where any permanent record of the examination could be impossible to interpret without expert advice.

Hagen (2012) reviewed cross-sectional imaging and emphasised that it is becoming more and more necessary in order to assist in diagnosis of pathology and the planning of surgical techniques. Both CT and MRI are used for this purpose. Cross-sectional imaging permits reconstruction of the anatomical situation in any desired image and also three-dimensionally. It has great potential in forensic work.

Safety is always a consideration, especially when radiography is used. X-rays are dangerous, and excessive or unwarranted exposure to them may itself be grounds for legal action.

A particular word should be said about fluoroscopy. This x-ray-based technology that produces 'real-time' images, single images or a combination of both is becoming very popular in veterinary medicine and there is an increasing trend to use it to produce real-time images during such procedures as angiography or administration of barium. It has considerable potential in forensic investigations involving live animals. However, use of fluoroscopy rather than radiography extends the screening time and, consequently, exposure to radiation – both the primary x-ray beam and from scattered radiation which is produced when the beam interacts with the patient. There are various ways of reducing the amount of scattered radiation and of protecting personnel who are involved in performing fluoroscopy.

Ultrasound examination has benefits over other imaging modalities – such as endoscopic examination or radiography – because often anaesthesia or sedation is not required. The machines and transducers used must have a high enough degree of resolution for use on small animals, and both abdominal ultrasound and echocardiology are commonly employed. Echocardiography will not be discussed in this chapter, although thoracic ultrasound can reveal interesting diagnoses.

Valuable guides to imaging of 'exotic species', which can include captive, injured or recently captured wild animals, are the books by Rubel et al. (1991) and Krautwald-Junghanns et al. (2011). It is not only animals' organs that can be scanned but also eggs of birds and reptiles, using x-rays or ultrasound, but the quality of the latter is dependent on the shell's level of calcification. In chelonians there is usually a hyperechoic shell with some acoustic shadowing, but lizard and snake eggs can be assessed in more detail due to their more flexible exterior.

Normal Anatomy

A difficulty encountered in both clinical medicine and forensic science is interpretation of images in unfamiliar species (Figure 8.6). There is need for more publications on this on the lines of the work by Smith and Smith (1991). These authors provided a reference for xeroradiographic and conventional radiographic anatomy of the bobwhite quail, using this as a representative of the order Galliformes. The heads, bodies, wings and pelvic limbs of four adult birds were radiographed using xeroradiographic and conventional radiographic techniques. Nine xeroradiographs and their corresponding

Figure 8.6 A reference radiograph – the threatened Rodrigues fody, now recovering as a result of conservation measures.

conventional radiographs were selected, and the xeroradiographs labelled to illustrate the normal anatomy of these regions.

Emesis and Gastric Lavage

Emesis

In cases where oral poisoning is suspected (see later and Chapter 2), stomach or (in some birds) crop contents are valuable for examination and analysis (see also Chapter 9). In live animals it may be desirable to induce vomiting (emesis) or to remove gastric (or crop) contents (lavage) – in order both to attempt to treat the affected animal and to recover the toxic material for investigation. Emesis may be attempted if the animal is conscious and not convulsing. Vomiting cannot be induced in certain species such as equids, rodents or lagomorphs. The vomitus should be retained frozen for analysis but in addition, prior to freezing, some can be examined grossly and microscopically using both reflected and transmitted light.

Emesis in mammals that regurgitate/vomit may be attempted by

- Manual methods
- Placing washing soda crystals on the back of the animal's tongue
- Administering such substances as mustard and water, syrup of ipecacuanha (diluted 1:1 in water), apomorphine (given subcutaneously, not in felids), xylazine (intramuscularly in carnivores; its sedative action may, however, potentiate the poison's central nervous system [CNS] depressive action)

Use of such substances may, of course, complicate the situation where noxious substances may have been given prior to the forensic examination and, indeed, may be the very reason why the animal is presented for such investigation.

In an emergency emesis may also be induced by administering a solution of saltwater, copper sulphate or hydrogen peroxide, but these methods are not generally recommended by veterinarians because of the possibility of adverse effects; saltwater can cause severe electrolyte imbalance; copper sulphate may lead to copper poisoning; and hydrogen peroxide is corrosive.

Emetics and Stomach/Crop Flushing (Lavage)

Emetics and flushing can also be tried. In the context of avian research, Sutherland stated that emetics and stomach-flushing 'do not usually kill the bird, but occasionally they do and these methods are usually considered too invasive'. This has to be borne in mind when dealing with a legal case. According to Sutherland, $0.8\,cm^3$ of 1%–1.5% antimony potassium tartrate per 110 g of body mass is used as an emetic for birds. It is administered by syringe through a lubricated narrow flexible plastic tube that is pushed gently down the oesophagus. The bird is then placed in a dark box, with a carpet of absorbent paper and released 15–20 min later.

Euthanasia

When dealing with a live animal that is part of a forensic investigation, it may prove necessary to kill (euthanase) it. There are usually two indications for this:

- Humanitarian – the animal is suffering excessively
- Practicalities – if, for example, dozens of frogs are found alive, but paralysed, in a pond and are believed to have been poisoned, examination of a small number of them may provide vital information

However, before embarking on killing an animal that may be part of a legal case, the following must be considered:

- Is permission to euthanase the animal required from the owner (if captive), the landowner, the police, the court or others?
- Can euthanasia be carried out humanely and discreetly under the circumstances that often prevail at a crime scene?

Special techniques may be needed to ensure (as far as is possible) that certain species are killed humanely – reptiles and amphibians, for example (Cooper, 1987b).

Retention of Wildlife in Captivity/Hospitalisation

Occasionally, an animal that is the subject of a legal case has to be hospitalised for treatment or retained as evidence. Campbell-Malone and Bogomolni (see 'Marine Mammals: A Special Case' section in Chapter 13) give examples in the context of marine mammals that present with pneumothorax or haemothorax.

Any live animal that is possibly a forensic case should be hospitalised in a secure place where it cannot be moved or interfered with by any person. The cage or enclosure should be locked and careful custody ensured for the keys. Distinguishing marks or means of identification should be noted (see earlier) so as to minimise the risk of a suggestion in a court case that transposition may have occurred. When a batch of animals is hospitalised – for example, oiled seabirds—it is wise when feasible to mark them individually (see Table 8.1) since each or any one of them may form the subject of a legal action and then accurate identification will be essential.

If a wild animal dies while retained in captivity, the details should be recorded and appropriate investigations carried out immediately. This may entail a *post-mortem* examination, and it is wise to have the necropsy performed by someone with appropriate specialist skills rather than the veterinary clinician who was responsible for the live animal.

Monitoring (Screening) of Wild Animals Prior to Release

An animal may need to be returned to the wild before a court case or after it. Whatever the timing, in addition to a clinical check (Figures 8.8 and 8.9), samples from the animal should, if possible, be subjected to laboratory-based health-monitoring. The purpose

Figure 8.8 The mountain gorilla depicted earlier is observed carefully before it returns to its group.

Figure 8.9 One of several green iguanas, stoned and collected by youths in Trinidad, in order to collect the reptiles as bushmeat, is checked prior to release.

of this is to be as confident as possible that the release does not compromise the welfare of the animal or carry the risk of introducing pathogenic organisms which may affect the same or other species of wildlife, contaminating the environment. Any screening is better than none. It is important to follow a protocol and to keep careful records. Whenever it is possible to monitor the animal after release, this should be done as thoroughly as is feasible.

What follows is a simple screening programme that has been used by the author to screen birds of prey prior to their return to the wild.

Screening of Birds of Prey Prior to Release

Observation and Examination

1. Presence or absence of
   a. Clinical signs of disease
   b. Injuries or external lesions
   c. Ectoparasites
2. a. Bodyweight
   b. Measurements
   c. Condition score
3. Gross appearance of
   a. Faeces
   b. Urine/urates
   c. Pellets (castings)

Laboratory Tests

1. Presence/absence of protozoan and metazoan parasites in faeces
2. Presence/absence of parasites or cellular abnormalities in blood smears

Additional investigations, if personnel and facilities permit

1. Bacteriological examination of swabs from
   a. Trachea
   b. Cloaca
2. Haematological examination of blood
3. Retention of serum for serological investigation

Supporting Investigations – Laboratory Tests

As the screening programme given above suggests, forensic examination of live animals should also always involve the taking of samples and the performance of laboratory tests – see Box 8.8.

BOX 8.8 LABORATORY TESTS ON LIVE ANIMALS

- Haematology
- Clinical chemistry
- Parasitology
- Microbiology
- Histology
- Cytology
- Electron-microscopy

Such samples for laboratory analysis will need to be taken. As a general rule, non-invasive or minimally invasive methods should be used, whether these animals are captive or free-living. This will reduce the need to handle and will minimise stressors (Cooper, 1998). Examples of non-invasive sampling include the collection of naturally voided

- Faeces (rather than rectal or cloacal swabs) for examination for parasites, for food contents or for corticosteroids
- Urine or urates (rather than samples taken by catheterisation) for testing for toxins or abnormal metabolites

An example of a minimally invasive method of sampling is the taking of buccal swabs from small animals for DNA analysis, rather than having to kill them in order to remove blood or organs – in amphibians, for instance (Pidancier et al., 2003).

Some forensic investigations will, however, inevitably, necessitate intervention, for example the taking of blood or biopsies.

Care must always be taken when sampling, regardless of how minor the procedure. Some techniques may cause pain or distress – see, for example, the study by Kiska et al. (2010) on individual and group behavioural reactions of small delphinids to remote biopsy sampling. If during sample-taking an accident occurs and an animal is injured or dies, there may be legal consequences and a necropsy must be performed (Cooper and Cooper, 2007).

So, what samples should be taken? The answer is, as always, 'it depends', and in legal cases it depends very much upon the charges and the evidence that is required. Often, however, it is a matter of taking a selection of samples – in part to be prepared for any eventuality, in part, to obtain as much information about the animal that may assist in its care and possible return to the wild. As an example, if a raptor is confiscated pending charges, it is wise to take a selection of samples as soon as possible, consistent with the welfare of the bird. The reason for doing so promptly is because samples taken soon after the animal comes into captivity may yield particularly valuable information about its origin and recent history – on what it has been feeding, for example. A suggested list for sampling such a bird of prey is shown in Box 8.9.

Many forensic samples, including some that would be classified as 'trace evidence', come from live animals (see Table 8.5). All, regardless of their identity or destination, must be handled with care and packed correctly (PAW, 2005).

Record-keeping is important. Laboratory reports should be attached to the paper records and/or added to the electronic files. The reference numbers must match.

Specific Examination

There are, of course, many organ systems that are important when examining forensic cases – for example, the integument, which is so frequently damaged when animals are attacked – but it is not possible to discuss more than a few of these in any detail in this chapter. There is a need for texts that are specifically orientated towards examination of live wild animals that have either (a) been injured as a result of human action or (b) themselves caused injury or death to humans.

> **BOX 8.9   SAMPLES FROM CONFISCATED BIRDS OF PREY**
>
> *In a clean bottle and kept at 4°C*
>
> - A fresh faecal sample (mixed)
> - A recently regurgitated pellet
> - Any dropped feathers
> - Blood in EDTA for haematology – initially in fridge, then to laboratory without delay
> - Blood for biochemistry (if sufficient)
> - Blood for DNA
> - Blood smears (at least two) fixed in alcohol
> - Ectoparasites (lice, hippoboscids, etc.) – alive if possible, otherwise in alcohol
>
> *All samples from birds of prey should be accompanied by information, including:*
>
> - History, if known
> - Weight at time of submission or seizure
> - Comments on the bird's condition, plumage, etc.
>   Beak – normal/long/worn/recently clipped?
>   Talons – normal/long/worn/recently clipped?
>   Soles of feet – evidence of wear, scabs or discoloration?
>   Feather condition
> - Evidence of drooping or abnormal wing
> - Evidence of lameness or favouring one leg
> - Legs – rings, jesses, bewits, changes in scalation, wounds, feather loss
> - Feet and cere – colour, damaged/normal
> - Behaviour – tameness, fear, screaming, etc.

Central Nervous System

A variety of clinical signs may be seen in CNS damage in animals. Investigation has advanced greatly in recent years, but usually necessitates the use of sophisticated equipment (especially imaging – see Chapter 8) and cautious interpretation. Often in wildlife the injuries prove fatal – in which case it is the pathologist who will investigate (see Chapter 9).

Eyes and Ophthalmoscopy

A thorough ophthalmic examination (Williams, 2012) of the animal is to be recommended in all forensic cases. Eyes are especially significant for two main reasons. The first of these is the importance of sight in most wild animals in terms of survival. For example, as pointed out in Chapter 8, raptors will only catch prey adequately when both eyes are intact – an important consideration when an injured bird of prey has to be released. On the other hand, snakes that have damaged eyes can often be rehabilitated successfully because they depend more on olfactory stimuli, using their Jacobson's organ, than on eyesight.

Table 8.5 Samples for Laboratory Analysis from Live Animals

Sample	Investigations	Comments
Faeces/vomitus, regurgitated material, pellets (birds of prey)	Parasitology. Microbiology. Chemical tests (e.g. for fat). Toxicology. DNA technology.	Often readily available. Stomach/crop contents can be obtained in some species by inducing emesis or lavage (see text).
Blood (or haemolymph – certain invertebrates)	Haematology. Clinical chemistry. Parasitology. Serology. DNA technology.	Many tests can be carried out on blood so long as sufficient is available and it has been taken and stored correctly. Limited investigations can also be performed on the haemolymph of invertebrates.
Urine/urates	Toxicology Microbiology Clinical chemistry Cytology DNA technology.	Many tests can be carried out on urine or urates so long as sufficient is available and it has been taken and stored correctly.
Pus	Microbiology Cytology.	Pus may be liquid or (e.g. reptiles) caseous. Pus-like material from any species should be sampled and examined. Both aerobic and anaerobic cultures are advisable.
Semen	Direct microscopy DNA technology.	Important in alleged bestiality cases.
Tissues (biopsies, scrapings, washings, brushings, etc.)	Histology (histopathology and histochemistry) Scanning electron-microscopy Transmission-electron microscopy Cytology Chemical analysis Toxicology Immunological tests (e.g. for detecting snake or scorpion venom) Radiography (plus/minus whole carcase) DNA technology.	Biopsies and other tissue samples can be used for a whole range of investigations that are applicable to forensic matters, such as alleged trauma, poisoning or spread of zoonoses.
Hair, feathers, scales	Low-power examination with hand-lens. Various other procedures as earlier.	Important samples because they may be shed naturally (trace evidence) or can be obtained readily, using minimally invasive techniques.

Source: Adapted from Cooper, J.E. and Cooper, M.E., *Introduction to Veterinary and Comparative Forensic Medicine*, Blackwell, Oxford, UK, 2007.

The second reason for stressing the value of ophthalmic examination relates to animals where non-accidental injury (NAI) is suspected.

It is widely recognised in human forensic examination of children who may have been subject to physical abuse (and, in particular, 'shaken baby syndrome' – see later), that retinal bleeding is an important sign (Morad et al., 2002; Riffenburgh and Sathyavagiswaran, 1991). A similar situation may pertain in animals where shaking can lead to retinal haemorrhage or even detachment (Serbanescu et al., 2008) although significantly more evidence needs to be provided before one can be sure that NAI in animals can be substantiated by retinal pathology.

The Importance of the Oral Cavity

Special attention is paid now to the examination of the oral cavity because of its forensic significance:

- Bite wounds caused by wild animals, especially mammals, are often a cause of injury, infection or death in humans and other species. The effect of a bite is influenced by both the morphology of the oral cavity and the microorganisms that it harbours.
- Orofacial trauma is sometimes a feature of abuse of animals, as it is of children.
- The oral cavity can provide other information pertinent to a legal case – for example, about an animal's diet or health status.
- Teeth can be examined in order to identify certain *species*, for example marine mammals (cetaceans), and sometimes to identify *individuals*. Teeth can be used to determine age (see also Chapter 9).
- Dental material often persists when other tissues have decomposed or been predated and can therefore remain a source of valuable forensic evidence.

Investigation of the teeth should be considered essential because much can be learned from them about the age, sex or health status of an animal (see 'Dentition of Mammals' section in Chapter 13). Dental examination can be carried out either in the live animal or *post mortem*. In this book much of the subject, with the exception of oral examination and bitemark analysis (both in the following), is covered in Chapter 9.

Examination of the Oral Cavity and Teeth

The principles of oral examination of a live animal for forensic purposes are essentially similar to those followed in routine veterinary practice, but the following points (Cooper and Cooper, 2007) are especially important:

- The mouth must be properly opened, for a reasonable period of time. This may be difficult because of species differences, especially those that hamper clinical examination or *post mortem* (see Chapter 9) because of the rapid development of rigor mortis.
- Meticulous records of observations must be kept. Use of a dental chart is advisable – a diagram of upper and lower jaws, with drawings of teeth, upon which comments can be superimposed. When investigating soft tissue changes (e.g. periodontitis) 'six-point charting', as used by (human) dental surgeons, can be helpful. Photography is important.
- The procedure must be carefully planned beforehand. The animal may need to be sedated or anaesthetised. Appropriate equipment (and assistance) is essential, especially if foreign material has to be collected, swabs taken or casts made.
- The assistance of a dental surgeon, or an odontologist, especially one with comparative interests, may be of value.

The detailed examination of teeth is an important part of the investigation, especially if there is a suspicion that the animal may have bitten humans, other animals or inanimate objects. In addition, however, as stressed earlier, the teeth can provide supplementary information (see 'Dentition of Mammals' section in Chapter 13). A dental assessment chart, similar to those used for domesticated animals in veterinary practice, can be easily devised in order to mark lesions and to record such points as absence or duplication of specific teeth.

Dental disease can be painful and a consideration in welfare cases. As William Shakespeare reminds us 'there was never yet philosopher That could endure the toothache patiently'.

Although the information above is orientated primarily towards work with mammals, in wildlife forensic investigations other taxa may also need to be examined. All vertebrates have an oral cavity and mammals, most reptiles, some amphibians and most fish have teeth. Extrapolation is often possible from mammals to other species, but there are also useful publications about oral anatomy and pathology of non-mammalian animals.

Damage to teeth may be a particularly significant finding in forensic cases. There are various types of fracture – in order of severity: enamel infraction, enamel fracture, crown fracture, crown-root fracture and root fracture. The last three can then be further divided into those fractures that are uncomplicated and those that are complicated. In captive wild animals – and sometimes those that are free-living but fed unnatural foods (e.g. monkeys around tourist hotels) – periodontal disease, ranging from gingivitis to advanced periodontitis, may be significant. Other changes of possible relevance to forensic cases include dental caries and various stages of tooth resorption. A specialist veterinary odontologist – for example, one holding the Diploma of the European Veterinary Dental College – can usefully be consulted.

Bitemarks

In human and animal forensic cases, bitemarks in living or non-living material at the scene of the crime may be valuable evidence. Bitemarks on skin are affected by distortion of the tissues during biting and by changes in the skin following a wound or after death. Therefore, critical observations must be made on the marks *in situ* and not after the lesions have been handled (live animals) or skin excised from the body (dead animals). As stressed by Cooper and Cooper (2007), a full photographic record is the first requisite, with a millimetre scale in the same plane as the marks. The bites often present better in ultraviolet or infra-red light.

An impression of the marks can be made with a rubber or silicone-base material that can flow over the skin without distorting it. Subsequent observations, such as comparison with the plaster casts of dentitions, can be made with a positive cast prepared from the impression, with the photographs as an additional check.

Samples of saliva may be taken from the bitemarks for DNA and other tests, but must be done *before* impressions are taken.

Lip-print identification (cheiloscopy) is not widely used in animal forensics but has potential. It has parallels with (e.g.) the well-established identification of gorillas using their unique noseprints (Figure 8.3).

Even if properly set, traps, nets and snares can cause injury, stress and pain to animals. The lesions they produce are discussed in Chapter 9.

In many countries traps of certain types can be used legally to catch pest species (see later). For example, in the United Kingdom there is an Open General Licence (WML/GEN L08), under the Wildlife and Countryside Act 1981 (as amended), which permits landowners, occupiers and other authorised persons to carry out a range of otherwise prohibited activities against certain species of wild birds.

In the context of trapping, the General Licence permits the use of a cage trap to take birds such as certain corvids. Amongst species that may be used as live decoys in cage traps are carrion crows and the monk parakeet. The Open General Licence stipulates that all relevant animal welfare legislation must be followed, including providing decoy birds with adequate food, water at all times, appropriate shelter and a suitable perch 'that does not cause discomfort to the birds' feet'. These requirements are defined in more detail in the Explanatory Notes to the licence: for example, 'adequate food' means 'sufficient, palatable food, which is of a type suitable for the decoy species, to meet the needs of bird(s)'. Water must be 'free from chemical additives and changed regularly to ensure that it is clean'.

The use of traps and snares is widespread in much of the world. Their deployment in many poorer countries is linked with a desire to obtain food (bushmeat – see Chapter 2), to protect crops and property, and/or to make money. One example cited earlier was the report by Hyeroba et al. (2011) describing the immobilisation and treatment of a speared chimpanzee in Kibale National Park, Uganda. Human-inflicted injuries are estimated to affect nearly 25% of the chimpanzee population in Uganda, the most common form of such injury being snaring (Stokes and Byrne, 2000; Waller, 1995). Snares or traps that are used are normally intended to catch other species such as small antelopes (Ohashi and Matsuzawa, 2010). A decision to intervene or not to intervene when a chimpanzee is injured depends upon on a number of issues such as whether the injury is human-inflicted or is a natural process (see 'Mountain Gorilla Disease – Implications for Conservation' section in Chapter 13), the severity of the injury, the risk that an intervention might present to the injured chimpanzee and field staff, the resources available and the attitude of local people.

These examples relate to Africa, but forensic investigators elsewhere in the world will encounter animals where detailed assessment of injuries is essential before a course of action commences. The veterinarian/investigator must accurately describe (count, measure) the injuries and should ensure that they are photographed, in colour, with a scale and unique reference number.

Although traps may kill their victim, a wild animal may be rescued from a snare or trap before death supervenes, in which case attention will need to be paid to treating or euthanasing the animal. In cases where clinical intervention is needed, this may assist the affected animal in terms of its health and welfare and can also provide important forensic information about the incident. For example, victims of shooting (see Chapter 9) may well need surgery; the removal of a bullet or shotgun pellets to treat the patient can often also yield evidence about the projectile and the animal's wounds that might be used in a trial. This, however, is not always an easy task; for instance, the position and location of shotgun pellets in an animal will relate to the type of weapon used and the distance from which it was fired.

Road-Traffic (Vehicular) Accidents

These are an important cause of injury and are discussed in Chapter 2.

Effects of Aircraft

Aircraft can have various effects on wildlife:

Disturbance: This can result in the dispersion of animals and deserting or neglecting young and eggs.

'Physiological' effects: These are poorly understood and generally referred to as 'shock' or 'stress'.

Physical: Collisions (see later), possible damage by shock waves and sound waves.

Secondary effects from some of these: For example, a tethered bird of prey may repeatedly 'bate' from its perch and damage itself.

Any of these – or a combination of them – may prove lethal or non-lethal.

Collisions between birds and aircraft have resulted in the instigation of bird control programmes on many large airfields throughout the world (PAW, 2005) (see 'Investigation of "Aircraft Wildlife Strike" Using Forensic Techniques' section in Chapter 13).

Wind-Farms

There is increasing concern about wind-farms and their possible adverse effects on wildlife (see 'Bats and Wind Turbines' section in Chapter 13). Martínez et al. (2010) studied the potential effect of 'wind-energy developments' planned in Spain in an area incorporating both golden eagle and Bonelli's eagle territories and concluded that some adverse interaction was likely. Long et al. (2010) investigated possible effects on bat mortality.

Abuse

The question of abuse (infliction of NAIs) is discussed here because it is not only dogs and cats that may be abused. As Table 8.6 shows, a wide range of taxa of animal, vertebrate or invertebrate, may be subject to such ill-treatment. This topic was introduced in Chapter 2 and is discussed here in more detail. Brief mention is made of types of lesions in Chapter 9.

Evidence that 'child abuse' might be transmitted down the generations in wild animals was claimed by Walker (2011) who reported studies on the Nazca booby, a South American seabird. In crowded nesting grounds, bouts of violence often take place. Visiting adult birds are aggressive to chicks and may try to copulate with them. Researchers Martina Müller and David Anderson found a strong correlation between the number of times that chicks had been attacked and the frequency with which they themselves were aggressive to other chicks when they reached adulthood. It was suggested that abuse earlier in life 'conditions' birds for life by raising levels of stress hormones that later trigger aggressive behaviour.

Table 8.6 Examples of Abuse of Captive and Free-Living Animals

Species	Type of Abuse	Comments
All	Physical assaults, e.g. beating with sticks, gouging of eyes. Cars may be driven at, and over, various species, either maliciously, or for 'sport'. Use of fire. Exposure to fireworks. Intentional starving, deprivation of water, neglect.	Care must always be taken in assessing suspected cases. For example, attempts to kill a pest species (e.g. a rat) or to euthanase a casualty may sometimes be mistakenly interpreted as abuse.
Mammals	Cutting-off tail, snipping patagium of wing (bats), pulling-out spines (hedgehogs), etc.	—
Birds	Pulling-off wings of young birds.	—
Reptiles	Disembowelling, cutting/pulling-off tail.	Predators, e.g. crows, may also cause disembowelling. Some lizards shed tail as a natural defence process (autotomy).
Amphibians	Disembowelling. Removal of/damage to limbs by cutting or treading on them. Placing toads in a fire to see them 'explode'.	As above.
Fish	Putting irritant chemicals in water.	—
Invertebrates	Various, including placing worms in a fire, pulling-off wings of moths.	Sometimes defended by perpetrators who claim that invertebrates cannot feel pain.

Source: Adapted from Cooper, J.E. and Cooper, M.E., *Introduction to Veterinary and Comparative Forensic Medicine*, Blackwell, Oxford, UK, 2007.

As recounted by Cooper and Cooper (2007), suspicion of the presence of NAIs in pet animals is usually prompted by such matters as unusual distribution or appearance of wounds, untreated lesions, an inconsistent history, suspicious behaviour on the part of the owner(s) and meekness of the animal. These may or may not, however, be a feature of abuse of wild animals.

Sexual abuse of domesticated animals by humans also takes place. Cooper and Cooper (2007) categorised sexual assaults under three headings – human/animal, animal/human and animal/animal. Any of these can involve wild animals, especially (but not exclusively) when in captivity. Those involved in forensic investigations of injured wildlife, especially those in zoos and private hands, should be aware of the possibility of sexual abuse, but should exercise caution and not jump to conclusions without careful investigation.

NAIs in animals are not necessarily linked with violence to humans (see, e.g. Piper [2004]). At a symposium held at the Royal College of Pathologists in January 2011, speakers from a wide range of disciplines (medical, veterinary, sociological) discussed this topic and examples given of where a link clearly did exist and others, some of which prompted draconian measures by the authorities, where no association could be detected between trauma inflicted on an animal and a propensity to cause harm to humans. In the authors' opinion, great care has to be taken, and caution exercised, in instances where there might, or might not, be such a link. In addition, pressure for veterinarians to report suspicious cases raises important dilemmas concerning client confidentiality – relevant to wildlife work as well is to domesticated animal practice.

Starvation

A simple, but eminently practicable, definition of starvation is the one used by Newton (1998), in the context of birds – 'loss of body condition to the point of death'.

The term 'inanition' has a slightly different meaning. Inanition is a condition resulting from lack of not just food but also water and/or a defect in assimilation of nutrients. Like starvation, it is characterised clinically by profound physiological and metabolic disturbances.

'Cachexia' implies general ill-health and malnutrition (see later), marked by body weakness, emaciation and secondary diseases.

Many wild species can tolerate long periods of markedly reduced food intake so long as they have had an adequate fluid intake. As Cooper and Cooper (2007) pointed out, reptiles can often survive weeks without food and spiders months. Some animals may refuse food or restrict its intake during times of, for example, courtship, egg-laying or when feeding their young.

Starvation is, however, a common cause of death in free-living wild animals and a major factor in population dynamics. For example, approximately 70% of birds of prey die in their first year of life, the majority of them on account of failure to obtain sufficient food (for references, see Cooper [2002]). Free-living wildlife may starve for a variety of reasons, often relating to unavailability of food but also competition and infectious disease. It may be seasonal. Birds, for example, tend to die of starvation in the winter on account of food shortage (Newton, 1998).

Charles Darwin and Alfred Russel Wallace drew attention to the part played by starvation in living organisms in their so-called struggle for existence. Both men were influenced in their thinking by the 'Theory of Population' put forward by Thomas Malthus (1766–1834), which anticipated the current problems of world food shortage. Malthus argued that disease, hunger and death due to starvation were 'nature's way' to control populations.

Limited food supplies will result in weight loss, sometimes leading to emaciation. In Northern climates, such emaciation may be the norm during the winter, and it may cause the death of substantial numbers of wildlife (see earlier, regarding birds). Animals in tropical areas may also undergo long periods of starvation, sometimes resulting in marked weight loss, during periods without rain, when food is scarce. Although these scenarios are generally well recognised, a definitive diagnosis of starvation/inanition in free-living animals can be difficult and necessitate interdisciplinary studies.

Notwithstanding the aforementioned points, starvation in wildlife can be a cause of legal action, particularly when it occurs in captive animals.

Animals that are alleged to have been starved may be examined/seen alive, necessitating clinical examination, or presented dead, requiring *post-mortem* examination. In view of the importance of the subject, starvation is considered both here, as a clinical matter, and in the succeeding chapter, under *post-mortem* examination.

Starvation and inanition can be due to a variety of causes, not only omitting to provide food.

Starvation in captive wild animals, whether fatal or not, can be the result of one or more of the following:

- Failure of the keeper to provide any food
- Failure of the keeper to provide sufficient food in terms of quantity and quality
- Failure of the animal to accept food, in part or in whole, even though food is being offered
- Failure of the animal to masticate and swallow food, even though food has been offered and, perhaps, taken initially
- Food not remaining within the stomach (on account of regurgitation or vomiting)

- Food not being adequately digested and absorbed in the intestinal tract (because of malabsorption, increased gut transit time, internal parasites, etc.)
- Food that has been ingested, digested and absorbed not being fully utilised because of metabolic or other disorders

As Cooper and Cooper (2007) pointed out, clinical examination of an animal in which starvation is suspected should incorporate the following:

- The animal must be weighed accurately and appropriate measurements taken (see earlier).
- A proper condition score should be used, rather than vague statements such as 'in poor condition' or 'clearly emaciated'.
- The condition score should be explained to make it comprehensible to a layman and, if appropriate, a reference to its use given.
- Appropriate comment should be made about clinical observations that might be relevant to an allegation or starvation – for example, 'poverty lines' on the sides of the tail of reptiles, 'hunger traces' on the feathers of birds (see Chapter 10).
- It must be made clear in any report as to whether a full clinical examination of the animal was carried out (the ideal situation) or, perhaps because of limitation of time or the circumstances (often relevant in work with wildlife), only certain points could be addressed. There are specific features that should be considered in any possible case of starvation (see later) and these should be enumerated and explained in the statement.

Important records in any circumstance where starvation of an animal is suspected or alleged are set out in Box 8.10.

Video-recording of the animal is helpful – in recumbency, if that is how it has been found, or while it is standing/walking.

If wild animals are seized, serial weight comparisons are useful evidence. One person should perform such measurements, using the same weighing scales. Variability in balances and scales can be significant when dealing with very small animals, such as passerine birds, reptiles and invertebrates, and may be a cause of contention in court.

BOX 8.10 SIGNS OF STARVATION—RECORDS

- Bodyweight (mass)
- Morphometrics
- Condition score
- Other features that may be relevant to condition and dietary intake (e.g. skin, hair, pressure sores)
- Assessment of subcutaneous fat, either manually or using ultrasound
- Assessment of internal body fat, as above
- Muscle mass and evidence of weakness or other locomotor deficiency
- Behavioural features that may add weight to or refute a diagnosis of starvation, for example marked hunger
- Change in bodyweight over time following provision of an adequate diet

Scoring systems for screening and assessing nutritional/metabolic status in humans are well established; some have been applied to domesticated and laboratory animals.

Care must be exercised in stating how long it takes for clinical signs of starvation to develop, especially if this is to be linked with allegations of 'cruelty' or 'suffering'. The food requirements of animals vary greatly and must be taken into account in any calculation. Endothermic species with a high metabolic rate (usually the smaller species) need to eat frequently and to ingest a substantial proportion of their bodyweight as food each day in order to maintain energy balance. Thus, a vole that has been deprived of food for 8h may be on the brink of hypoglycaemia and death, while a similar period of food deprivation will have little effect on a rabbit and probably none at all on a fox or a deer.

As Cooper and Cooper (2007) pointed out, publications on the speed of weight loss are few because such studies usually necessitate depriving animals of food and this raises legal and ethical questions (see Chapter 2). However, some work has been reported on the fasting of animals for veterinary or scientific purposes, for example prior to the administration of an anaesthetic agent or to study metabolic changes. There are rodent 'models' for the study of fasting on, for instance, the immune system.

The *post-mortem*, histological and metabolic effects of starvation are discussed in Chapter 9.

Dehydration

This may be concurrent with starvation or separate. It is discussed in Chapter 9.

Obesity

Obesity can be defined as being when accumulation of adipose tissue reaches a stage where there are pathological changes as a result (Cooper and Cooper, 2007).

In humans and domesticated animals these can include osteoarthritis, cardiovascular compromise, increased anaesthetic and surgical risk, increased risk of hyperthermia, diabetes mellitus and hepatic disease (lipidosis) – and similar clinical signs are seen in many overfed wild species, of different taxa.

An animal that is given or ingests too much food, or food of the wrong quality, or is under-exercised, can 'suffer' – which may result in legal proceedings. Criteria for assessing and scoring obesity are available for dogs and cats and for a few other species and it may be possible to adapt these to non-domesticated animals.

Overfeeding of some species can rapidly prove fatal – fish, for example. This is usually due to other effects such as hypoxia or toxicity, but can be grounds for legal actions.

Shock

This is an inadequate perfusion of the tissues and can be a sequel to many different insults. It is not always easy to differentiate *post mortem* from autolytic changes (Rutty, 2004).

Neglect

Defined in the Oxford Dictionary as 'a condition when a parent or guardian fails to provide minimal physical and emotional care for a child or other dependent person', neglect of wildlife is largely applicable only to captive animals. Neglect in such cases can be either intentional or accidental.

Examples of where neglect of an animal may be alleged to have occurred are summarised in Box 8.11.

Pollutants

Many pollutants can affect the health of humans, animals and plants. They are to be found in the air, in water and in terrestrial environments.

As Cooper and Cooper (2007) pointed out, insidious insults due to 'pollutants' can cause sub-clinical effects in individual animals or long-term sequelae in populations. In wildlife they include

- Toxic chemicals, such as lead from cartridges and anglers' weights
- Chemicals, such as soaps and detergents, that have an adverse effect on, for example, the plumage of birds but are not directly toxic
- Materials such as metal, wire and plastic that primarily cause physical damage or result in impaction and/or toxicity

---

BOX 8.11   CAUSES OF NEGLECT

Failure to

- Feed
- Provide water
- Clean a cage or enclosure
- Groom
- Keep at an appropriate temperature
- Provide environmental enrichment

Signs of neglect in wild animals may include

- Loss of weight/condition, leading to starvation
- Dehydration
- Soiling/matting of coat
- Myiasis
- Unclean environment

These were discussed in some detail by Cooper and Cooper (2007).

Pollution by oil has long been recognised as an environmental problem. Birds, turtles and marine mammals (see Chapter 2 and 'Marine Mammals: A Special Case' section in Chapter 13), so often the subject of public concern, are only the tip of the iceberg. Many other marine animals and plants are affected by oil, sometimes with long-term ecological effects. Oil contamination is one consideration in the relatively new field of 'environmental forensics' (Morrison, 1999; Murphy and Morrison, 2002) – see Chapter 2.

The effects of oil on birds have been well documented over the past few decades and will not be discussed here in any detail. An understanding of the pathogenesis of oiling is essential if one is dealing with live animals that have been affected by it. A summary of the pathological effects of oil on birds is given in Chapter 9.

Poisoning

As explained in Chapter 2, toxicology is the study of poisons and poisoning, intentional or accidental, and is an important part of wildlife work (PAW, 2005).

Poisoning is discussed elsewhere in this book in the context of types of wildlife investigation (Chapter 2), working with dead animals (Chapter 9) and dealing with samples (Chapter 10). It can occur on a large scale – for instance, the decline of vultures in India, Nepal and Pakistan on account of ingestion of diclofenac (see Chapter 2) and the intentional killing of pest species such as the use of organophosphate chemicals to control *Quelea* in Africa (see Chapter 2).

From a clinical point of view, as a rule of thumb, it can be useful to divide the clinical and other effects of poisons into two categories, based on whether they are chronic toxicants or acute toxicants.

Chronic poisons, such as arsenic, mercury and lead, generally produce a slow death, are relatively easy to detect in the body, even after decomposition or cremation, and (important in forensic cases) are often easy to acquire.

Acute poisons, on the other hand, often cause a rapid death, are capable of decomposing (thus not so easily detectable) and are often, not always, more difficult to acquire.

Hyperthermia

Overheating of captive animals can easily occur if shelter and housing are inadequate or if animals are overcrowded. Hyperthermia can occur as a result of chase and capture, especially when an animal is pursued or struggles and when the ambient temperature is high. Capture myopathy is often also a feature. Large, obese, animals are more susceptible to hyperthermia than are small, well-proportioned ones.

Over the past few years various reports have been published in veterinary journals and elsewhere about climate change and the effect that higher temperatures may have on the environment, on biodiversity, on the spread of infectious diseases and on the health of both domesticated and wild animals. A useful summary of current thinking was provided by Summers (2009).

Some species and scenarios predispose to hyperthermia – for example, a stranded marine mammal can rapidly become overheated if not kept cool. Ectothermic animals are particularly susceptible if they are unable to lower their body temperature by behavioural means (see earlier). In some reptiles with a low 'critical high' temperature, even modest heat can prove fatal. Amphibians and fish may die or show clinical signs of distress.

Heat stress can be acute or chronic. Acute effects can include panting or other signs of overheating, and hyperthermia may be detectable. Tachycardia is characteristic. In severe cases of sunburn, cutaneous erythema and other skin lesions are present, especially on unpigmented skin. The animal frequently is prostrate, in a stupor or coma and very dehydrated; it tends to seek cool or sheltered places. Birds show similar panting respiration and droop their wings. Chronic heat stress is more subtle in its effects and animals may show only poor growth and high embryonic mortality. Equine anhydrosis, an inability to sweat, occurs in some domesticated horses in the tropics and may affect wild equids.

Hyperthermia following exertion or capture appears to produce vascular collapse (shock), necrotic changes in the liver and other organs, and hyperkalaemia. Acute heat stress in domesticated animals can produce similar effects to burns but may, in addition, cause largely irreversible damage to kidneys, liver, heart and brain and changes in serum enzymes and electrolytes. Chronic heat stress (e.g. when cattle are taken from temperate countries to the tropics) depresses metabolism and production of milk and meat. High temperatures can affect spermatogenesis and produce faetal abnormalities.

The circumstances of overheating can be influenced by such factors as

- Relative humidity
- Air movement
- Dehydration
- Insulation by hair or feathers
- Stocking density/proximity to other animals

Wild animals, like their domesticated counterparts, have a 'critical high' temperature at which irreversible changes occur, especially in the brain. Death is the usual sequel. In any legal case where overheating is alleged, it is important to be familiar with this concept and to understand how different species thermoregulate.

Hypothermia

Extreme cold is a familiar danger to both free-living and captive wild animals. Many species inhabiting or passing through very cold or high altitude regions are adapted behaviourally and physiologically to such extremes; for instance, some hibernate or go through periods of torpor. Adverse weather can have significant effects on many species (see Chapter 9). Research on how weather can affect the behaviour and survival of animals has helped throw light on the circumstances under which morbidity and mortality may result from cold weather (see, e.g. Elkins [1983]).

Frostbite is frequently seen in captive animals when exposed to low temperatures and unable to move to a warmer location. Captive birds are particularly susceptible when kept in aviaries. Leopold, the 'father of game management', studied the effects of low temperatures, including ice and snow, on a number of species. Mortality can occur on a large scale as a result of the trapping underground of ground-roosting birds and rabbits by sleet or crusted snow.

Chilling is an important stressor, especially when animals are young, and can prove fatal. A sudden drop in temperature in the spring in temperate countries will kill large numbers of passerine birds. Storms can also have profound effects – see Chapter 9. Hypothermia in water can be exacerbated if the animal's feathers or hair are damaged.

Chilling will cause a drop in body temperature. Body metabolism is reduced and the animal becomes weak and unable to feed; hypoglycaemia ensues as carbohydrate reserves are exhausted. More severe exposure to cold can result in frostbite. In mild chilling the animal may appear cold to the touch, depressed in appearance and shiver; often no other clinical signs are present. In the case of freezing, the body surface initially feels cold and appears pale; later it may become erythematous and oedematous with the animal exhibiting pain. As with burns, pruritis may accompany healing. In severe cases the affected area becomes ischaemic and sloughs; this has been reported in free-living deer and opossums and in captive mammals and birds (Cooper, 1996).

Ectothermic animals may or may not show clinical signs or specific lesions as a result of cold. Much depends upon the species, its preferred temperature range and the circumstances of the exposure. A few ectothermic species produce cryoprotectant chemicals.

Burning

Burns can be caused by ultraviolet light, heat, chemicals or friction. Sunburn is due to overexposure of the skin to ultraviolet rays from the sun.

Bush and grass fires are a common cause of death from burning and suffocation in wild animals (Figure 8.11). The effects on wildlife have been studied by many authors – see,

Figure 8.11 Bush fires can kill a range of wild species, vertebrate and invertebrate.

for example, the paper by Harty et al. (1991) on direct mortality and reappearance of small mammals in grassland after a prescribed burn.

Over the past 20 years, uncontrolled fires across Indonesia and other countries have had devastating effects on wildlife, human health, climate and economies (Harrison et al., 2009) and, as they usually are started by humans, are likely to lead to legal action in due course.

Chemical burns can occur as a result of contact with either man-made or natural areas of acid or alkali – for example, flamingoes in the alkaline lakes of East Africa sometimes accumulate deposits of soda on their feet (Cooper and Cooper, 2007).

Burning by friction can be caused by snares or nets or, in captive animals, by a leash or collar that abrades the skin.

The clinical effects of burning are associated with tissue damage – see Chapter 9. Burns vary in severity. First-degree burns involve only the outer layer of the epidermis and cause pain, heat and oedema. Second-degree burns affect the whole epidermis and vesicles are characteristic. In third-degree (full-thickness) burns both epidermis and dermis are destroyed. There is severe tissue damage, and because nerve endings are lost, pain is not a major feature. There may be dehydration and the animal is lethargic and depressed as a result of electrolyte imbalance and circulatory shock. Inhalation of hot smoke causes dyspnoea, coughing, changes in vocalisation, expectoration of blood and sometimes carbon particles.

Explosions

Wild animals can be injured and/or killed in explosions and these may be accidental or illegal or, in the case of land mines, detonate them and damage humans, livestock or property (Cooper and Cooper, 2007).

SEM examination of synthetic material following exposure to an explosion may demonstrate characteristic 'clubbing' of nylon fibres: similar changes may be seen on birds' feathers exposed to heat (Cooper, 2002a).

Electrocution and Drowning

These are both often fatal and therefore covered in some detail in Chapter 9. When the animal is still alive, clinical signs may assist in diagnosis and these are discussed in the following.

Electrocution

Captive wild animals are usually electrocuted as a result of exposure to electrical circuits and free-living animals on account of contact with power lines. Lightning can sometimes be significant. Power lines are of more importance than lightning to wild birds and can cause physical damage due to collision as well as electrocution (Mañosa, 2001).

Electrocuted animals may be comatose, paralysed or show involuntary muscle contractions. There may be apnoea, ventricular fibrillation, dehydration, burns, haemorrhages and even fractures.

Drowning

Drowning is the inhalation of water in sufficient amounts to cause severe injury or death.

Whether a wild animal survives drowning or not hinges upon three factors – the duration of submersion, the water temperature and the speed of resuscitation. Young animals are generally more tolerant of submersion in water than adults, aquatic species more than terrestrial and ectotherms more than endotherms. The diving reflex of some mammals occurs in cold water and can permit survival after prolonged submersion; bradycardia occurs and oxygenated blood is shunted to the heart and brain.

The integument of immersed or drowning animals is usually wet and there may be water in the buccal or nasal cavities. If drying has occurred, the hair or feathers are matted. If water has been inhaled, respiratory signs predominate and careful examination and manipulation of the animal may reveal fluid sounds.

Ionising Radiation

Wild animals may be exposed to natural or anthropogenic sources of radiation. The effect of the radiation will depend on the total dose received, dose rate and organs and area affected. As a general rule, wholebody doses of 200 rad or more are likely to prove fatal; however, species vary in susceptibility. At doses of over 50 rad, many cells die and the multiplication of other cells is inhibited.

Within 1–3 weeks of exposure to radiation, an affected animal is likely to show leucopoenia, thrombocytopoenia and anaemia. Bacterial infection is a common sequela. The epithelium of the intestine is affected. The germinal cells of testes or ovaries may be badly damaged, producing infertility and possibly genetic changes. Fetal tissues can be similarly affected. The nervous system is usually only slightly damaged unless exposed to massive doses (2,500–12,000 rad).

The lesions induced by radiation depend on the dose and the area affected and may include petechial haemorrhages, enteritis (resulting in diarrhoea and severe loss of fluids and electrolytes), aplasia of lymphoid tissue and bone marrow destruction. Clinical signs depend upon the length of exposure and the dosage. In acute radiation syndrome, immediate clinical signs may be slight, but vomiting usually occurs within minutes or hours of exposure. Later, cellular changes begin to take effect and such signs as diarrhoea, anaemia, petechiation, increased susceptibility to infection and alopecia occur.

More chronic exposure to radiation produces fewer clinical signs, but reduced fertility, anaemia, leucopoenia, alopecia and neoplasia are well recognised in humans and in some species of animal.

As Cooper and Cooper (2007) pointed out, legal action relating to irradiation is likely to be brought on such grounds as

- Incorrect or unauthorised use of x-rays – routine radiography or fluoroscopy on a captive wild animal (see earlier)
- Radiation leakage from a power station or laboratory
- Exposure of animals to 'natural' radiation in the soil, air or water

Working with Dead Animals

9

JOHN E. COOPER

Contents

Introduction	238
Aims of the Necropsy	239
Before *Post-Mortem* Examination Commences	240
Determination of Death	240
When Should the *Post-Mortem* Examination Be Performed?	241
Chain of Custody	246
Personnel	247
Performing the *Post-Mortem* Examination	248
Stages of a *Post-Mortem* Examination	249
Background History and Confirming the Identity/Recording the Provenance of the Specimen	251
Receipt of Carcases	251
Record-Keeping	252
Health and Safety	252
Post-Mortem Technique	254
Mistakes and Omissions in Necropsy	255
Full or Partial Necropsy?	257
Skinning	258
Dissection Techniques	259
Group Examinations	261
Gross (Macroscopical) Examination of Material	262
Identification of Whole Animals or Their Parts	264
Differentiation of Human and Animal Material	266
Description and Recognition of Animals	267
Morphometrics	268
Weighing of Animals and Organs	268
Organ Size and *Post-Mortem* Change	270
Gender Determination (Determination of Sex)	270
Determination of Age of Animal at Death	271
Post-Mortem Imaging	272
Laboratory Investigations	273
Histological Examination	275
Water-Testing	275
Questions That May Need to Be Answered by the Necropsy	275
Post-Mortem Change and Estimation of Time of Death	275
Description, Interpretation and Ageing (Age Determination) of Lesions	281
Cause of Death	284

Circumstances of Death	286
Assessment of Reproductive Activity	287
Sudden or Unexpected Death	287
Making a Diagnosis	288
Special Investigations Necessitating Necropsy and/or Dissection	288
Examination of the Gastrointestinal Tract (GIT)	288
Presence of Fat	289
Examination of Teeth	289
Poisoning	291
Some Specific *Post-Mortem* Findings and Considerations	293
Starvation	293
Dehydration	294
Hyperthermia and Hypothermia	295
Drowning	296
Shock	297
Traumatic and Allied Injuries	297
Collisions, Road-Traffic (Vehicular) Accidents and Aircraft Incidents	300
Traps, Nets and Snares	301
Wounds from Weapons	301
Firearms and Unexploded Ordnance	307
Injuries due to Strangulation and Hanging	308
Asphyxia	309
Trauma to the Central Nervous System	310
Transportation Injuries	310
Electrocution	310
Burns	311
Bones and Skeletal Lesions	311
Retention of Tissues	320
Museums and Reference Collections	321
Conclusions	323

> 'How long, O lion, hast thou fleshless lain?
> What rapt thy fierce and thirsty eyes away?
> First came the vulture: worms, heat, wind, and rain
> ....
> The thunders of thy throat, which erst were wont
> To scare the desert, are no longer there;
> Thy claws remain, but worms, wind, rain, and heat
> Have sifted out the substance of thy feet'.
>
> **Charles Tennyson Turner**

Introduction

The importance of *post-mortem* (necropsy) examination in forensic work cannot be overstated. Dead wild animals, like dead humans, attract interest and concern. Whatever the reasons for the death ('natural causes', attack by another animal, traffic accident, assault by a human, etc.), there may be enquiries, and some cases lead to legal action of one kind or another. Often it is members of the public who report the finding of a carcase. This has to be borne in mind when dealing with such cases.

A well-performed *post-mortem* examination can also yield data on morphometrics, organ weights and organ/body weight ratios, and on the gross and histological appearance of tissues. Quite apart from their evidential use, such data can be of great value in research on wildlife species. Every effort should be taken, therefore, to record relevant information and, when circumstances permit (not always easy in legal cases), to collect material for study and future reference. However, priorities have to be considered. While a full necropsy would normally include some routine recording, such as weight and certain measurements, extensive collection of information on morphometrics and (e.g.) pelage may be counter-productive in cases where these data are of little or no direct relevance to the main objective of the forensic necropsy (cause of death/time of death, etc.). This is why the pathologist must first ask 'why is this *post-mortem* examination being performed?' (see later). In the case of rare or endangered species, especially when time is short, a balance may have to be struck. It may be worthwhile inviting a zoologist with knowledge of the species that is being examined to attend the necropsy, to advise on any special morphological features and to record biological ('non-pathological') data. Even if such attendance is not possible, the pathologist should consider contacting a suitably qualified person before or during the *post-mortem* examination in order to ascertain which 'zoological' data should, if possible, be recorded.

It is because necropsies are such a significant part of wildlife forensic work that they need to be carried out well, by an appropriately experienced person. Experience of, and training in, such work can be most important. The particular features of the forensic *post-mortem* examination were described and illustrated by Munro and Munro (2011), and they advocated that such investigations should be performed by a specialist pathologist, not even a veterinary practitioner. Wobeser (1996) also argued that the investigation of wildlife crime where dead animals or their derivatives have to be examined should involve the services of a specialist veterinary pathologist. It is certainly the authors' experience that biologists doing necropsies on wildlife are inclined to be simplistic in their interpretations – for example, sometimes declaring the presence of infectious disease on gross findings alone.

Notwithstanding this, there are occasions when an initial *post-mortem* examination of a wild animal has to be performed by someone who is less experienced or trained than is desirable; this is particularly the case under field conditions, when time is priority because of climatic or security factors (see Chapter 7). This situation as discussed under 'Personnel' later in this chapter.

Aims of the Necropsy

Post-mortem examination (necropsy) of a dead wild animal or its parts is an important component of wildlife forensic work. A necropsy can provide information on the following:

- Causes of death
- Causes of ill health
- Underlying abnormalities/pathology
- Specific points, such as the sex and age of a raptor or the reproductive status of a red deer

In most cases, a full necropsy is essential. Occasionally, an incomplete examination is necessary, either for legal reasons or because the rarity of the species means that the carcase is required for research or display. This aspect is discussed later in the chapter.

The necropsy may not be complete for other reasons – in particular, because in wildlife work portions of a carcase may be found or only certain derivatives are submitted for examination.

Before *Post-Mortem* Examination Commences

Important questions should be asked before embarking on a *post-mortem* examination of any wild animal:

- Do I have permission to carry out this examination?
- Do I know enough about the species to be confident to perform this necropsy?
- Is it clear as to what questions I am being asked?

If necessary, because of doubts about the aforementioned, the *post-mortem* examination should be delayed.

Determination of Death

A consideration in wildlife forensic necropsy work is whether the animal destined for necropsy is indeed dead. This subject cannot be discussed in detail in here, but the investigator should be aware that (1) some animals can go into a state of hypothermia and hypometabolism, during hibernation for example (Heldmaier and Rug, 1992; Osborne and Milsom, 1993), and (2) ectothermic species (reptiles, amphibians, fish and invertebrates) can appear dead because of low ambient temperatures and consequently (as in hibernating animals) a lowered metabolic rate.

Traditional indicators of death, as used in human and veterinary medicine, are useful but not always totally reliable and even they can be challenged in court. For example, it is generally true to state that in a carcase the heart has stopped beating and therefore the animal cannot bleed. However, ectothermic species that are 'dead' may still have a beating heart (Frye, 1999), which means that they can bleed, and even in a mammal or bird, blood sometimes comes out of a haematoma or a well-vascularised wound, especially if the animal also had a clotting defect.

Remarkably little appears to have been published about determination of death in ectothermic species. An exception was the work of Frye (1999), referred to earlier, who emphasised that even when 'clinically dead', these animals can exhibit residual cardiac contractions. Frye outlined key '*post-mortem* indicators' including the presence or absence of blood flow or myocardial contractions (using ultrasonic Doppler devices and electrocardiography), the ability of skeletal muscle to respond to galvanic stimulation, intra-ocular pressure and deep-core body temperatures.

Table 9.5 (continued)   Equipment for *Post-Mortem* Examination of Wildlife Species

Species	Essential	Useful Additions	Special Precautions	Comments
Invertebrates	As above (all). Micro-instruments (e.g. ophthalmic) are often needed. Examination may necessitate the use of dissecting microscope or magnifying loop. Some aquatic species need to be kept damp or even immersed in saline or water (to prevent desiccation or collapse) during examination.	As above (all).	As above (all).	As above (all). Care must be taken with toxic species. The small size of many invertebrates can make detailed examination difficult. Fixation in *toto* may be advisable, followed by histological examination.

Source: Adapted from Cooper, J.E. and Cooper, M.E., *Introduction to Veterinary and Comparative Forensic Medicine*, Blackwell, Oxford, UK, 2007.

Background History and Confirming the Identity/ Recording the Provenance of the Specimen

The background history will consist mainly of information provided on the submission form (see Appendix D), together with any additional data provided by the person or organisation requesting the necropsy.

The general guidance regarding background information about a *post-mortem* case is as follows:

- Always try to obtain some history before embarking on the necropsy – however brief, inadequate or unreliable this may appear to be.
- Heed the history but do not rely upon it.
- When you have the opportunity to take notes about the history yourself, do so – rather than depending on others who may have a less objective approach to the case.

Important features of a good history include

- Species of animal
- Correct dates and times of incidents or findings: it can be very significant whether an animal was found sick or dead early in the morning or later in the day
- The circumstances regarding the animal(s)
- Photographs or drawings

Receipt of Carcases

Strict hygienic precautions must always be followed when handling and unpacking carcases, regardless of the history, because of the possible transfer of pathogenic organisms or poisons (see Appendix F). Gloves should always be worn and full protective clothing, including goggles and facemask, should be available. This and other procedures will be influenced by the relevant risk assessment.

Material submitted for forensic *post-mortem* examination should arrive packed in suitable containers, appropriately sealed and labelled. Insulation for the journey can be provided by wrapping the sealed package (the carcase wrapped in two plastic bags) in layers of newspaper. Any paperwork accompanying the body (e.g. information on history) should have been wrapped and sealed separately and not enclosed in the same plastic bag(s) or container as the carcase.

Wherever possible, dead animals should be delivered to the laboratory by hand. Whatever the means of delivery, the relevant legislation and regulations should be followed (see Chapter 3 and Appendix I).

Record-Keeping

Carcases should be labelled (Figure 9.3). Standard *post-mortem* forms are available for most domesticated species and have also been designed for some wild animals such as reptiles, birds of prey, eggs and embryos and marine mammals (Cooper, 2002a; Kuiken and Hartman, 1992; see Appendix D).

Although standard necropsy forms are useful, they usually concentrate upon routine diagnostic work. In many forensic cases, more specific questions are likely to be asked; there is merit in compiling a specific form which can subsequently be used as the basis for a report.

A specimen *post-mortem* form is given in Appendix D, together with a model draft report.

Health and Safety

Health and safety are important during all *post-mortem* examinations and, in the case of those of a forensic nature, also help to ensure that systems are in place which will minimise

Figure 9.3 Scarlet ibis, subject of a court case, received for laboratory examination. Appropriate tags or other methods of identifying the specimen should be attached to the carcase, not just the outer wrappings (Courtesy of Ravi Seebaransingh).

Skinning of the carcase is usually an important part of a forensic necropsy, in order to detect injury to the integument and superficial tissues. Most pathologists tend to leave removal of the skin until after the internal examination is completed, but Wobeser (1996) advocated skinning the entire animal at the outset. Ultraviolet light may assist at this stage, especially if photographs are to be taken (Barsley et al., 1990).

There are several reasons for skinning, of which the following are perhaps the most pertinent:

1. Haemorrhage and/or oedema may be detected, both often indicative of traumatic injury.
2. Wounds due to traps and snares (see later) are often more readily detected in the deeper (dermal) tissues than on the furred, feathered or scaled surface. Snare injuries in mammals are usually especially common on the limbs.
3. Bullets and shot that fail to exit the body often become lodged under the skin. The presence of such projectiles should be suspected if subcutaneous haemorrhage is present with no skin perforation is found on the side of the body apposite a puncture wound (Wobeser, 1996).
4. Skinning permits assessment of subcutaneous fat and the appearance of the musculature.

Dissection Techniques

Useful guidance on certain aspects of forensic *post-mortem* examination can of course be obtained from medical texts (see, e.g. Burton and Rutty, 2001) but animals, especially the rich variety of species that come under the heading of 'wildlife', present many specific challenges (Cooper, 2003). There are many texts available describing *post-mortem* techniques for domesticated species, and certain 'exotic' species and a few relate specifically to forensic cases (see, e.g. Munro, 1998 and Wobeser, 1996). Necropsy methods suitable for animals ranging from monkeys (primates) to millipedes (invertebrates) were reviewed by Cooper and Cooper (2007). Subsequent to that, Cooper (2008) published a paper on forensic *post-mortem* techniques in the herpetological field, drawing attention to the special features of reptiles and amphibians.

Post-Mortem Procedures for Wildlife Veterinarians and Field Biologists by Woodford et al. (2000) provides useful general information for wildlife veterinarians and biologists who work in the field. One omission in that book, which might usefully be rectified in future editions, was the mention of the importance of saving material, fixed and fresh, for DNA and museum studies.

Nevertheless, necropsy methods do not differ substantially from species to species in terms of the need for standard protocols, appropriate equipment and proper record-keeping. In this context, regardless of the species involved, all dissection must be carried out with sensitivity and care. For example, incisions should be made cautiously, especially in cases where there is already traumatic injury or, perhaps, a ruptured aneurysm is suspected and therefore rough handling may damage important evidence. Throughout, a forensic necropsy one must remember that follow-up examinations may be necessary either by the pathologist himself or herself or by someone else called by the prosecution or defence.

There are various specialised *post-mortem* techniques that can – and should – be applied to forensic investigations. For example, examination of the hooves of ruminants is difficult because deep, concealed, lesions are protected and hidden by the overlying horn shoe. The latter can be removed by immersion in hot water at 60°C–65°C for 15–20 min (small ruminants) and 60 min for the hooves of equids. Once the outer, impervious, horn shoes have been removed, the underlying tissues can be fixed and usually provide good-quality samples for histology (Ossent and Lischer, 1997).

Small specimens such as certain fish, amphibians, embryos, eggs, fetuses and neonates need special care. Neonatal pathology is an increasingly well-developed speciality in veterinary and laboratory animal work. Examination of such material ('mini-necropsy') necessitates the use of special small instruments, such as those used by ophthalmologists.

Dead aquatic animals, especially fish and amphibians, warrant immediate necropsy as such species tend to autolyse rapidly. If available, some dying but live specimens should also be taken as these will often yield more reliable laboratory results. Often it is wise to carry out some initial necropsies in the field and to take appropriate samples. Carcases that need to be taken transported to the laboratory must be kept cool and well-wrapped. Samples of water should also be taken and subjected to a range of tests as appropriate, including standard water testing (see later).

Carcases of wildlife submitted for examination have sometimes been exposed to conditions that lead to desiccation and dry preservation. In some cases the animal's morphology may have become distorted (Figure 9.4). Truly mummified remains often attract an unusual and rather restricted range of invertebrates (Bourel et al., 2000).

Techniques for the examination of dried, desiccated and mummified carcases are poorly-documented in the veterinary and zoological literature, but some of the methods used to examine human mummies are applicable to the desiccated remains of animals,

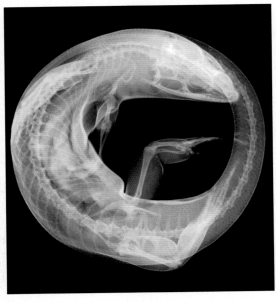

Figure 9.4 Whole body radiograph of a free-living Round Island skink (in a truly round posture!) found dead, dried by the sun, pending rehydration for necropsy and histological examination.

> **BOX 9.5   INVESTIGATION OF MUMMIFIED CARCASES**
>
> - Careful examination of the environment where the animal was found, including measurements of ambient temperature, relative humidity and airflow
> - Equally careful examination of the exterior of the carcase, to include any newspaper or substrate that is attached to it
> - Collection of skin/hair samples for toxicological and microbiological/parasitological investigations
> - Rehydration in saline for 72–96 h of selected portions of skin and internal soft tissues prior to their being fixed, embedded and sectioned for histological study
> - Microbiological culture of selected tissues, either rehydrated or ground with a pestle and mortar
> - Meticulous examination of bones (see 'Miscellaneous and Techniques' section in Appendix G)

especially the precautions that have to be taken both to conserve the specimen and to protect those involved in the dissection (Gill-Frerking et al., 2011). Advice can also be sought from museum personnel who often have to 'resurrect' dry and apparently featureless material in pots and drawers (see 'The Preparation and Fixation of Specimens for Museum Display' section in Appendix G).

Forensic examination of a mummified carcase may be requested because it has been suggested that an animal found in such a state is likely to have died under suspicious circumstances (often not the case!). Correct investigation of such material is important, nevertheless, and can be carried out as depicted in Box 9.5.

There is scope for enterprising students of comparative forensic medicine to examine the many thousands of mummified animals in Egypt and in the world's museums. Following radiography of such specimens, minimally invasive endoscopic examination is often a wise step. This both minimises damage to the material and reduces exposure of the prosector to possible pathogens – see Appendix F and paper by Irvin et al. (1972) on the possible health hazards associated with the collection and handling of *post-mortem* material.

A number of *post-mortem* techniques can likewise be applied to a mounted specimen or a study skin – hair or feather investigation, for example.

Group Examinations

In a forensic case involving large numbers of animals – for example, flamingos found dead in hundreds alongside an alkaline lake in East Africa – it may be impossible to examine all the animals that are available for necropsy. Under such circumstances it is usually acceptable to take and to examine a *sample* of them. It is important, however, to record how the selection was made, in order to counter any criticisms later. Sometimes a random sample is taken, at other times a chosen mixture of, for example, severely affected and apparently normal specimens. The number of animals sampled is also an important consideration. As a general rule, the larger the sample size the better (i.e. the greater the chance of making a diagnosis, or detecting a pathological change), but this has to be balanced against such factors as time, cost, practicalities and the possibility that another expert may want to examine the material. Nevertheless, regardless of the number of carcases sampled at the time, as many others as possible should be stored, at least for a few days – to be made available to other experts, if required, or (if new questions or theories arise) to be examined later by the original pathologist.

An alternative approach to large numbers of wild animals is to use more than one protocol so that some carcases are subjected to full necropsy while others are only partly examined, usually with particular attention being paid to target organs and a search for specific lesions. This formed the basis of a study of dead flamingos in Kenya; three different protocols were used: (1) Full/Comprehensive, (2) Fast track and (3) Super fast track (Cooper, Deacon and Nyariki, in preparation).

When examining large numbers of dead (or live) animals, there is merit in recording the main findings on a special 'group or flock' form (see Appendix D). This enables findings in different animals to be compared and contrasted rather more readily than if individual forms are used. The data can be hand-written or prepared as a computer spreadsheet.

In some forensic investigations, where large numbers of carcases are available, it is useful, possibly essential, to estimate how many animals have died or been killed. This is often the case, for instance, when investigating seabirds found dead on the beach or mudflats. They may have died as a result of natural factors (e.g. food shortages, algal toxins) or the deaths may be anthropogenic (e.g. due to oil pollution). The usual approach is to calculate the extent of the mortality event on the basis of counting numbers of carcases. This, however, is not always an easy task. It can involve painstaking search by humans, including volunteers, or (when on a large, extensive, scale) the use of dogs or helicopters. Various publications have addressed the question. For example, Van Pelt and Piatt (1995) argued that using counts of beached bird carcases in order to estimate total mortality is an uncertain science because (a) only a fraction of birds that die at sea are washed ashore, (b) comprehensive or repetitive beach surveys are rarely logistically feasible and (c) birds tend to be cast-up on, and removed from, surveyed beaches at variable rates because of changing deposition patterns and on account of scavenging or physical processes that remove carcases. As a result, many researchers have called for better estimates of persistence (Philibert and Clark, 1993). Van Pelt and Piatt (1995) monitored the deposition and subsequent disappearance of 398 beach-cast guillemot carcases over a 100 day period. Scavenging appeared to be the primary cause of carcase removal, followed by burial in beach debris and sand. They presented an equation that estimates the number of carcases deposited at time zero from beach surveys conducted some time later, using non-linear persistence rates that are a function of time.

Gross (Macroscopical) Examination of Material

Gross examination is of great importance in wildlife forensic work. Particularly if linked with light microscopy (see Webb, Chapter 12), it offers an inexpensive and simple way to identify, to sex and to age certain species. In addition, it often yields sound, scientifically based, evidence relating to such issues as cause and circumstances of ill-health or death and the provenance of animals and their derivatives. Gross examination may answer questions that cannot be resolved using DNA technology – for example, how long an animal has been dead, the sequence of events leading to death, whether it has been moved from one location to another and the contacts it may have had *ante mortem* with the same or other species. The examination may also throw light on the health status of the animal and whether or not it suffered pain or distress during life (see later). An example of its value is in marine mammal work, when looking for gross evidence of human-induced mortality in small cetaceans (Read and Murray, 2000).

Gross examination, especially if coupled with histological investigation, may provide useful date as to whether an animal may have been in pain during life – for example, pressure on nerves or tension on peritoneum. Gregory (2011) explored the relationship between pathology and pain severity in humans and animals. He emphasised that pain can be extremely difficult to recognise in animals if based only on behavioural signs. He listed the wide range of conditions known to cause pain in people and suggested that similar effects must occur in animals. In humans, pain is usually assessed using some form of visual analogue scale. However, developments in MRI may help to establish quantitative relationships between pathology and pain.

Gross examination clearly plays a key part in forensic pathology. However, it is vital to know the normal appearances of the various organs, and this can be difficult in wildlife work as species vary substantially. Experience is needed. The importance of recognising abnormality from normality – and also of having an enquiring mind about pathological changes – is illustrated in the writings of John Hunter, who on 9 February 1760, wrote of a human autopsy:

> The liver was somewhat contracted, and harder than common, and very irregular, both on its surface and internal structure …. I took a piece of the liver home with me, to try some experiments with it.

A strong light is of value in gross *post-mortem* work. Tissues should be examined using both reflected and transmitted light. Transillumination needs a strong focussed source of light, such as an operating light, a torch (flashlight) or an endoscope. In the tropics, the sun provides an excellent source of illumination (see Chapter 7).

Description is the mainstay of gross pathology and provides the basis of a *morphological* diagnosis, such as polycystic kidneys, as opposed to an *aetiological* diagnosis – which often consists of two words, for example staphylococcal dermatitis, and generally depends upon supporting laboratory tests.

Not all causes of death manifest themselves as gross *post-mortem* findings. Often laboratory investigations (histology, bacteriology, toxicology, etc.) will be needed. Examples of diseases in which no specific macroscopical lesions are likely to be visible include rabies and tetanus. In some conditions, for example electrocution, lesions may or may not be detected, depending upon the circumstances. In a few cases (e.g. certain toxicoses) the diagnosis may even be made as a result of noting a characteristic odour rather than observing gross lesions (see Poisoning, later).

Identification of Whole Animals or Their Parts

The identification of the remains of whole animals is generally facilitated if a whole range of photographs is taken at the outset, before handling or disrupting the material in any detail. This is particularly true of autolysed, badly bloated or predated problematic species, such as marine mammals. Fish and aquatic amphibians usually decompose very quickly after death.

Derivatives, such as the antlers of deer or the skin of a badger may be recognisable morphologically. Or, more precisely, they can often be linked with the carcase from which they originated, using DNA-profiling. However, as Linacre (2009) points out (see later), the success of this approach depends upon the quality of the samples. This is why it is wise, wherever possible, to use other, non-molecular techniques, in addition to DNA-profiling, when this is feasible (see later). Linacre stated

> The unambiguous identification of species can be determined by a number of tests depending on the sample type. If there is gross morphology, then visual identification may be possible and sufficient. This might be for eggs or protected birds, turtle shells of protected species or skins from protected big cats or bears. If visual inspection is not possible due to the incomplete nature of the sample, but hairs or feathers are available, then microscopic inspection is an option.

Photography, essential regardless of the method of identification being used, should be followed by collection of morphometric data and the collection of material for reference.

The various methods of identifying animal products are summarised in Table 9.6. It should not be assumed that only one method can be used to identify derivatives of animals.

Table 9.6 Identification of Animal Products

Method	Applicability	Comments
Gross examination	Some animal skins, horns, bones, eggs, feathers, wings, elytra, etc. of invertebrates, some tissues and meats. Depends upon characteristic markings, often with comparison with known reference material.	Cheap, applicable in countries and circumstances where more sophisticated techniques are impossible.
Microscopic examination (hand-lens, dissecting/ light microscope)	Hairs, feathers, scales, pieces of bone or eggshell, invertebrates, some tissues and meats.	As above. Microscopy is always wise initially, regardless of what other tests are planned.
Scanning electron-microscopy (SEM)	Some of the above.	Detailed SEM can reveal important taxonomic features.
Electrophoresis including gel diffusion	Some tissues, including meats.	Species identification.
Karyotyping	Various tissues. Many species: species identification, sexing, detection of chromosomal abnormalities.	Species determination and determination of sex, requires knowledge of normal chromosome patterns.
DNA techniques (various)	Blood, saliva, other body fluids or tissues. Identification of species and individuals, parentage, sexing.	Samples must be handled and stored well, to prevent damage and contamination (see Chapter 11).

Source: Adapted from Cooper, J.E. and Cooper, M.E., *Introduction to Veterinary and Comparative Forensic Medicine*, Blackwell, Oxford, UK, 2007.

Often in a court case it strengthens the argument if a number of techniques have been applied. For example, in the case of *shahtoosh* (the wool of the Tibetan antelope), both microscopical and DNA techniques can be used (PAW, 2005).

The identification of bird remains can be important evidence in forensic work, including the determination of avian species involved in aircraft accidents ('bird strikes') (see 'Investigation of 'Aircraft Wildlife Strike' Using Forensic Techniques' section in Chapter 13). Prast and Shamoun (1997), from the University of Amsterdam and Tel Aviv University, respectively, compiled a CD-ROM in close co-operation with both the Royal Netherlands Air Force and the Israeli Air Force. This includes detailed descriptions of the feather barbule structure for each species, accompanied by light microscope images of the structure of plumulaceous (downy) barbs at magnifications between ×40 and ×400, as well as scanning electron-microscope (SEM) photographs made at ×500 and ×1000 magnification.

Familiarity with comparative anatomy can be of great help in distinguishing species, sexes and age groups on gross morphological grounds. This, though, may need to be supplemented with a thorough search of the literature, if necessary going back many years. Thus, for examination of the larynx, the seminal work on the comparative anatomy of this structure by Negus (1929) is invaluable. Microscopical descriptions of the normal appearance of even common species of animal are often lacking. Thus, for instance, Liping et al. (1989) studied the histological structure of the digestive organs of the ring-necked pheasant and compared it with that of the chicken and quail.

There is a need for many more such publications, covering other structures and a wider range of species, but basic anatomical and histological studies are nowadays often pushed to one side, in the mistaken belief that only molecular research is of relevance to twenty-first century forensic investigations.

There are some problems in gross examination, however. For example, species identification can be complicated by geographic variation. Busack and Pandya (2001) investigated this in *Caiman crocodilus* and *Caiman yacare*, economically valuable and protected species, that, between 1758 and 1955, had been classified as five distinct populations. They examined 13 external morphological characteristics in 360 *Caiman* species from known localities distributed throughout the range of these taxa. Using covariance, principal component and discriminant function analyses, they could find no defensible basis for the partition of *C. crocodilus* into sub-species, nor did they feel that considering *C. yacare* a sub-species of *C. crocodilus*, as proposed by some workers, is warranted.

Hybridisation can present challenges in identification, not least because some molecular techniques, for example use of mtDNA, with its maternal inheritance (see Chapter 11), cannot identify hybrids. Karyotyping can sometimes be used to confirm that an animal with intermediate characteristics is indeed a hybrid – as was demonstrated by Griffiths et al. (1987) in the case of a natural hybrid newt, *Triturus helveticus* × *Triturus vulgaris*, from a pond in mid-Wales.

As Linacre said, microscopy is a useful tool in the identification of animal material and in determining the taxon from which it came. Examples are given in Table 9.7.

Stomach (and, in birds, crop) contents can provide much valuable data, especially concerning diet and other ingesta (including poisons – see earlier and Chapter 8). As far as identification of prey species is concerned, gross examination – following flushing and removal of the contents with water or saline – will often yield the required information.

Table 9.7 Roles of Gross and Microscopical Techniques in the Species Identification of Material from Wild Animals

Specimen	Gross	Microscopical	Comments
Hair	Yes	Yes	See text.
Feather	Yes	Yes	See text.
Scale	Yes	Yes	See text.
Other keratinous structures, for example horns, hooves	Yes	Yes	See text.
Tooth	Yes	Yes	See text.
Elephant ivory	Yes/No	Yes	See text. When used for ornamental purposes ivory may change, intentionally or accidentally, in appearance (see Chapter 10).
Bone	Yes	Yes	See text.
Viscera (liver, kidney, etc.)	Yes	Yes	See text.
Contents of GIT (stomach contents, faeces)	Yes/No	Yes/No	GIT contents may provide clues to the animal's identity (types of parasite, e.g.). Primarily of use in determining food items and answering more specific forensic questions, for example presence of poisons.

Where gross examination fails, molecular techniques may succeed; for example, Walter et al. (1986) described the use of ELISA in the identification of fish and molluscan prey from the unrecognisable stomach contents of marine birds.

Even inorganic material found in the gastrointestinal tract (GIT) may be relevant in a legal case. For example, Gudmundson (1972) was able to use grit as an indicator of the overseas origin of certain birds occurring in Iceland.

Differentiation of Human and Animal Material

Those involved in wildlife forensics may face the dilemma of not knowing whether remains are human or non-human. Human tissues may be indicative of homicide or another crime: they may also be subject to special controls because they are important culturally (as in the case of aboriginal peoples in Australia and elsewhere). Specific legislation can apply to material from *Homo sapiens,* as with the United Kingdom's Human Tissue Act.

It is important to be able to distinguish human material from that of other animals, even if the latter cannot be identified down to species. The task is often not easy. For instance, any vertebrate bone may present initial difficulties, especially if it is small, damaged (e.g. burned) or is in an 'assemblage' of bones from diverse species. Domesticated animal tissues sometimes confuse anthropologists and forensic investigators (less so veterinarians), but the main areas of concern regarding misidentification are (1) apes and large monkeys and (2) bears and other large carnivores.

This is an area in which links between the medical and veterinary professions – together with others, such as zoo-archaeologists – are important and where a comparative approach can be of great value. Methods and criteria used to distinguish human material from that of other animals are listed in Box 9.6.

> **BOX 9.6 DIFFERENTIATION OF HUMAN MATERIAL FROM THAT OF OTHER SPECIES**
>
> - Gross morphology – bones, viscera
> - Light microscopy – hair, blood (non-mammalian species)
> - Electron-microscopy – muscle (certain species)
> - Gel diffusion and electrophoresis – tissues, body fluids
> - DNA analysis – most samples

A more detailed discussion as to how the bones (as opposed to the soft tissues) of humans can be distinguished from those of wild animals is to be found later in this chapter.

A whole carcase of a domesticated or familiar species usually presents few problems insofar as identification is concerned, but wildlife can still be problematic. Difficulties often arise when only portions of an animal are presented. Methods for distinguishing carcases (or parts thereof) of horse and ox, cat and rabbit and various other familiar species are well-documented in books covering meat inspection and pathology (see, e.g. Saxena et al., 1998).

As hinted earlier, bones of small mammals and birds often present difficulties, especially if they are damaged. There are, however, published papers on some species, including detailed studies on certain bones or structures – for example, the work on the pelvis of British rodents (Clevedon Brown and Twig, 1969), intended to assist those analysing 'osteological assemblages', such as those in owl pellets.

An investigator may be asked to 'match' portions of carcases – for example, in cases where a number of wild animals have been killed, butchered and then had their parts distributed. A useful guide to such work, concentrating on wild mammals, is Stroud and Adrian (1996). Various DNA techniques are now being developed to distinguish meats, even those that are canned (see, e.g. Rosel, 1999). The inspection of meat and other food of animal origin is a specific discipline, covered in most countries by legislation, but it can involve the wildlife forensic pathologist if there are allegations of fraudulent practice, illegal disposal of condemned carcases and parts, inadequate inspection or failure to recognise and take action over signs of disease in carcases. Texts that cover the inspection of fish, crustaceans, molluscs and game meat are helpful; for the last of these the excellent volumes in German by Dedek and Steineck (1994) and Ippen et al. (1995) are recommended.

Gel diffusion methods, using prepared antisera, continue to be standard techniques for differentiating species in many parts of the world (William and John, 2001).

Description and Recognition of Animals

Accurate description and recognition of an individual dead animal are essential. Failure to be able to confirm that an animal is the one that was examined or sampled earlier can prove disastrous (and embarrassing) in legal proceedings – as was stressed in Chapter 8. Description and recognition are distinct from identifying an animal's species, but the two needs may sometimes coincide if features are present that both

Table 9.8 Description of Individual Dead Wild Animals

Method	Comments
Characteristic external markings such as colour patterns.	Of limited value. Easily masked by *post-mortem* change.
Characteristic morphological features, for example shape of horns or wear of hooves	Often of value, especially if linked with other characteristics. Horn and hooves (keratin) often persist long after soft tissues have decomposed.
Presence of external collars, chains, ear tags, sometimes tattoos and other human-introduced devices	Can be surprisingly helpful. Mainly applicable to captive wildlife.
Characteristic (sometimes abnormal or readily recognisable) appearance of foot pads, the plastral pattern of tortoises, nose prints of gorillas, etc.	Comparable to the taking of fingerprints from human cadavers as part of individual identification (DiMaio and DiMaio, 2001).
Presence of an internal device, for example a transponder (microchip) – correctly termed a radio frequency identification device (RFID)	Very reliable but detection not possible without the appropriate instrument or detailed dissection (see Chapter 8 and 'Identification Systems in Hunting Falcons in the Middle East' section in Chapter 13).
Dental features, including presence/absence/stage of eruption of teeth and presence of abnormalities	Very valuable. Will also often help in other respects such as identification of species and age determination (see 'Dentition of Mammals' section in Chapter 13).

distinguish an individual *and* help to determine its species – characteristic antlers, for example.

Photographs are an important part of description. It is also useful to include on *post-mortem* forms drawings (outlines) of animals – left and right, dorsum and ventrum – on which relevant features, including lesions (see earlier), can be drawn.

When dead animals or their remains have to be described, the methods listed in Table 9.8, adapted from Cooper and Cooper (2007), can be employed. Reference should also be made to the section on identification and recognition of individual animals in Chapter 8, where methods used in live animals, many of them applicable to carcases (some mentioned in the following), are detailed.

Facial reconstruction, from skeletal remains, has become a skilled part of human forensic work, but is at present of little relevance to wildlife forensic medicine other than for taxonomic and palaeontological studies on, for example, the evolution of primates.

Morphometrics

Weighing of Animals and Organs

Despite its importance, in terms of both individual and species recognition, sexing and assessment of condition (see later), weighing is sometimes forgotten or records are lost or erased. In such cases, measurements become even essential, especially as some can be used to estimate the weight. There are various methods, based on formulae, that provide radiographing is possible and often desirable. However, use of an x-ray machine always carries with it responsibility for the safety of operators and bystanders (see Chapter 8).

Other imaging techniques may also be helpful (see also Chapter 8). Neuropathologists in the United Kingdom faced with the dilemma of whether or not consent is likely to be given for traditional autopsy are increasingly using *post-mortem* MRI (Cohen and Whitby, 2007). Jeffrey et al. (2007) discussed the role of CT in forensic investigations.

Geometric morphometrics is an increasingly used method of quantitative analysis in some disciplines, including anatomy and osteology. It can be used on organisms to investigate shape mutations, phylogenetic parameters of size and relationships between ecological factors and organism shape. In many recent evolutionary studies, geometric morphometrics has been used to investigate island radiation, where relatively small (but significant) morphological changes may result from much-reduced 'micro'-evolutionary time-scales. In terms of a forensics application, geometric morphometrics can be used to verify the species (or sub-species) status of a skeletal sample, for example a confiscated skull.

The principles of geometric morphometrics cannot be discussed here, but essentially (Ben Garrod, personal communication) morphological 'landmarks' are captured using a NextEngine 3D Laser Scanner (with a precision of 0.25 mm), using associated ScanStudio HD software, and are analysed with specialist morphometrics software Landmark and MorphoJ. Such software allows shape-based data to be analysed in conjunction with molecular data. The software permits standard multivariate analyses such as principal components, discriminant analysis and multivariate regression. When sampling skulls and mandibles, for example, the NextEngine 3D Laser Scanner (Positioning: 360°; Divisions: 6–8; Points: HD; Target: Neutral; Range: Macro) is able to generate a viable scan when at least 50% of the specimen is intact.

Imaging clearly has much to commend it in forensic work but should not be used in isolation. For example, radiographic examination of the contents of the rumen and reticulum of an animal may reveal foreign bodies. However, radiography should never be used as an excuse for not also examining the ruminal contents by hand – grossly, with a magnifying lens and with both reflected and transmitted light, a traditional technique that may also yield findings that are missed by x-rays.

Laboratory Investigations

Forensic *post-mortem* examination often permits a whole range of techniques to be employed that are not usually part of the repertoire of the routine diagnostician.

A small laboratory, adjacent to the clinical or *post-mortem* area, is of great value. It enables certain tests to be carried out immediately, and these may yield information while the animal is still being examined: cytological ('touch preparation') samples are an example. The proximity of the laboratory also minimises the risk of sensitive samples drying—for example, buccal swabs from birds that may be harbouring motile *Trichomonas gallinae* organisms. A list of basic items to equip such a small laboratory is provided in Appendix B.

In wildlife forensic work, standard (traditional) laboratory tests such as histology, haematology and multiple investigations on urine may provide valuable supporting information (see later and Chapter 10). The selection of samples depends upon the circumstances and the questions that are being asked (see Chapter 10).

Samples from dead animals that need to be submitted for toxicological analysis often present particular challenges, and it is wise to consult the relevant laboratory before taking material. Samples are not only taken for biomedical investigation. Some, such as explosives, ammunition and weapons, may need to be examined chemically by specialists in forensic and other laboratories (Wallace, 2008).

These and other tests that are useful in routine gross *post-mortem* investigation are listed in Table 9.10. They are discussed in more detail in Chapter 8.

Table 9.10 Some Additional Investigations

Technique	Comments
Direct examination with hand-lens, magnifying loupe or dissecting microscope	Recommended initially for any tissue or sample.
Radiography (part or whole body)	Dental or low kv x-ray machines are ideal for small animals or structures. Digital kits are increasingly being used, but check on admissibility in court.
Histology	Samples (lung, liver, kidney plus abnormalities) should be taken routinely. Buffered formol saline (BFS) is the standard fixative.
Bacteriology	If a bacterial infection is suspected.
Mycology	If a fungal infection is suspected.
Virology	If a viral infection is suspected but virology is neither easy nor cheap.
Parasitology	A useful routine (ecto- and endo-parasites) in all necropsies. Combine faecal examination with cytology below. 'Ecological tagging' uses helminths to identify the stock from which fish originate. The presence of parasites may help to pinpoint the origin of other species.
Cytology–specific organs, intestinal contents, exudates, etc.	A whole range of simple and rapid tests that can provide useful information in legal cases.
Haematology	Small quantities of blood may be obtained from the heart or elsewhere of freshly dead animals.
Biochemistry	As above.
Toxicology	See text – important in many forensic investigations.
Karyotyping	Identification and sexing. See, for example, Gonzalez and Brum-Zorrilla (1995).
DNA studies	Identification, sexing. Valuable. Appropriate material should always be stored.
Scanning electron-microscopy (SEM)	Of value in specific cases. Expensive.
Transmission electron-microscopy (TEM)	As above.
Bone preparation, examination and analysis	Bone density and other studies.
Digestion studies	Studies on soft tissues.
Ashing and mineral analysis, including whole body studies	Studies on hard tissues.
Radio-isotope studies	See text – age determination of remains.
Ballistic and allied investigation	For bullets, shot, cartridges, explosives, ammunition and weapons.

Source: Adapted from Cooper, J.E. and Cooper, M.E., *Introduction to Veterinary and Comparative Forensic Medicine*, Blackwell, Oxford, UK, 2007.

In most cases only a few additional investigations are likely to be justified initially. However, it is always good practice to retain material in case further investigations prove necessary or tests are requested by a court; in the case of tissues, for example, liver in glutaraldehyde for TEM and frozen at 20°C for toxicology.

As stressed in Chapter 10, correct, secure, storage of material taken from a forensic case is vital – as is confidentiality. Samples must be handled carefully because of health and safety considerations and indelibly labelled or marked so that they cannot be misplaced or transposed.

Histological Examination

Histological examination is probably one of the most important laboratory tests in forensic pathology and therefore a section is devoted to it here. There is, however, a dearth of literature on forensic histopathology of animals, even domesticated species, and the pathologist must extrapolate as best he or she can from text and papers on diagnostic and research topics. Sometimes the most helpful work dates back many years – as in the case of the studies on the histology of the skin of the elephant by Smith (1890) and the tome about avian integumental anatomy (Lucas and Stettenheim, 1972). As far as contemporaneous texts are concerned, the book by Aughey and Frye (2001) is particularly recommended for its broad comparative approach.

There is also some useful information in medical texts – for example, the Atlas by Cummings et al. (2010), which offers an excellent set of high-quality colour photomicrographs and covers a wide range of *post-mortem* diagnoses and topics.

Following a 'mini-necropsy' (see earlier), small specimens can be embedded and serially sectioned for histological examination (Cooper, 1989; Reichenbach-Klinke and Elkan, 1965).

As explained earlier, in the context of mummified carcases, it is perfectly feasible to apply laboratory techniques to a mounted specimen or a study skin – hair or feather investigation, for example. Portions of skin and other tissues can be rehydrated in saline for 72–96 h prior to their being fixed, embedded and sectioned for histological study.

Water-Testing

This is an integral part of any laboratory investigations concerning dead or alive aquatic or, indeed, non-aquatic animals that have been found dead in water.

It is therefore mentioned here. The most appropriate reference tests for such water-testing are often those produced for aquaculturists – for instance, the book by Stirling (1985).

Questions That May Need to Be Answered by the Necropsy

Post-Mortem Change and Estimation of Time of Death

Accurate differentiation of *ante-mortem* and *post-mortem* changes is important in all forensic necropsies. Often one has to include analysis of the history and circumstances since death (Archer, 2004), coupled with use of laboratory tests, for example histology.

DiMaio and DiMaio (2001) referred to enzyme tests that can be used in humans both to demonstrate that a wound was *ante mortem* as well as to date it.

An example of how *ante-mortem* lesions can be confused with *post-mortem* changes was given by Rutty (2004), who drew attention to pathological changes, in the lung, for example, that can be mistaken for autolysis.

An understanding of *post-mortem* change is not only of importance in terms of helping assess what might have occurred before or after death but also in assessing how long an animal might have been dead (see later) and/or how it has been preserved or stored since then (see earlier). It is also relevant to all necropsy practice. Sometimes in standard diagnostic veterinary work, a laboratory may state that the carcase 'was too decomposed for a *post-mortem* examination', but such a verdict must never be applied to a forensic necropsy. A dead wild animal may be reduced to a liquefied 'soup' or a skeleton, but must be properly investigated if evidence is not to be lost.

Progression of *post-mortem* changes can depend upon many factors (see Box 9.8). Useful clues to the animal's history, time since death (see earlier) and/or the environmental conditions to which the carcase has been exposed may be provided by

- Presence or absence of rigor mortis, livor mortis (hypostasis, pooling of blood, etc.) and algor mortis (cooling)
- Appearance of organs and tissues
- Evidence of predation/scavenging by other animals such as dogs: this may be *ante mortem* or *post mortem* (see earlier)
- Infestation by maggots, carrion, beetles, etc. (see later)

The 'ageing' of tissues in remains in terms of how long ago the animal died can be even more difficult. This determination is often requested, however, and may include the estimation of 'age' of bones found during road maintenance or the clearing of a forest. In human forensic medicine such 'ageing' can be important as remains of over a certain vintage are unlikely to prompt criminal investigations, on the grounds that anyone culpable will now be deceased.

Remains should be photographed as soon as they are found, preferably *in situ*, as their appearance may change subsequently as a result of many factors (see earlier). Descriptions are essential too. The scoring of *post-mortem* change helps in assessing tissues, especially

BOX 9.8   FACTORS AFFECTING *POST-MORTEM* CHANGE

- Species of animal
- Health status before death
- Presence or absence of ingesta in the GIT
- Manner of death
- Body temperature at time of death
- Environmental temperature and humidity
- Location and position of the body
- Handling and storage following retrieval of the body

those that are of forensic importance. A very basic approach, better than none at all, is to describe *post-mortem* change as (e.g.) (1) minimal, (2) moderate and (3) marked. It is preferable, however, to draw up criteria that relate more accurately to the species and circumstances in question. For example, the following proved helpful and practicable in a study of dead flamingos in Kenya (Cooper, Deacon and Nyariki, in press):

1. Intact, including eyes. Plump body shape, neck curved. No change in colour. Lice sometimes present. A very fresh carcase may be flaccid or show early rigor mortis.
2. Not intact. Eyes dull, convex. Plump body appearance but beginning to decompose, skin losing elasticity but feathers still soft. Early discoloration. Lice sometimes present. Slight smell (depends on hydration of the carcase).
3. Beak friable. Body shape contracting and skin drying. Eyes convex. Feathers losing softness. Moderate smell (depends on hydration of the carcase).
4. Maggots visible. Skin from legs peeling. Eyes not readily visible. Abdomen becoming obviously concave. Marked smell (depends on hydration of the carcase).
5. Internal organs missing. Maggots mature, on ground as well as body. Other invertebrates, such as beetles, often present. Some part of skeleton visible. Feathers missing. Eyes missing, orbit apparent.
6. Bones and dry skin only, often scattered.

The ageing of animal remains that date back hundreds or thousands of years has traditionally been the preserve of the zoo-archaeologist or palaeontologist and a variety of techniques is available, many based on measurement of isotopes.

The expert must always be cautious when initially presented with a bone or a skin of an animal and asked to state when the animal had died. Many factors can influence the appearance and feel of such specimens. Storage conditions or an unusual combination of preservation factors may have acted on a carcase, complicating interpretation as, for example, in the very wide range of ages, diverging between tens and thousands of years, estimated in the literature for seal carcases found far inland in the dry valleys of Antarctica (Richard Norman, personal communication).

The 'ageing' of the remains (parts of the body) of an animal may be requested if such residua might provide evidence relevant to a cruelty case (e.g. for how long was a zoo animal that starved dead in its cage?) or contravention of conservation legislation (e.g. were feathers in a collection imported before or after the enactment of CITES legislation?)

'Skeletonisation' is a term used to describe the disappearance of soft tissues of the carcase and the appearance of parts of the skeleton. It is not the same as mummification; the latter is associated with desiccation. In humans and in many – but not all – animals the cranium is the first part of the body to skeletonise. This is probably because of the accessibility to invertebrates of the various orifices. Trauma to the head, which is not uncommon in forensic cases, will often predispose to decomposition. When describing a carcase, or the remains of it, the investigators should specifically mention the areas of the body that have skeletonised (see later – Bones) and take photographs to illustrate this.

Organs and tissues undergo *post-mortem* autolysis at different rates, starting with these organs that rapidly show breakdown and finishing with those that show relatively little decomposition (see Box 9.9).

> **BOX 9.9   RATES OF AUTOLYSIS**
>
> - Retina
> - Brain
> - Testis
> - GIT
> - Pancreas
> - Liver
> - Kidney
> - Skin
> - Skeletal and heart muscle
> - Uterus

There are also species differences. For example, the skin of fish and amphibians tends to autolyse sooner than does that of a mammal or bird.

Estimation of time of death is always a challenge to the medical pathologist (DiMaio and DiMaio, 2001; Shepherd, 2003) and such is also the case in wild animal work. The need usually is to determine the *post-mortem* interval (PMI or time since death estimations) – as important in many wildlife forensic cases as it is in work with humans and domesticated animals.

Studies on *post-mortem* interval (PMI) have largely been carried out on humans (DiMaio and DiMaio, 2001; Haglund and Sorg, 2002), but other species investigated have included rats (Seaman, 1987), poultry (Munger and McGavin, 1972) and beagle dogs (Erlandsson and Munro, 2007). Remarkably little appears to have been published about *post-mortem* change in ectothermic animals (see later).

In Seaman's book on *post-mortem* histological change in the rat, referred to earlier, sections of a range of tissues were examined microscopically and appearances were described and photographed. Significant differences were noted between, for example, the tongue (where only minor changes were seen over the 16 h *post-mortem* interval) and the colon (where epithelial loss and saprophytic bacterial invasion occurred early).

The results obtained by Seaman are of some, but limited, relevance to forensic studies because the conditions under which wild animals die and decompose under 'natural' conditions are very different from those that apply in a research institute. The differences include

- Variation in ambient temperature
- Carcase(s) often not intact – if, for example, there are *ante-mortem* wounds or *post-mortem* predation
- Variability in body size and condition
- Species other than *Rattus norvegicus* are often involved

In wildlife forensic work there is a need for studies similar to Seaman's that take into account the aforementioned points (and others) but which, nevertheless, are scientifically sound. This proves far from easy, as explained in Cooper and Cooper (2007), citing a project at a Rift Valley soda lake in Kenya, East Africa, measuring *post-mortem* change in lesser flamingos. The study used dead chickens (domesticated fowl) as models.

Oates (1992) and Oates et al. (1984) published information for those working in the field who are likely to be asked to determine the time of death of various wildlife such as racoons, rabbits and waterfowl. Kienzler et al. (1984) provided useful data on temperature changes in dead white-tailed deer.

Post-mortem changes in mallards were studied by Morrow and Glover (1970) with a view to developing techniques for estimation of time since death in birds killed by hunters. Birds were trapped, killed and held for observation in controlled-temperature rooms. Body temperature was found to be the most reliable indicator of time since death. Ambient temperature had the most pronounced effect on rate of cooling; weights of the birds had the least effect. Soaking in water after death caused a rapid decrease from the live body temperature. The electrical resistance of muscle tissue fell rapidly after death, then more slowly until a base level was reached after 7–11 h. Initially, resistance was greater across the fibres than along them, but the difference decreased steadily after death. Eye changes, particularly configuration of the globe and colour of the pupil and iris, generally indicated time since death. The peak of rigor in muscles of the jaw, neck and legs was commonly reached by one hour after death; rigor development in the flight muscles required about 1.5 h. Ciliary activity of the tracheal lining declined gradually after death and was largely absent after 32 h. Body temperature, eye appearance, muscle rigor and tracheal ciliary activity appeared to prove useful as field methods of estimating time since death.

There remains a particular death of published data on *post-mortem* change in ectothermic animals, and investigators have often had to extrapolate from data from mammalian and avian studies. Work on quality changes in fish, mainly directed towards promoting food hygiene (see, e.g. Huss, 1995), may or may not be relevant. Cooper (2012) regretted that studies on PMI in reptiles and amphibians have been so sparse (Figure 9.7). He pointed out that reptiles and amphibians present a range of challenges in terms of accurately assessing

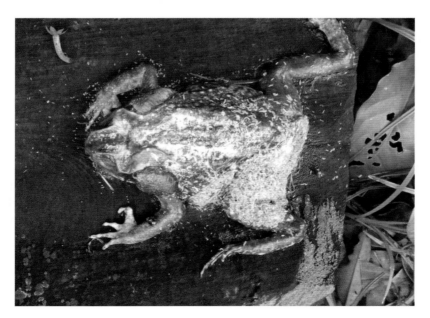

Figure 9.7 Part of a study on PMI in amphibians. This toad has been dead for 72 h in a hot and humid tropical environment. Flies have laid eggs on much of the caudal (posterior) part of the body and larvae have emerged.

those changes that are due to autolysis and decomposition, on account of their variability in morphology and lifestyle. In particular, there are effects due to ectothermy, the very different anatomical features and marked variation in body size of diverse taxa and seasonal fluctuations in subcutaneous and internal fat content. Eggs, embryos and fetuses of reptiles and amphibians present particular challenges and the presence of a larval stage in amphibians is a further complication. Cooper said that specific research is needed and advocated that this, to involve both amateur and professional herpetologists. should include the keeping of properly recorded accounts of changes in dead reptiles and amphibians, especially those kept in captivity or which can be carefully monitored, such as road kills. Experimental studies are also needed, carried out in collaboration with others, such as pathologists.

Frye (1999) made very useful contributions to our understanding of PMI in reptiles and amphibians. He listed changes that could be used in the assessment, such as rigor mortis, lividity (hypostasis), bile-staining, chromatophore relaxation and expansion, and the development of a characteristic odour. Frye's paper and a sequel (Frye, 2008) remain the key guides to the criteria that can be of value in assessing PMI in reptiles and amphibians, but neither provides any specific indication of the time it takes for such changes to occur.

Insofar as invertebrates are concerned, zoo-archaeologists sometimes use environmental cues to assess when the death of an animal may have occurred, such as the tidal lines on marine molluscs (Davis, 1987).

Various specialised methods have been developed in order to try to assess the PMI in humans (DiMaio and DiMaio, 2001), including temperature measurements, changes in the vitreous humour, muscle studies and evaluation of DNA degradation. Other methods tried have included changes in morphology of sweat glands, the cellular content of cerebrospinal fluid, quantification of melatonin, photometric measurement of colour changes in livor mortis, *post-mortem* activity of lactate and malate dehydrogenase in liver, nerve conduction patterns, compound muscle action potential analysis, immunohistochemical detection of glucagon in pancreatic α-cells and insulin in pancreatic β-cells, analysis of the cardiac protein troponin I, application of biochemical and x-ray diffraction analyses and Fourier transform infra-red spectroscopy. None of these methods is widely accepted.

Recently, Mao et al. (2011) attempted experimentally to estimate PMI using electric impedance spectroscopy. The spleens of rats were studied *post mortem* at 10°C, 20°C and 30°C. The results obtained demonstrated that PMI negatively correlated with the absolute value of Im Z//(capacitive reactance component) in electrical impedance, suggesting that electric impedance spectroscopy may be a sensitive (and rapid) tool to determine the PMI.

The rate of cooling of a carcass is recognised as being dependent on many different factors (Shepherd, 2003). In the case of wild animals it is greatly influenced by such variables as size (mass), surface area, insulation – presence/absence of subcutaneous and internal fat – and environmental factors including ambient temperature, wind, rain and whether the body is in water, mud or dry land. Autolysis is generally more rapid at higher temperatures and when the carcase is wet (Cooper and Cooper, 2007).

Munro (1998) described the particular difficulties for the animal pathologist because, as implied earlier, he or she is likely to be dealing with species that differ in size, shape and structure. He suggested that general estimates are usually all that can be given in animals, and advocated for them the use of the following broad categories:

- Recent death – less than 48 h
- Several days

> **BOX 9.10   ASSESSING TIME OF DEATH**
>
> - Cooling of the carcase
> - Rigor mortis
> - Desiccation
> - Discoloration of skin and internal organs
> - Maggot and other invertebrate activity
> - *Post-mortem* change
> - Scavenging patterns (see later)
> - Circumstances of death and storage thereafter

- Weeks
- Months
- Years

To conclude, it appears that at present there are relatively few methods of assessing time of death readily available to the wildlife forensic pathologist other than the basic criteria depicted in Box 9.10.

Description, Interpretation and Ageing (Age Determination) of Lesions

Lesions should be described using standard headings (see Box 9.11).

The shape and size of a lesion can be described in words and measurements (Thomson, 1978) or photographed, drawn or traced. Images should be in colour, with a non-reflecting scale and other relevant information, particularly the case number, date and (where appropriate) 'left' and 'right'.

All senses (except perhaps taste) should be used when examining a carcase – eyes, ears (crepitation, etc.), nose (odours associated with infections, parasites and poisons) and touch. The last can be very important: a slight change in the surface morphology of an organ can be best detected by gently running an index finger over it with the eyes closed – as if one is blind. Proper description of the 'feel' of organs or lesions is important.

> **BOX 9.11   STANDARD HEADINGS FOR THE DESCRIPTION OF LESIONS**
>
> - Numbers and distribution
> - Location/orientation
> - Raised, flat or depressed
> - On surface only or in deeper tissue
> - Shape
> - Size (cm and mm)
> - Colour
> - Consistency
> - Odour

Cooper and Cooper (2007), extrapolating from others, suggested that a useful approach is to relate this to parts of the human body so that, for example, 'hard' is likened to the feel of the forehead, 'firm' to the tip of the nose and 'soft' to the lip.

The colour of lesions can be important and changes may play a significant part in making a diagnosis. It is wise to standardise descriptions employing a colour key rather than using terms such as 'light green' or 'dirty grey', which may be very subjective, or relating the colour seen to a supposedly familiar object such as a plum (Cooper, 2009).

The appearance of lesions can be affected by the presence of pigment, especially melanin but also (e.g. in fish) guanine and xanthine. This primarily concerns the skin, but in many ectotherms the liver contains melanin or melanomacrophage foci and in a few species (e.g. certain African reptiles) the peritoneum may also be deeply pigmented.

The colour of organs or tissues may change *post mortem*. Some colours are indicative of specific changes, such as bile-staining that results in a green colouring in most tissues but may cause others, for example the yolk sac of young turtles, to appear black. Pseudomelanosis can occur – for example in kidneys – because of contact between the organ(s) and the alimentary tract, with the formation of iron sulphide.

Any changes in appearance following sampling should be noted as should the behaviour of samples when placed in saline or fixative. For instance, do they float or do they sink? Colours in carcases and tissues may be more easily seen 12–24 h after death or even following immersion of the tissue in saline or formalin.

Hypostasis (the gravitational pooling of blood in certain areas during or after the animal's demise), also termed livor mortis or 'lividity', will produce colour changes and can occasionally be mistaken for bruising. It may be of relevance in ascertaining the circumstances of death. If hypostasis is in an unexpected position (e.g. the reindeer is lying on its left but the right kidney and right lung show hypostatic congestion), this may mean that the carcase has been moved. Differentiation of hypostatic congestion of the lung and pneumonia is often important in forensic cases. Sometimes both are present.

A pink coloration to the tissues can be indicative of carbon monoxide or cyanide poisoning but may also be attributed to *ante-mortem* hypothermia or *post-mortem* refrigeration, especially at near-freezing temperatures; differentiation can be very important in a forensic case.

The ageing (age determination or dating) of skin and soft tissue wounds, contusions and fractures can be of great importance in forensic cases, in both live and dead animals. Two examples of the 'forensic' value of such information are (1) in a live animal, to be able to estimate when a skin wound was inflicted, in a possible cruelty case for instance; and (2) in a dead animal, to be able to age fractures or internal haemorrhage, in an alleged road-traffic accident for example. However, as medical forensic texts testify, the topic is fraught with difficulties.

Few studies have been carried out on the progression and thus the ageing of lesions of animals, in contrast to the situation in humans – bruises, for example (DiMaio and DiMaio, 2001; see Chapter 8). Assessments are even more problematic when body size and endothermy/ectothermy are brought into the equation. Some general rules are

1. The metabolic rate of endotherms (mammals and birds) is inversely related to body mass. A wound is likely to heal more rapidly in a lemming than in a lion.
2. Repair and/or regeneration takes place more rapidly at a high than at a low ambient temperature. Garter snakes (ectotherms) heal more rapidly at 24°C than at 20°C (Smith et al., 1988).

> **BOX 9.12 IMPORTANT QUESTIONS WHEN AGEING LESIONS**
>
> 1. With what species (or taxon/group) of animal am I dealing?
> 2. Is it an endotherm or an ectotherm?
> 3. If an endotherm, what is its likely body temperature? Was it hibernating/aestivating (and thus likely to be at a lower metabolic rate)?
> 4. If an ectotherm, at what temperature was it living/being kept?
> 5. Armed with the aforementioned information, plus any extra relevant history and clinical/*post-mortem* observations, can I estimate the age of the wound/contusion/fracture on the basis of
>   a. Published or unpublished data on this or allied species?
>   b. My own or colleagues' experience?
>   c. Extrapolation from human forensic medicine?
> 6. Do I have to make allowance for other factors that may have had an effect, for example the presence of infection or excessive movement, both of which can retard healing?
> 7. Do I have relevant photographs, drawings, radiographs, other images that can be used in a court case and/or sent to experienced colleagues for a second opinion?

Recognition that certain species heal more rapidly than do others is not new. Over 200 years ago, John Hunter demonstrated the speed of callus formation (and thus of fracture repair) in birds.

The important questions that anyone who is asked to age lesions in animals should ask are given in Box 9.12.

Approximate ageing of lesions in mammals is feasible based on published work in comparative pathology journals and medical texts (DiMaio and DiMaio, 2001). The criteria are listed in Box 9.13.

> **BOX 9.13 CRITERIA FOR AGEING LESIONS**
>
> - Organisation of blood clots
> - Inflammatory infiltrates – acute versus chronic, numbers of pyknotic/karyorrhectic cells, presence or absence of phagocytosed organisms or other material
> - Fibroplasia
> - The healing of abrasions, which takes place in humans in four distinct stages (DiMaio and DiMaio, 2001)
> - (In fractures) presence of callus, its degree of ossification, etc.

Even the criteria for such ageing of lesions may prove unreliable, because of such variables as

- The species, age, sex of the animal – some of this a result of differences in metabolic rate, as discussed (and see Chapter 2)
- Whether a lesion is infected or subjected to movement/trauma, which can delay 'normal' healing

Bruises are important clues in any *post-mortem* examination but the detection of bruising (contusion) in animals can be difficult because of the presence of hair, feathers, scales or pigmentation. The ageing (age determination) of bruises is discussed in Chapter 8.

Cause of Death

Determining why an animal died can be crucial in a legal case and yet in wildlife work, especially when dealing with free-living species, this can often be difficult to ascertain. As was pointed out earlier, a wild animal is, by definition, likely to die in the field, away from human habitation, where there is a strong possibility that it will be scavenged by other animals. Its body may well be found partly consumed, possibly some distance away from where it actually died. It therefore can be much harder to discover the cause of death of wildlife than it is in human or domesticated animal forensic medicine.

Analysis of cause of death in some free-living species is particularly difficult because of bias: thus, for example, animals that are hit by cars are more likely to be found than those that die in woodland. As mentioned in Chapter 8, animals wearing transmitters yield fresher carcases than those that die unrecorded.

Cooper and Cooper (2007) suggested that dividing cases into two categories, 'anthropogenic' and 'natural', was useful. It is important to bear in mind that sometimes the cause of death is multifactorial – not always an easy concept to explain in court and there often open to rigorous questioning in cross-examination.

Quite apart from any legal considerations – in some countries, such as the United Kingdom, only a registered veterinarian can make a diagnosis – caution should be taken in deciding who defines the cause of death in a forensic report. This is illustrated in respect of wild birds in Newton (1998) where there is a table of 'Causes of mortality as determined by post-mortem analysis of birds found dead', summarising *post-mortem* findings by various authors from a range of disciplines – some biologists, some veterinary pathologists. Newton's table includes disease as one of its headings. A footnote states that 'This includes infectious diseases and other disorders; not all diseases, notably viruses (*sic*), were checked for in all studies'. This statement, indicating a broad-brush approach to the diagnosis of diseases, without properly defining them, is an important reminder that analyses of the causes of death must always be interpreted and explained very carefully. This is especially so if those performing *post-mortem* examinations do not have a background in pathology and if, as in some of Newton's cases, what would be standard supporting tests in a veterinary laboratory, such as histology and bacteriology, have not been done. It also emphasises the need for far closer collaboration between veterinarians and biologists who are working on wildlife, as emphasised by Cooper (1993).

Some neonatal deaths in humans and domesticated (occasionally wild) animals are due to intentional infliction of trauma – abuse (see Chapter 8). In human infants there is teeth show ridges over a large area of crown surface. Enamel hypoplasias are caused by many factors, including nutritional deficiencies, metabolic disorders, infectious diseases and localised trauma. They are well recognised in humans and also seen in non-human primates (Moggi-Cecchi and Crovella, 1991).

Dental microwear analysis can help distinguish between diet patterns by measuring the different fracture properties of teeth. The technique has been used in extant primates and in fossils of extinct humans, chimpanzees, bonobo and their predecessors (Scott et al., 2005).

Poisoning

Poisons can cause clinical disease (see Chapter 8) and/or death. Toxic substances may either have a directly toxic effect on animals (e.g. causing damage to the liver) or they may injure superficial or exposed tissues such as skin, mucous membranes, alimentary system or respiratory tract (see later).

Many substances can prove toxic to animals and cause illness or death. Poisons that can kill animals may be ingested, inhaled or absorbed through the skin and range from arsenical compounds and herbicides to zinc phosphide and insecticides. Acids and alkalis are still used commonly, by mouth or by topical application, to kill or to injure animals and can sometimes form part of a pattern of animal abuse (see Chapter 8).

Caution is necessary when approaching and dealing with a carcase where poisoning is suspected. In many countries the emergency services have at their disposal a variety of probes and sensors for toxins (e.g. cyanide) that can be used at the scene of a suspicious death. These can be useful but their use does not mean that appropriate toxicological analysis can be omitted, nor that there is no risk of release of poisonous material during dissection.

The aim in toxicological studies in forensic necropsy cases is (1) to eliminate other causes of death, and (2) to demonstrate the presence of sufficient quantities of a poison to account for death. During the course of a *post-mortem* examination, this may require examination of the digestive tract or submission of relevant tissues, such as kidney or liver, for specific analysis.

The pathologist can play a key role in the diagnosis of poisoning, even if only by excluding other causes of death and then submitting appropriate samples (which may include dead animals and possible sources of poisoning, e.g. baits) to a toxicologist (Shepherd, 2003). Although poisoning cannot usually be diagnosed with accuracy on macroscopical examination, there may be significant clues. Sometimes a distinct odour is detectable – a reason during necropsy for carefully *smelling* a carcase. The remains of a toxicant, such as the leaves of poisonous plants, may be found in the GIT. Raptors that have died from ingesting carbofuran or alphachloralose (the commonest substances seen in the United Kingdom) may have their oesophagus and crop stuffed full of poison bait – in such cases they have probably absorbed a fatal dose through the wall of the upper GIT.

Other clues to poisoning may be more subtle. For instance, internal tissues may be found not to have decomposed to the extent that would normally have been anticipated. Small maggots are unexpectedly dead rather than alive and motile and may, in certain circumstances, have 'spared' an area of GIT or elsewhere, suggesting the presence there of toxic material.

Although it is difficult and potentially dangerous to generalise, a summary of some of the main gross and histological changes that may be seen in poisoning of wildlife follows:

- *Destructive*: ulceration of skin, mucous membranes, damage to capillary walls
- *Proliferative*: (usually secondary and long-term) teratogenic effects, neoplasia
- *Cellular*: disturbance of enzyme systems, leading to gross histological and ultra-structural changes in tissues
- *Local or systemic, or both*: Visible macroscopically, microscopically or only using biochemical, physiological or molecular techniques

The oiling of birds and other species is a form of poisoning and is discussed briefly in Chapter 2. As stressed in Chapter 8, an understanding of the pathogenesis of this syndrome is essential if one is to deal adequately with animals that have been affected by it. Much work has been carried out over the past few decades on this subject, particularly in birds, and only a summary will be given here. Useful background reading includes Leighton (1995).

A summary of the pathological effects of oil on birds is given in Box 9.14.

In addition to the direct effects of oil ingestion, detailed in Box 9.14, physiological derangement and 'stress' render birds more susceptible to infections.

The pathological changes in the intestine of oiled birds are often very marked and significant. Ingested oil affects the mucosa initially, causing inflammation, haemorrhage, necrosis (ulceration) and sloughing and, sometimes, acute catarrhal or haemorrhagic enteritis. If oil is inhaled, the respiratory tract – lungs and air sacs – is affected. One manifestation may be lipid pneumonia, with severe inflammation, often associated with pulmonary oedema and haemorrhage.

In long-term survivors other changes may be seen. For example, the spleen is sometimes enlarged due to amyloid deposition.

Crude petroleum oil is also toxic to embryos in birds' eggs (Lusimbo and Leighton, 1996). Oil is readily absorbed by the feathers of swimming or diving birds and can then be transferred to the shell of eggs during laying or incubation in sufficient quantities to cause overt toxicity and teratogenicity in embryos of various avian species.

BOX 9.14 EFFECTS OF OIL IN BIRDS

External effects. Destruction of the waterproofing of the plumage because the oil disrupts and damages the barbs and barbules and allows water to penetrate the plumage. This feather derangement is accompanied by a loss of insulation (necessitating rapid use of energy resources to maintain body temperature) and a reduction of feeding efficiency. Very heavily – oiled birds may drown.

Internal/systemic effects. The chilling and starvation induce physiological changes, including 'shock' and 'stress'. This is followed by internal effects as the bird ingests oil in an attempt to preen. Changes may include a drop in white blood cell count, dehydration, disturbance of endocrine activity, fatty change in the liver, renal damage, necrosis and/or oedema in various organs and metabolic disorders.

Wild animals in zoos, laboratories, rehabilitation centres or perhaps held pending a court-hearing are particularly susceptible to poisoning. If they have not previously been in captivity, wildlife may be vulnerable because they encounter, in (e.g.) the grounds of a zoo, materials with which they were not familiar in the wild – zinc from the wire of cages, for example. They may also be exposed to other, novel, sources of zinc, as illustrated by the recent report of zinc toxicity in dogs associated with the ingestion of identification tags (Adam et al., 2011).

Even common salt (NaCl) may poison wildlife. Duff et al. (2011) described mass mortality of great crested newts, a protected species in the United Kingdom, on ground treated with road salt. A definitive diagnosis was not made, but circumstantial evidence indicated that road salt application to a car park surface was the probable cause.

Some Specific *Post-Mortem* Findings and Considerations

This book is not the place for detailed pathology but some guidelines and comments will be provided on gross findings that are often of relevance in legal, insurance or other cases. Reference should also be made to Chapter 8 for description of conditions of forensic importance that can cause clinical signs as well as death.

Starvation

Starvation – defined by Newton (1998) as 'loss of body condition to the point of death' (see Chapter 8) is a significant finding in both free-living and captive wildlife. It is often an important forensic finding. Gross *post-mortem* examination plays an important part in diagnosis of cases where death has supervened. An emaciated appearance, with absence of subcutaneous fat and an empty GIT, are strongly suggestive of prolonged starvation but not necessarily. Starvation can be due not only to a reduced food intake but also other factors, both non-infectious and infectious. This can present difficulties in legal cases, where the certainty of diagnosis may be questioned. Often a diagnosis of starvation depends upon circumstantial evidence as well as gross *post-mortem* findings. In some taxa there may be important, possibly pathognomonic features, as illustrated later.

The effects of an insufficient food intake are well recognised. Animals go into negative energy balance and depend upon their own fat reserves. When dietary restriction occurs over an extended period, various changes take place in internal organs, including a reduction in relative organ sizes and in total protein content. Lipid reserves become smaller, including bone-marrow fat. Nevertheless, many wild species can tolerate long periods of 'starvation' so long as they have had an adequate fluid intake.

In starvation organs may be reduced in size and weight and fat deposits depleted. In ectothermic vertebrates the characteristic fat bodies may be absent. A reduction in size of the liver has long been seen as a sign of starvation in reptiles by those who eat these animals; André (1904) reported that the liver of *Geochelone* tortoises in South America 'shrinks in proportion to the length of time during which its owner has been deprived of nourishment'.

Moore and Battley (2003) employed a novel method to determine condition before death (starvation) in brown teal in New Zealand, using the wing fat content. They extracted the lipids from four outer wing components of 17 intact brown teal carcases and the lipid content of each component reflected the birds' nutritional condition (based on body mass and size and visible fat).

Bollo et al. (1999) studied the liver histology of Alpine chamois and showed in that species a significant decrease in the mean size of the nuclei of hepatocytes of emaciated chamois in contrast to normal animals. More such morphometric studies are needed and might assist in the diagnosis of emaciation and the detection of metabolic stress during winter as a result of inadequate food intake. Most studies relate to populations rather than changes in individuals; for instance, Gunnarsson et al. (2012) looked at the direct and indirect effects of winter harshness on the survival of mallards.

Starvation and hypothermia due to maternal separation was diagnosed in a neonatal harbour porpoise by Bartges and Osborne (1995). Over-inflation of the lungs and the presence of rib imprints on their caudal surfaces suggested that live-stranding had occurred.

Fatty acid analyses may provide insight into the progression of starvation among squamate reptiles, as demonstrated by McCue (2008). Analyses of fatty acid methyl esters (FAMEs) conducted on the total body lipids of various snakes and lizards revealed that these species exhibited similar characteristic changes in the quantities of body lipid and levels of circulating fatty acids during starvation stress.

Many of the criteria outlined in the preceding chapter, relating to clinical examination of suspected starvation cases, apply also to necropsy examination and will not be repeated unnecessarily here.

As will be clear from the aforementioned, a competent *post-mortem* examination can assist in diagnosing starvation so long as internal organs are investigated both macroscopically and microscopically. Macroscopical examination will provide an opportunity to assess the size of organs and their appearance – for instance, whether the liver is pale and therefore possibly fatty. At the same time, the pathologist can examine and weigh any internal body fat. The GIT can be dissected in detail for the presence of ingesta, lesions or parasites and other possibly relevant abnormalities.

Microscopical (histological) examination enables the pathologist to assess more accurately the appearance of organs. This should indicate, for example, whether they contain lipid or show other significant changes and will permit the measuring of hepatocytes or assessment of zymogen granules in pancreatic acini (sometimes used as an indication of impaired nutrient intake).

Malnutrition may contribute to starvation, or the sequelae to it, or be significant in a legal case in its own right as evidence of neglect or cruelty. An important example is metabolic bone disease (MBD), usually caused by a calcium/phosphorus imbalance, which remains common in captive primates, birds and reptiles. It can cause severe clinical signs, culminating in death, and its presence is not infrequently considered grounds *de facto* for allegations or charges of cruelty (under the [English] Animal Welfare Act, for example – see Chapter 3).

Some metabolic diseases are difficult to diagnose accurately in life and present even more problems *post mortem*. Little appears to have been published on the investigation of metabolic diseases after death in most wild animals, with the possible exception of exertion (capture) myopathy (Cooper and Cooper, 2007), but Olpin and Evans (2004) provided useful ideas from the human perspective.

Dehydration

Dehydration can be a very important finding in legal cases, especially when deprivation of water is alleged.

Dehydration in many live mammals can be assessed using a combination of clinical investigations and laboratory tests, but this does not apply to all taxa and is often inapplicable to ectothermic animals. Even in mammals apparent dehydration can be a misleading sign. In a freshly dead specimen, skin tenting and other 'clinical' assessments may still be helpful but PCV estimation, even on apparently still liquid heart blood, is unlikely to be reliable because of clotting.

Dehydration in wild animals can be associated with cold weather. Duff et al. (2011) described water bird mortality in the United Kingdom associated with prolonged freezing weather in December and January.

Drought is an important cause of dehydration and starvation in free-living wild animals. In addition to its direct effect on animals, it can predispose to poaching (see Chapter 2 regarding the effect of severe, prolonged drought on elephants and other species in Kenya and to the effect of climatic extremes on many species, particularly birds). In recent years, there has been much international concern about the prolonged droughts that have affected humans and animals in certain parts of the world, such as East Africa, and the extent to which climate change and/or the technological need to utilise water in large quantities to produce the consumables that richer parts of the world demands may be contributing to this. The UN Food and Agriculture Organization predicted that by 2025 two-thirds of the world's population could be suffering 'water stress'.

At the level of individual animals and groups of wildlife, there have been many studies of the impact of drought – for example, Dunham (1994) looked at and analysed the effect of drought on the large mammal populations of Zambezi riverine woodlands.

One of the many possible effects of drought can be inhalation of dust. Not only dust but various other substances may be inhaled, including different types or soil, sand, diatomite and silicates. Pneumoconiosis, important legally in humans, may be seen in animals – either 'naturally' (as in burrowing or fossorial species, e.g. moles) or accidentally (as in animals kept in urban zoos). The presence of inhaled material, identifiable in terms of its origin, sometimes provides useful evidence in a court case.

Hyperthermia and Hypothermia

These are discussed in detail in Chapter 8. Some aspects of their pathology were described by Cooper (1996). Hypothermia is further discussed below.

Like dehydration, hypothermia can be associated with bad weather. Duff (2007) reported suspected wild bird mortality in Britain due to stormy weather and hailstones. An unusual combination of widespread internal haemorrhage in the absence of skin penetration indicated traumatic injury. Circumstantial evidence strongly suggested that this could have been associated with flying in a violent storm front, causing the birds to drop precipitously from a considerable height.

In chilling there are often only mild, non-specific, pathological changes. The lungs are often congested and an excess of clear fluid may be present in the body cavities; the liver may appear pale and the stomach is usually empty. Prolonged exposure to cold weather may predispose wild animals to diseases associated with stress – for example, gout, amyloidosis and myocardial infarction in waterfowl (Karstad and Sileo, 1971).

In freezing there may be areas of hair or feather loss and specific lesions, such as oedema. Tissues may appear avascular or necrotic, and in severe cases have been sloughed or removed by the animal. The pathogenesis of frostbite is mainly dehydration, disruption of cells by ice crystals, thrombosis and disturbed metabolic activity resulting in ischaemia of the superficial tissues that can progress to gangrene and sloughing.

Diagnosis of earlier exposure to cold may be important from a forensic point of view, but is not easy. Cipriano (2002) detected cold stress in captive great apes by examining incremental lines of dental cementum. While the dental cementum of free-ranging apes was regularly structured into alternating dark and light bands, four of five captive animals showed marked irregularities in terms of hypomineralised bands. These could be dated to the year 1963 (characterised by an extremely cold winter) when all four had been kept in a zoo located in the Northern Hemisphere. So-called 'cold stress' is a calcium-consuming process and the lack of available calcium in newly forming cementum could be responsible for the hypomineralisation.

Attempts to diagnose exposure to severe cold on the basis of biochemical and histological changes have not proved easy. Bradley et al. (1996) looked at biopsies of paw skin from Alaskan sled dogs. No significant histopathological changes were noted in any of the specimens, but in some dogs there was elevation of 2,3-dinor TxB2. Wet snow at higher environmental temperatures caused more 'paw stress' than did hard crusted snow at lower environmental temperatures.

Drowning

Drowning usually implies the inhalation of water in sufficient amounts to cause severe injury or death. Medical pathologists sometimes differentiate between 'dry drowning' (causing vagal cardiac arrest – see earlier) and 'wet drowning' (inhaled and ingested water). Laryngospasm can cause asphyxia. In the case of 'wet drowning', vomiting may occur, especially if seawater is swallowed. If water is inhaled, gaseous exchange is reduced as bronchioles become blocked, causing anoxia. An osmotic effect is produced – haemolysis in freshwater, dehydration and pulmonary oedema in hyptertonic seawater.

Although there are no pathognomonic findings (DiMaio and DiMaio, 2001), the pathogenesis of drowning in humans is well understood (Camps and Cameron, 1971).

Post-mortem diagnosis of drowning in animals is based largely on criteria similar to those for humans mentioned earlier (Cooper, 1996). Some aquatic and marine species have anatomical and physiological mechanisms devised to reduce the risks of inhalation of water. In fatal cases of drowning in most (non-piscine) vertebrates excess frothy fluid is found in the respiratory tract, including the air sacs of birds, and often in the oesophagus and stomach. In addition, the lungs may be distended and congested, with sub-pleural petechiae. The presence of water in the respiratory tract is usually considered always significant, as can be the finding of inhaled diatoms and other aquatic flora. However, caution must always be exercised. In marine mammals, for instance, the aforementioned observations are not at all conclusive (Kuiken et al., 1996), and water and organisms may be inhaled or ingested *post mortem* (Larsen and Holm, 1990). Studies on inhaled material, coupled with analysis of seawater, have been used to predict the location of death place of dead, drowned, green sea turtles (Cenigiz et al., 2002).

- (Usually wild birds) They are the cause of an accident (see later).
- Animals on the ground are killed or damaged as a result of a crash, or falling debris or frightened ('stressed') following the passage of a noisy or low-flying aircraft (Brown et al., 1999; Grubb and Bowerman, 1997; see Chapter 2).

Although much has been published about the pathological effects on humans of depressurisation and exposure to low temperatures in aircraft, little appears to be available, certainly in the literature, concerning aircraft and animals.

'Birdstrike' (collision between birds and aircraft) has long attracted interest and research, primarily on account of the dangers to humans and the costs that such encounters incur. As Johnson (Chapter 13) points out, a better term is 'wildlife strike' because bats, terrestrial mammals and reptiles may also be involved.

Injuries and deaths due to railway trains appear (with the exception of already debilitated animals, such as rabbits with myxomatosis) to be rare – or possibly under-reported.

Traps, Nets and Snares

These are an important cause of death in wildlife and relevant to both animal welfare and species protection (Cooper, 1996). They are also mentioned in Chapter 8, in the context of live animals, because various injuries may result from them.

Careful examination, gross and microscopical, is essential when investigating wounds that may have been inflicted by any device or implement that is being used to catch animals. This includes fish hooks, which can cause damage or death in both target species and other (protected) animals (Borucinska et al., 2002).

Injuries from traps and snares are usually characterised by 'patterned injuries' (also shown in vehicular and other damage). Traps may cause crushing lesions, nets and snares strangulation or impairment to blood flow. A trapped animal may succumb to strangulation, blood loss, ischaemia, septicaemia, hyperthermia or hypothermia or be predated by another species.

Pitfall traps can cause falling injuries and may impale animals if, as is frequently the case when creatures are being caught for meat, stakes are placed in the bottom. Drowning may occur if pit traps become flooded or, if a research worker fails to protect or check his vertebrate or invertebrate traps regularly. A trapped animal will be stressed and, if left long enough, is likely to succumb to dehydration, starvation or death.

Wounds from Weapons

These will initially be discussed under two main headings – knife (stab) wounds and gunshot wounds. Various other weapons can be used to wound or kill wild animals, ranging from hammers to garden forks. In poorer countries or more rural locations, non-ballistic weapons such as machettes (cutlasses, *pangas*), spears and arrows are often used to kill or to maim wildlife; the offending items usually producing a characteristic 'chop wound' (DiMaio and DiMaio, 2001). Adrian (1992) is one of many authors who has published advice about the differentiation of arrow, gunshot and other wounds; however, few European or American authors are able to cover adequately the range of weapons that are used to kill or injure wildlife in such parts of the world as South America, Africa and Asia.

Knife (stab) wounds are a frequent cause of injury in humans and animals. Their dynamics have been studied using human cadavers (Knight, 1975; O'Callaghan et al., 1999) as have the gross and microscopical features of different types of wound (Humphrey and Hutchinson, 2001; Tucker et al., 2001). Studies on naturally occurring firearm lesions in animals were reported by Corradi et al. (2004); Nozdryn-Potnicki et al. (2005) and Urquhart and McKendrick (2003).

Williams et al. (1998) advocated that, when describing stab wounds, the following should be included:

1. Position and angle of the wounds
2. Their dimensions
3. Tissues or organs affected

Similar criteria apply to knives and other cutting weapons, but here the examination and interpretation of wounds must take into account the sharpness and structure of the weapon, particularly its shape and the number of cutting edges (e.g. serrations) that may be present.

As a general rule, a stab wound is deeper than it is long while a slash wound is longer than it is deep. The wounds produced by such weapons may be incisions or lacerations, or sometimes a combination of the two.

Firearms play an important part in the physical assault of humans in certain countries and can produce gunshot wounds of medical and legal importance (Cooper and Cooper, 2007; DiMaio, 1999; DiMaio and DiMaio, 2001). Guns also contribute to injury and death in animals, both domesticated and wild (Nozdryn-Potnicki et al., 2005; PAW, 2005; Stroud and Adrian, 1996; Wobeser, 1996), but the prevalence varies from country to country.

The prevalence of gunshot wounds in live or dead wild animals depends upon a number of factors. In some countries, such as the United States, guns are widespread and hunting is popular. In others, such as the United Kingdom, relatively few people own a firearm and neither legal nor illegal killing of mammals and birds is on a comparably large scale. In many other parts of the world firearms, ranging from airguns and shotguns to high-velocity rifles and machine guns, are used against wild animals. Some such animals are killed, others are injured; a few survive, but with residual pellets, shot or bullets left in their bodies.

Some knowledge of ballistics is essential for anyone involved in forensic work with wildlife. Both pathologists and clinicians need to understand how guns – and other weapons – cause damage and the sort of injury they are likely to produce. A useful summary of gunshot wounds was provided by Wobeser (1996).

In poorer parts of the world, arrows, where guns are difficult to obtain, or too expensive, bows and arrows, spears and catapults are often employed to kill animals. This is usually for food but sometimes for other illegal purposes, such as the killing or wounding of protected adult animals so that their young can be captured and sold (e.g. gorillas). In North America, arrows (fired with a conventional bow or crossbow) are frequently used, legally and illegally, to kill deer and other species.

The situation in the United Kingdom is not typical, but will be briefly discussed. A useful précis was the article by Saunders (2004). The weapons used most frequently to injure or kill

It has been claimed many times over the years that the fossil of *Archaeopteryx* in the (London) Natural History Museum is a fraud. Despite the work of Charig et al. (1986) that showed unequivocally that the specimen is genuine, Hoyle and Wickramasinghe (1986) attempted to re-ignite the debate and to link a claim of forgery with their own theory of how birds and mammals might have evolved on Earth. Both authors were eminent astronomers who had received many academic and national honours and, as the review of the book by Norman (1987) pointed out, it is difficult to believe that scientists of their calibre, albeit from a different discipline, could contemplate so amateurishly trying to influence palaeontological history.

The techniques used when investigating allegations of fraud involving mounted museum specimens might usefully be applied in wildlife forensics.

While museums provide a rich source of information for the wildlife investigator, one must never assume that all specimens are what they claim to be. As in all forensic work, the words of Epicharmus are apt 'A judicious distrust and wise scepticism are the sinews of the understanding'.

Cooper and Cooper (2007) stressed the value of reference collections in forensic work. They pointed out that collections relating to wildlife crime are being established, citing the USFWS National Fish & Wildlife Forensics Laboratory in Oregon, United States and smaller official or private collections in the United Kingdom and Australasia. The establishment and use of reference collections have been standard amongst some zoological and conservation bodies for many years – see, for example, Cooper et al. (1998) and Munson (2002). Reference collections are not confined to animal and plant material. Webb and Cooper (Chapter 10), in the context of diet analysis, advocate that the wildlife investigator should collect as many reference samples as possible of ingredients, contaminants and adulterants, which should be catalogued, dried and stored carefully in air-tight containers, away from the light.

Conclusions

The forensic examination of material from dead animals is a key part of many wildlife investigations. All tissues must be saved and stored after the necropsy. They constitute 'evidence' (see Chapter 12) and may need to be made available for a second *post-mortem* examination or as an exhibit in court.

Forensic necropsy requires experience and expertise and a degree of 'lateral thinking'. The words of Sir Arnold Theiler, the eminent South African veterinarian, when teaching osteology to students in the 1930s, exemplify the approach:

> Gentlemen … you may think it is only an old bone. You have to put your soul into that bone.

Dealing with Samples

10

JOHN E. COOPER

Contents

Introduction	326
Laboratory	326
Samples	329
Value of Multiple Investigations	330
Collection of Specimens	331
Laboratory Techniques	332
Initial Examination: Use of Hand-Lens	335
Use of Light Microscope	335
Electron-Microscopy	335
Transmission Electron-Microscopy	336
Scanning Electron-Microscopy	336
Parasitological Examination	336
Types of Sample	337
Pellets (Castings) from Birds	337
Background	337
Practical Considerations	338
Examination	338
Animal Diets and Foodstuffs	338
Faeces	343
Feathers	345
Background	345
Biology of the Feather	345
Collecting Samples	346
Examination	347
Gross and Low-Power Examination	347
Light Microscopy	347
Scanning Electron-Microscopy (SEM)	348
Transmission Electron-Microscopy (TEM)	348
Microbiology	349
Uses of Feathers	349
Hair	350
Collection and Handling of Hair Samples	352
Examination of Hair	352
Microscopical Examination of Hair	353
Other Investigations Using Hair	355
Hair Analysis	356
Example of the Forensic Value of Examination of Hair	356

Scales	357
Eggs and Eggshells	357
Teeth, Including Ivory	359
Other Samples	360
Poisons	361
Sampling in the Field	361
Identification of Invertebrates	362
Special Treatment of Samples	362
Movement and Tracking of Samples	362
Storage of Samples	363
Foreign Bodies and Contaminants	364
Legal Controls on Samples	364
Samples and the Chain of Custody	364
Samples and Data	365
Retention of Samples	365
Methods and Pitfalls in Laboratory Examination	366

'This very evening I was going to write to you, when behold a basket came with pea-fowls, lizards, and birds' legs'.

Letter from John Hunter to Edward Jenner (4th March 1781)

Introduction

Laboratory investigations are an important part of many forensic cases (Jones and Williams, 2009). Some relevant disciplines are almost entirely laboratory-based – for example, toxicology and DNA technology – but in wildlife work, it is often necessary to use the laboratory and its facilities in combination with preliminary investigations of the samples carried out in the field.

It is clearly not practicable to cover all laboratory methods. There are plenty of relevant publications that will guide the wildlife forensic investigator in this respect. Some are referred to in this chapter and others are included in the References and Further Reading at the end of the book. Here, the importance of laboratory work is emphasised and examples given, both of techniques and of samples and specimens that may need to be examined. Some of the equipment that may be needed is listed in Appendix B.

A useful review of criteria for dealing with wildlife forensic material in the United Kingdom was published by PAW (2005) and this drew attention to the importance of valid methodology, quality assurance, accreditation, quality control, staff competence and practices complying with good forensic practice. These points are discussed by Harbison (Chapter 12).

Laboratory

The main requirements of a wildlife forensics laboratory are listed in Box 10.1.

The facilities are a critical part of ensuring the high-quality forensic investigation. Ogden (2012), writing in the context of wildlife DNA, pointed out that various laboratory models exist and that the most appropriate solution will vary with the situation. He cautioned,

> **BOX 10.1   REQUIREMENTS OF A WILDLIFE LABORATORY**
>
> - Reliable
> - Appropriate facilities, including containment where necessary (see later)
> - Properly trained staff (see later) who are familiar with the legal process and such concepts as a chain of custody
> - Willing to be involved in forensic work, which will include the production of appropriate reports and possible appearance in court
> - Aware of the higher standards, extra costs, inconvenience and sometimes vulnerability (possibly needing additional security) likely to be incurred when dealing with material in legal cases
>
> *Source*: After Cooper, J.E. and Cooper, M.E., *Introduction to Veterinary and Comparative Forensic Medicine*, Blackwell, Oxford, UK, 2007.

however, that 'basic forensic requirements cannot be compromised'. This is absolutely correct but the statement has to be seen in the context of the wide range of forensic wildlife work that is carried out, in diverse situations, much of which does not concern DNA. Ogden described four types of facility that he believes reflect the diversity of laboratories currently in use. He is highly critical of two of these – (a) the university or institutional research facility with dedicated forensic laboratory space, and (b) the multi-use research laboratory. His verdict in this respect should not, however, be taken to be true of all types of wildlife forensic laboratory investigation. Scientists in other disciplines may not require the same level of containment of samples – for example, if their speciality is the morphological and histological examination of skeletal material, and there is no question of applying molecular techniques to the studies. This is not to suggest that non-DNA forensic work necessarily requires a lower standard of containment of samples. On the contrary, those working with micro-organisms from wildlife, especially highly dangerous viruses of primates, will need a laboratory that is equipped, both legally and practically, both to contain these organisms and to protect those working with them and to ensure high-quality forensic science – quite regardless of the cost involved. The important thing is to relate the needs to the type of forensic work being performed; ideal facilities for an entomologist or an ultrasonographer may be totally inadequate or unusable as far as a molecular biologist is concerned – and *vice-versa*.

The need in wildlife forensics for rigorous quality management (QM) applies to all types of work, regardless of the disciplines involved. Harbison (see 'Why Is Quality Management Important?' section in Chapter 12) defines QM as 'the policies, procedures, behaviours and practices (management) of an organisation or individual, that provide a guarantee (assurance) of a degree of excellence (quality) in a service or product' and states that, 'to be effective, QM generally requires endorsement from a recognised independent testing body, such as an accreditation agency, which in turn relies upon regular audits to ensure conformance'.

Almost all accrediting agencies base their requirements on the ISO/IEC 17025:2005 standard.

In most countries, there is a long way to go before forensic laboratories that deal with wildlife samples are properly and consistently accredited. In the United Kingdom, for example, the government's Forensic Science Regulator is embarking upon a programme of assessing and accrediting facilities, including laboratories, but is starting with those establishments that deal with human material and it will be some time until wildlife forensic practice is likely to be considered.

Those involved in wildlife forensics should have a list of laboratories that meet their own requirements and expectations. Whichever laboratory is used, the relevant investigators should ideally have visited the premises, met the staff, discussed the sort of work that might be needed and carried out a mock chain of custody and ascertained that the laboratory's standards are adequate, especially in respect of Good Laboratory Practice (GLP) and Standard Operating Procedures (SOP).

The two key factors in performing laboratory forensic work are (a) chain of custody, and (b) quality control. Chain of custody is discussed later and widely elsewhere in this book. Quality control is part of QM and is primarily covered in Chapter 12.

A more detailed assessment of the capability and suitability of a laboratory that might deal with wildlife material should address the matters set out in Box 10.2.

The need for highly trained staff in forensic work has never been in doubt. In wildlife investigations this can raise questions because scientists and technicians may not have worked with such species before. In most cases, this makes little difference in terms of laboratory investigation; the bacteriological culture of a swab from a degu in a zoo is no different from the culture of a swab from a dog in a kennel. Nevertheless, some training may need to be specifically tailored to needs. In the past, many scientific and most technical staff of medical and veterinary diagnostic laboratories would have had experience of working in a wide range of laboratory-based disciplines as part of their responsibilities. It could be assumed that a good technician, in particular, had reasonable familiarity with histological, microbiological, haematological and clinical chemical techniques, and in the context of each of these was familiar with good working practice and adherence to SOPs. Now, however, this is not necessarily the case. Many of the people working in laboratories are specialised in their knowledge and application. A scientist or laboratory assistant

---

BOX 10.2   CHOICE OF A LABORATORY – IMPORTANT QUESTIONS

- Can the laboratory ensure continuity and security of evidence?
- Will the work be carried out to the desired standard?
- Are the laboratory staff competent in wildlife laboratory forensic work?
- Do laboratory staff understand such principles as the unique marking of exhibits, the correct labelling of photographs and radiographs and the need to retain all material, including wrappings, correspondence, notes and documents?
- Are risk-assessment and appropriate safety codes in place and are they relevant to the range of wildlife material that might be received?
- Is a specific person available to be consulted when necessary, including weekends and public holidays?
- Are there facilities for the long-term retention of wildlife forensic material or for safe disposal of animal by-products?

Source: Adapted from Cooper, J.E. and Cooper, M.E., *Introduction to Veterinary and Comparative Forensic Medicine*, Blackwell, Oxford, UK, 2007.

Dealing with Samples

Table 10.3 Laboratory-Based Investigations That May Provide Useful Forensic Information

Discipline	Investigation	Comments – Information That May Be Obtained
Parasitology	Identification of ecto- and endoparasites, including blood parasites (haematozoa and microfilariae)	Species present may give clues as to geographical origin of host, time of death or contact with other species, including prey or a predator.
Haematology	Examination of blood smears	Presence of anucleate red cells will confirm that the blood came from a mammal. Other features may pinpoint the species more closely.
	Haemoglobin estimations	Elevated values may indicate that an animal has originated from a high altitude.
Histology	Standard methods	Many uses, e.g. anthracosis may indicate an industrial origin, a giant cell focus a reaction to a foreign body.

Initial Examination: Use of Hand-Lens

All samples should first be examined with the naked eye and a hand-lens, magnifying loupe or dissecting microscope. This work must be carried out carefully, using both reflected and transmitted light.

Use of Light Microscope

In his book about wildlife forensic investigations, Linacre (2009) stated

> The use of microscopy in forensic science is paramount and a standard tool in species identification. If this simple technique can answer the question asked, then there is no reason to proceed with more expensive tests. It may be that the type of material is inappropriate for microscopy, and if the species present is required to be identified, then molecular techniques can be employed.

The value of the microscope in forensic work is emphasised by Webb (see 'The Value of Microscopy' section in Chapter 12). Its use, including the importance of proper maintenance, is covered in many textbooks (see, e.g. Bell and Morris, 2009). A useful review of some particular microscopic techniques and their application to wildlife forensic science was provided by Sahajpal and Goyal (2009), in Linacre's book referred to earlier. They described the types of microscope that they use in their work, stereo, comparison and scanning electron (SEM), and detailed how these instruments can be employed to investigate mammalian hair. Their techniques and others that can be used to examine hair (and other keratinous structures, such as feathers and scales) are discussed elsewhere in this book.

Electron-Microscopy

Both TEM and SEM have an important role to play in certain types of forensic investigation. The correct preparation of material for electron-microscopy is all important and is therefore covered in some detail.

Transmission Electron-Microscopy

Rapid fixation of small fresh pieces of tissue is essential for good results. The fixative of choice is 2% buffered glutaraldehyde but a freshly prepared buffered paraformaldehyde solution may be substituted if glutaraldehyde is unavailable. The ideal specimen is a biopsy or comparably tiny piece of tissue. Larger specimens should be cut immediately into very small blocks (1 mm^3) prior to fixation. Tissues should be rapidly fixed as soon as possible after death. Small (1 mm^3) pieces of tissue require at least 1 h in the fixative whilst larger specimens will require up to 4 h fixation before transference to a buffer solution.

Scanning Electron-Microscopy

SEM has many uses in wildlife forensic work, particularly the examination of keratinous structures such as hairs, hooves, feathers and scales and other surface structures such as mucosa. A number of techniques have been developed as part of veterinary or zoological studies and can be applied to forensic investigations – for example, the work on microwear of teeth and important research on nutrition. In the past such studies, essentially the examination of scratches, were carried out using a stereo-microscope (see 'The Value of Microscopy' section in Chapter 12). SEM uses higher magnification and provides better differentiation of microwear parameters. A seminal paper on microwear in wild animals was that by Young and Robson (1987) on the molar teeth of the koala.

Tissues that are to be examined by SEM should be thoroughly washed in saline/buffer to remove mucus, plasma, urine, fluid, etc. before fixation in 2% glutaraldehyde for at least 1 h for small (1 mm^3) specimens and up to 4 h for large (>10 mm^3) specimens, following which they can be transferred to a buffer solution.

Specimens for both transmission electron-microscopy (TEM) and scanning electron-microscopy (SEM) can be stored in buffer at 4°C before being submitted to the laboratory for investigation.

All samples for EM must be adequately packed (screw-top bottles containing solutions should be sealed in a plastic bag before packing). The Post Office and other regulations regarding pathological material must be followed (see Chapter 3).

Parasitological Examination

Parasites have been mentioned many times in this chapter and elsewhere in the book. Parasites are not just of importance in terms of the health of animals but they can also provide information of forensic significance. In captive wildlife, large numbers of parasites may be relevant to the health and welfare of the animal, and reference to them can form part of necessary evidence pertinent to appropriate, especially welfare, litigation.

Space does not permit detailed discussion of parasitological techniques but these are amply covered in many other texts, including those orientated towards veterinary diagnosis (see, e.g. Taylor et al. [2007] and Elsheikha and Khan [2011]).

Determining the presence and identity of parasites is always important in forensic cases. This information may help to pinpoint the origin of an animal (see Table 10.3) because some parasites are strongly host-specific – for example, fleas and monogenean trematodes – and because parasites can help to determine such things as the origin of nesting material or even a piece of animal tissue.

Types of Sample

Samples for laboratory examination in wildlife forensic work can range from whole carcases to tiny portions of tissue. In this section some of those samples, most very specific to work with wild animals, will be discussed with a view to illustrating the wide range of laboratory tests that are available and demonstrating the different ways in which they can be applied to diverse specimens.

Pellets (Castings) from Birds

Background

A remarkable number of species of bird can produce pellets (castings), in addition to the familiar raptors, gulls and crows. For example, Gillham (1967), writing in the Bulletin of the (now defunct) International Bird Pellet Study Group, described pellet ejection by thrushes. Internet searches reveal many websites with information about finding and recognising pellets: see, for example, that of the Royal Society for the Protection of Birds, www.rspb.org.uk/youth/makeanddo/pellet/recognising-pellets.asp.

Research on birds of prey (Order Falconiformes and Strigiformes), in particular, often employs pellet examination in order to obtain data about the feeding habits and diet of these species (Cooper, 2002; Marti et al., 2008). Such techniques can also be used to investigate species that are lower in the food chain, such as seed-eating birds.

Bird pellets can yield much useful scientific information, quite apart from providing visual evidence of what a specific bird has been eating. They have even proved valuable in archaeological studies. In one study, two concentrations of animal bones, almost exclusively from small mammals and wild birds, were found within the destruction debris of a Roman bath complex in Sagalassos (Turkey). They were found to be those of the European eagle owl. Radiocarbon-dating indicated that eagle owls lived in the collapsing bath complex during the second half of the sixth to the beginning of the seventh century AD, before the final abandonment of the town. In another report, Holdaway and Worthy (1996) reported the investigation of fossil deposits of degraded pellets of the endemic, now extinct, strigid owl *Sceloglaux albifacies*, in New Zealand. They described techniques they used for collection and processing. Using radiocarbon-dating they were able to determine the age of the degraded pellets as more than 10,870 years. Examination of pellets has also yielded information on neoplasms in the bones of frogs (Tyler et al., 1994).

Pellets often have much to offer in legal cases, but their examination is rarely covered in forensic texts. As was pointed out by Cooper and Cooper (2007), pellets may contain the remains of prey species and this can be of great evidential value since it provides information on recent diet and/or parasites or vegetation that may be characteristic of a particular location. In addition to this, ejected pellets attract moths and other scavengers, especially when the castings are a few days old, and these too may be relevant to both scientific and forensic enquiries. Other examples of their potential forensic value are (a) the detection of ingested poisons, or of prey that contains poison (Coeurdassier et al., 2012), and (b) helping determine when a bird was captured or killed. A simple example of the latter was given by Knight (1968) who showed that pellets produced by little owls in the United Kingdom in May or June are very likely to contain the remains of chafers.

Practical Considerations

Recognition of pellets is not always easy. They can usually be differentiated from faecal masses by the fact that they (pellets) do not usually smell or compress easily (Sutherland, 2004).

For optimal forensic investigation, pellets should be obtained as soon as possible after ejection. Older samples voided days (sometimes weeks) earlier may reveal items of note, such as parasites, but physical damage and invasion by invertebrates can complicate interpretation.

Gloves should be worn to protect both the pellet and the handler. The glove can be turned inside-out and sealed in order to contain the specimen. Transportation and storage are important. Pellets are relatively robust but there are exceptions, such as those from kingfishers (Knight, 1968). They can be transported in plastic or paper bags; the latter are less prone to condensation and moulds. Pellets can be stored by drying or freezing with little effect on gross contents, but the effect of such methods on laboratory investigations has, apparently, yet to be evaluated.

Examination

Gross and microscopical examination remains the mainstay of pellet examination. One of the best and most straightforward protocols devised was that described by Knight (1968), for naturalists, and this can readily be modified for forensic laboratory investigations.

Although pellet examination is a valuable tool, it has its limitations and drawbacks, including bias (Marchesi et al., 2002), as was recognised by some workers 80 years ago (Brooks, 1929). Although numerous studies of predatory birds have reported dietary proportions based on analyses of large numbers of pellets, such analyses are often markedly biased and hence unquantifiable because some prey remains are more conspicuous or persistent than others. Simmons et al. (1991) investigated this bias for the bird- and small mammal–eating African marsh harrier, using an essentially independent measure of diet – that is, observed prey deliveries to the nest. Observations showed that bird prey, particularly large wetland species, was over-represented almost threefold among remains. Small mammals were under-represented, whilst fish, frogs and eggs were marginally over-represented. Analyses using pellets were also biased but in the opposite direction from that of remains. In a similar context, Simmons et al. (1991) showed that by combining pellets and prey remains (collected with equal effort), accurate estimates of overall diet could be achieved. This was confirmed using month-by-month comparisons of small mammals, in which proportions derived from pellets and remains never differed by more than 10% from those established from direct observations.

When identifying the contents of pellets, reference should be made to appropriate texts – for example, the work by Clevedon Brown (1969) on the recognition of the pelves of British rodents.

The key steps of any investigation are listed in Box 10.3.

Animal Diets and Foodstuffs*

Legal and insurance cases relating mainly to captive wildlife, for example in zoological collections, can be concerned with the quality of food that is provided. Foodstuffs (feed) can be the cause of infectious disease or malnutrition in animals and the manufacturer or supplier may be held liable.

---

* A joint contribution by Jill Webb and John E. Cooper.

refractive index. An acidified chloral hydrate–glycerol solution, also known as Hertwig's solution, serves this purpose extremely well. However, use of chloral hydrate is controlled in some countries and the investigator should establish its legal status in a country before attempting to transport it there.

It is important that, during sampling, all substance types present are represented. The investigator should screen with stereo-light microscope or, if that is not available, with a magnifying glass (hand-lens). Two important aspects are (a) to look for homogeneity/heterogeneity and (b) to isolate small volumes from each different substance type. Pelleted feed should be allowed to soak and then dispersed in water.

Other important guidelines are to smear/squash semi-solid material, to tease/section solid materials (note if it is fibrous, as in, e.g. muscle or tendons or if it is solid tissue, e.g. liver, kidney); to see if it consists of vessels or alimentary tract; to look for seeds and isolate them separately; to look for grain, flowers and leaves, bones and fibres/hairs. It is also important to search for foreign materials – pebbles and rocks, fragments of wood, metal/bullets, wire/synthetic line and mould hyphae and spores.

Captive wildlife, like domesticated animals, may be fed processed foods. Those designed for carnivores (dog/cat foods) will have highly processed contents indicated by gelatinised starch. Ungulates may be fed grain (and/or pellets) but this is likely to be rolled.

In any investigation it is useful to collect examples of seeds and flowers from the local environment.

Various methods have been used to retrieve and to examine the stomach contents of wild animals, especially birds (see Chapters 8 and 9) (Sutherland, 2004). Logistically, screening the stomach contents of a well-fed, large ruminant is physically far more demanding than screening that of a much smaller carnivore, for example a cat or a bird.

The investigation of bird pellets is covered earlier in this chapter.

Faeces

The faeces of animals are potentially valuable evidence in their own right. They also attract invertebrates (Cooper and Cooper 2007). Scarab beetles are the best-known and documented examples, but there are many species of Coleoptera and other taxa that are attracted to faeces, and play a part in their breakdown and thereby can provide information.

Numerous studies have been carried out over the years on the value and drawbacks of using faecal analysis to study the biology of mammals. Examples include research on otters (Carss and Parkinson, 1996; Carss and Elston, 1996) and on Arctic foxes (Hersteinsson and Macdonald, 1996). The latter is one of many that includes useful information on how samples were collected and processed; such data can often be applied to forensic cases. In the case of the Arctic fox faeces (scats), for instance, each sample was individually numbered, dried in an oven at 50°C and weighed. It was then fragmented, individual food items were removed and their relative volume was estimated.

The examination of faeces for forensic purposes was discussed briefly by Cooper and Cooper (2007). Descriptions need to be as accurate as possible, not only for professional reasons but also because, like any other evidence, they may be re-examined or checked by someone else. Quantification is important and some scoring system is recommended when there is a need to quantify. Colour codes are useful (Cooper, 2009a); objectivity is enhanced if reference can be made to colours on a chart (such as those used by paint companies).

The consistency of faeces can be difficult to describe in an unambiguous way. Where appropriate, use should be made of the grading systems (usually 1–5, liquid–solid) advocated by some of the animal food companies. It must be borne in mind that the consistency of any sample may be affected by a number of different factors. Thus, for example, loose faeces may be a feature of any of the following:

- A change in diet or a diet that is very rich contains much fluid
- Ill-health – diarrhoea (loose faeces), which may or may not be an indication of enteritis (also see Appendix A)
- Stress, usually acute – mountain gorillas, for example, will void very loose faeces if they are startled by humans or predators

Changes in the appearance of the droppings of birds may also be significant. Sometimes they are indicative of disease, but often they reflect diet or other factors. In birds of prey, for example, green coloration to the faecal component of the dropping may be seen when the bird has fasted, reflecting bile in the intestine (Cooper, 2002). Tan colour may be associated with the feeding of dead day-old cockerel chicks to birds, usually readily confirmed by the finding of tiny, characteristic, chick feathers both in faeces and in regurgitated pellets. These findings can all be an important clue to the history of a bird of prey, especially if it is alleged to have come in from the wild, where cockerel chicks are not a regular part of a raptor's diet.

Droppings of birds and reptiles are a combination of faeces, from the rectum, and urates, from the kidneys. It is important when dealing with droppings of birds and reptiles that the faeces have been separated from the urates (the latter can be examined independently) – see Figure 10.3.

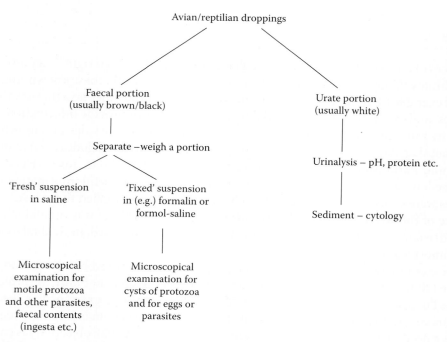

Figure 10.3 Appearance and examination of avian/reptilian droppings (faeces/urates).

Sample age and storage method are not too critical in standard tests on faeces. Indeed, parasites can often be detected in very old samples (coprolites) (see, e.g. Luiz et al., 2002). However, there is evidence that delay and suboptimal storage may adversely affect DNA yield and microsatellite amplification – as shown in the work on baboon faecal samples by Bubb et al. (2011).

Two examples of the value of faecal examination follow: both of them date back some years, but they illustrate methods that have potential in forensic work with wildlife.

The chemical composition of elephant dung was measured by Weir (1972). This information was used to show that the sodium budget of an elephant may be precarious, and that the biology, behaviour and distribution of elephants may be affected accordingly. Samples of this dung were ashed, extracted in acid, and the sodium, potassium and calcium content determined by flame photometer.

There are many species of carnivore whose behaviour and ecology is still very imperfectly understood, especially amongst the smaller members of the order. Food selection of aardwolves was studied by Kruuk and Sands (1972), who were able to report on the role of termites in the diet of this species and the interactions between aardwolves and their prey.

Feathers

Background

In the past, investigation of feathers attracted less attention than did hair despite the importance of the plumage of birds in terms not only of their health, welfare, appearance and performance but also the information feathers can yield that is relevant to forensic investigations.

A considerable amount of information follows about feathers. This is intentional, not only because of the paucity of information about these structures in most other forensic publications but also because certain of the techniques discussed later can be used in whole or in part for other keratinous structures, including hair of mammals and scales of reptiles and other species. A few of the methods described are also applicable to the exoskeleton of invertebrates, but it should be noted that chitin is different from keratin in many important respects.

Little has been published on the laboratory examination of feathers other than relatively brief descriptions in some textbooks (Cooper, 1985) and more detailed accounts in papers dealing specifically with viral infections (Pass and Perry, 1984) and chemical or toxicological studies (Bortolotti and Barlow, 1985; Gerlach and Leipold, 1986; Goede and De Bruin, 1984). The paucity of published literature on the subject as a whole, however, means that many forensic investigators, as well as veterinarians, ornithologists and research workers, are unaware of the diagnostic tests that can be carried out on feathers and ignorant of how samples should be taken and submitted.

Biology of the Feather

An understanding of the biology of keratinous structures is an important prerequisite to working with such material and understanding its value in forensic investigations.

A number of authors have described the structure and development of feathers (Ginn and Melville, 1983; Lucas and Stettenheim, 1972; Spearman and Hardy, 1985), and some have also discussed the various abnormalities and disorders that may be seen in different species (Cooper, 2002a).

Feathers vary considerably in strength, hardness and resilience to trauma, and this may be relevant in legal cases where the appearance and physical status of a feather is a matter of dispute. Factors that may influence strength, hardness and resilience include

- The type of feather, for example flight or down
- The species of bird, for example a ratite versus a passerine
- The age of the bird
- The health of the bird
- The age of the feather
- The colour (pigmentation) of the feather, including extraneous coloration (Brown and Bruton, 1991)

Normal feathers are strong and resilient. Very large feathers, for example those of geese, may be difficult to cut and impossible even to bend by hand. Breaking a feather shaft is not easy unless the feather is old and brittle or of poor quality (or if a feather is naturally moulted and then exposed to environmental influences). When a normal feather shaft is bent, it tends to become dented and pliable at that point and bends easily. A brittle feather *can* be broken but when this happens, the ends are not flat and relatively plane (smooth) as in snapping a dead twig, but jagged, with points and tags of keratin remaining – as when a tree is felled without sawing completely through the bole.

Feather shafts, largely regardless of the various aforementioned factors, can be cut clearly with a scalpel blade or razor blade. If the cut is complete, the two pieces separate and expose flat, clearly incised, ends. If the cut is incomplete, the two pieces remain attached initially. They can be separated by moving them backwards and forwards repeatedly but, as with breaking (earlier), tags of keratin remain, giving a jagged appearance that is clearly visible microscopically.

In the laboratory, the keratin of feathers can be softened for processing using phenol. This will facilitate studies on damage caused to feathers by traumatic insults.

Collecting Samples

Examination of the live or dead bird is important before feathers are removed for laboratory tests. In the case of a live bird, preliminary observation is important and should precede clinical examination (see Chapter 9): a bird that is unaware that it is being watched may show significant clinical signs such as excessive preening, scratching or muscle fasciculations.

Samples for laboratory investigation will include dropped feathers, plucked feathers, or portions of feather removed by clipping, biopsies of feather follicle, swabs, parasites and other material.

The choice of feathers to be sampled will primarily be dictated by the demands of the forensic case, but other factors may need to be considered. For example, when looking for feather and quill mites in parrots, it is important to remember that different species of mite tend to be found in distinct areas of the body (Perez and Atyeo, 1984). Some feather specimens can be taken from live birds without touching it; others may involve minimal contact,

Scales

Examination of the scales of reptiles (and sometimes of birds) plays an important part in species, identification, but can also provide other information of forensic importance. Martin (2012) detailed the identification of reptile skin products using scale morphology.

Reptile skins may consist of full-thickness skin together with overlying scales, as in the case of prepared material, or the shed (sloughed) skin that comprises only the outer keratinous layers. The latter, the sloughed skin, is produced quite naturally by reptiles, particularly snakes and lizards, as part of their growth process. Sloughs are used by herpetologists and veterinarians to assess the health of reptiles and can be used for similar purposes in forensic investigations. Shed skins should be secured in separate plastic bags, properly labelled. It is important that these bags are transparent so that the slough can be examined through the bag before being removed; in this way, it may be possible to detect such items as parasitic mites that might otherwise be missed.

The examination of a slough, complete or partial, comprises examination with the hand-lens followed by low-power microscopy. Transmitted light (transillumination) is vital and will help to reveal lesions such as scars from burns and sites of attachment of ticks. Pieces of slough can be mounted in normal saline and examined under a microscope. They can also be sectioned for histological examination.

Complete skins of reptiles are usually presented for examination in a prepared state. They still consist of full-thickness dermis and epidermis, but the material has been treated in order to preserve it. This can complicate examination and interpretation of findings. Skins that have been dried, with or without salt or other temporary preservative, can be rehydrated for examination and a whole range of laboratory tests, including histology, can then be performed.

Amphibians do not have scales, but they, too, shed the outer layers of their skin and this presents as thin, soft, transparent material which is delicate and must be handled with care. Such material, which can often provide useful information about the amphibian and its environment, is best handled and processed in normal saline in a Petri dish. A range of investigations can be carried out, including those described for the slough of reptiles above. Transillumination is, again, vital. Amphibian material should be cultured for bacteria and fungi in cases where infection is suspected or forms part of the grounds for prosecution.

The exoskeleton of arthropod invertebrates is made of chitin. It too is sloughed and can be used to assess the health of an invertebrate or, sometimes, to provide information relevant to a legal case. These exoskeletons should be dealt with in a very similar way to the shed skins referred to the aforementioned. They are generally more robust.

Eggs and Eggshells

The examination of whole eggs or parts from them can yield information that may be useful forensically. In particular, the failure of eggs of birds or reptiles to hatch may be the subject of a legal case because either (a) the animals were free-living and there is a suggestion of disturbance to the eggs which resulted in their not hatching, or (b) the animals were in captivity and it is alleged that mismanagement or malicious interference with, say, an incubator caused the eggs to fail to hatch.

Abnormal findings in eggs and their possible significance are tabulated in Table 10.6. Whilst primarily relating to birds, much of the information can be applied also to the eggs of reptiles.

Table 10.6 Some Findings in Eggs and Their Possible Significance

Finding	Possible Interpretation(s)
Infertile egg.	Parent birds not of same species; of one sex; one or both immature; incompatible (e.g. because of physical injury) of courtship or copulation.
Soft-shelled, thin-shelled or abnormal egg (infertile or fertile).	Calcium deficiency; oviduct abnormality or old age (hen bird). Elevated levels of chlorinated hydrocarbon insecticides.
Profuse growth of bacteria from contents of infertile egg.	Contamination in nest/incubator. Prolonged/poor storage.
Profuse growth of bacteria from contents of fertile egg. More than one species of organism, not recognised avian pathogens.	As mentioned earlier.
One (possibly two) organisms in pure culture, recognised avian pathogens.	Embryonic death possible due to infection.
Dead embryo, late incubation, no evidence of infection, unusual position in egg.	Death possibly due to inability to hatch (e.g. malpositioning).
Dead embryo, no evidence of infection or malpositioning. Abnormalities, e.g. hydrocephalus, duplication of digits.	Death possibly due to genetic or other factor causing abnormality. Radiographic examination may assist.
Dead embryo, no evidence of infection, malpositioning or abnormalities.	Incorrect incubation temperature or relative humidity, faulty turning in incubator; prolonged/poor storage; infection; nutritional deficiency.
Embryo dead-in-shell at hatching stage.	Low relative humidity prior to hatching, trauma (e.g. human interference).
Elevated levels of chlorinated hydrocarbon insecticides, heavy metals or other chemicals in egg contents or eggshells.	Infertility/death of embryo, possibly due to poisons *or* elevated levels indicative of high environmental pollution.

The investigation of eggs from a veterinary, diagnostic, point of view was discussed by Cooper (2002). The basis of such work is depicted in the following:

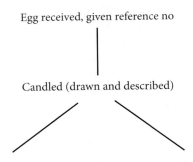

Egg received, given reference no

Candled (drawn and described)

Probably infertile
Weighed, measured, exterior drawn and described. Cleaned with methanol/ethanol. Opened, observed, then examined, drawn and described in situ. Placed in Petri dish, samples taken as necessary for histology, microbiology. Contents frozen for toxicology and DNA studies. Shell dried, weighed and retained.

Probably fertile
Weighed, measured, exterior drawn and described. Cleaned with methanol/ethanol. Opened, observed, then examined, drawn and described in situ. Placed in Petri dish, samples taken as necessary for histology, microbiology, Contents fixed/frozen (embryo retained) for toxicology and DNA studies. Shell dried, weighed and retained.

Dealing with Samples

Figure 10.9 An avian embryo submitted for forensic necropsy. There are no obvious external abnormalities.

Reference should also be made to the document entitled Forensic investigation of eggs/embryos (Appendix D).

Various components of the egg may be needed as evidence, ranging from albumen and eggshell to embryonic tissue. Eggshells can usually be kept dry but other components should be refrigerated, frozen or fixed, depending upon the circumstances. Mikhailov (1997) produced an excellent atlas, depicting the different features of eggshells, visible on SEM, and Solomon (1991) described the appearance of eggshells and contents as they related to quality. Both texts have forensic application.

The measurement of avian eggshell thickness may be necessary in forensic investigations. It is often expressed by means of an index based on the length, breadth and weight of the shell. Reduction of eggshell thickness in raptors may indicate a nutritional deficiency or imbalance (Cooper, 2002) or can be an effect of environmental pollution (see Chapter 2).

Shell thickness measurement can be difficult unless specially modified micrometers are available. The eggshell thickness index, first described by Ratcliffe (1967, 1980) in connection with his studies on shell-thinning associated with chlorinated hydrocarbon pesticides in birds of prey continues to be refined (Maurer et al. 2012).

The importance of examination of embryos was discussed by Cooper (2002), who provided detailed information and a number of references. In a legal case it may be necessary to determine at what stage an embryo died or was subjected to adverse stimuli (Figure 10.9). Developmental abnormalities may be detectable with the naked eye, dissecting microscope or on radiographic examination. There remains a need for more baseline information on embryonic development in order to assist those involved; only a few publications provide such data – for example, Pisenti et al. (2001) on embryonic development of the American kestrel.

Teeth, Including Ivory

As emphasised elsewhere in this book, the teeth of mammals often provide valuable forensic evidence, in large part because they are made of durable materials and generally survive better than do soft tissues and even bone. A useful review of mammalian teeth is provided by Chapman (see 'Dentition of Mammals' section in Chapter 13).

Despite their robust nature, teeth must be handled carefully and metal instruments used sparingly. Careful inspection of the surface of the surfaces of crown and root is essential. Note should be made of colour. The presence of calculus may be significant in legal cases, especially those relating to the nutrition and care of captive animals. However, in some species the presence of deposits on the tooth surface may be normal – for example, black tartar is a feature of most adult mountain gorillas (Schaller, 1963). Miles and Grigson (1990) commented that calculus on primate teeth tends to be detached after a time and 'may even have been cleared off when the cranium was prepared', an important consideration when assessing teeth forensically.

The specialised dentition of elephants was described by Spinage (1994). As Chapman (see 'Dentition of Mammals' section in Chapter 13) pointed out, elephant tusks have, in addition to concentric rings, an engine-turned appearance. Ivory from elephants has long been prized and used for a variety of purposes ranging from carvings to billiard balls and piano keys. Miles and White (1960) recounted some of this history in a paper that also described specimens, both normal and pathological, based on ivory held by the Royal College of Surgeons of England (RCS). The RCS material provides a valuable reference collection for forensic investigators.

In recent decades the demand for ivory has been the stimulus for conflict, social unrest and widespread poaching of elephants (see Chapter 2). Possession of certain ivory samples is permitted prior to a particular date. For instance, the European Union permits the trading of genuine antiques if the samples date back to before 1 June 1947. There are few methods that can accurately differentiate ivory prior to 1947 and post-1947, but Linacre (2009) alluded to techniques relating to the types of radio-isotopes released by nuclear weapons, as these would only be in the environment from 1945 onwards.

The recognition of elephant ivory and its differentiation from superficially similar substances, such as cow horn, is therefore very important. Various forensic tests are available to assist in such determinations. Ivory can be identified by microscopical examination or by elemental analysis. The former depends upon the amount of material present and whether the structure of the original ivory has changed. There are methods for the isolation of trace amounts of DNA from ivory, including from ivory statues, in which case species identification is possible.

Other Samples

In a wildlife forensic case, it may be necessary to examine various specimens, from different species of animal and often in varying degrees of preservation. These can include, for example, blood, urine, cerebrospinal fluid, exudates, transudates and pus. Various laboratory investigations that are performed routinely in veterinary (and medical) laboratories can be applied.

Water samples for bacteriological analysis must be collected under sterile conditions to minimise the risk of contamination by organisms. Water for investigation may be taken from an open water source, such as a lake or river, a well or a tap. There are standard techniques for sampling each; these are described in publications for hydrologists, aquaculturists and laboratory technologists.

Genetic Methodologies in Wildlife Crime Investigations

11

LOUISE ANNE ROBINSON

Contents

Wildlife Crime and Forensic Genetics	367
Use of Mitochondrial DNA	369
Advantages of Mitochondrial DNA	369
Drawbacks of Mitochondrial DNA	369
Methods of Species Classification	370
Species Identification	371
Universal Primers	371
Nested PCR	372
Comparative Sequence Analysis	372
Contaminated DNA Samples	373
Species Detection	374
Species-Specific Primers	374
Restriction Fragment Length Polymorphisms	375
Alternate Methodologies of Species Detection	376
Geographic Location	377
Individual Identification	377
DNA-Sexing	378
Conclusions	378

Wildlife Crime and Forensic Genetics

Recognised as the second-most prevalent crime worldwide after drug trafficking (Linacre, 2008), wildlife crime encompasses a range of criminal activity such as poaching (Caniglia et al., 2009), animal cruelty (Merck, 2007), the use of derivatives in traditional medicines (Alves et al., 2010; Peppin et al., 2008) and trafficking of animals (Oldfield, 2003) (see also Chapter 2). Not only is trafficking an issue due to the reduction of endangered populations but the translocation of new species to an environment can have devastating effects on the ecosystem by introducing infectious pathogens (Crowl et al., 2008; Pedersen et al., 2007; Smith et al., 2009) and non-native species (Jenkins and Mooney, 2006; Manchester and Bullock, 2000) which may be able to out-compete native species (Smith et al., 2009a) (see also Chapter 2).

Wildlife investigations occur when it is suspected that laws regarding issues such as trade or the protection of endangered species have been breached (see Chapter 3). Wildlife legislation from the European Union (EU) and Council of Europe (CoE) is implemented by national laws (Cooper and Cooper, 2007). In the United Kingdom, the primary legislation for the protection of wildlife and their habitats is Part 1 of the Wildlife and Countryside Act, 1981. To assist the enforcement of regulations, the Partnership for Action against Wildlife Crime, United Kingdom (PAW UK) brings together over 100 organizations representing a wide range of interests, such as the Wildfowl and Wetlands Trust, The British Association for Shooting and Conservation, DEFRA, the police and RSPCA (http://www.defra.gov.uk/paw/about/partners/), to help combat wildlife crime by improving awareness and regulation (see Appendices E and J). In the United States and Canada, the majority of wildlife legally belongs to the state and permits or licences are required to kill or possess wild animals (Stroud, 1998).

Species considered to be threatened or endangered are listed in the appendices of the Convention on the International Trade in Endangered Species of Wild Fauna and Flora (CITES) from which stem the national or European Union laws that regulate the international trade in wildlife (Cooper and Cooper, 2007). Any movement of a CITES species or its derivatives from one country to another requires permits from a national CITES Management Authority. It is the lack of these documents or the incorrect declaration of species that is commonly seen in trafficking offences. For more information on wildlife legislation, see Chapter 3.

It is often stated that the illegal wildlife trade is estimated to value more than $20 billion but this figure, often referenced to Interpol (Alacs et al., 2009), appears to be a fabricated value (Christy, 2010). As limited data are available to make these evaluations, vast discrepancies have been found to occur between estimations based on CITES and US Customs data (Blundell and Mascia, 2005). With few wildlife crime cases leading to prosecution (Johnson, 2009), coupled with low penalties and the high value of prized species (Linacre and Tobe, 2011), pressure is placed on endangered wildlife in much of the world as this trade continues to grow and populations are over-exploited (Alacs et al., 2009).

Genetic methodologies within wildlife forensic investigations often seek to answer questions regarding species identification, geographic location, individual identification and occasionally the sex of the species of study. This chapter will discuss in detail one of the most frequently sought pieces of information within a wildlife crime investigation, species identification, and the current methods in place for detection and identification. All other areas will be covered briefly for the information of the reader but will exclude botanical forensic methods (Koopman et al., 2011).

There is currently no standardised methodology for the identification of non-human samples (Tobe and Linacre, 2007). In view of this lack of standardisation, good laboratory practice is encouraged and guidelines have been suggested for quality control and assurance (Ogden, 2010). With few specialised wildlife laboratories in the world (see Chapter 12), the first of which was established by the US Fish & Wildlife Service in Oregon, the United States (http://www.lab.fws.gov), many non-human investigations are conducted by academic institutions (Budowle et al., 2005; Ogden, 2010). Methodologies approved by the International Society for Forensic Genetics (ISFG) include nucleotide sequencing (Carracedo et al., 2000; Linacre et al., 2011a) and the use of validated genes such as Cytochrome *b* (Cytb) following Technical Working Group on DNA Analysis Methods (TWGDAM) (Branicki et al., 2003). Many of the procedures relating to wildlife crime scene investigation and the storage and transport of materials mirror those in human investigations as these methods have been established to ensure that evidence presented in court has not been contaminated or tampered with (Cooper et al., 2009).

Use of Mitochondrial DNA

Genetic evidence in human crime cases most frequently refers to the identification of an individual through their genetic fingerprint. Variable number tandem repeats (VNTRs) are used within a person's nuclear DNA (nDNA) which can discriminate between all but identical twins. However, although individual identification is sometimes required in wildlife investigations (to be discussed later), often identification is only required to the level of species. For this we turn to maternally inherited extranuclear mitochondrial DNA (mtDNA) as the level of inter-species variability is often sufficient to separate closely related species.

Advantages of Mitochondrial DNA

The number of mitochondria within a cell (containing two copies of nDNA) varies depending on cell type, with each mitochondrion containing between 1–15 copies of mtDNA (Linacre, 2009). Having a higher copy number, mtDNA is more suitable in cases where DNA may be old or degraded (McDowall, 2008). Its overall rate of degradation is also slower than that of nDNA as the organelle is protected from bacterial activity by a protein coat. In addition, the structure of mtDNA is circular rather than linear as seen with nDNA, which is thought to make it less vulnerable to exonucleases attaching to free ends of the DNA (Lee et al., 2011; Seutin et al., 1990). The rate of mutation within the mitochondrial genome is far greater than that seen in the nuclear genome, assisting the identification of species through inter-species variation. This higher mutation rate occurs because the specialised correction mechanism for copy errors present in nDNA is absent in mtDNA.

The mitochondrial genome is 15–20 kb in length and variation between species normally occurs due to the size of the non-coding region called the displacement loop (D-loop), also known as the control region. With no introns, mtDNA contains 37 genes, 13 of which code for sub-units of the electron transport chain (protein coding genes), 22 transfer RNA (tRNA) and 2 ribosomal RNA (rRNA) genes (Ballard and Whitlock, 2004). All vertebrate mtDNA contains the same number of genes which can differ in their arrangement (Mindell et al., 1998). The standard avian mtDNA arrangement can be seen in Figure 11.1.

Drawbacks of Mitochondrial DNA

Because of its maternal method of inheritance, one of the most important limitations of mtDNA to consider when conducting species identification is its inability to identify hybrids. If it is suspected that a species may be susceptible to hybridisation, mtDNA may not be the most suitable avenue for establishing identification and nDNA techniques should be utilised to provide classification.

In addition, amplification of nuclear mitochondrial insertions (NUMTs) is possible when working with mtDNA. These occur when mitochondrial sequences transpose to the nuclear genome (Sorenson and Quinn, 1998) and the data obtained are therefore not representative of the genome region required. This can cause problems in forensic investigations when sequence data are used for identification purposes (Goios et al., 2006). However, measures can be taken to overcome this issue as short sequences of mtDNA are usually transposed. NUMTs can therefore be identified by routine examination as they will frequently co-amplify with the correct mtDNA sequence, contain unusual mutations and produce mismatched sequencing. Overlapping polymerase chain reaction (PCR) products

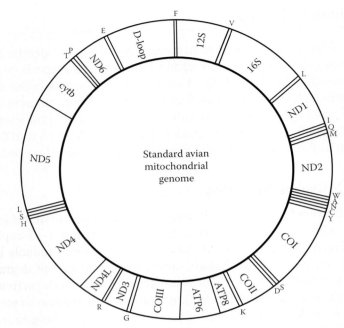

Figure 11.1 Standard avian mitochondrial genome demonstrating gene order. Protein coding genes can be seen on the internal circle (excluding the non-coding region of the D-loop and the 12S and 16S rRNA genes). The positions of all 22 tRNA genes are shown by their one-letter amino acid symbols.

assist the verification of sequencing as it is unlikely that a NUMT region will amplify with two different primer sets (Sorenson and Quinn, 1998).

Methods of Species Classification

When the full carcase of an animal is available (or species-identifying characteristics present) it may be possible for species classification to be conducted by a forensic veterinarian or similar professional by assessing distinctive morphological characteristics (see Chapters 9 and 10). However, if these features are indistinguishable or missing, as may be the case if they are removed by hunters (Linacre, 2009) or chemical processing of materials has occurred, genetic analysis is required to establish identification.

As no unanimous method of species identification has been determined, a number of techniques are currently used for classification. However, care must be taken to define between methods of species detection and species identification (Ogden et al., 2009). In essence, detection methods use known variation within the species' mitochondrial genome to create methodologies that can detect DNA of a target species. Identification methods use nucleotide sequence data to provide either a comparison match to reference sequences or create a phylogenetic tree to classify the species (Linacre, 2009).

Single nucleotide polymorphism (SNP) is the term given to variation seen at a single base which can be used in many methods of species classification. SNPs can be employed specifically to perform species identification (Bento et al., 2009; Kitpipit et al., 2009; Köhnemann and Pfeiffer, 2010) but are most commonly used within other methodologies.

They are the basis of species-specific primers, incorporating SNPs of the target species in primer design, restriction fragment length polymorphism (RFLP) analysis where SNPs can alter the number of recognition sites within a sequence, or in mitochondrial sequence data comparison where their accumulation enables separation of species by similarity scores.

Species Identification

Universal Primers

If no information is available regarding the origin of a sample in a forensic investigation, the most common practice is to amplify a region of DNA with universal primers for comparison with a known data set and establish identity by sequence similarity. The most frequently used mitochondrial region for species' identification is that of Cytb. This gene is validated for use and is the most widely represented mitochondrial gene within nucleotide databases due to its use in taxonomy (Branicki et al., 2003; Linacre and Lee, 2005; Parson et al., 2000). Universal primers are created by aligning sequence data from many different species and identifying regions of conservation within the alignment. If suitable with regard to primer design issues such as GC content, these regions are then used to create forward and reverse primers which will amplify DNA from a high number of species. Some of the most commonly used Cytb universal primer sets are those described by Kocher et al. (1989) often combined with Irwin et al. (1991) primers which have been used to identify rhinoceros horns (Hsieh et al., 2003), snake species (Wong et al., 2004), petrels (Parik et al., 2008) and have been tested on multiple species (Hsieh et al., 2001; Tsai et al., 2007).

More recently, the primer combination mcb398/mcb869 was created by Verma and Singh (2003); this amplifies a larger region for comparison than many of the Kocher et al. (1989) and Irwin et al. (1991) combinations. This longer sequence size improves the similarity score and therefore the confidence level of the identification whilst remaining a suitable size for use in forensic casework. This primer set has shown to be successful for a wide variety of taxa and has been used in a number of cases, such as the identification of a stray carnivore (Verma et al., 2003), a small Indian civet (Sahajpal and Goyal, 2010), bear bile (Peppin et al., 2008) and the identification of tissue samples from a slaughtered Indian peafowl (Gupta et al., 2005).

Genes also validated for use are 12S and 16S for which universal primers have been designed (Kitano et al., 2007; Mitani et al., 2009) and species identification performed (Cawthorn et al., 2011; Melton and Holland, 2007). These regions are more commonly used when very little inter-species variation is expressed. It has been shown that more species-specific sites may be present in these genes, allowing clearer identification (Arif et al., 2011). In some cases, the lack of variation is such that the non-coding region (which is therefore not restricted to mutation and evolves more rapidly) is used for species identification (Dalebout et al., 1998). The control region is also commonly used when dog breed has to be established (Baute et al., 2008; Webb and Allard, 2009) or to identify sub-species (McGraw et al., 2012). Multiplex reactions may be used to perform species identification utilising primers within a number of genes to assist classification (Bellis et al., 2003).

With the introduction of DNA barcoding in 2003 (Hebert et al., 2003a,b) and the creation of the barcoding initiative and bar code of life database (www.barcodinglife.org; Ratnasingham and Hebert, 2007), it was claimed that 648 bp of the cytochrome *c* oxidase sub-unit I (COI) gene could provide an identification system for animals. Although shown to be effective (Dalton and Kotze, 2011; Dawnay et al., 2007; Webb et al., 2006), the method can be problematic when trying to identify closely related species (Wilson-Wilde et al., 2010). Fewer sequences are publicly available for COI than for Cytb, which can be an issue when trying to identify an unknown sample against reference material. With respect to which gene (Cytb or COI) more accurately provides species identification, Tobe et al. (2011a) found that Cytb contains more variable sites than COI within a shorter sequence length and provides a higher level of congruence in species' assignment by phylogeny. Recent studies have compared these gene regions for the purpose of identification and in each it has been demonstrated that Cytb contains more inter-species variation and can therefore more easily perform species assignment (Boonseub et al., 2009; Tobe et al., 2009).

Nested PCR

In some instances, DNA quantity may be too low to perform amplification using standard universal primers. When this occurs, it may be possible to use nested PCR in order to obtain DNA at quantities suitable for further analysis. When performing nested PCR, a primer pair is initially used to amplify a larger region than that which is required for analysis (e.g. 600 bp of Cytb) in order to increase the quantity of the target sequence present in the starting material. Using a second round of PCR, primers which anneal within the previously amplified region are used to obtain the required fragment (perhaps 450 bp of Cytb). The higher level of starting material in the second cycle provides a greater quantity of final product on account of the exponential amplification process of PCR. This method has been used to identify tortoise and turtle shells (Hsieh et al., 2006; Lee et al., 2009), bulb mites (Li et al., 2010) and rhinoceros horn (Hsieh et al., 2003).

Comparative Sequence Analysis

Regardless of which primer pair has been used to produce amplification, sequencing is subsequently performed to obtain data for analysis. In wildlife studies this process is sometimes termed "forensically informative nucleotide sequencing" (FINS) (Bartlett and Davidson, 1992; Blanco et al., 2008; Sahajpal and Goyal, 2010; Wen et al., 2010) and is often completed by Sanger sequencing. The sequence data obtained are compared to a database of DNA sequences to provide identification by similarity. This may be a database created by a research group if the target species is not represented within the public domain but most frequently the Basic Local Alignment Search Tool (BLAST) available in the National Center for Biotechnology Information (NCBI) database of genetic information termed 'GenBank' (www.ncbi.nlm.nih.gov/genbank/) is used. In August 2009, the database reached 108 million sequences (Linacre and Tobe, 2011) with over 3800 taxa being added every month (Benson et al., 2011). BLAST compares the target sequence fragment to the data deposited in GenBank and produces a similarity score of the 100 most closely matched sequences. Species identification, therefore, occurs by the obtainment of a high similarity score (99%–100%). A 100% similarity match is often found from a BLAST search, which in combination with a match of less than 97% with another species (Linacre, 2008), provides a high level of confidence in the identification, depending on the length of the sequence compared.

There is no consensus on how many SNPs define intra- and inter-species variation, but it is generally accepted that one base variation per 400 bp sequence is acceptable intra-species variation (Linacre, 2009). Studies have since been performed to provide threshold values and present statistical confidence in species identification. In these studies, inter-species variation was examined by aligning 217 mammal species and intra-species variation was compared in three model species (Tobe et al., 2010a, 2011a). It was found by K2P (×100) calculated distances that intra-species variation ranged from 0 to 1.16; therefore, a value of 1.5 or lower obtained in a comparison can be classed as intra-species variation. The majority of inter-species comparisons (99.99%) presented a K2P value greater than 2.5, therefore setting the threshold for species assignment. It is possible that values that are obtained between 1.5 and 2.5 could be accounted for by a higher level of intra-species variation than would be expected and therefore analysis of further loci would be required to classify species (Tobe et al., 2010a, 2011a). Although useful as a guideline for species identification, this research cannot be accepted as a general rule because some closely related species may not conform to these figures. The size and loci of the sequenced fragment may also cause variation in the value which is calculated.

One of the drawbacks of database comparison is the possibility that sequences deposited within GenBank are not representative of the species. Although voucher specimens (see Chapter 14) are recommended for genetic study, these samples are not always available to researchers and therefore reference samples that are obtained to represent the species may be misclassified. Sequences deposited in GenBank do not undergo any validation process and therefore misclassified data may be present. It is also possible that the species being investigated has not been previously studied and therefore sequences are not present within GenBank to produce a match (Coghlan et al., 2012). In these circumstances, classification is reduced to the most closely-related species present in the database. Universal Cytb primers have recently been designed for mammal species to amplify the complete gene region. This assists sequence-based species identification by generating reference material to be deposited in GenBank for comparative purposes (Naidu et al., 2012). In addition, the use of 454 sequencing (Rothberg and Leamon, 2008) may have implications in future wildlife forensic studies as the short sequence reads which are appropriate for ancient DNA (Leonard, 2008) are suitable for use in forensic investigations. Full genome sequences, therefore, have the potential to be generated and deposited in GenBank for comparative analysis. This may in turn reduce the problem of partial sequence comparisons which may not be sufficient for reliable species identification.

Contaminated DNA Samples

When working with forensic samples that are contaminated, it is not possible to use methods such as universal Cytb amplification and sequencing analysis. However, methods of detection that are specific to a target species can identify DNA within a mixture, using techniques such as species-specific primers (Črček and Drobnič, 2011; Tobe and Linacre, 2008a, 2009) and/or RFLP analysis. These detection methods can be used to identify species based on banding patterns but additionally the species-specific bands seen by electrophoresis can be excised from the gel and sequenced. The level of human contamination within a wildlife DNA sample can be quantified (Tobe and Linacre, 2008b) but sequencing cannot occur without a method of separation (Bataille et al., 1999). Size variation seen in the control region can also be used to assist species assignment by the creation of different band sizes between species using universal primers (Fumagalli et al., 2009; Pun et al., 2009).

Species Detection

Species-Specific Primers

If a species has been previously investigated and sequence data obtained, it is possible to use this information in an alignment with closely-related and morphologically similar species to design species-specific primers. Unlike universal primers, which look for regions of conservation between sequences, species primers are designed using areas of variation that occur only in the target species (Tobe and Linacre, 2010). The higher the number of SNPs incorporated within the primer, the more likely that amplification will only occur in the target species. Primers should terminate at the 3′ end with a species-specific mutation, thereby preventing extension in non-target species. SNPs incorporated elsewhere within the primer assist annealing to the target sequence. Figure 11.2 demonstrates the creation of a species-specific primer for cheetah using Cytb data available in GenBank for a jaguar

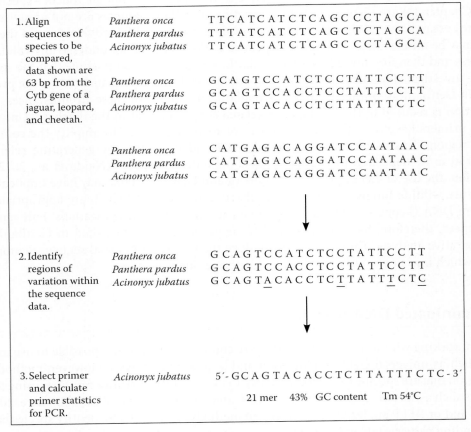

Figure 11.2 Demonstration of species-specific primer selection using Cytb sequence data obtained from GenBank for jaguar (*Panthera onca*), leopard (*Panthera pardus*) and cheetah (*Acinonyx jubatus*). Selection of a species-specific forward primer for cheetah is performed using point mutations seen only in the target species.

and leopard and cheetah. GenBank accession numbers used for the comparison were HM107682, NC_010641 and NC_005212, respectively. This diagnostic PCR process has been used to detect species such as tigers (Wan and Fang, 2003), bats (Zinck et al., 2004), alligators (Yan et al., 2005) and beetles (Cassel-Lundhagen et al., 2009).

Species-specific primers can be designed as a standard pair which anneal only to the target species or used in a multiplex reaction in which at least three primers are incorporated, thereby creating a difference in band size between species when separated by electrophoresis. If universal primers are included, this alleviates the possibility of false negatives through insufficient DNA quantity. Multiple species primers used in combination with a single universal primer has previously been used to separate Arctic fox from red fox and wolverine (Dalén et al., 2004). If few SNPs are incorporated into a species-specific primer and mutation occurs by chance, it may be possible that closely related non-target species are amplified. However, if used as an initial detection method to eliminate more distantly related taxa (such as distinguishing between domestic cattle and endangered species), this method can reduce the cost of investigations.

Species primers have been successfully developed for pygmy rabbits (Adams et al., 2011) and to identify endangered species used in traditional medicines (Tobe and Linacre, 2011). A second species detection method, such as RFLP analysis, can also be used to provide further evidence of species classification (Ravago-Gotanco et al., 2010) to increase the confidence of the findings before conducting further analysis. To confirm species identification if primers have not been validated, samples producing a positive result in these assays are sequenced and comparative analysis is performed using BLAST.

In addition, microsatellite markers have been shown in some cases to be species-specific, allowing identification from a single marker (Singh et al., 2004). Primers have also been created within nDNA using a sequence characterised amplified region (SCAR) identified with random amplified polymorphic DNA (RAPD) markers where species-specific bands are extracted and sequenced to create primers. The SCAR method, although uncommon, has been used to identify snake species in traditional medicines (Yau et al., 2002) and common fly species collected in death investigations (He et al., 2007).

Restriction Fragment Length Polymorphisms

RFLP analysis uses restriction enzymes isolated from bacteria which identify a specific sequence of 4-, 5-, 6- or 8-bases and digest at this recognised site (Cronin et al., 1991). By using RFLPs, it is possible to separate species that differ by a single-point mutation within an enzyme's restriction site. There are many endonucleases available for the digestion of a DNA strand and choosing the correct enzyme to allow detection is definitive to the success of the method. Not only can this technique be used on pure DNA amplifications but it can also be performed on contaminated samples amplified by universal primers.

Selection of endonucleases can be conducted experimentally or by using online software such as Restriction Mapper (Blaiklock, 2002; Brakmann, 2002). Enzymes are selected which distinguish the target species in a restriction digest by the presence or absence of restriction sites. Figure 11.3 shows how the restriction enzyme *Eco*RV with recognition site 5′…GAT▾ATC…3′ would digest a DNA fragment containing this sequence (represented as a single strand but occurring in double-stranded DNA). New methodologies are fast replacing the use of RFLPs, but the technique can still be of use when closely related species cannot be separated by species-specific primers and in laboratories with

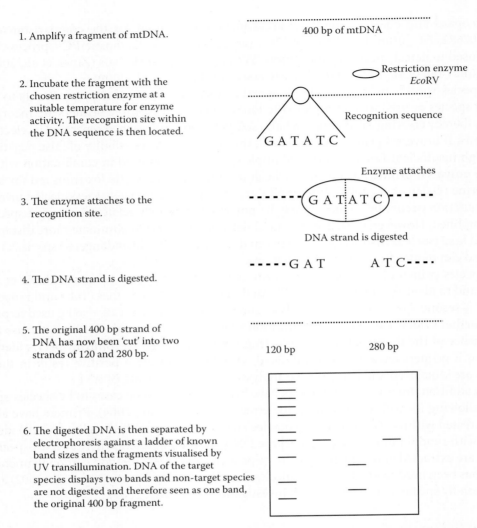

Figure 11.3 Representation of a restriction digest in a fragment of DNA containing the recognition site for the enzyme *Eco*RV subsequently separated by electrophoresis using agarose or polyacrylamide gel.

only basic equipment. Previously, RFLPs have been used to detect domesticated species (Bravi et al., 2004), bushmeats (Malisa et al., 2006; Ogden and McEwing, 2008), blowflies (Preativatanyou et al., 2010), bats (Ferreira et al., 2005), eagles (Rudnick et al., 2007), tench (Lajbner and Kotlík, 2011) and turtles (Moore et al., 2003).

Alternative Methodologies of Species Detection

High resolution melt analysis (HRMA) has been previously used to identify species (Morgan et al., 2011). Using this technique, an amplified fragment is heated within a temperature range of 50°C–95°C in which the double-stranded DNA separates. This separation is identified with a fluorescent dye which only binds to double-stranded DNA; therefore, when the strands separate, the fluorescence dramatically reduces. This separation depends on the composition of the DNA fragment and therefore can be predicted, creating species-specific melt curves.

The single base extension (SBE) method can also be used in which a primer extends by only one base which is fluorescently labelled and detected by a DNA sequencer (Belfiore et al., 2003; Nakaki et al., 2007). The presence of this base and its subsequent detection by fluorescence therefore identifies the species.

Pyrosequencing has been performed in a limited number of wildlife identification studies (Balitzki-Korte et al., 2005; Karlsson and Holmlund, 2007) but is limited to sequencing up to 500 bp and is outside the scope of many laboratories.

Geographic Location

With respect to mtDNA, its alternative use within wildlife crime investigations is to link an individual to a location or a population by its mitochondrial haplotype. This has been used to track hammerhead sharks in the fin trade back to their original location (Chapman et al., 2009) and detect the species and population origin of seahorses used in traditional medicines (Sanders et al., 2008). This analysis is however more commonly performed by microsatellite analysis, using short tandem repeats (STRs) within the nuclear genome. As populations become isolated, genetic material cannot be exchanged which therefore leads to the accumulation of genetic markers that are unique to a population, known as discrete variation (Ogden et al., 2009). Assignment tests can be performed to relate a sample back to its original population to establish the origin of derivatives, to identify regions which are poaching 'hot spots' and to ensure that captive stocks are not being supplemented from wild populations (Alacs and Georges, 2008) as occurs at times in the farming of turtles (Haitao et al., 2007). It has been suggested that a DNA register of legal stocks should be created so that investigators might more easily establish whether wildlife has been illegally taken (Palsbøll et al., 2006).

Microsatellites have been used extensively in investigations of the ivory trade (see Chapter 2) to determine the region of origin of seized goods (Wasser et al., 2004, 2007, 2008) and were also used to identify the flock to which a poisoned lamb belonged after its consumption led to the death of an endangered Egyptian vulture (Rendo et al., 2011).

In addition, isotope composition which is defined by environmental factors such as diet (Bowen et al., 2005) can be used to trace the origin of individuals. Different tissues incorporate and replenish isotopes at different rates and therefore the sample obtained must be representative of the time period required. Bone can be used to study isotopes over a lifetime, whereas tissues such as the liver have a high turnover; hair and nail provide temporal data and feathers become inert after the growth phase (Bowen et al., 2005; Linacre, 2009; see also Chapter 10).

Individual Identification

It is sometimes necessary in a case of alleged wildlife crime to identify a specific individual through analysis of STRs. However, this can only be performed in instances where a comparator is available and markers are known or can be created (Dawnay et al., 2008; Ogden et al., 2012; Pillay et al., 2010). In some cases, microsatellite markers can be used for cross-species amplification of closely related species – for example, feline markers have been used to identify tigers (Xu et al., 2005) and human markers to identify snub-nosed monkeys (Liu et al., 2008). Microsatellites vary in size between individuals and therefore by amplifying and analysing a number of sites, identification of an individual can be performed through the creation of a DNA profile. The number of markers used for this analysis has to be sufficient to create a high statistical probability that this profile is unlikely to occur in another individual. In human crime investigations in the United Kingdom, 10 loci plus a sex marker are required for full DNA-profiling but no figure has been established for wildlife profiling. SNP analysis can also be used within the non-coding regions but many more markers are required to perform the identification (Linacre, 2009), and multiplex and interpretation are more difficult than when using STRs (Butler et al., 2007). This ability to identify an individual allows investigators, for example, to monitor populations (Rudnick et al., 2005), to ensure that the correct problem animal has been culled in fatal attacks (Frosch et al., 2011), to investigate poaching incidents (Gupta et al., 2011; Lorenzini et al., 2011) and to relate derivatives back to a known individual (Caniglia et al., 2009; Gupta et al., 2010).

Microsatellite markers have been successfully developed for species such as Eurasian badgers (Ogden et al., 2008a), Northern European brown bears (Eiken et al., 2009), boars (Caratti et al., 2010), birds of prey (Dawnay et al., 2009), foxes (Weissenberger et al., 2011; Wesselink and Kuiper, 2011) and the ivory gull (Yannic et al., 2011).

DNA-Sexing

Hunting regulations sometimes require that capture of only one sex of a species is permitted; therefore, it may be necessary to establish the sex of a sample (see also Chapter 10 – Feathers). In mammal species the female is homozygous (XX) and the male heterozygous (XY), in avian species the sex chromosomes are homozygous in the male (ZZ) and heterozygous in the female (ZW). The presence or absence of the Y and W chromosomes can therefore distinguish the sex of a species. Although this information may be used alongside other data in wildlife investigations, it can be the foremost important piece of information in crimes relating to sex-specific poaching.

The hunting of female pheasants is illegal in Korea and therefore carcases suspected to be of female origin can be investigated (An et al., 2007). Male Asian elephants are hunted for their ivory; therefore, decomposing carcases found in the field can undergo testing to identify the sex of poached animals and to monitor the population of male elephants (Ahlering et al., 2011; Gupta et al., 2006).

Conclusions

The current absence of standardisation within the field of wildlife DNA forensics emphasises the importance of good laboratory practice, following TWGDAM and ISFG guidelines, when conducting conventional DNA methods. Until standard procedures are brought about that are specific to wildlife investigations, these guidelines should be upheld to ensure the validity of a wildlife crime study. The development of threshold values to bring some form of standardisation to species identification is an important step, but there are ranges of values for which classification is reliant on the knowledge of the researcher. A universal value is unrealistic and although many species may comply, closely related taxa may be an exception to the rule. Knowledge of the species of study and its most closely related relatives may therefore be required in many wildlife investigations so as to ensure correct classification. Validated techniques are not always appropriate to an investigation and the method

Table 11.1 Summary of Wildlife Forensic Methodologies

Analysis Required	Conventional Methodologies
Species Identification	
Origin unknown	Universal primer amplification and FINS
Origin speculated	Species-specific primers, RFLP and FINS
Contaminated sample	Species-specific primers, RFLP
Possible hybrid	Microsatellite analysis
Low DNA quantity	Nested PCR and FINS
Low inter-species variation	12S, 16S or D-loop sequencing, SNP analysis and FINS
Individual identification	Microsatellite analysis
Geographic origin	Microsatellite analysis, mitochondrial haplotype or isotopes
Sex identification	Chromosome analysis

chosen may therefore depend upon a number of factors such as the information required, the sample available and the capabilities of the laboratory conducting the analysis. A summary of conventional wildlife molecular biology methods is given in Table 11.1.

As a result of the marked increase in wildlife publications over the past decade, thousands of sequences are being deposited daily in reference databases. Data are therefore rapidly becoming available for the development of genetic markers and for use in species identification by sequence comparison. Future use of rapid techniques such as 454 sequencing, enabling the generation of whole genome data, may be of great benefit for comparative analysis by improving the reference data which are available. Continued interdisciplinary collaboration is essential in tackling wildlife crime in order to increase awareness of techniques that are available and to assist enforcement officers whenever possible.

level of scrutiny increases, practice in maintaining composure under cross-examination is required. Many a forensic scientist has found the court experience a difficult one, whilst others relish the challenge (see Chapter 15).

Conclusions

By recognising the need to follow QM practices, and by making use of readily available online resources, wildlife forensic practitioners must work together and take practical steps to provide an assurance of quality science that reflects well on the discipline, the scientists and – most importantly – the wildlife with which it is concerned.

The Value of Microscopy

Jill Webb

> No human has fulfilled his manifest destiny of joy and awe in this life if his eyes have never looked through a telescope or microscope.
>
> **William Beebe**

Introduction

Ever since the end of the sixteenth century, when the father and son, Hans and Zacharias Janssen, invented the first compound-light microscope, the microscope has been pivotal to our understanding of the shape and function of living things. The original was a simple device illuminated by light, comprising two lenses mounted in tubes, one inside the other and which could be slid closer and further apart to permit focus. Over the next centuries and continuing to this day, the Janssens' microscope was developed and improved by others, to provide higher magnification with improved clarity. Notable early contributors to microscope development date back to the seventeenth and eighteenth centuries and include Antonie van Leeuwenhoek, who described bacteria and protozoa and Robert Hooke, who established the term 'cells', concepts so fundamental to our understanding of life today. The microscope has become a scientific tool that is as versatile as it is indispensable and this is as true in modern times as it was for the early pioneers and is especially so for forensic investigations.

Λ The ability to distinguish two points (known as Airey disks) as separate entities is known as 'resolving power' and the minimum resolvable distance is the closest distance between two points where they can be distinguished as separate features; this is commonly loosely referred to as 'resolution'.

Over the years, the microscope manufacturers have learned to correct most of the lens defects during design and machining and it is now possible to produce microscopes that come near to the theoretical limit of their performance. The theoretical limit that affects the minimum resolvable distance is imposed upon us by the physical properties of light. Essentially, the shorter the wavelength of the illumination radiation, the smaller that resolvable distance is and it equates to a little under 200 nm (0.20 μm), when good-quality, high-magnification oil-immersion objective lenses are used. To enable effective imaging of smaller features, a shorter wavelength radiation is required and this is why electron-microscopes have been developed and, assuming a suitable specimen, these have a resolving power of less than 1 nm.

Types of Microscope

The stereo-light microscope might be considered an extension of one's magnifying glass. A good quality and robust instrument designed with a relatively long working distance and supplied with a versatile illumination system is invaluable for examining surfaces and for enabling dissections and delicate manipulations to be carried out accurately and with precision. In an ideal world where anything larger than a small rodent is to be examined, more than one instrument would be available, one standing on a mobile floor stand with the microscope mounted and extendable arm capable of travelling over the largest of specimens and additional bench-mounted instrument(s) where the examination of excised parts can be undertaken. In the practical world, the floor stand version will not be available in any but the largest laboratories.

Arguably, the most widely used of the range of microscopes available today is the compound-light microscope (LM), routinely used to examine the structure of tissues and other items. LMs range from simple pocket microscopes that can be carried on field trips and costing a few tens of pounds to large, research instruments that can cost tens of thousands of pounds, depending on the quality of the lenses, the configuration and how many accessories are included with them. This is the instrument that enables the donkey work of screening and examining the countless histological and histopathological specimens that are generated every day. With selective staining of the specimen, different structures can be indicated, based on their chemical or physical interaction with the stain. With appropriate attachments, it can be operated in different 'modes', which include brightfield, polarising, phase contrast, Nomarski DIC, reflected and fluorescence. These serve either to provide additional contrast or to demonstrate physical or chemical properties associated with the specimen. In short, the LM can be considered an analytical instrument in its own right.

Stereo- and compound-light microscopes are instruments that commonly equip vehicles that serve as mobile investigative laboratories. The laboratory is then available for travelling to the specimen at the site of investigation rather than have the specimen travel to it.

We mentioned previously that the physical properties of light used for illumination limit the fine detail light that microscopes are able to distinguish. Scanning electron-microscopes (SEMs) and transmission electron-microscopes (TEMs) provide that extra definition. The SEM can be considered an extension of the stereo-light microscope, whereas the TEM continues where the compound-light microscope leaves off. Apart from revealing the ultrastructural detail in tissues, the TEM plays a fundamental role in virus identification (see Chapter 10).

Bombarding any material with an atomic number of 4 or more with high energy electrons, as occurs in an electron-microscope, can result in the production of x-rays (along with other signals), essentially as a by-product of the electron-microscopy process.

Each element has a unique x-ray spectrum. It follows, therefore, that if a device capable of measuring the energy or the wavelength of x-rays is attached to the electron-microscope, the elemental composition of the feature being irradiated can be determined on the basis of detected x-rays. So, in x-ray microanalysis we have a powerful analytical tool and this is capable of generating at least semi-quantitative data from regions as little as $1\,\mu m^3$ to help characterise the chemical composition of a material.

The varieties of microscope described earlier are by no means exclusive. It is evident that, rather than the microscope being an instrument used solely for assessing the physical structure of the sample under investigation, it will provide chemical information. The capability of the microscope has been exploited further in its application to vibrational spectroscopy. Modified light microscopes can be interfaced to Fourier Transform Infrared (FT-IR) spectrometers or to Raman spectrometers. This effectively converts them to instruments with a micro-spectroscopy capability, thus enabling the acquisition of chemical data from tiny samples; loosely, these instruments are termed FT-IR microscopes and Raman microscopes. Traditionally, a spatial resolution of 10 μm has been accepted for an FT-IR microscope. However, advances in FT-IR analysing and imaging techniques have given rise to claims of obtaining information from areas as small as 3 μm. Confocal Raman spectroscopy complements FT-IR. Functional groups that give strong Raman bands often give weak infra-red signals and *vice-versa*.

Over recent years, digital microscopes have developed into a range of useful instruments. A digital microscope has no means by which an operator can look down an eyepiece; the image is captured on a sensor, much like that in a digital camera, and is transmitted directly to a PC screen. Current systems do not compete with the best of the conventional microscopes but are useful for low-resolution imaging; some are very simple. They are an extension of a digital camera operated in macro mode. However, they are being developed constantly and more sophisticated versions will undoubtedly become available.

Recording Images

The ability to record good quality, digital micrographs has revolutionised the capture of routine micrographs, for both light and electron-microscopy. Gone is the need to try multiple exposures in the hope that one will be satisfactory; gone is the wait for the darkroom staff to process the film and print the images or the wait for Polaroids to develop; gone is the worry about the expense of Polaroids. Now, in true WYSIWYG (what you see is what you get), the image captured after viewing it on the preview screen is what you end up with. Then a secure image archive and proper file management will ensure the original data cannot be deleted or updated/overwritten (e.g. following 'enhancement' of the original file) and will stand up to challenges of whether or not they are admissible as original evidence. However, routine digital images do not have the fine detail available on the best film images.

Microscopes in Action

Microscopes and associated preparation techniques provide the following important features:

- Importantly, microscopes reveal information that no one knew was there. This type of information may be present only in traces but can be crucially important in the understanding of events that have led to the particular investigation.
- Microscopes enable relevant selected materials to be isolated for analysis by other disciplines.
- Once material can been found, it can be examined. Microscopes can generate analytical data that will positively identify foreign materials associated with the subject of the investigation.
- Microscopes enable the investigator to make observations so he or she can express an opinion about the condition of tissues and materials that are the subjects of the investigation.

Microscopy in Wildlife Forensics

Investigating the origin of an alleged illegally held wild animal illustrates the importance of microscopy in collecting and examining any evidence available (see Chapters 8, 9 and 10).

Following a thorough examination by eye, likely to be helped by a magnifying glass (see Chapter 10), and the recording of initial photographic and x-ray evidence, investigation by stereo-light microscopy can provide a wealth of information. Locard's principle (see Chapter 5) states that whenever there is contact between two objects (whether either is a living thing or not), there is a transfer of material between them. This is fundamental to the approach to every forensic investigation. In support of this view, the stereo-light microscope can be applied to the examination of the finer features of a specimen to look for evidence of small lesions as well as other damage; detection of deformation is permitted by using a stereo-light microscope. During examination, dusts, insect activity and other foreign materials may become evident, giving the investigator the opportunity to subsample and retain for further investigation. All of those foreign materials will have come by contact with something from somewhere, be it from brushing past foliage whilst walking through undergrowth or maybe as a result of forcible contact with something/someone.

The majority of 'dust' found on an animal will have originated naturally, from exfoliating skin cells. However, brushing past plants in flower will inevitably result in the release of clouds of pollen, some of which is liable to become trapped under hairs and scales or become packed into mud lodged into crevices in the foot. Similarly, small insects, larvae and insect eggs can be brushed off the plants and trapped along with the pollen and stones incorporated in the trapped mud (see Chapter 6).

Pollen grains can be identified on screening isolated dust by compound-light microscopy and also by SEM, revealing the varied and beautiful shapes of pollen specific to different species. Having speciated (identified) the pollen, a picture of the type of countryside in which the animal has been living or travelling can be started. Invertebrates help to colour the picture and their speciation is based on the shapes and size of different body features, almost all of which will be determined by stereo-light microscopy and SEM. Many species inhabit quite specific locations and will also help to pinpoint the region the animal has occupied (see Chapter 10). All of this can be reinforced by examination of mud and stones trapped in the animal's feet. Geological samples are supported in resin and then polished to reveal flat surfaces for examination with a variation of a compound-light microscope known as a petrological microscope, followed by examination by SEM and determination of its elemental composition with x-ray microanalysis. By determining the rock and soil type, it may be possible to narrow its origin down to a particular region, thus adding another piece to the jigsaw puzzle.

The presence of fibres other than the animal's own hair possibly indicates the use of cords/ropes for restraint and of sheeting used as a cover, as well as brushing against human clothing. After identifying the presence of extraneous fibres by stereo-light microscopy, compound-light microscopy is used in the first instance to determine if the fibre is synthetic or natural. Assuming it is not a mineral fibre, a positive material identification can be made using FT-IR microscopy. Compound-light microscopy is used to characterise further cellulose (plant fibres) and, for example, to distinguish cotton from flax.

Particles of metal, glass and residues of paints or lubricants all indicate contact with materials that are associated with human activity and, following their isolation, different aspects of microscopy play an important role in their characterisation.

The classic application of SEM examination and x-ray microanalysis in forensic science is in the analysis of finely-divided metal that turns out to be gunshot residue (see Chapter 9). When fine residues are lifted from the specimen on to an adhesive tape and shown to have an elemental composition consistent with that of gunshot residue, this is proof that the subject of the examination has at some time been in close proximity to a discharging firearm. The elemental composition characterises the shot used and can distinguish between some manufacturers – for example, shot typical of that produced in the West (North America/Western Europe) as opposed to shot produced by the former Eastern Bloc (Soviet Union).

The application of x-ray microanalysis is not restricted to metallic specimens; it also enables identification of glass type (often in combination with refractive index measurement) and it contributes to the characterisation of paint flakes, frequently in conjunction with FT-IR or Raman microscopy.

X-ray microanalysis works in synergy with FT-IR microscopy and compound-light microscopy when tackling the identification of unknown materials. For example, selective staining (toluidine blue, Sudan IV and iodine) and light microscopy might indicate that a whitish deposit probably comprises agglomerates of proteinaceous material, lipid and a glycan/proteoglycan but no starch; FT-IR definitively confirms it contains protein, the lipid is a triglyceride and carbohydrate is present; x-ray microanalysis demonstrates the presence of calcium and phosphorus, which in combination with the other materials is characteristic of milk. The deposit is dried milk.

Considerable insight into an animal's recent activity is derived from looking at what it ate at its last meal. This involves screening such samples as stomach contents and regurgitated bird pellets (see Chapter 10).

Individual components of ingesta can be identified and isolated with the aid of a stereo-light microscope. Hand-sectioning/teasing/squashing, followed by simple staining techniques and examination by compound-light microscopy can identify the types of food. Fibrous tissue with cross-banding is easily identified as skeletal muscle (meat) but possibly more interesting are hairs, feathers, bones and scales associated with it (see Chapter 10), which can be used to identify the source of the meat. Where sufficient details remain, plants can be characterised – based on cell shape and the shape and size of leaf hairs. The different sizes and shapes of starch granules are specific for different plants. Starch exhibits a classic Maltese Cross-type birefringence when viewed by light microscopy using polarised light and because this shows up as bright features in a dark background, the features are easy to recognise and screening is fast. Starch also turns a characteristic blue/black colour when stained with iodine. However, thoroughly cooked starch loses its birefringence but retains its staining properties with iodine, a state in which processed food fed to an animal in captivity or perhaps stolen during a dustbin raid would be expected to present itself.

Pathological Changes

The previous paragraphs have discussed the role of the microscope in the identification of some of the types of extraneous material that can provide information about an animal and its environment. However, the microscope plays a pivotal role in the diagnosis of disease and this too can be relevant to animal legal cases (see Chapter 2). Although much of the microscopical examination involves the use of transmitted LM and haematoxylin and eosin (H&E)-stained sections of tissue fixed in formalin, 'special' staining can also be carried out and this may reveal additional information. For example, the time-honoured Gram stain that dates back to 1884 distinguishes between two major classes of bacteria – Gram-positives (e.g. streptococci) and Gram-negatives (e.g. coliforms). More modern developments (again, a microbiological application) include fluorescence demonstration of the bacilli that are the cause of tuberculosis, the mycobacteria, a technique that dates back to 1950.

Immunological and genetic techniques are increasingly being used in the diagnosis of human diseases. However, they depend on the production of specific antibodies and probes and the restricted market presented by wildlife forensics usually does not make them financially viable.

Summary

This account has skimmed the surface of how microscopy contributes to forensic investigations involving wildlife. Emphasis has been laid on the pivotal role played by microscopy in its various forms. Stereo- and compound-light microscopes should grace the benches of every forensic laboratory.

Special Considerations and Scenarios

13

JOHN E. COOPER
MARGARET E. COOPER
NORMA G. CHAPMAN
ALEXANDRIA YOUNG
REGINA CAMPBELL-MALONE
ANDREA BOGOMOLNI
REBECCA N. JOHNSON
STUART WILLIAMSON
JAIME SAMOUR
MADHULAL VALLIYATTE
JÁNOS GÁL
MÍRA MÁNDOKI
MIKLÓS MAROSÁN
MAURICE ALLEY
GLADYS KALEMA-ZIKUSOKA
CHLOE V. LONG

Contents

Introduction to the Chapter	397
Dentition of Mammals	399
Introduction	399
Methods for Assessing Age	399
Eruption	399
Assessment of Wear	400
Incremental Lines in Cementum	401
Cusp Height	402
Pulp Cavity	402
Teeth as Indicators of Stress	402
Teeth as Biomarkers	402
Environmental Factors in the Developing Tooth: Fluoride and Isotopes	402
Identifying Foods Eaten from Abrasion of Teeth	403
Non-Pathological Departures from Normal Dentition	403
Oligodontia	403
Advances in Scavenging Research	403
Marine Mammals: A Special Case	404
Introduction	404
The Importance of Marine Mammal Forensic Analysis	405
Human Activities Can Threaten Marine Mammal Populations	405

Marine Mammals Are Important Bio-Sentinels for Ecosystems and Human Health	409
Marine Mammals May Be Especially Sensitive to the Effects of Climate Change	409
Specific Challenges	412
Special Equipment in the Marine Mammal Forensics Toolkit	419
Conclusions and Summary	421
Acknowledgements	421
Investigation of 'Aircraft Wildlife Strike' Using Forensic Techniques	**421**
Introduction	421
Methods – Wildlife Strike: An Example of Applied Wildlife Forensics	422
Methods – Wildlife Strike, from Morphology to Molecular	423
Conclusion – Wildlife Strike: An Example of Applied Wildlife Forensics	424
Dealing with Illegal Wildlife Trade: A Case Study of Coral Imports into New Zealand	**425**
Background	425
Case Study	426
Identification Systems in Hunting Falcons in the Middle East	**429**
Welfare Issues of Captive Asian Elephants	**432**
Introduction	432
Objectives	432
Materials and Methods	432
Interaction with the Mahouts, Camp Staff and Veterinarian	432
Questionnaire	434
Field Survey	434
Body Condition Index	434
Veterinary Inspection	434
Results and Discussion	434
Management Practices	434
Elephant Housing Systems	435
Feeding	435
Health Status	436
Mahout Welfare Issues	436
Management Protocols Followed for the Tourist Camp Elephants of Jaipur (2003–2009)	437
Protocols Followed	437
Salient Findings	437
Lessons Learned and Recommendations	438
Conclusions	439
Acknowledgements	439
Forensic Expert Work in Wildlife Medicine in Hungary	**439**
Areas of Wildlife Forensics in Hungary, with Practical Examples	**441**
Criminal Offences	441
Killing of Wildlife	441
Persistently Damaging the Health of Wildlife	445
Illegal Trading or Keeping of Wildlife, Resulting in Health Damage	447
Other Cases of Interest	449

the tooth. For each molar, the mesial cusps wear before the distal cusps, and their mesial slopes before their distal slopes. A graph of age versus score allowed scores from animals of unknown age to be assessed, but possible differences between populations must be borne in mind if the graph is used for deer from elsewhere (Brown and Chapman, 1990, 1991b). However, applying a score allows comparison between individuals within any population.

A refinement of the Brown–Chapman scheme was used to score the wear of the molars of wild known-age red deer from Scotland, aged up to 16 years (Dudley Furniss-Roe, 2008). The total score predicts the age of the deer in months without recourse to a graph. Again, because wear can depend upon a number of factors, the scheme will not provide an accurate absolute assessment of age for specimens from a different environment but should offer a reasonably accurate relative assessment between different animals. The measure of reliability between observers or the same observer on different occasions was very high.

In France, visual (but not scored) assessment of wear in roe deer was generally within a year of the true age for more than 50% of animals up to 7 years but errors ranging from −5 to +6 years occurred. For a 16-year-old, the maximum error recorded was −10 years. Hewison et al. (1999) concluded this was not an adequate method for age-related and behavioural studies. Some researchers have reported so much variation that wear is not a useful indicator of age, for example American mink in coastal United Kingdom (in Harris and Yalden, 2008).

Incremental Lines in Cementum

Incremental lines in cementum have been found in almost every group of mammal and have became a routine tool for assessing ages. A common interpretation was that white lines (cellular cementum) alternating with translucent lines (acellular cementum) represented a period of rapid growth (e.g. summer) followed by a period of slow growth (e.g. winter), thus together indicating 1 year. Grue and Jensen (1979) studied 52 species (3700 individuals) of terrestrial mammals and realised that there was no simple way of relating the number of lines to conditions in the life of an animal. They even observed differences in free-living fox siblings which had been ear-tagged as cubs. Formation of the lines could relate to several factors, of varying importance in different species, and resorption of cementum can also occur to allow teeth to move through the mandible. For a dense population of white-tailed deer in an environment with contrasting seasons, McCullough (1996) reported ease of interpreting cementum lines, but when the population level was reduced the lines became obscure or absent. In the European hedgehog, the cementum lines were too tightly packed to count but an alternative method of assessing age was found by sectioning the mandible and counting periosteal growth lines (Morris, 1970). Cipriano (2002) reported the recognition of cold stress in captive apes recorded in incremental lines of dental cementum (see later).

Incisors and premolars have been utilised but usually, for large herbivores, the first molar (first permanent tooth to complete development) is selected to be sectioned longitudinally through the root bifurcation to reveal the inter-radular pad of cementum. Thin sections are prepared (Azorit et al., 2002; Dudley Furniss-Roe, 2008) and viewed by transmitted light; a simpler method is to polish the surface of the half tooth and view under a dissecting microscope by reflected light. The position on the cement pad at which the lines are counted is critical as has been thoroughly demonstrated by Dudley Furniss-Roe (2008) who discussed the pitfalls of using cementum to estimate age and stated that the method is not applicable to material that has been bleached by chemicals which make the lines indiscernible. In her study of known-age red deer teeth from the Scottish Highlands, she found that thick sections (i.e. polished half tooth viewed under reflected light) gave a more accurate assessment than thin sections viewed under polarised transmitted light, but reliable repeatability for different observers or one observer on different occasions was poor. She concluded that scoring wear was more reliable and accurate. Omar (1992) who also studied known-age red deer (113 from the Scottish island of Rum) found that the majority could be assessed from thin sections within 2 years of actual age but some assessments were out by up to −8 or +9 years. Anyone considering utilising cementum for age assessment would benefit from reading Dudley Furniss-Roe's Report (2008) and also Berkovitz et al. (1992), which explained the complexities and classification of cementum. The biology of cementum was also discussed by Lieberman and Meadow (1992).

Cusp Height

The use of cusp height of molars has been used to attempt to assess age in large herbivores including African buffalo (Spinage and Brown, 1988) and waterbuck (Spinage, 1986). With roe deer, Aitken (1975) found extensive overlap between animals of similar age, especially older ones, as did Miura and Yasui (1985) for Japanese serow. For known-age Scottish red deer, the correlation between age and cusp height was very poor in Dudley Furniss-Roe's specimens (2008), and for white-tailed deer, Gee et al. (2002) found this long-used method to be unsatisfactory. For horned species, counting annual rings on the horns (Goss, 1969) can be a useful cross-check against any other method of assessing age, for example Japanese serow (Miura, 1985).

Pulp Cavity

In cross sections of an upper canine of Arctic foxes, Russian workers recorded the ratio between the external diameter and the diameter of its pulp cavity. The results allowed separation into different groups but age could not be determined in the absence of information on the rate of deposition of secondary dentine (see Morris, 1972).

Teeth as Indicators of Stress

Linear enamel hypoplasia, an indicator of metabolic stress, affects only that portion of the crown that is in the process of forming (Ritzman et al., 2008). Early studies on this deficiency of thickness of enamel, caused by the slowing or stopping of the secretory phase of amelogenesis, were on humans, modern and ancient. However, it has been applied to a range of non-human primates (Collett and Teaford, 2008; Guatelli-Steinberg and Skinner, 2000; Newell et al., 2006; Skinner, 1986) and other species with brachyodont molars (e.g. *Sus*, Dobney et al., 2004 and *Giraffa*, Franz-Odendaal, 2004).

For genera with hypsodont molars which continue to grow after the tooth has come into wear, a complete unworn crown is never available; so problems arise when trying to correlate hypoplasia with a causal event. This is thoroughly discussed by Upex (2009) in her very detailed investigation of present-day sheep and goats from Iceland, Orkney and Kenya.

Teeth as Biomarkers

Environmental Factors in the Developing Tooth: Fluoride and Isotopes

Fluoride occurs naturally in the environment but is released in large amounts during many industrial processes; so both the natural environment and local or downstream industrial sources must be taken into account. Its high affinity to calcium causes accumulation in bone and teeth. Most investigations have utilised the teeth of large game species or the antlers of deer, for example Kierdorf and Kierdorf (1989a, 1999, 2000a,b).

Structural changes in fluorised tissues are very well-illustrated by Kierdorf et al. (1996b, 2000) and Schulz et al. (1998), the latter also showing changes to bone. The normal structure of enamel for red, roe and fallow deer was shown by Kierdorf et al. (1991). Defective mineralisation of dentine has also been reported for deer living in a fluoride-polluted region (Richter et al., 2010).

Different geological terrains contain varying ratios of strontium isotopes ($^{87}Sr/^{86}Sr$) which, via the food chain, leave a signature in bones and enamel, the latter being the preferred tissue for archaeological investigations (Price et al., 2002). Analysis of samples from a molar from each of two fallow deer from a Roman site in England showed that one deer had originated elsewhere but the other was born at the site (Sykes et al., 2006).

Stable carbon isotopes, together with dental morphology, were used to distinguish sheep and goats in grassland environments (Balasse and Ambrose, 2005). The ratios of stable isotopes of carbon and nitrogen have been used to investigate diet (O'Regan et al., 2005). Analysis of stable isotopes of carbon, nitrogen and oxygen of teeth of white-tailed deer living around 1000 years Before Present (BP) in Canada is being assessed to determine the seasons at which the deer had access to maize (Morris et al., 2011) (see also Chapter 2).

Identifying Foods Eaten from Abrasion of Teeth

Microwear patterns (pits, scratches) of a molar were examined to indicate the types of food eaten by a population of roe deer and the findings have potential application for other herbivores, including archaeological specimens and extinct species (Merceron et al., 2004).

Non-Pathological Departures from Normal Dentition

Examples of non-pathological variations in teeth as well as diseases were described in the book by Miles and Grigson (1990).

Extra, absent, misaligned and rotated teeth are mostly known from game species because the mandibles are often retained for age assessment but cases have been reported from a range of animals.

Oligodontia

In a survey of *Ovis*, the incidence of absent teeth was higher in the New World species of wild sheep than in the Old World species (Hoefs, 2001). Amongst the cervids, a reduced dentition has been reported in several genera.

Polydontia also has been reported in many species of deer and less commonly in badger (in Harris and Yalden, 2008), red fox (Ratcliffe, 1970) and camel (Gauthier-Pilters and Dagg, 1981) but no doubt most examples are not recorded (see also Miles and Grigson, 1990).

Advances in Scavenging Research

Alexandria Young

Techniques used within the field of wildlife forensics usually focus on identifying the animal as the victim and the human as the perpetrator; however, comparable techniques can readily be applied when these roles reverse. Animal attacks occur not only on living humans but also on deceased individuals. The effects of scavenging on human remains are of great interest to a variety of fields of study such as zooarchaeology, palaeoecology, behavioural ecology, archaeology, wildlife forensics, forensic archaeology and forensic anthropology.

The ways in which various animal scavenger species manipulate a human body greatly differ due to species' size, environment, behaviour and the condition of the human remains. It is important that all of these factors are considered in an investigation.

In order to understand how mammalian scavenger species modify human remains, it is essential to employ a multidisciplinary approach to the investigation. A widely used method within forensic studies for understanding the decomposition and scavenging of the human body is to use animal carcases as human proxies. In North America, the species most frequently employed as a human proxy is the domestic pig on account of its comparative body size, fat content and lack of dense hair (France et al., 1992; Kjorlien et al., 2009; Morton and Lord, 2006; VanLaerhoven and Hughes, 2008; Wilson et al., 2007). However, deer have also been used in various scavenging studies and are currently the taxon of choice in Britain due to regulations by DEFRA (see Appendix E) relating to the deposition of carcases of domestic livestock out-of-doors that might encourage the spread of infectious organisms (Haynes, 1982; Willey and Snyder, 1989). Unfortunately, scavenging studies within North-Western Europe are limited and, as a result, many investigations rely on data from North America, where there has been a large amount of research on scavenging using pig, deer and human carcases. The majority of the North American studies focus on large canid scavenger species such as domestic dogs, coyotes and wolves, which is appropriate for that region where these species are the most common wild canid scavengers (Haglund et al., 1989). Nevertheless, those North American models of scavenging by large canids are now being applied to a wide range of contexts encompassing different regions, environments and scavenger species (Moraitis and Spiliopoulou, 2010; Ruffell and Murphy, 2011). In particular, employing North American models of scavenging by large canids cannot be used satisfactorily to explain scavenging within Britain where the largest wild canid scavenger is the red fox. This animal is smaller than North American carnivores (e.g. wolves, coyotes) and it displays different scavenging behaviour and patterns. Likewise, the canid model cannot be applied to Britain's second most common wild scavenger, the Eurasian badger, the dental morphology and scavenging behaviour of which differ greatly from those of canids.

Forensic investigations would greatly benefit from species-typical and region-specific scavenging studies that assist in the correct identification and interpretation of a scavenger species and in more efficient search and recovery of key skeletal elements for victim and trauma identification.

Marine Mammals: A Special Case

Regina Campbell-Malone and Andrea Bogomolni

Introduction

The term 'marine mammals' includes a diverse group of animals from several distinct taxa that are each reliant on the ocean as a habitat and primary food source. This group includes cetaceans (whales, dolphins and porpoises); pinnipeds (seals, sea lions and walruses); and polar bears, sea otters and sirenians (manatees and dugongs). Although populations of most marine mammal species are characterised by a finite geographic distribution, marine mammals can be found in every ocean basin around the globe.

Marine mammals are in a precarious position. While valued by humans as a classic example of 'charismatic megafauna', they are also prized as a source of entertainment, sport and food. Human utilisation of marine mammal parts dates back to the Upper Palaeolithic time period in Western Europe (17,000–9,000 years Before Present)

Special Considerations and Scenarios

Figure 13.7 Male manatee carcase exhibiting external gross skin lesions (multifocal pustules and bleaching of epidermis) associated with cold stress syndrome in manatees. (Courtesy of the Florida Fish and Wildlife Conservation Commission, Panama City, FL, USA.)

Special Equipment in the Marine Mammal Forensics Toolkit

The forensic examination of remains requires not only training and careful execution of standardised methods, but also a host of tools (equipment) (see Chapter 7 and Appendix B). The following is a list of specific items that, when available, may be utilised during marine mammal necropsy.

Marine Mammal-Specific Necropsy Guide. Several groups have put together detailed and informative guides for performing the diagnostic necropsy of marine mammals. These guides are complete, with safety considerations, field-tested best practices from over 30 years of necropsy experience and most also contain datasheets specifically designed for marine mammals. Familiarity with the appropriate data sheets (and several copies, including some made on waterproof paper or laminated!) will facilitate data collection. Here is a list of some of the recent and most useful guides available.

Cetaceans and Pinnipeds
Pugliares et al. (2007). Marine mammal necropsy: An introductory guide for stranding responders and field biologists. Woods Hole Oceanographic Institution Technical Report WHOI-2007-06, Woods Hole, MA, USA. This guide focusses primarily on pinnipeds, small cetaceans and large whales http://darchive.mblwhoilibrary.org:8080/handle/1912/1823.

Young, N., Ed. (2007). Odontocete salvage, necropsy, ear extraction and imaging protocols. http://www.nmfs.noaa.gov/pr/health/noise/docs/protocols.pdf

Geraci, J.R. and Valerie, J.L. (2005). Marine mammals ashore. This text provides a thorough treatment of stranding response for cetaceans, seals, sea otters and manatees tailored to a wide audience from beachcomber to scientist. http://www.aqua.org/research_marinemammalsashore.html

Raverty, S.A. and Joseph, K.G. (2004). Killer whale necropsy and disease testing protocol. http://www.vetmed.ucdavis.edu/whc/pdfs/orcanecropsyprotocol.pdf

Dierauf, L.A. (1994) Pinniped forensic necropsy and tissue collection guide. NOAA Technical Memorandum NMFS-OPR-94-3, USA. http://www.nmfs.noaa.gov/pr/pdfs/health/pinniped_guide.pdf

Thijs, K. and Hartmann, M.G. (1991). *Proceedings of the First European Cetacean Society Workshop on Cetacean Pathology: Dissection Techniques and Tissue Sampling*, Leiden, the Netherlands. Available from the European Cetacean Society (ECS) Special issues/ECS Special Publication Series: http://www.europeancetaceansociety.eu/forms/ecs-order_2009.pdf

Sea Otters

Burek, K. et al. (2005). (Burek et al., 2008) A pictorial guide to sea otter anatomy and necropsy findings. USFWS unpublished report, USA. For more copies of this guide contact Verena Gill at Marine Mammals Management, U.S. Fish & Wildlife Service, MS 341, 1011 East Tudor Road, Anchorage, AK 99503, USA or e-mail verena_gill@fws.gov
http://alaska.fws.gov/fisheries/mmm/seaotters/pdf/Necropsy_photo_guide.pdf

Sirenians

Lightsey, J.D. et al. (2006). Manatee necropsy method/mortality paper.
http://www.nmfs.noaa.gov/pr/interactions/injury/pdfs/day2_costidis_lightsey.pdf

Rommel, S.A. et al. (2007). Forensic methods for characterizing watercraft from watercraft-induced wounds on the Florida manatee (*Trichechus manatus latirostris*). *Marine Mammal Science* 23(1), 110–132.

Eros, C. et al. (2007). *Procedures for the Salvage and Necropsy of the Dugong (Dugong dugon)*, 2nd edn. Research Publication No. 85. Published by the Great Barrier Reef Marine Park Authority, Australia. http://www.reefhq.com.au/data/assets/pdf_file/0020/21971/dugong_necropsy_2ndedition.pdf www.gbrmpa.gov.au/corp_site/info_services/publications

Polar Bears

AZA Bear Taxonomic Advisory Group (2009). (AZABearTAG,2009) Polar bear care manual. Association of Zoos and Aquariums. http://www.aza.org/uploadedFiles/Animal_Care_and_Management/Husbandry,_Health,_and_Welfare/Husbandry_and_Animal_Care/PolarBearCareManual.pdf

Equipment for Moving Carcasses. Specially designed wildlife stretchers, carts, pontoons and able-bodied volunteers are the primary vehicles for moving small marine mammal carcasses. Though harder to come by, hand-operated cranes, vehicle-mounted cable winches, block and tackle pulley systems, and heavy machinery make all the difference when it comes to moving larger carcasses and post-necropsy waste. The Ez-Lift foldout crane and winch has proved to be a reliable tool thanks to its nearly 1000 kg capacity and slim storage profile. A front-end loader and (more importantly) a skilled operator can reduce the work involved in flensing skin and blubber from whale carcasses weighing 80 t by providing critical tension for knife-wielding prosectors.

Blades. Like the animals themselves, the dissection equipment utilised for marine mammals also range in size from scalpels and small knives to large carving knives and flensing knives (pole-mounted blades that measure 45–60 cm in length). The thick pelts of the furred marine mammals (seals, sea lions and otters); the rough skin of walruses and manatees (Kipps et al., 2002); and oft-present sand and sediment on the external surfaces rapidly dulls the sharpest of blades. Hilted knives prevent one's hand from sliding up the blade when cutting with force. Blade sharpeners and buckets for holding used knives are highly recommended.

Hooks. Flensing hooks are important when working with large marine mammals. They are used to hold the skin and blubber in tension so that it can be removed with relative ease. They are also useful for manually moving small carcasses or large 'sheets' of skin and blubber for disposal.

Scanning Equipment. Various forms of scanning or imaging (see Chapters 8 and 9) provide important diagnostic data that can greatly assist in a forensic investigation. Metal detectors have been used to scan marine mammal carcasses for evidence of bullet fragments and metal shot (Neme and Leakey, 2010).

As emphasised elsewhere in this book, medical/veterinary diagnostic imaging, including plain film x-ray (radiography), ultrasound, computed tomography (CT) scanning and MRI, permit the non-invasive visualisation of internal anatomical structures, including the auditory pathway, fractures, lesions, parasites and foreign bodies (Campbell-Malone, 2007; Cranford et al., 2008; Montie et al., 2007; Zucca et al., 2004). Prior to opening the carcase, imaging can be used to observe evidence that may be missed or destroyed during a necropsy, such as bullet fragments, fractures, fistulae or gas bubbles. Gas bubbles have been imaged (MRI and CT) in marine mammal tissues and have been correlated with gross and histological findings that implicate barotrauma or gas bubble injury as a likely factor in marine mammal morbidity and mortality (Moore et al., 2009; Van Bonn et al., 2011).

Conclusions and Summary

Marine mammal carcasses, like all remains, may provide important data regarding their lives and the circumstances of their death. Here, we have provided a summary of the details needed to obtain access to and interpret that information in light of the specific obstacles faced when investigating and interpreting marine mammal remains. The rewards are, nevertheless, gratifying – in spite of the often challenging circumstances that one may encounter when studying marine mammal remains.

Acknowledgements

We thank Drs Nadine Lysiak and Michael Moore (Woods Hole Oceanographic Institution, Woods Hole, Massachusetts, USA); Brandon Bassett and Dr Martine de Wit (U.S. Fish & Wildlife, Vero Beach, Florida, USA); Eric Montie (University South Carolina, Beaufort, South Carolina, USA); Dr Jarita Davis (National Marine Fisheries Service, Woods Hole, Massachusetts, USA); and Heather Pettis and Amy Knowlton (New England Aquarium, Boston, Massachusetts, USA).

Investigation of 'Aircraft Wildlife Strike' Using Forensic Techniques

Rebecca N. Johnson

Introduction

Birdstrike to aircraft has been an issue for concern since humans were first able to take to the skies in the earliest modes of air transport. However, a more accurate term for 'birdstrike' is 'wildlife strike' since flying mammals, terrestrial mammals, reptiles and even some flightless birds have been involved in significant aviation events. The risk of wildlife was starkly evident to all air travellers in 2009 during the so-called Miracle on the Hudson where U.S. Airways Airbus A320–214, carrying 155 passengers and crew, experienced a double engine failure following a wildlife strike during take-off from La Guardia Airport in New York and had to make an emergency landing on the Hudson River.

Wildlife strikes are a serious safety hazard to the aviation industry. Since the advent of air transport, hundreds of human lives, in both civilian and military aircraft, have been lost in incidents directly attributable to wildlife strike (Allan, 2006; Dolbeer et al., 2009). In the event of a strike to an aircraft, the collision impact force between the aircraft and a flying vertebrate (such as a bird or bat) is proportional to the animal mass and the square impact of speed (Canada, 2004). For example, a bird commonly reported to be involved (ATSB, 2010) is the Australian white ibis, with an average bird weighing approximately 1.8 kg. If a single bird collides with an aircraft travelling at 250 knots, the impact force of a bird of that weight will be approximately 56 t over a relatively concentrated area. That same bird colliding with an aircraft travelling at 400 knots increases in collision force to over 140 t.

In addition to the serious risk to safety that wildlife strikes can pose, even a seemingly minor strike can lead to costly flight delays and expensive damage to aircraft and engines, requiring repair. The cost to the aviation industry is calculated to be several $ billion dollars annually with each serious incident estimated to cost approximately $100,000–$150,000 (Dolbeer et al., 2009). Further, as the number of aircraft movements grows and documentation of wildlife strike improves, it follows that the reported frequency of wildlife strikes is increasing year by year (ATSB, 2010; Canada, 2004; Dolbeer et al., 2009).

Methods – Wildlife Strike: An Example of Applied Wildlife Forensics

Wildlife strikes, and also wildlife near-misses, are now reported to national airline safety regulators and in many countries this reporting is mandatory. Analysis of reported strike data revealed that in approximately 50% of reported wildlife strikes it was not possible to identify the species involved (ATSB, 2010; Canada, 2004). However, accurate species identification after a wildlife strike is extremely important for the following reasons:

- If the species can be identified, then the size and weight of the animal can be approximated, allowing investigators to assess possible effects to the aircraft, even in the absence of visible damage.
- If the species can be identified, investigators can then assess the likelihood that more than one individual was involved in the strike (e.g. whether it is a solitary species or a flocking species).
- Prevention of wildlife strikes is desirable to both the industry and the animals involved; therefore, accurate data of the species commonly involved will enable more effective management of the habitats surrounding aerodromes by targeting known problem species.
- Knowledge of the species involved will enable closer monitoring of these species present in the aerodrome environment so as to manage the variation in risks due to the seasonal or climactic changes in population.
- Knowledge of the high-risk species, their weights, etc. can be used to design new and improved aircraft components, particularly those parts frequently hit (engines and windscreens) and damaged by flying wildlife.

Special Considerations and Scenarios

In common with many other wildlife forensic applications, the analysis of wildlife strike on aircraft can interface with the legal system, especially in the event of a significant strike where lives are at risk or significant damage caused. The legalities of this area are complex but there is a general acceptance that the onus is on the providers of airport space (and surrounds) to assess potential risks and have procedures in place to minimise them (Allan, 2006).

Methods – Wildlife Strike, from Morphology to Molecular

Techniques now in common use in other areas of wildlife forensic analysis have proved extremely useful in the field of wildlife strike analysis. This is particularly relevant for the identification of animal remains from a wildlife strike that cannot be identified morphologically because they are mutilated due to the extreme forces of the impact.

Since the 1960s, the Smithsonian Feather Identification Laboratory (Dove, 2000; Feather Identification Laboratory, 2011) has employed the highly specialised skill of 'direct feather morphology analysis' on the remains of birds involved in wildlife strikes. However, analysis of species using feather morphology can be inconclusive if the feather is damaged or lacking in informative characteristics and is entirely inapplicable, of course, if the animal is non-avian. DNA-analysis is now routinely used in the laboratory study of wildlife strikes in many parts of the world and is a reliable and accurate tool for species identification.

Following a wildlife strike, the most critical information required is accurate species identification; however, DNA analysis can also be used to provide individual identification, assignation of an individual to a specific population, and it is even possible to determine the gender(s) of the animal(s) involved. Figure 13.8 depicts a typical sample obtained after a collision with an aircraft, demonstrating how challenging identification from visual cues can be.

DNA-analyses are now commonly used in many cases for the forensic analysis of smuggled animals or animal parts where the samples being analysed are of unknown species

Figure 13.8 An example of the remains after a collision with an aircraft. There are very few morphological features that would allow identification using visual cues alone. DNA-analysis revealed this material to be from a mammal, a little red flying-fox.

(see Chapter 11), and the techniques employed in those circumstances are easily applied to fragmentary remains following a wildlife strike. In these cases, the so-called "universal DNA regions" which are often mitochondrial DNA regions, are a good starting point for analysis (Dawnay et al., 2007; Parson et al., 2000). Mitochondrial DNA sequence analysis is well suited to many wildlife forensics applications because it is present in high copy number (1–15 copies per mitochondrion) compared with nuclear or chromosomal DNA (where there are two copies per nucleus for a diploid genome). In addition, the mitochondrial genome is circular, a shape that is more resistant to the changes of autolysis so that DNA-analysis is often possible even if the sample is degraded or damaged. Finally, the mitochondrial genome is thought to evolve up to 10 times faster than the rate of the nuclear genome (Brown, 1981; Brown and Simpson, 1982; Brown et al., 1979; García-Moreno, 2004; Wolstenholme, 1992; Zink and Barrowclough, 2008), thus providing the level of variability essential to distinguish even closely related species with statistical significance. The mitochondrial DNA cannot, however, provide evidence as to gender as it is derived from the maternal line only.

Mitochondrial gene regions are now used routinely and have a high rate of success in aviation wildlife strike analyses (Dove et al., 2008; Kerr et al., 2007; Waugh et al., 2011; Yoo et al., 2006).

A typical work flow following receipt in the laboratory is as follows:

1. DNA is extracted from the unknown sample
2. Mitochondrial gene regions are amplified via the polymerase chain reaction (PCR)
3. The DNA is then sequenced to obtain a sequence from the unknown strike sample

The results obtained are then compared with the analyses held in an established DNA database such as the Barcode of Life Database (BOLD) (Hebert et al., 2004; Ratnasingham and Hebert, 2007), which has been populated with confirmed reference sequences obtained from collections such as those held in museums (see also Chapter 14). This confirmation of identity against a sample from a properly identified and curated collection is essential as an independently identified and validated reference source (Johnson, 2010). It is essential that any laboratory involved in this work maintains and documents strict forensic standards of specimen-handling and tracking or the results obtained will be legally challengeable (Case Study 13.4).

Conclusion – Wildlife Strike: An Example of Applied Wildlife Forensics

Over 90% of aviation wildlife strike incidents occur when an aircraft is travelling below 3000 ft (1000 m) and most occur during landing and take-off. In order to manage the risk of wildlife strikes to aircraft adequately, individual airfields have a responsibility to be aware of the species present in their local areas throughout the year and to manage them appropriately. If the morphological characteristics of a specimen (or specimen part) from a wildlife strike are lacking, then visual identification (to species, population or individual) can be difficult or even impossible and direct feather analysis (Dove, 2000) or DNA analyses (Dove et al., 2008; Kerr et al., 2007; Waugh et al., 2011; Yoo et al., 2006) will be necessary to ensure the accurate identification of the animal(s) involved. It is this identification that is the critical factor in ensuring the proper management of the species that are known to be at high risk of involvement in aviation wildlife strikes.

Wildlife forensic techniques now offer a suite of valuable tools to assist in identifying those species most commonly involved in wildlife strike of aircraft. With increasing

Special Considerations and Scenarios

In order to obtain the information needed to make a decision, the authorities address specific questions to the expert in the commissioning document; these are to be answered in the expert's report. The opinion of the forensic expert commissioned by the investigating authorities is considered to be final (compulsory) and cannot be redefined by other expert(s). In the event that the investigating authorities have commissioned a special or divisional expert, the injured party or the suspect (if one has been identified), or his or her legal representative, may request a fresh investigation by a forensic expert in wildlife medicine. In the majority of these cases, the forensic expert can only base his or her opinion on reports prepared by the investigating authorities after the inspection of the location and the consideration of all available evidence, as well as the expert's report prepared by the special or divisional expert. Thus, the investigating authorities strive to contact a forensic expert in wildlife medicine licensed by the MPAJ – at least in cases of serious criminal offence, as the expert's opinion will be decisive and final.

In a small percentage of cases, in the final stage of a trial, the law court commissions an expert to answer a specific question or to redefine the earlier opinion of a special or divisional expert. The court always contacts an expert at least one level up in the hierarchy, and when possible, a forensic expert recognised by the MPAJ. In such cases, the expert must rely exclusively on the available documents and reports in formulating his or her opinion in order to help the judge to arrive at a conclusion. The court may want to hear the expert's opinion, no matter at which level s/he is in the hierarchy, either as a witness or as an expert. This is usually the case when the report is prepared by the special or divisional expert and that produced by the forensic expert differ substantially on certain issues. The final conclusion is always that of the expert who is higher in the hierarchy.

Areas of Wildlife Forensics in Hungary, with Practical Examples

János Gál and Miklós Marosán

The work of a wildlife forensic expert is complex and, in order to function properly and to formulate a solid opinion, one needs to be proficient in more than just one field of science. Expert work basically relates to the following areas:

1. Criminal offences:
    a. Killing of wildlife
    b. Persistently damaging the health of wildlife
    c. Illegal keeping or trading of wildlife
2. Other issues of interest

The actual judgement of the aforementioned areas and the expert's opinion related to the given case always raise special questions depending on the circumstances. The individual categories are best illustrated by a practical example of each.

Criminal Offences

Killing of Wildlife

Wildlife may be killed, often in contravention of the relevant law, by inappropriate means or methods, or under unlawful conditions (see Chapter 3). Even if a given species can be legally pursued, its hunting is regulated by law – for instance, it can only be harvested in the open season, by specific means and with specific tools. All other instances are likely to be criminal offences. Most countries apply the equivalent of the term "poaching" to this category (see Chapter 2) and sanction it accordingly.

Example 13.1

A hunting company denounced the authorities on the grounds that illegal and unknown offenders killed a red deer stag on its hunting grounds, outside the open season. After taking note, police investigators located tracks at the site (crime scene) left by the denouncer. The culprit discarded offal on the hunting ground after eviscerating the intestines and other internal organs of the animal, together with its hooves. By thoroughly investigating the scene, detectives gained information about the vehicle used by the offender (wheel tracks left behind in the leaf litter) and data based on questioning eyewitnesses (people who were hiking in the area at the time when the offence took place). On account of this information (the tread pattern of the tyres was known and the eyewitnesses recalled the licence plate number of the car) the offender could be identified. An inspection and search at the suspect's house resulted in the confiscation of frozen, wild meat-like, substances in plastic bags, as well as the detection of traces of blood and animal hairs in the boot (trunk) of the suspect's vehicle. Based on these circumstances and the seized evidences, authorities launched a trial against the suspect for illegally harvesting wildlife outside the open season as well as for theft (poaching).

A wildlife forensic expert was called in at the trial, and the following questions were asked:

1. From what kind of animal (species) were the blood, hairs and frozen meat, as well as the offal found on the hunting ground, derived?
2. Can the market value of the wild meat be established?

The expert called to the trial was given the confiscated meat found in the suspect's freezer as well as the hair and blood sampled from the boot of his vehicle and the animal parts recovered from the (crime) scene for investigation.

After examining the viscera, one part of the first question could be relatively easily answered, as the organs undoubtedly originated from a large ruminant (cervid). The kidney was also present in the offal, the anatomical characteristics of which pointed to the animal being a red deer. Specific identification of the blood, hair and meat samples was more complex. From each sample, genomic DNA was isolated (Macherey-Nagel). The quantitative and qualitative analysis of genomic DNA samples was performed with the aid of a NanoDrop® spectrophotometer. Genetic testing was done by real-time PCR and restriction fragment length polymorphism (RFLP) methods with a Sartagen® MX30000 real-time PCR device (see also Chapter 11).

An answer to the second question was needed in order to categorise the crime. To establish its market value, the frozen meat had to be thawed and sorted in a similar fashion to that used when game meat is processed in abattoirs. Data collection was also necessary to obtain commercial prices for each type of meat and to calculate their market value according to the weight.

Special Considerations and Scenarios

The harvesting of a wild animal protected by law is always illegal in Hungary, irrespective of when and how it is being killed. Protected animals might be killed for several reasons – as follows:

- Deliberate killing for food (protected passerine birds are considered a delicacy in certain European countries)
- Deliberate killing so as to protect private property (small carnivores are often shot to protect livestock, especially in the countryside)
- Accidental killing during legal harvesting of unprotected wildlife, when the hunter confuses a protected species with an unprotected one (e.g. bean geese can be legally shot, whereas superficially similar greater white-fronted geese cannot)
- Accidental killing of a protected species (e.g. a protected bird crashes into the windscreen of a car or enters a public road within the breaking distance of a vehicle)

Example 13.2

During routine inspection of a vehicle at the country border, Customs officials found a large number of packed protected birds in the boot and seized them immediately. The passengers in the car were taken into custody. The carcases were sent to experts for examination. It was necessary to consult several experts from various fields, as two main questions had to be addressed:

1. Which species constitute(s) the confiscated consignment, and as such, what is its theoretical value? For an answer to this question, an ornithological expert had to be called in, who identified the birds according to their species and determined their theoretical value as specified by the national conservation law.
2. How and when exactly were the animals killed? For an answer to this, a wildlife forensic expert was consulted. As a first step, the birds had to be categorised. Early in the investigation, it became clear that the birds, which were all European common quail, a species protected within the European Union (EU), were properly eviscerated according to traditional hunters' practices. They were lacking gizzards and glandular stomachs as well as intestines. It was important to clarify in what way the birds were harvested; hence, they were subjected to radiography whereby radio-dense foreign bodies (small shot) could be clearly identified in almost all instances. These had to be collected and handed over to the authorities, who called in a ballistics expert to analyse the bullets. (It should be noted that a wildlife forensic expert is not entitled to perform ballistic analyses in Hungary.) Dissection of the area of bleeding (which is a vital reaction indicative that the birds were alive) around the bullets (shot) in the quail indicated that the lesions were fresh. It could thus be concluded that the birds were killed by gunfire. Hunting common quail, a protected species, is considered illegal and hence it is a criminal offence.

To establish when exactly (at least in which month) the birds were killed, two tests were performed. First, their gizzard contents were collected; these contained seeds of different plant species within which different phenological stages could be identified. These were submitted to another expert for bromatological (constituent) analysis. The investigation revealed the presence of plant seeds which ripen in a relatively short period each year. It turned out that the quail were in all probability shot at the height of summer, in July. Also, the birds could be assigned to age groups (juvenile or adult) on the basis of their plumage, and the reproductive organs (testes and ovaries) of the adults were closely examined. The size and activity of the latter suggested that the quail were pursued during the birds' reproductive season. The age range, in particular chicks that were less than 1-year-old present in the sample, pointed to the killing taking place in midsummer.

The three pieces of evidence – the results of the gizzard contents analysis, age composition and the activity status of the reproductive organs of adults – all confirmed that the birds were shot in July.

The aforementioned example shows that wildlife forensic work consists of much more than simply dissecting an animal and determining the cause of its death (see Chapter 9). It also includes the collection of evidence and consultation with other experts (Chapter 4).

An animal, whether protected or not, may be killed in Hungary by a legal method. If the animal is considered vermin, it may be killed in the open season only. According to Hungarian law, a protected or a non-protected animal may only be killed outside the open season (i.e. in the closed season) if it has an incurable disease. This fact can be reliably judged by its behaviour, general appearance and body condition. In such instances, after the animal is shot, it must be inspected by a veterinarian and, in the event of any controversy, both diagnostic dissection and other tests are required to be performed in order to formulate an expert opinion.

Example 13.3

The owner of a non-profit private enterprise that kept bears and other large carnivores reported to the authorities that two of his grey wolves, which were alive and in perfect health in the morning, were found dead by the end of the day, close to the fence of their enclosure. After taking note, the authorities transferred the animals' carcasses to a private veterinarian, who performed a necropsy using unconventional irregular methods and reported signs of injury on the animals. Subsequently, a wildlife forensic expert was called in by the authorities and the park proprietor. The incorrectly dissected bodies were given to the expert. The authorities formulated several questions, the main points being what caused the death of the wolves, and whether they could have been deliberately killed.

Upon receiving the carcasses for examination, the forensic expert had to document their immediate condition first – that is, all details of their defective dissection. Such documentation is inevitable and essential because it is possible that an improperly performed necropsy has altered the original situation and lesions may no longer be correctly evaluated, resulting in erroneous conclusions (see Chapter 9).

Because the animals showed no sign of any disease and had died unexpectedly, they were subjected to radiography prior to dissection. As a result, radio-dense foreign bodies were detected in the head of one of the wolves.

External examination of the carcasses revealed the presence of holes (bullet wounds) in the skin of both animals (Figure 13.18). One of the bullets penetrating the chest of one wolf left the body on the opposite side but the size of the exit wound could not be determined on account of the improperly performed necropsy. Portions of bullets detected by radiography (Figure 13.19) could not be recovered during dissection; they were therefore collected by preparing the skull, shredding the meat into small pieces, and subsequently forwarding these to a ballistics expert.

Special Considerations and Scenarios

Figure 13.18 Hole caused by a bullet in the temporal region of a wolf skull.

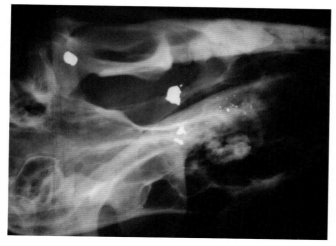

Figure 13.19 Bullet particles in the head of a wolf, detected by radiography.

It should be stressed here how important it is for a wildlife forensic expert to examine carcases and other animal body parts following clear-cut veterinary criteria (see Chapter 9) and to document the process photographically (see Chapter 6). Improper diagnostic work and dissections that do not meet professional standards may alter the final conclusion of an examination (see Chapter 14).

Persistently Damaging the Health of Wildlife

A problem often encountered during wildlife forensic work is whether or not the injury caused to a wild or exotic animal is the result of deliberate or accidental action (see Chapter 2). This question can often be answered only if all circumstances are known to the expert. A decision is needed because this will determine the legal status of a given situation.

Example 13.4

During a bean goose hunt, a species open to hunting in seasons defined by law in Hungary, a greylag goose was injured. One wing was pierced, as a result of which the bird was caught and treated: following radiography, the two pieces of lead shot wedged between the left radius and ulna were removed under general anaesthesia.

Several questions arose, one of which, whether the protected bird was deliberately shot, could not be unambiguously answered, even by knowing about the circumstances. Another question that was addressed to the expert related to repatriation (return to the wild) (see Chapter 2), which could only be answered after the goose was operated upon by the veterinarian. A test of the bird's flying ability helped make a decision.

The aforementioned example raises the issue of defining the degree of damage inflicted to wildlife, in particular to animals that are not fatally injured. Whether or not repatriation is possible can only be answered after long observation and repeated examination of the animals (see Chapter 8).

A central question in determining the degree of damage and the prospects of repatriation is for how long the animal would be able to survive in the wild – in other words, whether it would be able to feed, clean and hide adequately, especially whether it would be able to avoid predators and successfully reproduce back in the wild. To formulate an answer, specific aspects need to be considered. Raptors can orientate and catch prey adequately only when both eyes are intact (see Chapter 8); therefore, only birds with two fully functional eyes should be released. On the other hand, snakes that are blind can often be repatriated without problems because their orientation is based on tasting the air with their tongues and passing scent molecules to their Jacobson's organ; eyesight is not vital to these animals.

Some wild animals may become chronically injured on account of other human activities or industrial disasters. These are often the consequence of incompetent or negligent handling.

Example 13.5

In Hungary, on 4 October 2010, at a plant of the company MAL Zrt., where valuable aluminium oxide is extracted (digested) by the so-called Bayer or wet process consisting of washing alumina with a hot solution of sodium hydroxide and producing a strongly alkaline corrosive waste of solid impurities (red mud), an embankment of one of the company's holding ponds broke. The resulting flood of toxic sludge had profound ecological effects and substantial financial loss. Animals (both domesticated and wild) exposed to the red mud sustained severe necrosis on their body surfaces, skins, eyeballs and the mucous membranes of body orifices. In view of their serious injuries and poor prognosis, several animals had to be euthanased after a few days of treatment.

Experts were asked to determine whether the lesions had really been caused by the toxic sludge. A further issue was the prognosis of injured animals; that is for how long they might have been able to live a full life after recovery.

An important point, particularly in the case of wild animals, was the extent of damage inflicted to the eyeballs and corneae. Of the domesticated animals, the eyes of ducks were so heavily affected that the corneae perforated and the contents of the eyeballs (aqueous humour, lens) leaked out. In sheep, red mud sticking to the wool dried and the resulting dust was inhaled and the mucous membranes of the respiratory tract were damaged, causing massive hyperaemia.

Special Considerations and Scenarios 447

Table 13.4 Heavy Metal Content of Red Mud Collected from the Disaster Site

Heavy Metals Examined in Mud Sample	Arsenic (As)	Barium (Ba)	Cadmium (Cd)	Cobalt (Co)	Chrome (Cr)	Copper (Cu)
Concentrations of heavy metals in mud sample (mg/kg)	36.84	51.74	<0.1	19.5	248.04	42.12
Heavy Metals Examined in Mud Sample	Mercury (Hg)	Molybdenum (Mo)	Nickel (Ni)	Lead (Pb)	Zinc (Zn)	
Concentrations of heavy metals in mud sample (mg/kg)	<0.1	4.8	106.56	57.78	82.62	

In cases of environmental disasters, it is important to evaluate the mid- and long-term effects of contaminants in addition to their acute influence on living organisms. For this, a constant monitoring and observation of living organisms is necessary. Domesticated animals may serve as sentinels for wildlife (see Chapter 2). In the earlier example, the heavy metal content of the red mud was determined, with a view to serving as a basis for further wildlife forensic work in that area (Table 13.4).

Illegal Trading or Keeping of Wildlife, Resulting in Health Damage

An interesting area of forensic work relates to the illegal keeping and trading of wildlife, whereby animals are often kept improperly and therefore can suffer persistent health damage (see Chapter 2). Keeping and trading of wildlife is regulated by national and – as a result of legal harmonisation in wider regions, such as the EU – international laws (see Chapter 3).

Example 13.6

A private individual (hobbyist) breeding exotic animals as a hobby was transferring his captive-bred snakes to a southern European Member State within the EU, when he was halted at a border control. The animals were packed singly in boxes lined with an appropriate substrate. Special panels in the boxes ensured ventilation.

The keeper was able to present to the authorities papers (in English) signed by an official, licensed veterinarian in the country of origin. However, a transport licence signed by the Chief Veterinary Officer was lacking for the transporting vehicle and the driver was not properly qualified to transport animals (as specified by EU Animal Transport Regulation No. 1/2005). As a consequence, the consignment was not permitted to cross the border and the hobbyist was heavily fined.

As a special Hungarian rule, the keeping, breeding and trading of wildlife naturally occurring and protected by law in Hungary is illegal, even if it is permitted in other member states of the EU.

Example 13.7

Hungarian authorities refused the importation of goldfinches, that were bearing closed rings, that had been bred in a southern Member State of the EU. The question arose as to why birds proved to have been bred by recognised breeders within the EU and legally kept in their country of origin may not be imported legally if they also breed in the wild in Hungary.

Legal and fully authorised transportation may also result in health damage or death in animals. In such cases, a wildlife forensic expert's responsibility is to determine the causes of such health damage or death; that is to establish the state of affairs relating to warranty.

Example 13.8

On a day in January, at a Western European airport, Madagascan frogs, chameleons and day geckos were awaiting transfer to Hungary. The cargo shipping company 'forgot' about the consignment and the animals were left on the ramp for 10 h. During this time all the animals died. When the consignment was opened on arrival at its final destination, none of the animals, which were all properly packed, was alive.

A wildlife forensic expert was called in and was given the carcases for examination (Figure 13.20). Dissection revealed only hyperaemia in the internal organs. According to the accompanying documents (a veterinary report made-up before releasing the consignment from Madagascar), the animals were still alive when loaded; thus they must have died either on the plane or during transfer. Documents also confirmed that the animals spent nearly 10 h at the airport, out-of-doors. The forensic expert retrieved airport registration data including ambient temperatures, which revealed that temperatures dropped below −13.5°C on the day in question.

It is a well-known fact that reptiles generally tolerate a minimum of +4°C without suffering persistent or lethal ill-effects, but this is true only for temperate zone and mountain species, not tropical taxa.

In the aforementioned case, it could be proved (based on freight and airport documentation, as well as direct examination of the carcases) that the animals died as a result of improper handling during transport.

In this context, it is important to stress that wildlife forensic work must include an examination of the circumstances and documentation, when these are available, in addition to necropsy. A diagnostic dissection is often not sufficient for formulating a solid expert opinion (see Chapter 9).

Figure 13.20 Various individuals of *Mantella* sp., frozen during transportation.

Special Considerations and Scenarios

Other Cases of Interest

Wildlife forensic work includes several other areas of responsibility, in part related to keeping, breeding and trading wildlife. With regards to keeping wildlife that is not protected by law, it is mainly the owners rather than the authorities who ask a forensic expert for help. Usually, any cause of disease or death related to human error must be identified. Of course, expert work does not end with formulating an opinion in such cases but advice should be given about possible treatment and prevention of the disease (see Chapter 8).

Example 13.9

One game management enterprise produces pheasant eggs and sells them to incubator houses. The stock is kept in mobile cages, with a ground surface of 2.5 × 3.5 m, in which six hens are accompanied by a single cock. Birds are fed a diet of fixed composition *ad libitum*. Dry pine branches are inserted in one corner of each aviary as hiding and egg-laying places.

The breeder consulted the expert upon observing that some females and the males lost their feathers. The males' courtship activity was inhibited and hatcheries reported a large percentage of infertile eggs (see also Chapter 10—Eggs and Eggshells). The question was what the breeder could do to improve the situation.

Local inspection and food analysis revealed that the pheasants' diet was far too low in fibre. Females were mutilating each other, and, in particular, the males, in order to counterbalance this low-fibre diet. As a consequence, the sexual activity of cocks decreased while the number of infertile eggs significantly increased.

By adding plant fibre to the diet, both the destructive *pterotillomania* of the females and the limited courtship activity of the affected males could be treated with success. Instead of pine branches, the expert suggested placing black locust branches in the cages from which the birds could pluck fresh leaves and sprouts. This provides not only enrichment but, more importantly, also serves as a source of dietary fibre. During breeding periods, it is essential to have surplus males to hand so that cocks showing poor sexual activity can be replaced, and thus a constant (high) fertility maintained.

Example 13.10

A pheasant breeder/game manager contacted a wildlife forensic expert upon noting that a large number of eggs transferred to an incubator house 'fell out' at the first candling—that is showed no sign of embryonic development (see Chapter 10—Eggs and Eggshells). Eggs were being collected twice a day and kept at 18°C –20°C in a dark place before being transferred to the incubator house, located at a 3 km distance from the breeding farm.

The expert's task was to determine the cause of these losses and to recommend treatment.

The expert inspected the breeding farm, that is the aviaries and egg-laying sites, as well as the feeding and egg-collecting methods. Everything was found to be in order. Incubating conditions were in accordance with the technology employed at the hatchery (see Chapter 10). By examining the eggs discarded at the first candling, it was observed that the albumen was liquefied and turbid, whereas the chalazae were torn. The reduction in hatchability of the eggs proved to be the result of strong vibration during transport: they were being transferred in a vehicle that had a defective shock absorber. The eggs were subjected to trauma that resulted in the chalazae being damaged and the yolk losing its central position.

Warranty claims can also arise relating to the traffic of captive-bred wildlife. Very often pheasants reared for hunting are the subject of such disputes – for example, when hunting companies purchase chicks from a hatchery. Deaths in the first 5 days of life are of particular importance under these circumstances.

Example 13.11

A hunting company purchased 2500 one-day-old pheasant chicks from an incubator house. The chicks were received in a sterilised, clean room littered with mould-free straw and heated to 35°C–36°C. On the day of arrival 165, a day later 473 and on the third day 532, chicks died.

The carcases were sent to an institute for examination. During the investigation, three important questions arose: What has caused the death of these birds? How might the remaining chicks be treated and further losses avoided? Could the seller be held responsible for the disease and were there any warranty claims to be endorsed?

When examining pheasant chicks, their weight should always be measured, and both the umbilical ring (whether it has closed or shows signs of inflammation) and the contents of the yolk sac (the level of absorption and the quality of the contents) should be thoroughly inspected. Pulmonary mycosis and renal failure should also be ruled out.

In the aforementioned case, *Salmonella* bacteria could be cultured from the chicks' yolk sac; this infection was making the yolk thinner and adopting a reddish-brown colour. The umbilical ring was still open.

Thus, *Salmonella* infection was pinpointed as the chicks' cause of death. As the buyer took action within 5 days of purchase, and the time span between purchase and peak drop-off was shorter than the probable latency period of the pathogen causing the disease (see Chapter 2), an infection in the incubator house could be proved.

As stressed in Chapter 2, wildlife and agriculture often conflict – as when animals damage crops or are themselves affected by certain agricultural practices. It should be noted that such conflicts are rated and prevented differently in each country. In Hungary, damage to farmlands and forests (crops and woodland renewal) is an existing legal category (see later). Less frequently, wildlife can be harmed by feeding in an agricultural location.

Example 13.12

In the winter period, mid-January, sudden intense snowfall covered the fields with a 30 cm thick layer of snow, concealing the undergrowth.

Three days after the snowfall, roe deer started to die: each day 5–6 dead roe deer were found by hunters. Carcasses were sent for pathological examination because the animals were suspected to have died as a result of poisoning or infection.

Upon dissecting the deer, large amounts of frothy, vivid-green, alkaline (pH 7.8–8) matter was found in their stomachs (Figures 13.21 and 13.22). However, apart from hyperaemia, no other lesions could be seen. After performing the dissections, the area where the animals were recovered was visited. It was a rape field with some rape leaves protruding from the snow cover.

The alkaline pH of the ruminal content and the sudden death of the animals proved to be a consequence of consuming excess amounts of rape leaves, high in protein. The microflora in the rumen could not adapt to the high-protein diet in such a short period of time, and an increase in ammonia resulted in alkalosis.

Special Considerations and Scenarios

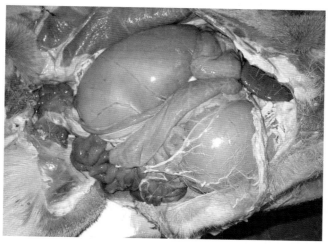

Figure 13.21 Distended rumen of a roe deer.

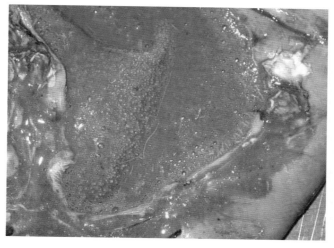

Figure 13.22 Thin, frothy ruminal content in a roe deer that had eaten excessive amounts of rape leaves.

The expert advice to the hunting company was to set-up feeders in the area well before food becomes scarce in winter, so that the roe deer could more easily survive drastic weather changes, especially snowfall.

Example 13.13

On the hunting grounds of a hunting company, just a few weeks before the onset of the open season, six extremely valuable red deer stags died in a single day. As no infectious disease was known from the area that might have resulted in such a large scale and sudden decline, the president of the hunting company called in a wildlife forensic expert to investigate and determine the cause. The expert examined the deer and noted that their bellies were distended and saliva was oozing from their mouths. Dissection revealed the presence of large amounts of chewed maize in the rumen which pointed to lactic acidosis. It turned out that the problem was caused by maize used to feed wild boar. The president of the hunting company told the expert that they fed their animals throughout the year and did not offer more maize than at any other time. However, a new employee had not spread the maize from the back of the pick-up truck with a shovel, as was usual, but emptied the sacks at once, thereby building heaps. The red deer thus ate much larger quantities of maize (rapidly soluble carbohydrates) in a single session; this resulted in multiplication of lactate-producing bacteria, a rise in lactic acid that caused a drop in ruminal pH and finally lactic acidosis.

The aforementioned examples clearly illustrate how important it is to evaluate the environmental circumstances in addition to performing a diagnostic dissection.

An interesting and rather problematic area of wildlife forensics relates to financial loss resulting from a collision between wildlife and vehicle (see Chapters 2 and 9). Besides evaluating traffic conditions, an important question to be answered is whether the impact damage suffered by the animal matches that inflicted on the car. In such cases, it is usually the insurance company that calls in a forensic expert.

Example 13.14

Following a collision between a red deer with a medium-sized car, authorities called in a technical and a wildlife forensic expert to investigate. The left front and middle sections of the vehicle were affected. The deer left the collision site, but died approximately 200–250 m away. According to initial data, it was believed that neither the driver nor the game manager could be blamed for the accident. The sky was overcast and it was raining at the time of the accident. The driver insisted that upon noticing the deer he had started to brake, the car came to a halt and the deer ran into the immobile car. He claimed that the unusual behaviour of the deer had caused the collision, as it was actually running into a standing vehicle. The two forensic experts examined both the cadaver and the car, and concluded that they were both moving when the accident occurred: the breast of the deer collided with the left front of the vehicle. The impact caused the animal to rotate and its pelvis crashed into the side of the car. Technical simulation proved that, in order to produce the damage on the car, the vehicle must have been moving at a speed of approximately 80 km/h. Consequently, it was concluded that the deer was not running into a standing car but the driver had caused the accident by driving too fast.

Predation of Endangered New Zealand Weka by Dogs

Maurice Alley

Introduction

Many of New Zealand's threatened species are successfully maintained on predator-free offshore islands where they are safe from attack by introduced stoats, ferrets and cats which are common on the mainland. The North Island weka or woodhen (*Gallirallus australis*) is one such flightless bird of the rail family that is particularly vulnerable to mammalian predators. It is endemic to New Zealand and was once very common in lowland forests throughout the North Island but is now confined to a few offshore islands and three mainland sanctuaries.

About 35% of the total population of North Island weka is held on Kawau Island in the Hauraki Gulf which is situated only 1.4 km off the coast of the North Auckland Peninsula. About 10% of this island is managed by the Department of Conservation (DoC) and consists of an historic reserve containing Mansion House, the residence of the first Governor of New Zealand, Sir George Grey, who bought the entire 5000 acre island to establish a botanical and zoological park. In addition to the reserve, there are now about 60 permanent human residents living on the remainder of the island. The weka live mainly in the grounds of the historic reserve which is also a home to endangered kiwi, a variety of peafowl and five different species of wallaby, some of which are endangered in Australia.

On the morning of 5 May 2009, the DoC ranger on the island was walking on a beach path near Mansion House when two dogs came up to him. One was a small Shitsu without a registration number on its collar and the second was a medium-sized mastiff-cross. Because the dogs were in a dog-controlled area and posed a threat to the wildlife, they were immediately seized and locked away. Inspection of the surrounding grounds revealed the bodies of 14 freshly dead weka, one dead peacock and a decapitated and disembowelled wallaby. Both dogs were found to be the property of a local resident. Later that morning the dogs were delivered to the local Council dog ranger who was asked to hold them until DNA samples could be taken but miscommunication allowed the release of the dogs into the custody of a friend of the dog owner before the samples could be taken.

Findings

The dead birds were taken to the regional DoC office and, after a brief examination, 12 were dispatched to Massey University's Wildlife Health Centre at Palmerston North for necropsy. All the birds examined in the laboratory were in good body condition and showed similar lesions. Nine of the twelve showed severe thoracic trauma with multiple fractured ribs and haemorrhage in the lungs and thoracic air sacs. Six of the birds showed severe trauma to the head and neck with extensive subcutaneous haemorrhage, puncture wounds in the skin, and three had penetrating wounds of the cranium. Seven of the birds showed leg and abdominal trauma. Three had fractured legs and severe muscle tears and associated haematomas while four had abdominal haemorrhage, tears in the abdominal wall and lumbar region, with maceration of lumbar and pelvic muscles.

Puncture wounds in the skin that penetrated the subcutis and extended into the underlying muscle were seen in the majority of birds. These varied in size between 5 and 20 mm diameter and often had irregular torn margins. A reasonably accurate measurement of the distance between the inner margins of paired puncture wounds was possible in three birds and this was 35–40 mm (Figure 13.23). The larger wounds seen in the underlying muscle are presumably the result of the dog shaking the carcase after biting.

Further investigation revealed that the same dogs had been identified in the reserve area a month earlier by a member of the public and been returned to the dog owner without anyone reporting the incident. A visit to the dog owner's residence revealed boundary fences that were not adequate to contain any dog. The defendant admitted that his two dogs had disappeared on the afternoon before the incident, were at large all night and that he had made no attempt to search for them. Nevertheless, he believed they were not capable of killing the weka.

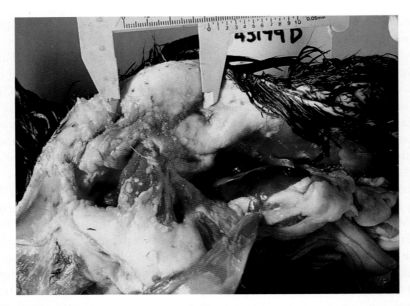

Figure 13.23 The lumbar region of a predated weka with the skin reflected dorsally and the leg caudoventrally. There are smaller canine tooth puncture wounds in the reflected skin and subcutis (callipers) and corresponding larger puncture wounds in the underlying muscle.

Questioning the Evidence

Following receipt of the pathology report, the defendant took his dog (the mastiff-cross) to a veterinarian who measured the upper intercanine tooth distances from the inner margins of tooth impressions on a piece of paper and found the distance to be 42 mm. The defence solicitor attested that this differed from the pathologist's report and may therefore not have been the same dog that killed the weka.

However, an examination of the skull of a mastiff-cross by the prosecution's pathologist revealed a significant angle of splay in both the upper and lower canine teeth (Figure 13.24).

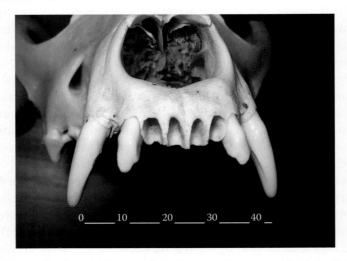

Figure 13.24 The skull of mastiff-cross dog showing the angle of splay of the canine teeth. Scale = mm.

Outcome

The defendant was charged with two charges of allowing his dogs to enter a dog-controlled area and one of owning a dog that attacked and killed endangered wildlife. The penalty for each of these offences was 12 months' imprisonment, a NZ$10,000 fine or both.

The accused pleaded guilty to the first two charges on the third morning of the trial, and after nearly 4 h of deliberation, the jury found the accused guilty of the remaining charges. The defendant's lawyer appealed for her client to be discharged without conviction, given the damage that had already been done to his reputation and self-esteem and that although he had not exercised the necessary precautions when walking his dogs, he had not intended any harm. However, the Crown prosecutor submitted that a sentence of community service, a fine and an order that the defendant not hold a dog licence for 2 years would be appropriate.

Although the judge showed leniency in sentencing the accused to 200 h of community work on the island, the publicity the trial received served as a warning for all dog owners to be more vigilant in controlling their dogs near areas where there are vulnerable wildlife.

Mountain Gorilla Disease: Implications for Conservation

Gladys Kalema-Zikusoka

Introduction

Mountain gorillas (*Gorilla beringei beringei*) are critically endangered, with half of the approximately 780 individuals residing in Uganda's Bwindi Impenetrable and Mgahinga National Parks and the remainder in Rwanda and the Democratic Republic of the Congo (DRC). The tourist site at Bwindi Impenetrable National Park, Buhoma, is in a tropical Afro-montane forest habitat at an altitude range of 1190–2607 m. Currently, 10 out of 28 gorilla groups in Uganda have been habituated for tourism and research purposes. Veterinary intervention in mountain gorillas is limited to habituated individuals, and then only in human-related or life-threatening disease conditions, to minimise the disturbance of natural behavioural patterns (see Chapter 2).

Before an intervention is performed, many stakeholders have to be widely consulted, on top of fellow veterinarians, wildlife managers or wardens in charge of the national park; biologists and ecologists also have to be involved in the debate in order to ensure that both welfare and conservation are not being compromised. For example, when two silverback gorillas fight during group interactions, it is considered inappropriate to treat them, however severe the wounds are, because this can be judged to be interference with nature. It may be time for another silverback to take over the group and to bring new genes into what is an already endangered population; this aspect has significant implications for conservation of the endangered sub-species.

The Uganda Wildlife Authority (UWA) set up a veterinary unit in 1996, primarily because mountain gorillas had been habituated for tourism and this put the animals at much greater risk from human disease (see Chapter 2); tourists visiting the gorillas may get as close as 5 m. As the first Veterinary Officer for UWA, I was called upon to attend to mountain gorilla clinical cases. Most of these did not need an intervention but some were unusual and there was a strong indication that an intervention should be performed. I learnt from senior wildlife veterinarians that the two most important subjects taught in veterinary school are anaesthesia and pathology. You have to justify why you have carried out an intervention on a wild animal, and pathology is probably the only justification, whether it is a *post-mortem* examination of an animal or treatment of a sick or injured one. Biologists or conservationists have asked why a *post-mortem* examination of an animal is done even though the cause of death is already known – for example, in the case of a gorilla that has died in a snare or as a result of 'old age'. I explained that, through conducting *postmortem* examination of all gorillas, we obtain valuable information about this endangered sub-species that helps us to manage their health and ensure their survival and population viability for future generations. This is particularly relevant in the case of mountain gorillas as disease is considered one of the most important threats to them. Although pathology was not my favourite subject at veterinary school, it has become the most useful of the disciplines I practise when dealing with wildlife and disease issues at the human/wildlife/livestock interface (see Chapters 2 and 9).

Clinical Cases

Two cases are briefly mentioned that resulted in significant implications for wildlife and public health management and could have been of forensic significance.

The first was a scabies outbreak in mountain gorillas in Bwindi (Kalema-Zikusoka et al., 2002). A provisional diagnosis of scabies (*Sarcoptes scabiei* infection) was made on the basis of clinical signs (see Chapter 8) and confirmed as a result of examination of a full-thickness skin biopsy. Initial treatment given was ivermectin and a long-acting antibiotic, both intramuscularly. Skin scrapings showed numerous mites resembling *S. scabiei*.

Subsequently, skin scrapings of a dead infant revealed numerous *S. scabiei* mites some of which were still alive and with eggs. The skin sections showed severe cutaneous acariasis associated with marked acanthosis, spongiosis, ballooning degeneration and hyperkeratosis, but few signs of inflammation – illustrating the importance of supporting laboratory tests (see Chapter 10). All the surviving gorillas improved and fully recovered after treatment. Follow-up health checks were continued for a year.

The second case was one of rectal prolapse in a mountain gorilla (Kalema-Zikusoka and Lowenstine, 2001). This female juvenile gorilla was apparently active and healthy other than for the presence of a 3–4 cm diameter mass protruding 10–12 cm from the anus. Liquid was released from the lumen when the gorilla occasionally squeezed the tissue. A diagnosis of severe, complete rectal prolapse was made.

Preparations were made for an intervention but, to our surprise, within the next 24 h the prolapse spontaneously resolved. No further health problems were noted in this animal. The field rangers from Bwindi reported that similar rectal prolapses had been seen in the past in two other juvenile male gorillas but had not been reported because they did not realise that it was unusual – a reminder of how much rangers, who work daily with these animals, know about their biology and natural history (see Chapter 8).

A second case was in a juvenile female gorilla that was observed with a persistent rectal prolapse of several days' duration. The day before the prolapse was noted, there had been a fight between a lead silverback and a solitary silverback. The prolapse progressively worsened. By the third day, it was approximately 15 cm long and trailing on the ground. The prolapse was shorter on the fifth day, estimated at 10 cm in length. On the sixth day, the gorilla's condition deteriorated rapidly, and the prolapse appeared necrotic and was malodorous from a distance of approximately 5 m. The gorilla became progressively weak, lethargic, anorexic, was surrounded by flies and appeared to be retching on the seventh day. The following day, the gorilla was moving with great difficulty from her night nest training, taking into account vegetation type and season, and identifying the likely size of carcases to be located. The overall corrected mortality rate (m) can then be calculated using the following formula, as described in Kerns and Kerlinger (2004):

$$m = \left(\frac{NC}{ktp}\right)\left(\frac{e^{I/t} - 1 + p}{e^{I/t} - 1}\right) \qquad (13.1)$$

where
    N is the total number of turbines at the site
    C is the total number of carcases
    K is the number of turbines surveyed
    t is the mean carcase removal time (days)
    p is the searcher detection probability
    I is the search interval (days)

However, this model does not allow for any additional error in searcher efficiency/scavenger rates.

Post-Mortem Forensics

Once carcase surveys have been made, it is logical for further analysis to be done on any collected carcases. Much information can be ascertained about the cause of death from a carcase (see Chapter 9), but in the case of bat bodies retrieved from wind turbine bases, not all is as it might appear. Initial assessment of such bats revealed that although many did show signs of direct trauma, such as broken wings and lacerations (e.g. Baerwald, 2008), some did not appear to display any outward signs of collision-type impacts, as might be expected to occur if the bat was struck by a moving turbine blade. It was hypothesised by Dürr and Bach (2004) (amongst others) that the cause of these mortalities could be linked with the variable pressure changes that occur around an operational turbine rotor (e.g. Bet and Grassmann, 2003). Further investigation to this effect was not published until 2008, when Baerwald and colleagues retrieved a sample of 188 bat carcases around turbines at a wind farm in Alberta, Canada. Of these carcases, 54% displayed injuries consistent with blade strike; however, those remaining had no obvious gross injury. Necropsy was performed on the bodies of 75 bats, revealing evidence of haemorrhage in 92% of cases (Figure 13.27).

Baerwald et al. (2008) also performed histological examinations on 17 of the bat carcases retrieved, focussing on the lung tissue where haemorrhage had occurred. In addition to haemorrhage, pulmonary lesions, lung collapse and small air bubbles were found, consistent with what might be expected from barotrauma. It is speculated that the pressure differential around the vortices of moving turbine blades is sufficient to cause barotrauma in susceptible mammal species, such as bats (Baerwald, 2008; Baerwald et al., 2008). However, more recent findings by Rollins et al. (2012) have indicated that pulmonary examination findings alone should be treated with caution, as lung tissue damage can also be caused by natural decay, temperature and freezing processes. Rollins and colleagues (2012) performed necropsy and histopathology examinations on 262 bat carcases retrieved from wind turbine bases in Illinois, USA, as well as 53 bat carcases retrieved from collisions with buildings, for comparison. Of the wind farm bats, 73% had traumatic injuries consistent with collision. Examination of the tympana inside the ear, rupture of which is an indicator of barotrauma

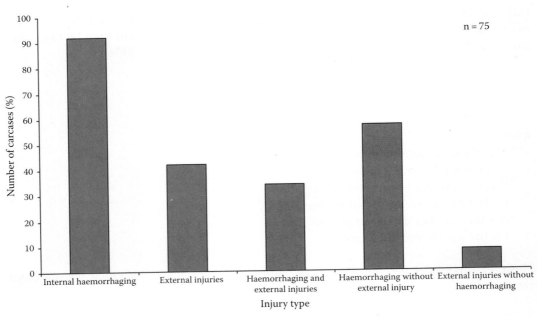

Figure 13.27 Necropsy findings from 75 bat carcases retrieved from wind turbine bases. (Adapted from Baerwald, E.F., Variation in the activity and fatality of migratory bats at wind energy facilities in Southern Alberta: Causes and consequences, MSc thesis, University of Calgary, Alberta, Canada, pp. 1–117, 2008; Baerwald, E.F. et al., *Curr. Biol.*, 18, R695, 2008.)

in humans, revealed that 20% of wind farm bats had evidence of rupture, but only 6% showed tympanal rupture without traumatic injury. Only 1% of the building collision bats demonstrated tympanal rupture. It was therefore concluded that barotrauma may be less of a significant cause of bat mortaility at wind turbine sites than was previously identified.

These findings demonstrate that it is likely that a combination of both blade strike (Horn et al., 2008) and barotrauma contribute to the high incidence of bat mortality at some wind installations. Examination of carcases is, therefore, a particularly important part of bat–turbine mortality investigation, which should include radiography (fractures and strike damage), necropsy (external and internal injuries particularly to the tympana) (Rollins et al., 2012) and histopathology (particularly of the pulmonary system) (see Chapter 10). However, it is not only important to carry out forensic investigation of mortalities caused by bat–turbine interactions but also to identify why bats are present in the vicinity of the turbine itself.

Behavioural Observations

To develop a better idea about why bats fly close to wind turbine rotors, it is useful to be able to monitor the active behaviour of the bat in the area. Such observations can be difficult to make, because bats are active at night, and often involve large-scale turbines, where the rotor region can be hundreds of metres from ground level. Because of the bat's echolocation system, it is possible to use sound to 'listen in' on bat activity. As well as emitting high-frequency sound to hunt insects and navigate, bats use social calls to interact with each other. Acoustic cues can be detected using audio equipment with an ultrasonic range; it is also possible to model the interaction of bat echolocation pulses with moving turbine blades to predict how bats might 'see' the operational rotor (Long et al., 2010). It is

Special Considerations and Scenarios

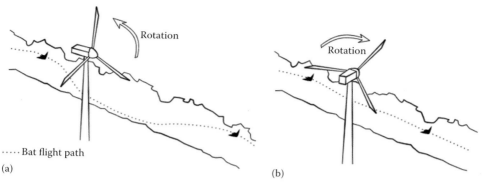

Figure 13.28 Demonstrating how the flight paths of common pipistrelles along a nearby (<10 m) hedgerow changes in accordance with rotor angle; perpendicular (a) and parallel (b). (Adapted from Bach, L. and Rahmel, U., *Bremer Beiträge für Naturkunde und Naturschutz*, 7, 245, 2004.)

also feasible to observe bat activity directly with thermal imaging cameras. Using thermal infra-red imaging, Horn et al. (2008) observed bat behaviour around operational turbine rotors, recording not only bats colliding with moving blades but also bats investigating and attempting to land on blades (both moving and stationary) and becoming trapped in rotor vortices. Other studies have noted more subtle changes to bat behaviour before and after turbine installation. Bach (2001) conducted pre- and post-construction bat activity assessments at a new wind farm site in Germany, finding that while the activity of certain species such as the serotine bat (*Eptesicus serotinus*) declined after turbine construction in the area, the activity of other species (most notably the common pipistrelle, *Pipistrellus pipistrellus*) increased significantly post-construction. However, the flight paths of the pipistrelles were seen to alter according to rotor position, suggesting that the turbines were influencing their foraging behaviour along the hedgerow (Figure 13.28).

Observations of bats attempting to land on turbine blades and towers, coupled with the peak in bat mortality frequently observed during the summer and autumn months, have even led to speculation that certain species of tree-roosting bats commonly involved in turbine interactions (such as the eastern red bat [*Lasiurus borealis*] and silver-haired bats [*Lasionycteris noctivagans*]) are attracted by the tall, tree-like turbine towers to use as part of their mating displays (Cryan, 2008). It is clear that not all bat species will be affected in the same way by any particular turbine, and therefore the species present in the local area should be identified before further investigation or assessment takes place. Using a mist-net to trap and record bat species in the area is an option, but since many countries, particularly in Europe, protect their bat species by law, official authorisation (often in the form of a permit or licence) is usually needed to do so (see Chapter 3). A more favourable, non-intrusive way to identify foraging bat species present in an area is to record their echolocation calls, many of which are unique to each species and may be identified during post-processing of recordings (e.g. Fisher et al., 2005).

Indirect Investigations

It must also be considered that wind turbines can have an indirect effect on local bat populations that contributes to the problem in a more subtle, complex manner than that previously described. A good example of this is the ultrasonic (above 20 kHz) emissions that can be produced by some turbine models during operation (e.g. Long et al., 2011b; Schröder,

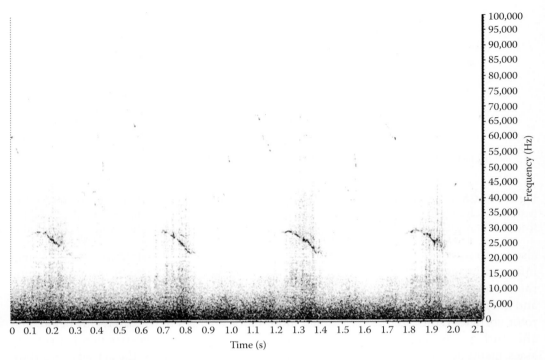

Figure 13.29 Fast Fourier Transform (FFT) sonogram of a recording from an operational 20 kW turbine with a blade defect. Periodic frequency modulated sweeps in the ultrasonic band between 20 and 30 kHz, caused by movement of the damaged blade, are potentially audible to bat species in the vicinity that use these frequencies for echolocation, such as the noctule (*Nyctalus noctula*). FFT produced using length 1024 bands, Hanning window, 75% overlap, 40% linear energy scaling.

1997; Szewczak and Arnett, 2006). Although not directly audible to humans, ultrasonic emissions may be detectable by certain bat species in the vicinity if the ultrasonic noise is of a similar frequency range to that used by the bat for echolocation (Figure 13.29).

Such noise has the potential to attract the curiosity of the bat, to interfere with the bat's echolocation when near the rotor region or even to be attractive to insects. Insect attraction to wind turbines is an interesting concept as this has the potential to be the root of the issue; if insects are drawn to turbines in large numbers, this in turn could be providing attractive foraging opportunities for insectivores such as bats. Other recent work (Long et al., 2011a) has looked into the possibility that the colour of the turbine rotor and tower could be attracting insects. The majority of wind turbines are painted in white or light grey, and it was found that these colours are significantly more attractive to insects than are other colours such as blue, red, green and purple. Although much more study needs to be done to confirm or refute whether wind turbines act as lures for hunting bats by attracting insects, it is clear that investigation into bat–turbine mortality should not be restricted to direct cause and effect alone.

In summary, the incidence of bat mortality at wind turbine installations is a phenomenon generating increasing concern, both for environmentalists and the wind industry. Forensic investigation of the problem is of critical importance in order better to understand the nature of bat–turbine interactions. Owing to the non-uniform locations, habitats and features of turbines, each case must be investigated consistently in ways that are appropriate to the different bat species present in each area.

Collection and Submission of Evidence

As stressed in Chapter 6 and by Cooper and Cooper (2007), standard crime scene precautions should be followed before evidence is handled.

Observation must be the first action before evidence is collected. If, for example, the case revolves around a wild bird found dead under unusual circumstances, there may be feathers on the ground. The position of these should be noted and recorded – both in writing and with photographs. The feathers should then be examined *in situ*, as far as this is practicable, with the naked eye and with a hand-lens available for more detailed investigation. If large numbers of birds are found dead, it will be important to count them – this may need to be repeated day after day if mortalities continue (see also Chapter 9 regarding groups of dead animals). Accurate counting of carcases of dead birds is surprisingly difficult, as pointed out by Philibert et al. 1993; various authors have covered this subject in papers subsequently.

However, potential problems exist when estimating turbine mortality rates based on carcase retrieval alone, owing to assessment error (Tuttle, 2004), inefficiency of searcher and removal of carcases by scavengers (Morrison, 2002). Some evidence may provide a strong indication as to the sort of activity that has occurred. For example, in the case of illegal poisoning of wildlife, the finding of baited carcases or several dead animals in a location is suspicious. Another suggestion of poisoning may be the presence of chicken eggs in unusual places; these may have been injected with a poison using a syringe.

When collecting evidence, bear in mind that some of it may have been 'planted' in order to confuse (Figure 14.6). This may not be apparent at the time but come to light later. Various clues may suggest such transposition. For example, the carcase of a dead protected

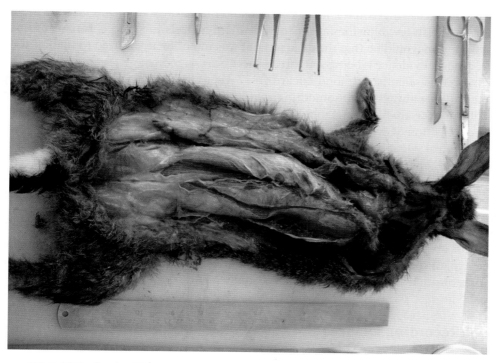

Figure 14.6 This dead rabbit was submitted for investigation because it was found under strange circumstances on a shooting estate where raptors are prevalent. There are straight-line incisions on the skin and in the muscles, indicative of human interference. The rabbit had been put out as bait, possibly laced with a toxic substance, so as to attract birds of prey.

Table 14.1 Some Actions That May Destroy, Damage or Corrupt Evidence

Action	Possible Results
Careless treading or walking carelessly.	Damage to footprints, tyre marks, vegetation.
Not wearing gloves.	Addition of fingerprints.
Wearing gloves but taking inadequate care in their use.	Obliteration of fingerprints.
Failing to wear adequate protective clothing.	Contamination of crime scene with (e.g.) foreign DNA. Exposure to pathogens, poisons, etc.
Smoking, eating or drinking.	Contamination of crime scene with cigarette ash, crumbs, drops of fluid, etc.
Using facilities at the site such as toilets, wash basin or telephone.	Contamination (see aforementioned). Addition or obliteration of fingerprints. Destruction (by flushing toilet or running water) of evidence.
Moving items or turning on/off taps or lights.	As mentioned earlier. Evidence that (e.g.) animals have not been fed, watered or kept warm/cool may be lost or damaged if the facilities are touched or used.

Source: Adapted from Cooper, J.E. and Cooper, M.E., *Introduction to Veterinary and Comparative Forensic Medicine*, Blackwell, Oxford, UK, 2007.

species may be put near to a public footpath so as to attract attention; careful examination of the carcase and its surroundings might indicate that that was not where the animal died. Other factors may also prompt doubts about the truth of the most obvious interpretation. A flyblown carcase high on the moors in Scotland may arouse suspicion because it is at an altitude at which few blowflies are seen and where myiasis in sheep is rare or unknown; the assumption would be that the dead animal was intentionally moved to that location.

Whenever evidence is being sought, it is necessary to remember the maxim of the great French forensic scientist Edmond Locard (1877–1966) that one always *leaves* something at the site and one always *takes* something away. Some actions that may destroy, damage or corrupt evidence are listed in Table 14.1.

The collection of evidence requires strict adherence to protocols and high standards of quality control. Health and safety are important (see other chapters). Wildlife cases may present special hazards such as unfamiliar zoonoses or poisons. Some of the dangers presented by live and dead animals are discussed in Chapter 2, but to them should be added physical injuries from rough terrain or poorly designed animal enclosures and assault by keepers of animals, activists, poachers or others. A risk assessment is a legal requirement in some countries (see Chapter 3). In the United Kingdom, government-appointed Wildlife Inspectors are issued with a letter that can be presented to the Inspector's medical practitioner, explaining that as a result of his or her duties he or she may come into contact with zoonoses or other health hazards (see also Chapter 2).

Use of Recording Devices

In wildlife forensics there is much in favour of using a tape-recorder, video or DVD, in addition to the time-honoured method of writing contemporaneous notes. Even if the recordings are not admissible as such in a court case, they serve as a useful aide memoire for the investigator when he or she is compiling a report. In certain circumstances, 'telemedicine' data that link live or dead animals or their derivatives with experts in other parts of the world may be acceptable evidence (Cooper and Cooper, 2007; Penzo and Pietersma, 2012; Wootton and Craig, 1999).

The other very important value of a tape-recording is when translation is necessary (see also later). For example, a dead tiger is found in India. Those who discovered the animal's body are interviewed, but they respond in Bengali or Hindi. An interpreter translates, but only he or she knows the accuracy of that translation. A recording allows review of both the original response and the interpretation.

There are some important considerations, however. A recording should not be taken without asking permission, or explaining what is proposed, beforehand. This in turn, though, may influence the response of those being questioned, or their interpreter, especially in countries with authoritarian regimes where there may be recriminatory action if 'mistakes' are made.

Marking of Evidence

In wildlife cases evidence may consist of live animals, dead animals or samples. Each of these will need to be reliably marked and identified.

The marking of live animals is a long-established practice and many methods can be used in wild species (Cooper and Cooper, 2007) (see also Chapter 8). The marking of dead animals is discussed in Chapter 9, and the marking of samples in Chapter 10.

In all forensic work, correct identification of live animals, dead animals, tissues and other derivatives is essential. Loss or transposition can occur by accident or by intent. The routine taking and storage of material for checking of identity by DNA identification is recommended as discussed elsewhere. Simpler and usually less-expensive methods range from the attachment of a tag, or tying tape or string to an appendage, to the implantation of a microchip (see 'Identification Systems in Hunting Falcons in the Middle East' section in Chapter 13). Dockets may also be present (see Chapter 10 and Appendix A).

An important reason for maintaining a proper chain of custody is to help ensure that evidence can neither be tampered with, nor replaced or transposed, prior to a court hearing. In wildlife forensic work, where carcases, parts of them or samples, may need to be sent to different people, there are opportunities for such tampering or replacement to take place.

The investigator needs to be alert to such possibilities and to take all practicable steps to protect the material. This is best achieved by using reliable methods to identify the material, including

- Photographs
- Detailed written descriptions
- Attachment of labels and tags that cannot easily be removed – or removal of which leaves recognisable damage (Figure 14.7)
- Always taking and retaining a small piece of tissue, for DNA and other tests, each time a carcase or parts of it change hands.

These measures should be routine in all wildlife work.

When an investigator receives material, he or she should not neglect to confirm its identity. Even if a specimen has passed through the hands of various experts, it is possible

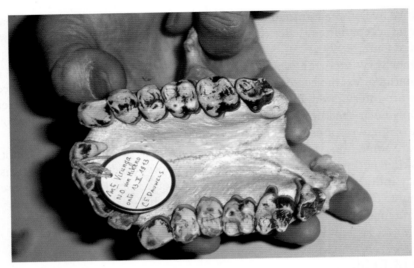

Figure 14.7 This specimen has an identification tag attached to it. Care must always be taken to ensure that such tags do not damage the evidence.

that it has not previously been correctly identified, or perhaps has been intentionally or accidentally transposed. It is easy to assume that, for example, a bone is a tibia, because that is the identity given on the paperwork by previous investigators. One should always examine it and assure oneself of its correct identity.

Storage of Evidence

As with other forensic work, all evidence must be saved and carefully stored before and during the hearing. It will be needed as an exhibit in court and may need to be made available to others (e.g. a second *post-mortem* examination).

It is a moot point as to whether or not samples should be stored by a non-governmental organisation that has an interest in the case. For example, in the United Kingdom, a body such as the Royal Society for the Protection of Birds (RSPB) or the Royal Society for the Prevention of Cruelty to Animals (RSPCA), involved in bringing a prosecution, may offer or be asked to keep evidential material in its store or deep-freeze. As a general rule, this should be discouraged as it might provide an opportunity for suggestions or allegations that the evidence could have been tampered with during the case.

Laboratory samples are often part of evidence and therefore may need to be retained for production in court or examination by another party (see Chapter 13). If this is not practicable, because, for example, the whole sample has been used, this should be recorded. Useful guidelines concerning the retention of samples are as follows:

- If in doubt, assume that the samples will be needed later, even if a legal case seems unlikely.
- Plan carefully how the samples are to be retained (+4°C refrigeration, standard deep-freeze, ultra-low temperature, fixed). In some cases (e.g. tissue that may later require toxicological, virological or electron-microscopical examination) it may be wise to divide samples and store them in different ways.

- Always retain original specimens at least until initial laboratory results are obtained.
- Designate a person to 'sign out' specimens when they are destroyed or sent elsewhere.
- Remember, in planning and budgeting, that the storage of samples can be expensive and take up valuable space.

Legal Controls Relating to Evidence

The collection, transportation and processing of samples that may form part of evidence are covered by legislation (see Chapter 3 and Appendix I).

Hazardous material (pathogens, poisons, radioactive or explosive samples, etc.) will be subject in most countries to strict regulation, even if being sent between laboratories. Sending forensic material from one country to another is not always easy.

Samples that may constitute evidence should be placed in separate packages and properly sealed so that their contents cannot be removed, changed or damaged without detection. Evidence seal tape should be used: the person sending the sample should sign/initial and date the area that is taped. A list of contents should accompany the consignment, with copies available for officials and for the receiving agent.

Perishable evidence will need special attention as it may have to be chilled or frozen *en route*. Various methods can be used and laboratories will advise on this.

Evidence may be contaminated and thereby present health hazards. When dealing with evidence that originates from live or dead animals, one should assume that wrappings, labels, dockets and relevant correspondence (even if in a separate envelope or bag) are contaminated and may present hazards to those handling such items. This is particularly likely when the material has been through the hands of several people: the docket should reflect such movement, but remember too that the docket and other items may have become contaminated. One cannot assume that everyone in the chain adheres to the same standard of hygiene.

This raises the following points regarding both storage and transferring the evidence elsewhere:

- A risk assessment may be appropriate.
- Whether or not a revised risk assessment is deemed necessary, full hygienic precautions should be taken.
- Protocols and SOPs should allow for hygienic precautions. This should include a change of gloves between handling 'dirty' paperwork (that has accompanied the specimen) and clean paperwork, tape-recorder, camera, etc. that are being used to document work. The best option is to have several people involved and to designate them as 'clean' or 'dirty' (see earlier and Chapter 7).

As pointed out by Cooper and Cooper (2007), wildlife material can present unusual, but significant, hazards; the fangs of dead snakes can still often cause envenomation if they penetrate human skin and rigid frozen material from any species may inflict damage through wrappings.

Reporting

The production of a final report is preceded by the preparation and collation of technical records. These are the written and electronic records of the steps that are taken, the findings made and the conclusions drawn from the examination and analysis of evidence. They are crucial to the preparation of the final report and to the expert evidence that may flow from it.

Technical records must be thorough enough to provide sufficient supporting data that another scientist could reasonably be expected to draw the same conclusion. Harbison (see 'Why Is Quality Management Important?' section in Chapter 12) points out in this context that the preparation of appropriate controlled SOPs to deal with the range of sample types and taxa involved in wildlife crime can be difficult to achieve.

The responsibilities of the expert who prepares a report are outlined in Chapter 15. As suggested, there is great merit in preparing initially a draft report that can be refined later. This provides lawyers and clients with an indication of the approach, opinion and conclusions of the expert at an early stage. This is not, however, to suggest that changes should be made merely to satisfy the wishes of those who have sought the expert's opinion. Responsibility in many, possibly most, jurisdictions is to the court, not to either party.

A report should be accurate, clear, unambiguous and objective. It must be based on information that is recorded in the technical notes and any expert opinion and interpretations drawn from the analysis should be clear. It is advisable to have reports reviewed by another expert. Some quality management systems require work to be reviewed in this way and this provides a further opportunity for correction and improvement.

It is important to write and spell correctly in reports. To illustrate this Cooper and Cooper (2007) related the Chinese proverb:

> If writing is poor then what is said is not what is meant. If what is said is not what is meant then what ought to be done is not done.

This applies to all written or computer-generated documents, including technical notes. Clear, unambiguous records will result in easily read reports and reduce the risk of misunderstanding. The general rule is to use plain language.

Not all reports, of course, are written in English. As emphasised in Chapters 3 and 7 and earlier, wildlife cases take place in many countries of the world and an expert may well be asked to produce evidence for a hearing that uses a different language. If possible, one should find a translator in whom one has confidence although use of an independent person is likely to incur extra costs. Preparation for appearing in a court case overseas needs to start early. The challenges that different languages and use of the vernacular can produce were discussed by Cooper and Cooper (2007) and will not be repeated here.

It is most important in reports – and on tape-recordings – to describe labels, tags and dockets correctly as they may be crucial to evidence. Photographs will help. A 'docket' is defined as 'any numbered and dated record of material that is recorded chronologically and, usually, signed by those receiving that material…'. The colour of the docket may be important – in the United Kingdom they are often yellow – and care should be taken to ensure that there are comparable dockets, each signed and dated for each piece of evidence.

The investigator should be very cautious of commenting on statistics and probability (see Glossary – Bayesian network). Reference is made in Chapter 15 to the case of Professor Sir Roy Meadow, an eminent medical man but not a statistician, who misguidedly gave an opinion in court as to the chance of two babies in the same family dying from cot death. If statistical analysis of findings or data is required in a wildlife case, expert advice should be sought.

Consulting the Literature

Remember to consult widely before writing a report or statement that would benefit from, or be strengthened by, reference to previous work on a subject. Be sure to carry out as full a search of the literature as possible, not just recent publications and not only books and articles written in English (Cooper, 2010c; Cooper and Cooper, 2007). If you have followed standard techniques in your investigation – for example, clinical ophthalmoscopy or toxicological analysis – it may be worthwhile (and helpful to the court) to put a reference to an appropriate book or scientific paper where the technique is described.

Although there is plenty of material on the Internet (see Appendix E), courts tend to be conservative and may give more credence to a citation from a textbook or a scientific paper than something on a website, especially if the latter is of uncertain provenance or not, apparently, subjected to scientific scrutiny or peer review.

Make any literature search as broadly based as possible. For example, if a case revolves around the feeding habits and associated features (e.g. pellet formation) of a species of bird of prey, remember that there may be relevant information in both ornithological and falconry texts. Studies on captive animals can be relevant to those that are free-living, and *vice-versa*, but be careful of extrapolating between the two.

Composite Reports

Elsewhere the advantages and disadvantages of working as a team in wildlife forensic investigations were discussed.

A corollary to this is the composite report – in other words, a report on a case, or an incident, that is compiled by more than one person. On the face of it, this presents no problems; each person writes up his or her own part and the different pieces are put together. In practice, however, this is not always so easy. Particular difficulties can arise when an investigation is prompted, and perhaps funded, by a body such as a conservation organisation which is anxious to mould the report, with all its constituent components, into a document that, adorned with the organisation's logo, can be presented to the media. Under such circumstances, the expert may find himself or herself under pressure to write in a way that matches the organisation's normal format and in which arguments are worded in order to promote the organisation's policy. The situation may be further complicated if all members of the team are invited to contribute wording and ideas to a section, especially where interpretation is a matter of opinion rather than science. Great care has to be taken in all such cases. The expert should be aware from the outset of the dangers of being a party to a report that, while meeting the public relations needs of the organisation commissioning it, may not stand up in a court of law. If the situation cannot be resolved by agreement, the expert should either (1) decline to be involved, (2) insist that his or her contribution is separate from the rest of the report or (3) ask to be listed as an 'advisor' rather than one of the authors.

Change of Opinion

The expert witness has an obligation to state if he or she has had a change of opinion over a matter (see Chapter 15). This is not always easy. Many people experience mental conflict (cognitive dissonance) when they are presented with evidence that their beliefs or assumptions are wrong.

Retention of Evidence after the Case

Legal cases may take years to resolve, especially if there is an appeal or if other actions (civil, public enquiries) are possible.

The retention of evidence is important, but presents problems of storage and, in some cases, deterioration. Biological material, especially samples, is particularly susceptible to deterioration.

It is difficult to know how long evidence should be retained. Many toxicological laboratories retain their records and materials for a decade or more. DiMaio and DiMaio (2001) suggested that toxicological samples from (human) medical cases should be kept for a minimum of 2, and preferably 5 years, after analysis. We propose that standard guidelines should be established for the retention and storage of wildlife forensic samples, coupled with the establishment of reference collections (see Chapter 9).

Some evidence will need to be destroyed after the court case on grounds of health or safety. Other items may need to be returned to the original owner. The fate of some may be decided by the court. Anything that remains should not necessarily be destroyed. Collections of reference material are much needed and can be very valuable in wildlife forensic work (PAW, 2005) (see Chapters 5, 8, 9 and below).

Reference Collections

Dedicated reference specimens and collections can be of great value in human and animal forensic work for teaching, research, and reference in court cases.

There are very few collections that relate specifically to animal forensics, and even fewer that contain wildlife specimens. The situation has improved in recent years with the development of electronic communication, whereby images of specimens or parts thereof can be sent all over the world. Nevertheless, depositories of original material that can be handled and even sampled are of great potential value. In the United States and Australia, these items are usually termed 'voucher specimens' (see, e.g. Griffiths and Bates, 2002). The word 'archival' is generally applied to stored laboratory material.

Collections relating to wildlife crime are slowly being established. The U.S. Fish & Wildlife Service (USFWS) has amassed large amounts of reference material at its National Fish and Wildlife Forensics Laboratory in Oregon, the United States (see Appendix E). Smaller official or private collections exist in the United Kingdom, New Zealand and Australia, and probably elsewhere. It has been argued (Cooper and Cooper, 2007) that, especially in view of the legal obstacles to moving specimens (see Chapter 3), each country or region should have a forensic reference collection.

Various criteria are important in the establishment and maintenance of reference collections and these are listed as follows:

- Legal acquisition of all material.
- Confirmation of provenance of all specimens; this is essential in order to validate forensic results used in case-work.
- Proper cataloguing, including unique reference numbers, and reliable labelling of specimens.
- Well-publicised information on how specimens can be studied or used.
- Strict adherence to national and international laws when material is sent elsewhere.

Relevant material may be available from other sources. In certain zoological societies and conservation bodies, the maintenance of reference collections has been standard for many years (Amato and Lehn, 2002; Bailey et al., 1997; Cooper and Jones, 1986; Cooper et al., 1998; Munson, 2002; Ryder, 2002). Davis (1987) emphasised the value of 'comparative collections' for use by zoo-archaeologists, and Gippoliti (2005) stressed the vital role of historical museums in documenting biodiversity. The same applies to plants. The Linnean Society of London has long held reference collections of 18th century material. There are gene banks where microorganisms, plant seeds and frozen embryos can be stored – for example, the Kew Millennium Seed Bank and the Frozen Ark (a joint initiative between the [London] Natural History Museum, the Zoological Society of London and Nottingham University). Johnson (see 'Investigation of "Aircraft Wildlife Strike" Using Forensic Techniques' section in Chapter 13) refers to the Barcode of Life Database – BOLD (Hebert et al., 2004; Ratnasingham and Hebert, 2007), which contains confirmed DNA reference sequences obtained from collections such as those held in museums (see also Chapter 14). All such collections have potential in wildlife forensic work.

Conclusions

Efficiently detecting and meticulously collecting and processing evidence are vital in any legal case. As readers were reminded by Weston (2004), 'Forensic evidence starts at the scene'.

Writing Reports and Appearing in Court

15

MARGARET E. COOPER
CHARLES FOSTER

Contents

Forensic Evidence in the Context of Wildlife	481
Legal Background	483
Levels of Law	483
Kinds of Law	483
Legal Systems	483
Judicial Systems	484
United Kingdom	485
What Is an Expert Witness?	486
Being an Expert Witness	487
General Duties of an Expert Witness	487
Accepting Instructions	488
Writing the Report	489
Conferences	490
Pre-Trial Questions Put to Expert Witnesses	491
Meetings between Experts	491
Appearing in Court	492
Training for Expert Witnesses	493
Finding and Assessing an Expert	494
Expert Witness Fees	494
Basic Advice for the Novice	495
Witness Immunity and Liability	497
Jones v Kaney	497
Other Countries	498
Conclusion	499

'It is a very easy thing to devise good laws; the difficulty is to make them effective'.

                                                             1st Viscount Bolingbroke, Henry St. John

Forensic Evidence in the Context of Wildlife

Many situations involving wildlife can give rise to legal action. Forensic evidence provided by an expert witness is frequently required to support the claims of parties to litigation. Most commonly, but not exclusively, this occurs in criminal prosecutions for offences that arise from legislation relating to unlawful interference with protected species or habitats, to the ill-treatment of animals. In addition, wildlife may cause damage to people or property for which compensation is claimed in the civil courts in the torts of negligence or nuisance, or under specific legislation that provides for strict liability, in some circumstances, for damage done by animals. Wild animals could also be the subject of a contractual dispute or may feature as part of the context of litigation – for example, when the presence of a protected species is used as a tool with which to object to the development of new buildings or roads. Other animal-related forensic evidence may be required in disciplinary hearings, particularly relating to veterinary surgeons (veterinarians) and veterinary nurses, insurance claims and professional negligence (malpractice).

Much litigation involving wildlife will involve the use of expert witnesses. This is essentially because it will turn on questions that are outside the tribunal's own knowledge. Judges can be expected to make up their own minds about whether or not it was negligent for a car to make a right turn in particular circumstances: they cannot fairly be expected to know sufficient about the dose of poison required to kill a sparrow-hawk, or what amounts to appropriate veterinary care of a badger's femoral fracture.

This chapter, although it refers to 'wildlife', will deal mainly with animals since this is the primary concern of the book (see Chapter 1). However, it is recognised that many plants and other organisms are legally protected and that habitat conservation, in law and practice, is vital to the survival of all living species.

Legislation and litigation can involve both free-living and captive wildlife. Common parlance frequently uses the word 'wildlife' without careful definition. It is often used generically to mean 'species of a sort found in the wild'. On this basis, some enclosed collections of animals use the title 'Wildlife Park'. The term wildlife in its generic sense will still be used in this chapter, particularly when no further definition is needed.

Such species are often called 'non-domesticated species' as a more concise alternative to 'wildlife' and also to distinguish them from domesticated livestock or pet animals. Animals covered by conservation legislation are often referred to as 'protected species'. Pet, or companion, animals may also be non-domesticated species, and in this case are often referred to as 'exotic species' or 'exotics'. It is important to define clearly whatever terminology has been adopted for the purposes of providing evidence in a particular case.

Prosecutions can occur when free-living protected species are taken from the wild, trapped, snared, injured or killed without authorisation (see Chapter 2). While many activities involving captive animals are legitimate, offences can also arise when such species, alive or dead, are illegally traded, imported, transported or kept in captivity, without proper authorisation, as pets, as breeding stock, for display (zoological collections, museums and live exhibitions), or for sport. The relevant law is discussed in detail in Chapter 3. Sometimes several offences can arise from a single incident for instance, where ill-treatment occurs in the course of smuggling or other illegal trade. Other areas of potential litigation have been mentioned earlier.

In many of these situations there may be an issue that calls for expert evidence. This is very often provided by a veterinary surgeon (veterinarian), behaviourist or biologist. Experts from other relevant disciplines such as animal welfare or management, environmental science, biological sciences, ecology, taxonomy, tracking, ballistics or taxidermy may be used as appropriate (see Chapter 4). Laboratory scientists provide evidence based on veterinary pathology, microbiology, histology, DNA analysis and other techniques (see Chapter 10).

All experts who present forensic evidence are, regardless of their discipline, subject to the rules and regulations of the court in which they give their evidence. They may also be required,

Writing Reports and Appearing in Court

or to follow particular ethical codes. These will vary from one country to another and from one court or legal system to another. An experienced expert will be aware of the judicial context in which he or she prepares and gives his or her evidence, but newcomers to the scene should take advice or undertake training or study. All experts need to remain up-to-date.

This chapter will provide some guidance for the expert witness. While it will be based primarily on the English judicial system and law, many of the general principles and guidance available are broadly applicable to giving evidence in other jurisdictions.

Legal Background

Levels of Law

While most law is made by the national parliament of individual countries (or by federal and state parliaments under federal constitutions), it may also be made at international or regional levels (see Chapter 3). For example, the legislation implementing the Convention on International Trade in Endangered Species of Wild Fauna and Flora is normally made at national level, but in the European Union (EU) it is implemented by regulations that apply directly in all member counties; only enforcement powers are provided by national legislation.

Kinds of Law

There are various kinds of law, but the most important to the expert witness in wildlife cases are criminal and civil law. The criminal law deals with the enforcement of laws by the state by way of prosecution. When a prosecution is successful, it results in punishment in the form of fines, imprisonment or alternative sanctions such as a suspended sentence or community service. The legislation may also provide powers to confiscate equipment used in a crime or to disqualify a person from keeping animals. Civil law deals with the rights between persons (individuals, companies and other legal entities). Claims have to be made in a recognised field of civil wrong such as tort (including negligence, professional negligence, nuisance or trespass) or breach of contract. Such provisions arise from a mixture of long-established rights, case law and legislation. The courts provide remedies for successful claims in the form of 'damages' (financial compensation) or injunctions to prevent future repetition of an unlawful act.

Legal Systems

Most countries have a legal system that falls into one of the following categories or, most often, a mixture of several:

- Civil law: Based on codes of law, many originally derived from Roman law. For example, France, Germany and many other continental European states, Latin American countries, China and Japan.
- Common law: Based on legislation and judicial decisions made in court. For example, the United Kingdom (except Scotland), the United States of America, Canada (except Québec) and many (British) Commonwealth countries.
- Religious law: Based on religious texts and interpretation, mainly Hindu, Jewish or Islamic (e.g. India, Israel, Iran and Saudi Arabia).

- Customary law: Derived from well-established customs and still used in some mixed systems.
- Mixed: Most legal systems, largely for historical reasons, use a mixture of sources of law. In some countries there may be separate civil and religious legal systems with their own courts. Some countries may have a primarily civil or common law legal system, but apply religious law to the personal and property rights of the members of that religion (e.g. Kenya and India).

Further information on national legal systems can be found at

- List of Country Legal systems. Wikipedia. http://en.wikipedia.org/wiki/List_of_country_legal_systems
- The French Legal system. About France.com. http://about-france.com/french-legal-system.htm
- CIA (undated). Legal System. The World Factbook https://www.cia.gov/library/publications/the-world-factbook/fields/2100.html
- University of Ottawa (undated). Juriglobe. http://www.juriglobe.ca/
- There are some comparative studies of legal systems such as Merryman and Pérez-Perdomo (2007) for Europe and Latin America, and Antoine (2008) for the Caribbean and Sapse and Kobilinsky (2011) for Europe and the USA

Judicial Systems

Functioning countries have judicial systems with courts and tribunals to interpret and enforce legislation – for example, criminal, commercial, personal or maritime. There are normally tiers of courts, from those that deal with simple cases at the lowest level to courts for more complex cases and appeals, and a final, supreme, court. These are regulated by legislation and rules of procedure. These may include specific requirements for expert witnesses. Many governments make legislation available on the Internet, including information specifically relating to expert evidence – for example:

- Australia: Practice Direction: Guidelines for Expert Witnesses in Proceedings in the Federal Court of Australia. http://www.fedcourt.gov.au/how/prac_direction.html
- Canada: Rules Amending the Federal Courts Rules (Expert Witnesses). Canada Gazette Vol. 144, No. 17 – August 18, 2010. http://www.gazette.gc.ca/rp-pr/p2/2010/2010-08-18/html/sor-dors176-eng.html
- The United Kingdom: see below and:
  civil jurisdiction: http://www.judiciary.gov.uk/about-the-judiciary/the-judiciary-in-detail/jurisdictions/civil-jurisdiction#headingAnchor1
  criminal jurisdiction: http://www.judiciary.gov.uk/about-the-judiciary/ the-judiciary-in-detail/jurisdictions/criminal-jurisdiction
- The United States: The Federal Rules of Evidence US Code Chapter 28 Part VII. http://www.law.cornell.edu/rules/fre/article_VII (Cornell University Law School). Detailed provisions are provided in the Federal Rules of Evidence 2012 (updated annually) (Federal Evidence Review, 2012)
- Many US state jurisdictions have also adopted these rules (to be found in the state legislation such as The Pennsylvania Code and Arkansas Judiciary)

Access to national laws is also available from the World Legal Information Institute http://www.worldlii.org/ which provides a worldwide collection of legislation databases.

Papers written by those who have acted as experts in a given country, such as Martin et al. (2003) (also see 'Forensic Expert Work in Wildlife Medicine in Hungary' section in Chapter 13), Hungary, and Vermylen (2010), Belgium, may be useful. Otherwise, expert witnesses should obtain advice from their instructing lawyers.

United Kingdom

The United Kingdom is a union of four countries: England, Wales, Scotland and Northern Ireland (http://www.parliament.uk/documents/commons/lib/research/rp2003/rp03-084.pdf)

They are all represented at the UK Parliament at Westminster, but the latter three also have their own national assemblies with 'devolved' legislative powers over specified matters, including wildlife conservation and animal welfare. (http://www.parliament.uk/documents/commons/lib/research/rp2003/rp03-084.pdf). Northern Ireland has a similar legal system to that of England and Wales while Scotland's is quite distinct, being based partly on Roman law. England and Wales have a joint judicial system, while those of Scotland and Northern Ireland have separate ones. The specific information provided in this chapter is based on the law and rules of procedure applicable in the English (and Welsh) courts.

The legislation of all four countries is available on http://www.legislation.gov.uk/. There are annotated versions that show amendments and where the law diverges for different countries.

Since 2011 the judicial system for England (and Wales) has been provided by Her Majesty's Courts and Tribunals Service. The hierarchy and functioning of the courts are explained on its website http://www.adviceguide.org.uk/england/law_e/law_legal_system_e/law_taking_legal_action_e/courts_of_law.htm. The lowest criminal courts are the Magistrates' Courts, which deal with minor criminal and some family matters. There are courts for serious crime and civil disputes together with various routes for appeals up to the Supreme Court (formerly called the House of Lords). There are various other tribunals for specialised areas of law and procedures for arbitration and alternative dispute resolution. Magistrates decide cases in the lower courts and judges (with a jury in most criminal trials) preside in the higher courts. The legal profession has two branches; solicitors who can appear as advocates in the lower courts and, if they have special certification, in the higher courts too, and barristers (Queen's Counsel and counsel) who can practise in any court.

There are procedural rules for all courts. The expert witness must understand and comply with the sections relating to expert evidence:

- Criminal courts: The Criminal Procedure Rules Part 33 Expert Evidence http://www.justice.gov.uk/courts/procedure-rules/criminal/rulesmenu/part_33
- Civil courts: The Civil Procedure Rules Part 35 Experts and Assessors http://www.justice.gov.uk/courts/procedure-rules/civil/rules/part35
- Practice Direction 35 Experts and Assessors http://www.justice.gov.uk/courts/procedure-rules/civil/rules/pd_part35
- The Civil Justice Council Protocol for the Instruction of Experts to give Evidence in Civil Claims must also be followed. http://www.xproexperts.co.uk/pdf/Experts_Civil%20Justice%20Council%20Protocol.pdf

What Is an Expert Witness?

Normally, evidence of opinion (as opposed to evidence of fact) is inadmissible. For present purposes, there is one important exception. That is, where the opinion is given by an expert who has the credentials to give helpful opinion evidence. What those credentials are will depend on the context of the case. There is sometimes argument about whether a purported expert is sufficiently expert to be heard. It is not necessarily a matter of formal qualification.

In the United States, Federal courts apply the 'Daubert Rule' established in *Daubert v Merrell Dow Pharmaceuticals* 509 U.S. 579 (1993). This provides that

> If scientific, technical, or other specialized knowledge will assist the trier of fact to understand the evidence or to determine a fact in issue, a witness qualified as an expert by knowledge, skill, experience, training, or education, may testify thereto in the form of an opinion or otherwise, if (1) the testimony is based upon sufficient facts or data, (2) the testimony is the product of reliable principles and methods, and (3) the witness has applied the principles and methods reliably to the facts of the case (Federal Rules of Evidence, Rule 701).

'Daubert motions' are common in the United States as a way of seeking to exclude expert evidence before or in the course of a trial. The courts in England and Wales have not codified the definition of an expert in a similar way, and such arguments do not occur. As already noted, the expertise of an expert is sometimes questioned in England and Wales as part of a contention that the expert evidence should not be heard, but the other matters encompassed in the Daubert rule would, in England and Wales, be the subject of cross-examination and later argument about whether a particular expert's evidence should be accepted. They would go, in other words, to the weight of the evidence, rather than to its admissibility.

While 'expert evidence' is generally understood by lawyers to be opinion evidence, a significant part of the job of an expert is to educate lawyers and judges. Thus, to say that blood passes from the lungs to the left atrium via the pulmonary vein is not a matter of opinion, but the judge may well need to be told this in order to understand the case. Normally, this sort of education is the background to the opinion evidence.

Where the case involves an adjudication of whether an individual (e.g. a veterinary surgeon) has met the standard required by the law (e.g. in a veterinary negligence claim), the choice of expert can be difficult. This is particularly so in veterinary negligence claims. Essentially, one is judged by the standard of one's professional peers: see later. But who are the peers of a veterinary surgeon in general practice, who divides his or her life equally and frantically between domestic pets, agriculture animals and horses? Should his or her management of an equine colic be judged by the standard by which one would judge a Newmarket colic specialist in comparable circumstances? Many would say not. But it can be argued that if one undertakes the management of equine colic at all, one is holding oneself out not merely as a practitioner as competent in the management of colic as one would expect a harassed veterinary general practitioner to be, but as a practitioner who is simply competent in the treatment of equine colic. There are insufficient decided veterinary negligence cases to know just how the English courts would approach the question.

The standard ubiquitously applied in professional negligence cases is the so-called *Bolam* test: a professional will not be held negligent if what he has done would be endorsed by a responsible body of opinion in the relevant speciality (see *Bolam* v. *Friern Hospital Management Committee* [1957] 1 WLR 582). *Bolitho* v. *City and Hackney Health Authority* [1997] 4 All ER 771 had the effect of putting in italics the word 'responsible' in the traditional formulation of the test. In order to be 'responsible', the endorsing body must be capable of being logically defended. Usually (the House of Lords emphasised in *Bolitho*), the fact that a particular practice is followed by respectable members of the professional community will indicate that it is logically defensible. But this is not necessarily so. To do something in a particular way simply because it has traditionally been done that way is not adequate justification.

The practice of evidence-based veterinary medicine increasingly limits the scope for argument about what amounts to responsible practice (Cooper and Cooper, 2007). If the literature conclusively demonstrates that procedure X is superior to procedure Y, and X and Y cost a similar amount of money, how can it be seriously contended that use of Y is acceptable?

Many matters that fall within expert witness territory, however, will not involve the assessment of the acceptability of a professional's management. Many will be questions of fact – facts about which the expert is qualified to opine. Examples might include: How long had the animal been dead? What was the probable cause of death? Suppose the case is a criminal case brought against a landowner in relation to an animal's death. The original examiner (a veterinary surgeon in general practice) concluded that the cause of death was poisoning by a substance found in the landowner's possession. You want to contest that conclusion. The appropriate expert is not a veterinary GP peer of the original examiner. The object is to challenge the examiner. You need a toxicologist/pathologist with intimidating credentials in the relevant niche.

Being an Expert Witness

General Duties of an Expert Witness

The rules governing the use of expert witnesses in civil and criminal proceedings in England and Wales have been consolidated, respectively, in the Civil Procedure Rules Part 35 and the Criminal Procedure Rules Part 33. The Civil and Criminal Rules are very similar. The Criminal Rules provide that

1. An expert must help the court to achieve the overriding objective by giving objective, unbiased opinion on matters within his/her expertise.
2. This duty overrides any obligation to the person from whom he receives instructions or by whom he is paid.
3. This duty includes an obligation to inform all parties and the court if the expert's opinion changes from that contained in a report served as evidence or given in a statement.

Crucially, both sets of rules emphasise that the expert's primary duty is to the court, rather than to any party. Long gone should be the days of the hired gun who will say whatever suits his or her paymaster. It follows from this, although it was always the case, that along with demonstration of relevant expertise, the greatest attribute of an expert is his or her ability to demonstrate objectivity. That has some important consequences for the way that experts write their reports and give their oral evidence.

Sometimes the court will direct that an expert is to be the only expert in a set of proceedings. That often proves to be a particularly stern test of objectivity, although in theory the absence of an 'opponent' should make no difference whatever to the contents of the report.

One of the corollaries of the expert's obligation to be objective is that it is undoubtedly a good thing for an expert to have experience on both sides of litigation: both prosecution and defence: both claimant and defendant. A reputation for appearing always and only for defendant insurance companies or the prosecution in protected species or animal welfare cases will not increase credibility. Likewise, an expert who is not dependent on giving evidence for his or her entire income is less likely to be challenged on impartiality. A related comment: experts should retire gracefully when they are no longer engaged in active practice. Professional organisations (such as the Royal College of Veterinary Surgeons) make it a professional obligation to keep up to date. That obligation is just as important when it comes to giving expert testimony (if not more so, since the consequences of getting one's testimony wrong can be disastrous). Keeping up to date does not just mean reading the relevant journals: it means being involved on a day-to-day basis with the relevant discipline.

There are a few cases in which retired experts have a crucial place. If the issue is the acceptable standard of professional practice in the 1960s, one will want an expert who was in active practice then.

Accepting Instructions

Instructions will usually come to an expert from a solicitor. Occasionally, they will come from a party acting in person.

If the instructing party is incompetent or inexperienced, the instructions may amount to little more than: 'Here is a bundle of papers: discuss'. That is never acceptable. Experts are entitled to a properly collated and indexed bundle of papers and a clear statement of what is needed. There will often have been a preliminary enquiry from the solicitor about the fees payable and the time scale for the production of the report. It is important to keep to both. Often the solicitor will have obtained approval for a specified expert fee from the client or an external funding body, and the time of production of the report may have been arranged in order to dovetail with arrangements made with counsel, the court, the other side or other experts. It is often not possible to give an exact indication of the likely fee or the time of production of the report: if there are significant uncertainties, declare them. Do not promise what cannot be delivered.

As soon as the instructions come in, the expert should check that

1. Everything necessary for the writing of the report has been provided. If there is anything missing, ask for it.
2. The matter really does fall within his or her own area of expertise.
3. It is clear what questions need to be answered.
4. There are no conflicts of interest.

Writing the Report

The Criminal and Civil Procedure Rules make similar stipulations about the contents of an expert's report. Here are the criminal stipulations:

1. An expert's report must
    a. give details of the expert's qualifications, relevant experience and accreditation;
    b. give details of any literature or other information which the expert has relied on in making the report;
    c. contain a statement setting out the substance of all facts given to the expert which are material to the opinions expressed in the report, or upon which those opinions are based;
    d. make clear which of the facts stated in the report are within the expert's own knowledge;
    e. say who carried out any examination, measurement, test or experiment which the expert has used for the report and
        i. give the qualifications, relevant experience and accreditation of that person,
        ii. say whether or not the examination, measurement, test or experiment was carried out under the expert's supervision, and
        iii. summarise the findings on which the expert relies;
    f. where there is a range of opinion on the matters dealt with in the report—
        i. summarise the range of opinion, and
        ii. give reasons for his own opinion;
    g. if the expert is not able to give his opinion without qualification, state the qualification;
    h. contain a summary of the conclusions reached;
    i. contain a statement that the expert understands his duty to the court, and has complied and will continue to comply with that duty; and
    j. contain the same declaration of truth as a witness statement.
2. Only sub-paragraphs (i) and (j) of rule 33.3(1) apply to a summary by an expert of his conclusions served in advance of that expert's report (Criminal Procedure Rules 33(3)).

The 'declaration of truth' in civil proceedings is as follows:

> I confirm that I have made clear which facts and matters referred to in this report are within my own knowledge and which are not. Those that are within my own knowledge I confirm to be true. The opinions I have expressed represent my true and complete professional opinions on the matters to which they refer. (Civil Procedure Rules, Practice Direction 35.3 (3)).

In criminal proceedings (see Appendix C) it is

> I confirm that the contents of this report are true to the best of my knowledge and belief and that I make this report knowing that, if it is tendered in evidence, I would be liable to prosecution if I have wilfully stated anything which I know to be false or that I do not believe to be true.(Criminal Procedure Rules 33(3)(j)).

These stipulations are fairly exhaustive and self-explanatory. A few words of caution:

1. Be careful not to stray beyond your own expertise.
   Sometimes counsel or solicitors will tempt you to do so (perhaps because they do not themselves recognise the limits of your expertise, and sometimes because they

would like to spare the money that they might otherwise have to spend getting a more appropriately qualified expert). Resist the temptation.

The most dramatic cautionary tale is that of Professor Sir Roy Meadow, a distinguished paediatrician who gave evidence at the trial, for murder of her two babies, of Sally Clark. Professor Meadow was asked about the chance of two babies dying innocently from cot death. He gave the figure of 1 in 73,000,000 – obtained simply by squaring the risk of one cot death in a middle-class, non-smoking, family. The figure was devastating for the defence, and the jury duly convicted.

Statisticians (and later the General Medical Council, the UK disciplinary body for the medical profession) were outraged at Professor Meadow's assertion. It was statistically utterly naive. He was not a statistician, and should not have trespassed on to statisticians' territory. The convictions were overturned on appeal although not, in fact, on this basis. Sally Clark's second (successful) appeal was on different grounds, although the court observed, a propos of the 1 in 73,000,000 figure, that '… if this matter had been fully argued before us we would, in all probability, have considered that the statistical evidence provided a quite distinct basis upon which the appeal had to be allowed'.

2. Cite authority for every significant proposition you advance.

The days of experts simply asserting in their reports: 'I do things this way, because I always have', were ended by *Bolitho (earlier)*. While this is now true of arguments about breach of duty, it has always been true for arguments about causation and other matters of scientific fact.

3. Write in simple language, not impenetrably technical language. And explain the basic science involved.

The report is for a lay audience. Assume nothing except bare literacy.

Conferences

Often experts will be invited to discuss cases with counsel and/or instructing solicitors. These are often pivotal meetings at which the fate of the litigation will be decided, strategy determined and the content of reports finalised.

Counsel will often put expert witnesses through their paces at such meetings. This is not, or should not be, a pre-trial rehearsal: that would be improper. It is an attempt to see how the expert witness is likely to fare under cross-examination from the other side, and to probe weaknesses. Sometimes the weaknesses will be shown to be fatal: sometimes they can be repaired by further evidence. While these testing exercises can be fun, experts should not get carried away with the game. The most important quality to bring to the meeting is frankness. If you have doubts about an argument that you have advanced in a report, or about the general thrust of the litigation, now is the time to air them. It is far, far better to know the real odds at this point than on day three of an expensive and traumatic trial.

Sometimes experts will discover that an argument that seemed compelling when the report was written looks fairly shabby under the forensic spotlight. If it does, the expert owes it to the court, the instructing party and himself to put the record straight.

Sometimes, as a result of the discussions, counsel or the instructing solicitor will ask for your report to be amended. If the amendment represents your considered view, and can be fairly squared with the stern declarations that the Rules require, make them. Otherwise, decline.

Pre-Trial Questions Put to Expert Witnesses

The Civil Procedure Rules make express provision for written questions to be put to experts (Civil Procedure Rules Part 35.6). Although the Criminal Procedure Rules make no such express provision, questions are often put. This usually happens where clarification is sought as to a matter dealt with in the original report, or to seek the expert's view on an issue relevant to the proceedings which has not been, or (in the questioner's view) not adequately been, addressed in the original report.

The answers should be regarded as an annexe of the report. They can be referred to at trial. All the comments made earlier about reports apply equally to such answers. The questions are sometimes framed in unnecessarily confrontational terms. Do not rise to the lawyers' bait by replying with corresponding intemperance.

Meetings between Experts

The Civil Procedure Rules provide for pre-trial meetings between experts in an attempt to narrow the issues:

1. The court may, at any stage, direct a discussion between experts for the purpose of requiring the experts to
    a. identify and discuss the expert issues in the proceedings; and
    b. where possible, reach an agreed opinion on those issues.
2. The court may specify the issues which the experts must discuss.
3. The court may direct that following a discussion between the experts they must prepare a statement for the court setting out those issues on which –
    a. they agree; and
    b. they disagree, with a summary of their reasons for disagreeing.
4. The content of the discussion between the experts shall not be referred to at the trial unless the parties agree.
5. Where experts reach agreement on an issue during their discussions, the agreement shall not bind the parties unless the parties expressly agree to be bound by the agreement. (Part 35.12)

The corresponding provision in the Criminal Procedure Rules provides

1. This rule applies where more than one party wants to introduce expert evidence.
2. The court may direct the experts to—
    a. discuss the expert issues in the proceedings; and
    b. prepare a statement for the court of the matters on which they agree and disagree, giving their reasons.
3. Except for that statement, the content of that discussion must not be referred to without the court's permission.
4. A party may not introduce expert evidence without the court's permission if the expert has not complied with a direction under this rule. (Part 33(6))

These meetings are usual in civil proceedings and common in criminal proceedings. Usually they follow an agenda drafted between the lawyers. A memorandum of agreement and disagreement is drawn up. It refines the issues to be decided at trial.

Often experts, after discussion with their counterparts, dilute or change the opinions they have expressed in their reports. The only advice that can be given is: be honest and remember the duty to the court.

Appearing in Court

Everything that has been said about writing reports, answering questions and participating in expert meetings applies equally to appearances in court. If there are new papers or other evidence supporting (or undermining) your position, you should disclose these at once to the lawyers instructing you.

If it is at all possible (and if the funds are available), you should hear at least the evidence of your expert counterpart, if not all the evidence in the case. It is not satisfactory for an expert to agree to come to court, give his or her evidence and go away immediately. The factual background is generally crucial to your expert testimony. Your evidence will carry much more weight if you have heard the factual evidence that was actually given, rather than merely reading factual witness statements prepared with the help of lawyers. The party instructing you will no doubt want you present in any event throughout the cross-examination of your expert counterpart, so that you can point out errors and suggest questions.

At the end of each court session (at lunchtime and at the end of the day), the lawyers will want to discuss things with you. Make sure that you arrange your diary accordingly. Once you have started giving your evidence, no one will be allowed to speak with you until you have completed it.

When you are giving evidence, be reasonable. This is not your fight. If you descend into the arena, and start fighting for 'your' side, it will be clear to everyone that you have forgotten the nature of your duty as an expert. No judge will want to rest a decision on the evidence of a partisan expert. Be prepared to make concessions. Do not doggedly defend the position adopted in your report if it is plain that the evidence does not justify it.

Be straightforward: if a question can be answered with a 'yes' or a 'no', just say 'yes' or 'no'. Only qualify such plain answers if failure to do so will misrepresent your position. Just answer the question: evasion looks sinister. Do not try to guess the agenda behind the question: it will lead to your answering a question other than the one you have been asked. Do not (in the witness box) call into question the expertise of the expert on the other side, or indulge in *ad hominem* attacks. It will just make you look like a combatant, which you must not be. If points about the corresponding expert's expertise can be made, the lawyers will make them.

While the Daubert principle (mentioned earlier) on the admissibility of evidence is not used as such in the English courts, there has been considerable concern for many years about the quality of expert evidence.

In the case *The Ikarian Reefer* [1993] 2 Lloyd's Rep 68, Cresswell J. set out the duties of expert witnesses. Later, following the enquiry and seminal report by Lord Woolf (1996), substantial changes were made in the civil court procedure and, subsequently, in that of the criminal courts.

As mentioned earlier, modern powers of the court in respect of case management allow the person(s) hearing a case to rely on the expert's report without the expert appearing, and to require experts from either side to meet and identify where they are in agreement and what is in dispute. The court may also appoint a 'single joint expert' to analyse the evidence available and identify the points in dispute. A new approach in Australia is 'concurrent evidence', colloquially known as 'hot-tubbing', in which all the experts are examined

Bite: injury caused by the mouth and/or teeth of a human or an animal. Often has a characteristic appearance and may contain indentations or bruises caused by the teeth which can be used to compare with the dental chart or dentition of the biter.
Blackening: the deposition of soot as a result of a discharge of a firearm. Its presence is used to determine the range of fire of a gunshot wound.
Blast: a short episode of high pressure usually caused by an explosion, commonly followed by a shorter period of lowered pressure. Can cause barotrauma (q.v.).
Bleeding: leakage of blood caused by damage to blood vessels (see also Haemorrhage).
Blister (vesicle): collection of fluid derived from blood below the outermost layer of the skin (epidermis). Caused by some infectious diseases and by many types of trauma, particularly heat and electricity. Also associated with hypothermia and barbiturate poisoning.
Blowflies: bluebottles, greenbottles and other sarcophagous flies in the order Diptera, class Insecta. Their larvae (maggots) and pupae can be used in the determination of the time of death in cases of decomposition (q.v.).
Blunt injuries: term used to describe the group of three types of injury caused by blunt objects; abrasions (grazes), contusions (bruises) and lacerations. Many objects cause blunt injury, ranging from the ground (a fall) to a human fist or a horse's hoof.
Brain death: complete absence of any functional activity in the brain. Diagnosed after a specified series of neurological tests have been performed and the effects of drugs, alcohol, hypothermia, etc. have been excluded.
Bruise: leakage of blood, often under the skin (see also Contusion).
Buggery: carnal intercourse by a male person with another person or an animal, consisting of penetration *per anum* (also Sodomy).
Bulla ('blister'): elevated, fluid-filled, lesion measuring more than 5 mm in diameter.
Bullets: projectiles fired singly from rifled weapons.
Burn: injury caused by either wet or dry heat. There are degrees of severity: first – outer surface of skin only, second – full-thickness of skin, third – deeper tissues. Chemical and friction burns also occur.
Bushmeat: non-domesticated animals (wildlife) used as food.
Cadaveric spasm: extremely rare form of rigor mortis recognised mainly in humans, of instantaneous onset, said to be associated with traumatic death following exertion.
Canker: a lay term for ulceration (q.v.), necrosis (q.v.) or proliferative lesions of oral mucosa.
Captive (animal): animal kept under some form of restraint (by humans), varying from permanent caging to general supervision.
Carbamates: synthetic organic insecticides that inhibit the function of cholinesterase.
Caries: demineralisation and loss of hard tooth tissue, leading to continued destruction of enamel and dentine, followed by cavitation of the tooth; may be caused by bacteria.
Carrier: an animal that harbours and may transmit the causative agent of a disease.
Case law: the body of law set out in judicial decisions, as distinct from statute law (q.v.)
Caudal: relating to the cauda or tail. Also used for 'Posterior' (q.v.).
Cheilitis: inflammation of the lips.
Child abuse: infliction of injury, usually over a period of time, on a child or infant. May be physical, sexual or psychological. Similar abuse can be inflicted on animals.
Chlorinated hydrocarbons: synthetic organic insecticides that block transmission of nerve impulses, have relatively low acute toxicity but are persistent in the environment.

Choking: accidental or deliberate obstruction of the upper airways leading to asphyxiation.
Chronic: describes diseases or clinical signs that may arise slowly and persist for a relatively long time. From Greek *'chronos'* – time.
CITES: Convention on International Trade in Endangered Species of Wild Fauna and Flora.
Civil law: a legal system set out in a written code. Also a type of law dealing with rights and obligations between individual or corporate persons (see also Tort).
Clinical signs: the features of a disease that can be observed, for instance, lameness, diarrhoea, nasal discharge, etc. (see also Symptoms).
Common law: a legal system derived originally from the English legal system, based on legislation, court decisions and custom. Also, rules of law developed by the courts as opposed to those created by statute.
Comparative forensic medicine: that branch of forensic science relating to different species of animals, including humans, and its application to the judicial and other processes.
Concussion: lay term for the effects of a head injury. No generally agreed medical definition exists. However, in humans a group of symptoms including amnesia, confusion and altered consciousness are referred to as Post Concussional Syndrome.
Congenital: defect present at birth; the effects may be immediately apparent or may not manifest themselves until later in life.
Congestion: excessive accumulation of blood in a tissue or organ. This is not bleeding or bruising as the blood is still in the blood vessels.
Contempt of court: (1) civil contempt: failure to comply with a court order; (2) criminal contempt: interference with the administration of justice (e.g. bribery/intimidation of judge, jury or witnesses, misbehaviour in court).
Contrecoup: injury to the brain on the opposite side to the site of an injury to the head. Some believe that contrecoup injuries indicate that the head was free to move when struck, others dispute this.
Contusion: leakage of blood from a damaged blood vessel, most commonly the smallest vessels – capillaries (see also Bruise).
Coroner: official who determines the cause of death of a person in sudden, unexpected or suspecious circumstances
Cot death: sudden Infant Death Syndrome (SIDS), also called Crib Death in the United States.
Cranial: relating to the cranium or skull. Also used for 'Anterior' (q.v.).
Crime scene: the place where the crime occurred. From Latin *'scaena'* (Greek *'skene'*) – the stage on which a play is presented, especially a Roman or Greek theatre.
Criminal law: legislation which imposes sanctions such as imprisonment and fines, *cf.* civil *law* (q.v.).
Cross-examination: the questioning of a witness usually by the lawyer for a party other than the one who called the witness. It may be used to obtain further information likely to benefit that party, or to try to discredit the evidence of the witness, or to challenge the overall credibility of the witness.
Cut: in popular parlance, a break in the surface of the skin. Correct forensic terminology depends on the cause of the skin injury; blunt trauma causes lacerations, sharp trauma causes incisions.
Cyanosis: blueness of the skin caused by reduced oxygen in the blood.

Appendix A

Damages: financial compensation awarded by a court as a result of a successful claim in civil law of tort (q.v.) or contract.

Death: the cessation of life. Usually ascertained by the absence of heartbeat and respiration.

Decomposition: decay of body after death (see also Autolysis). Very variable and exact type and speed of decomposition will depend on body type, environmental temperature, availability of water, etc.

Defence wounds: injuries received by victims while defending themselves. Most commonly seen with attacks involving a sharp weapon but any object can be the cause of defence wounds. In humans defence wounds are typically on the palm of the hand and/or outer border of the forearm: in animals they can be on the side of the neck or the rear end (depends on species).

Dehydration: lack of water in the body. Can occur in starvation, neglect, hyperthermia, etcv.

Demographics: study of factors that affect a population, such as birth and death rates.

Dental plaque: the accumulation of food debris and bacterial mass on the tooth surface.

Devolution: limited powers passed from a central, sovereign government to a lower authority. In the United Kingdom it refers to the conferring of limited powers of government and legislation by the UK Parliament upon the Welsh Government and the National Assembly for Wales and the Scottish Government and Parliament, respectively.

Diarrhoea: loose faeces.

Directive: legislation issued by the European Union (q.v.) that its member countries are required to implement through their national law or administrative measures (see also Regulation).

Disease: an unhealthy condition of the body or a part thereof, disordered state of an organism or organ. Anything that causes impairment of normal function. Diseases can be caused by infectious agents (e.g. viruses) or non-infectious factors (e.g. burns).

Disseminated intravascular coagulation (DIC): a complex disease process in which there is uncontrolled clotting of the blood inside blood vessels. It can be caused by many factors including septicaemia, trauma, etc.

Docket: any numbered and dated record of material that is recorded chronologically and, usually, signed by those receiving that material.

Domestic (animals): in this book used to describe animals that live in or around human habitation, not necessarily 'domesticated' (see below).

Domesticated: 'an animal that breeds under human control, provides a product or service useful to humans, is tame, and has been selected away from the wild type' (Mason, 1984). In biological parlance, domesticated means captive-bred for many generations with corresponding changes in appearance, morphology, behaviour, etc. from the wild progenitors.

Drowning: death caused by immersion in a fluid, usually water. Pathological features may be absent if death was rapid and they differ for fresh- and saltwater drowning.

Dysentery: blood in faeces, may be fresh (red) or partly digested (dark and tarry).

Dyskeratosis: abnormal keratinisation of skin.

Dysphagia: difficulty in swallowing.

Dyspnoea: difficult breathing.

Ecology: the study of the interrelationships of organisms and their environment.

Ecosystem: a dynamic complex of plant, animal and micro-organism communities and their non-living environment that serves as a functional unit.

Effusion: escape of fluid into a body cavity.

Electrocution: death or injury caused by the passage of an electrical charge. Effects depend on the voltage, amperage and the time for which the current flows. High voltages can result in great tissue damage while low voltages may leave only minimal marks.
Emaciation: excessive leanness; a wasted condition of the body.
Embolism: blockage of a blood vessel, usually an artery, by an extraneous object or material. In humans the commonest form is the blockage of the arteries of the lungs by fragments of blood clot which have become detached from areas of deep venous thrombosis (DVT) in the legs (see also Thrombosis).
Enclosed animals: usually used in respect of deer or similar species kept in a park; they are primarily fenced-in but in some cases, may have freedom to come and go.
Entomologist: one who studies insects.
Entomology: study of insects. Can be useful in forensic studies for the estimation of the time of death from blowflies, beetles, etc. present on the body and their stage of development (see also Blowflies).
Enzootic: the continual presence of a disease in an animal community.
Epidemiology: the study of disease in a population, including its prevalence/incidence, distribution and spread.
Erosion: focal loss of epidermis (see also Ulcer).
Ethology: the study of animal behaviour.
European Union: a political and legislative group of 27 European countries (2012).
Evidence: facts observed by a witness with his or her own senses, presented to a court under oath or affirmation by a witness (q.v.), that are intended to support the existence (or non-existence) of some other fact. The evidence must be admissible according to law and may be oral, written, documentary or real (physical). The evidence may be directly observed or circumstantial (from which a fact can be inferred) (see also Opinion evidence and Hearsay evidence).
Examination: the questioning of a witness (q.v.) in a court case. The lawyer presenting the case will examine the witness followed by cross-examination by the lawyer for the other side and, finally, re-examination by the presenting lawyer to clarify matters arising from the cross-examination.
Excoriation: linear, traumatic lesion resulting in epidermal damage.
Exhumation: recovery of a body from a grave or burial place.
Exit wound: site where a weapon, most commonly a bullet, leaves the body.
Exotic: strictly non-indigenous animals, often excluding introduced species already established in the wild in a particular country. The term 'exotic' is also used, especially by the veterinary profession, to describe unusual or non-domesticated species.
Expert witness: in court, a person with specialised knowledge, skills or experience who gives evidence and expresses an opinion on it in order to provide the court with information outside its normal scope of knowledge.
Exudate: a fluid, usually of inflammatory origin, with high protein content, a specific gravity of more than 1.020 and many cells. Often found in a body cavity or in interstitial tissue.
Farm animals: a generalisation for animals commonly used in commercial agriculture, being an alternative to listing separate species for convenience. Often termed "livestock".
Farmed or ranched animals: normally free-living species kept for animal husbandry or conservation purposes.
Fat embolism: blockage of blood vessels, particularly of the lungs and kidneys, by portions of fat. Seen following trauma particularly fractures.

Feral animals: animals (usually domesticated) living in a wild state after leaving captivity. Escaped wild animals are usually described as having 'reverted to the wild' or in 'a free-living state'.
Firearms: weapons that fire projectiles such as bullets and pellets.
Flotation test: the placing of tissues in water to see if they float or sink. Sometimes used to help to determine whether a baby or young animal was born alive: inflated lung floats but this is not a totally reliable test. Pneumonic lung sometimes sinks, sometimes floats, depending upon the changes present. Liver that floats may be fatty or contain air/gas.
Focus (plural 'foci'): a small, usually distinct, lesion such as a micro-abscess in the liver.
Food chain: a succession of organisms that eat other organisms and are, in turn, eaten themselves.
Forensic podiatry: the application of knowledge of the anatomy, function, deformities and diseases of the feet, lower extremities, and at times, the entire body, to a legal and/or criminal investigation. Foot-related evidence can include footprints, partial or complete pedal remains and the use of medical records to identify individuals. May also relate to gait.
Forensic veterinarians: veterinarians who have a particular knowledge and experience of a field of veterinary science, the recovery of forensic samples and the provision of opinion, preparation of evidence for court or the giving of oral or written evidence based on that opinion (see also Veterinary surgeon).
Forensic veterinary medicine: the application of veterinary knowledge to the purpose of the law. The term covers all situations in which verterinarians collect information or samples, give opinion on, prepare expert witness statements on or give evidence in court on any aspect of their professional fields of expertise. It can also be relevant to insurance claims and allegations of professional misconduct.
Free-living (animals): animals not living in captivity; this would normally relate to wild or exotic species (q.v.) but it may also be applied to domestic animals, although the adjective 'feral' (q.v.) is often used. Thus a feral cat is a domestic cat (*Felis catus*) living independently of human control whereas the (Scottish) wild cat is a separate species or sub-species (*Felis silvestris*) and usually occurs as a free-living creature, although it is also occasionally kept in zoological collections.
Frounce: a protozoon (*Trichomonas gallinae*) infection of birds of prey, characterised by necrotic lesion in the mouth, pharynx, oesophagus, crop and proventriculus (see also Canker).
Gingivitis: inflammation of the gums.
Glossitis: acute or chronic inflammation of the tongue.
Haematemesis: vomiting of blood.
Haematoma: used to describe a significant collection of blood outside blood vessels and in the tissues of the body (see also Contusion).
Haemoglobin: pigment in red cells responsible for carrying oxygen.
Haemoptysis: coughing-up of blood.
Haemorrhage: bleeding.
Hanging: form of ligature strangulation where the pressure on the neck is produced by the weight.
Health: a state of physical and mental well-being of the body (not merely the absence of disease).
Health and safety: legal obligations of employment to prevent harm to those in the work place.

Health-monitoring (screening): on-going evaluation of health status; may involve clinical examination, haematology, parasitology, etc.
Hearsay evidence: for England and Wales, "a statement not made in oral evidence in the proceedings that is evidence of any matter stated." Section 114 (1) Criminal Justice Act 2003. (Hearsay is generally accepted in civil law cases but restricted incriminal cases. The rules on hearsay vary with each country, thus in the US Federal Evidence Rules 2012 (and annually updated) hearsay is not allowed in either criminal or civil cases).
Herpetologist: one who keeps or studies reptiles and amphibians.
Herpetology: the study of reptiles and amphibians.
Hide and die syndrome: sometimes a feature of death from hypothermia where the individual (human or animal) hides away in a small space, e.g. cupboard, etc. either in an attempt to keep warm or through confusion induced by the low body temperature.
Histology: microscopic study of tissues and organs.
Homeostasis: constancy in the internal environment (the '*milieu interieur*' of the French physiologist, Claude Bernard) of the body, naturally maintained by adaptive responses that promote healthy survival.
Horizon scanning: the systematic search for potential threats and opportunities that are currently poorly recognised (see Sutherland et al., 2012).
Hyperaemia: increase in blood supply to an organ or tissue.
Hyperkeratosis: excess keratinisation of skin.
Hyperpnoea: increase in respiratory rate.
Hyperthermia: an increase in body temperature.
Hypertrophy: increase in size of a tissue or organ.
Hypoglycaemia: low blood sugar.
Hypostasis (adjective: hypostatic): the red-staining of the skin of a carcase caused by the settling of blood under the influence of gravity. Areas of the body that are in contact with surfaces will be unaffected and will remain white (if the skin is not heavily pigmented).
Hypothermia: decreased body temperature.
Hypoxia: low levels of oxygen in the blood and tissues.
Iatrogenic: an adverse reaction that results from treatment, induced by human intervention.
Ichnology: the study of burrows, tracks and trails.
Ichthyologist: one who studies fish.
Ichthyology: the study of fish.
Icterus (jaundice): yellowing of the skin or mucous membranes due to bile pigment.
In situ: conservation that takes place within the natural range of the species involved, as opposed to *ex situ* practices, usually involving captive-breeding, which may be carried out elsewhere, often in a different country.
Incidence: the number of new cases of a particular disease during a stated period of time.
Incision: break in the continuity of the skin caused by an object with a sharp edge.
Incubation period: the time between acquisition of infection and onset of clinical signs, *or* the period between a fertile egg (of a bird, reptile, etc.) being laid and hatching.
Indigenous (animals): animals that have always existed in a given geographical area.
Infarct: an area of tissue that has been deprived of a blood supply.
Infection: the entry of an organism into a susceptible host (human or animal) in which it may persist, but detectable clinical or pathological effects may or may not be apparent.

Infectious disease: a disease caused by the actions of a living organism (virus, bacterium, etc.), as opposed to (for example) physical injuries, endocrinological disorders or genetic abnormalities.
Inflammation: the normal response of the body to damage, whether biological, chemical or physical.
Injury: tissue damage (see also Wound).
Introduced (animals): non-indigenous species that are established in the wild where they have not previously existed.
Invertebrate: an animal without a vertebral column. The majority (89%) of living creatures, encompassing such groups as the insects, arachnids, crustaceans, molluscs and worms.
ISO 14001: international standards on environmental management.
ISO/IEC 17025: 2005: international standards on quality management.
Jaundice: see Icterus
Judge: a lawyer authorised to hear and decide upon disputes and other matters brought before the courts for decision. In criminal law courts, a judge hears certain cases with a jury (q.v.).
Jury: a group of people who, in a criminal law court, under oath or affirmation, after hearing the evidence, give a verdict (a finding based on the facts presented in the case). The jury sit with a judge who provides guidance and interpretation of the relevant law.
Kinetic energy: potential energy of a moving object.
Laceration: break in the continuity of the skin or organ caused by the application of blunt force. One of the triad of injuries associated with this type of force. Lacerations are usually characterised by jagged edges. They are not caused by sharp objects (which produce incisions).
Latent infection: an "inapparent" infection in which the pathogen persists within a host, but may be activated to produce clinical disease by such factors as stressors or impaired host resistance.
Lesion: an abnormality caused by the disease of a tissue; usually it is characterised by changes in the appearance of that tissue, e.g. a raised nodule on the skin.
Lichenification: thick, rough lesion of the skin with prominent markings, often as a result of repeated rubbing.
Ligature: a piece of material, such as thread, around an organ or hollow structure.
Ligature mark: a form of friction abrasion due to the effect of a ligature.
Lumen: cavity of a hollow organ, e.g. lumen of the stomach or uterus.
Maceration: destruction of tissues (or a carcase) caused by a combination of trauma and decomposition.
Macule: flat circumscribed, lesion distinguished from surrounding skin by its coloration.
Maggot: developing (larval) stage of some insects (see also Entomology and Blowflies).
Magistrate: 'justice of the peace' – an unpaid, non-lawyer who, normally with two others, hears cases in a magistrates' court in the United Kingdom other than Scotland. A stipendiary (paid) lawyer hears cases singly in some magistrates' courts.
Malocclusion: malposition of teeth, resulting in faulty meeting of teeth or jaws.
Mammology: the study of mammals.
Marine biology: the study of animals and plants in the marine environment.
Mechanical vector: an animal such as an insect which transmits an infective agent from one host to another but is not essential to the life cycle of the organism.
Melanocytes: cells that produce melanin pigment.

Microbiology: the study of microscopic or ultra-microscopic organisms such as viruses and bacteria.
Monitoring: the routine collection of information on disease, productivity and other characteristics. To maintain regular surveillance.
Morbidity rate: the proportion of clinical cases during a given time.
Mortality rate: the proportion of deaths during a given time.
Mucous membrane: the thin layer of mucus-secreting epithelium that lines areas such as the mouth, nose, eyes, vagina and rectum.
Mummification: a form of decomposition that is associated with warm dry conditions whereby the body dehydrates. Often results in remarkably good preservation. Usually involves the whole body but may affect only part, most commonly the extremities.
Münchausen syndrome: a complex psychiatric syndrome associated with a desire to obtain medical treatment (often invasive) by complaining of fictitious illnesses. May also involve the children or animals of the individual when it is called 'Münchausen Syndrome by Proxy'.
Necropsy: autopsy, *post-mortem* examination (of an animal).
Necrosis: death of a cell/tissue/organ.
Nodule: elevated, solid, lesion measuring more than 5 mm in diameter.
Oedema (American: 'Edema'): abnormal accumulation of fluid under the skin or elsewhere (e.g. pulmonary oedema).
Opinion evidence: a witness can only give evidence of facts but an expert witness (q.v.) can express an opinion on the facts that he or she presents to the court.
Organophosphates: synthetic organic insecticides that inhibit the function of cholinesterase and are acutely toxic but are generally not persistent in the environment.
Ornithologist: one who studies birds.
Ornithology: the study of birds.
Osteomyelitis: inflammation of bone and bone marrow.
Palaeontology: the study of fossil animals and plants.
Papule: elevated, solid, lesion measuring 5 mm or less in diameter.
Parakeratosis: abnormal retention of nuclei in stratum corneum of skin.
Parasitaemia: parasites present in the blood.
Parasites: organisms that live in or on another and benefit from it. Nowadays, especially by biologists, divided into 'macro-parasites', such as worms and ticks, and 'micro-parasites', such as bacteria and protozoa.
Parliament: the overall legislature of many countries.
Pathogen: an organism capable of producing disease. Can range from a microscopical organism, such as a bacterium, to a worm or a flea.
Pathogenesis: the *mechanism* of a disease, how a disease develops.
Pathology: strictly (from its Greek roots), the science of the study of disease. In common parlance tends to be related to *post-mortem* examination and related laboratory studies.
Penetrating injuries: injuries which pass through the skin.
Peracute: a disease with an exceedingly sudden onset and exceptionally short course.
Periodonitis: inflammation of the peridontium.
Periodontium: tissues investing and supporting the teeth.
Peritonitis: inflammation of the lining of the abdominal cavity and surfaces of the abdominal organs.
Petechiae (singular: petechia): pinpoint haemorrhages, due to rupture of small vessels.
Pharyngitis: inflammation of pharynx.

Poison (toxicant, toxin): a substance which can harm the body if taken in sufficient quantity. It is the dose that determines whether or not the substance is poisonous.
Poisoning (toxicosis, intoxication): disease produced by a poison.
Polychlorinated biphenyls (PCB): a group of synthetic industrial compounds that are toxic to animals and are persistent in the environment.
Posterior: the back of the body, limb, organ, etc. (see also Caudal).
Prevalence: the total number of cases of a particular disease at a given moment of time.
Primatology: the study of primates, including monkeys, apes and humans.
Privilege: in the law of evidence a witness or expert witness may have the right in specified circumstances to refuse to provide evidence, information or documents—for example, if it defined in law as confidential.
Prognosis: forecast of the probable course and outcome of a disease.
Propellant: the chemical substance used to provide energy to fire a projectile from a firearm, e.g. gunpowder.
Protected (animals): animals that have the protection of some form of law, usually conservation legislation. A specific term in the United Kingdom in the Animals (Scientific Procedures) Act 1986 and Animals Welfare Act 2006.
Pruritus (adjective: 'pruritic'): itching.
Public health: activities that promote the maintenance of health of the community, including food safety and sanitation.
Purulent: containing pus.
Pustule: discrete, pus-filled, raised area.
Putrefaction: decomposition.
Pyogenic: producing pus.
Quality assurance: the actions that are taken to make sure that policy and procedures are adhered to and that the work meets the required standard.
Regulation: European Union legislation that is directly applicable in each of its member states without further implementation by individual states. Note that some secondary (or subsidiary) legislation may be called regulations.
Reintroduced (animals): indigenous species that are established again in the wild in areas where they previously existed.
Reservoir: the host or hosts in which a pathogen is maintained in nature; often as a subclinical infection.
Rifled firearms: firearms that have grooves in the barrel that cause the bullet to twist as it leaves the barrel.
Rigor mortis: stiffening of the muscles after death. A very variable process that is affected by many factors and generally cannot be used to give an accurate time of death but may give some clue as to *ante-mortem* history.
Sample: in statistics is a subset of a population – and there are many types of statistical samples – in common parlance, and in much of biomedical science, a 'sample' is 'a representative specimen or small quantity of something' (see Chapter 10).
Scald: a type of burn due to hot liquid, usually water.
Scale: dry, plate-like, excrescence on the skin.
Screen: to test for evidence of the presence or absence of disease, quality etc., generally using tests that can be applied rapidly.
Septicaemia: multiplication of organisms in the blood, usually with pathological effects on organs.

Serology: the study of antigen/antibody reactions, especially the detection of antibodies in blood.

Shock: a physiological response to trauma, haemorrhage, infection, toxaemia and stressors, characterised by inadequate blood flow to the body's tissues (largely a deterioration of capillary perfusion) and sometimes life-threatening cellular dysfunction.

Smooth bore firearms: firearms that do not have rifling or groves in their barrels, e.g. shotguns.

Smothering: blockage of the external air passages (nose and mouth).

Sodomy: see Buggery.

Solicitor: a lawyer who may represent clients as an advocate in the lower courts in the United Kingdom and also provides a wide range of legal advice direct to the public.

Speciation: either (1) the evolutionary process by which new biological species arise or (2) the first description of a new species. Sometimes incorrectly used to mean the determination (identification) of an animal or plant in (e.g.) forensic evidence.

Squamous epithelial cells (keratinocytes): these are the majority of epidermal cells and the major mechanical barrier; also a source of cytokines, which regulate the cutaneous environment.

Stab: a penetrating injury caused by a sharp object that is deeper than it is wide.

Starvation: lack of food (but see also Chapters 8 and 9).

Statute: act of parliament (see Act); primary legislation passed by the legislature (often called a parliament (q.v.)).

Statute law: legislation made by a legislature (e.g. a parliament).

Statutory order: secondary (or subsidiary) legislation made under the authority of primary legislation (see Statute).

Stillbirth: birth of a dead human baby after 24 weeks' gestation or of an animal (mammal) late in gestation.

Stomatitis: inflammation of the buccal cavity.

Strangulation: application of pressure to the neck resulting in obstruction of the airways with or without obstruction of the blood vessels.

Studbook: detailed record of births, deaths and genetic relationships and other biological data.

Sudden death: no specific definition. Usually death without premonitory signs. Some medical authorities state that the time period should be less than a few minutes. Must not be confused with 'unexpected death' where the human or animal may have been unwell for some time.

Sudden infant death syndrome (SIDS): sudden unexpected death of a child of less than 2 years (in some definitions, 1 year) which is unexplained despite extensive investigation and tests (see text).

Suffocation: asphyxiation caused by obstruction of the external airways or lack of oxygen in the inspired air.

Superior: describing the upper part of the body, limb, organ, etc.

Symptoms: the features of a disease that are experienced and can be recounted by an affected human, for instance, giddiness, abdominal pain. The term should not be used for animals (see Clinical signs).

Syndrome: a group of symptoms, clinical signs or pathological changes that consistently occur together (Greek – 'to run together').

Systematics: the processes used to describe species, which encompass three disciplines: description of species (identification), taxonomy (q.v.) and description of relationships amongst and between taxa (phylogenetics).

Systemic disease: affecting the whole body (*cf.* local disease).

Tachypnoea: rapid breathing.

Taphonomy: from Greek '*taphos*' (grave); the *post-mortem* fate of biological remains. Forensic taphonomy is the application of such studies to assist legal investigations. Taphonomy is the study of fossilisation or, more broadly, anything that happens to a body after death, ranging from decomposition to responses to environment or climate.

Tartar/dental calculus (plural 'calculi'): mineralised plaque, yellowish film formed of calcium phosphate and calcium carbonate, food particles and other organic matter, deposited on teeth by saliva.

Taxon: a group of organisms such as species, genus or family.

Taxonomy: the science of classifying and naming organisms.

Thrombosis: blockage of a blood vessel by a blood clot (distinguish from *post-mortem* blood-clotting) (see also Embolism).

Throttling: see Strangulation.

Tolerance: a toxicological term; the ability of an animal to respond less to a specific dose of a poison on a subsequent occasion: this is attributable to acquired, not innate, resistance.

Tort: a civil law (q.v.) wrong (negligence, nuisance, trespass, defamation) for which compensation can be claimed in the civil law courts.

Toxaemia: the presence of a toxin in the blood.

Toxicity: the amount of poison needed to produce an adverse effect.

Toxicological pathology: the pathology of poisoning. The study of the pathogenesis of poisoning – the structural and functional changes in cells, tissues and organs that are induced by toxins.

Toxicosis: any disease condition due to poisoning.

Translocation: deliberate movement of wild animals from one area to another.

Transudate: a fluid with low protein content and a specific gravity of less than 1.012 and few cells.

Trauma: injury.

Traumatic asphyxia: asphyxiation caused by restriction of the movements of the chest wall, e.g. crushing (see also Asphyxia).

Trophic: the trophic level of an organism is the position it occupies in a food chain (q.v.). The number of steps an organism has from the start of a food chain is a measure of its trophic level. Food chains start at trophic level 1 (primary producers, such as plants), then to herbivores at level 2, to predators at level 3 and conclude with carnivores or apex predators at level 4 or 5. The path along the food chain can form a one-way flow (chain), or a food 'web.'

Tumour: a neoplasm, growth of tissue which is uncontrolled or abnormal.

Ulcer: focal *complete* loss of epidermis: may include dermis and subcutaneous fat.

Vagal inhibition: slowing or stoppage of the heart caused by stimulation of the vagus nerve, can be caused by pressure on the neck. Sometimes sudden death, in the absence of specific signs, is attributed to vagal inhibition.

Vector: a carrier of a disease or parasite, often an insect or other arthropod.

Vertebrate: an animal with a vertebral column (mammals, birds, reptiles, amphibians and fish).

Vesicle ("blister"): elevated, fluid-filled, lesion measuring 5 mm or less (see also Blister).

Veterinarian: one trained and qualified in veterinary medicine. In Britain and some other, mainly Commonwealth, countries veterinarians are usually, for historical reasons, called 'veterinary surgeons'.

Veterinary forensic medicine: that branch of forensic science relating to animals.

Veterinary surgeon: a term for "veterinarian" (q.v.) used in Britain and certain other countries. Elsewhere implies a veterinarian with particular skills in surgery.

Viraemia: the presence of a virus in the blood.

Wad: felt, cardboard or plastic material that is used to separate shotgun pellets from the propellant.

Wheal: erythematous, elevated area resulting from dermal oedema, usually pruritic.

Whiplash: type of injury to the neck usually caused by sudden backwards and forwards movement(s).

Wild animals: free-living, non-domesticated, animals. Also found in captivity. Defined by the common law and also given specific meaning (as to wild birds and protected wild animals) in the (UK) Wildlife and Countryside Act 1981.

Wildlife crime: activities that threaten wild animals, plants or their habitats and which constitute a breach of national, regional, or international law.

Witness: in court, a person who gives evidence (q.v.) of what he or she has seen or heard or experienced.

Wound: a breach of the integrity of the skin or other surrounding structure (see also Injury).

Zoology: the study of animals.

Zoonosis (adjective: zoonotic): an infection or disease that is naturally transmitted between vertebrate animals and humans (see Appendix F).

Zoophilia: sexual activity between a human and an animal (see also Bestiality).

Appendix B: Facilities and Equipment Lists

JOHN E. COOPER

Facilities

Facilities appropriate to forensic work with wild animals can range from a fully equipped laboratory with a national or regional role, through a small, specially designated, area of a veterinary practice, animal welfare centre or wildlife unit to a temporary station in a vehicle, a trailer or the field (see Chapter 7). However, the basic tenets of design and system apply to all facilities, large or small, permanent or temporary, and these relate primarily to

- Security
- Safety
- Chain of custody
- Accuracy of reports and reporting

Security is a vital prerequisite. The forensic laboratory, unit or station must, whenever possible, be lockable, with limited access, as should any reception area and storage facilities. There should be only one key for sensitive locations and surveillance cameras are a wise precaution.

The needs and features of a large forensic facility, including site building and laboratory design, a mechanical plumbing and electricity systems checklist and extensive guidance on construction and occupation, were described in some detail by the US Department of Justice (1998); that publication also contains information that is relevant to smaller or improvised units, such as those needed in the field.

A small, specially designated, animal forensic unit is likely to comprise a building or room that contains the following:

Basic

- Reception area
- Refrigerator and deep-freeze
- Computer
- Storage facilities for fixed or dry material, paperwork and back-up computer disks

As Necessary

- Appropriate lighting, including ultraviolet, both portable and ceiling-mounted
- Stainless steel examination table(s), with hydraulic lift system (when space is limited, folding tables, attached to a wall, are particularly valuable for forensic work)
- Laboratory bench and sink with appropriate facilities for disinfection, disposal of clinical waste and ventilation (see also in the following)

- Clinical facilities, including x-ray machine and viewer, heavy gauge steel dangerous drugs cabinet, endoscope(s), surgical instruments, etc. These may be part of an adjacent veterinary practice, in which case they need not be incorporated into the forensic unit as such
- Scavenging equipment (for anaesthetic gases, fumes, etc.)
- Laminar air flow systems or containment facilities, such as safety cabinets

Appropriate arrangements, in accordance with relevant laws and codes of practice (see Chapter 3), must be made for the provision of water and gas and for drainage. In temperate countries central-heating may be required; in the tropics or subtropics, air-conditioning. If these cannot be provided, full use should be made of natural ventilation, insulation or shade, as appropriate. Much has been published on the design of laboratories in isolated areas, especially in tropical regions, and advantage can be taken of this when planning forensic facilities.

Equipment

If finance is not an obstacle, no expense should be spared to ensure that the correct and best equipment is obtained. Most items needed for work with animals can be purchased from veterinary suppliers but for pathological investigations it is wise to consult also catalogues from companies that supply tables and other equipment for mortuaries and for human forensic work; the range that is available is usually extensive and often better tailored to the needs of a forensic investigation than are standard veterinary items.

Equipment – General Items for Forensic Work at the Crime Scene

Recommended equipment for work at the crime scene is listed next. More specific items for working the field – for instance, when investigating wildlife crime – are to be found later in this appendix.

- Protective clothing including boiler suits (coveralls), gloves (surgical, thick kitchen, long rubber, etc.), masks, goggles, boots and overshoes
- Barrier tape, flags, markers, cones and other crime scene security items
- Appropriate gloves and clothing, barrier tape and tools (such as tongs and hooks) for handling hazardous animals and materials
- Disinfectant and deodoriser to neutralise smells from carcases, etc. (both to be used with caution at crime scenes as they may destroy trace evidence)
- Pre-packed collection kits for taking samples for laboratory investigation, including microbiology, haematology, toxicology, DNA, etc.
- Appropriate labels, tamper-proof tags, evidence seals, etc.
- Marking pens
- Scales/balances, callipers and micrometer for weighing animals, tissues and samples and rules, tapes, centimetre scales, etc. for measuring (see Chapters 8 and 9)
- Equipment for taking casts of dentition, imprints in bones, animal tracks, etc., including plaster-of-Paris, other powders and silicone-based materials and waxes plus retention frames or stiff cardboard
- Trace evidence collection equipment

- Evidence packing, e.g. bags, boxes, tubes, envelopes and other supplies for packaging and storing evidence
- Photographic kit, including (as appropriate) digital/still cameras, video cameras, magnifier, night vision equipment, aerial camera system, camcorder (see Chapter 6)
- Binoculars/field glasses
- Torches (flashlights)
- Blue-Light kits and supplies
- Magnifying glasses/hand-lenses and/or magnifying loupe or dissecting microscope
- X-ray viewing screen
- Egg candler
- Clipboards and record sheets, plus pens and pencils
- Chalk and crayons
- Elastic bands and string
- Tape-recorder and tapes
- Evidence seals/tape
- Kits and systems for collecting and preserving delicate crime scene evidence (see text)
- Mobile devices, e.g. PDA, GPS, Smartphone, Netbook/Notebook computer and appropriate software (Cooper, 2013), including bar-code tracking systems for property, evidence and crime scene reconstruction. Back-up systems, such as a desk-top computer and/or an external hard-drive, for storing information (see Chapter 7)
- IT hardware and software for use outside the office or in the field. This will need to be portable, rugged, appropriate to the climate, terrain, easy to install and operate. Alternative power sources may be required, including replacement or rechargeable batteries, solar power chargers, portable generator and power storage and inverters
- Communications equipment, such as mobile phones, two-way phones and short-wave radio are important for safety, logistics and for transmitting and backing-up data collected in the field. The choice of equipment will depend upon the availability of mobile phone signals and network coverage in the location (Cooper, 2013)

Equipment: When Working in the Field

The detailed contents of field kits will depend upon the location and the type of investigations being carried out. A detailed description of a general diagnostic kit, suitable for various types of veterinary work, was provided by Frye et al. (2001) and some data from that paper are reproduced here, with permission.

In addition to photography, video-taping has much to commend it in the field but may have to be approached with care in some parts of the world where local sensitivities exist (see Chapter 7).

For a list of equipment that has proved useful for manipulating marine mammal material during necropsies the reader should refer to 'Special tools from the Marine Mammal Forensics Toolkit' in Campbell-Malone and Bogomolni (Chapter 13). Equipment for entomological (invertebrate) collection is covered in Chapter 6.

Metal detectors may be needed to scan carcases for evidence of bullet fragments and metal shot, including marine mammals (Neme and Leakey, 2010).

The following items are recommended in the field for (1) clinical, (2) *post-mortem* and (3) laboratory diagnostic work, respectively. In each case the list should be supplemented with general items, as mentioned earlier.

1. Clinical equipment – when live animals have to be examined in the field:
    - Pen torch (flashlight)
    - Spare bulbs and batteries
    - Spring balance(s) or battery-operated scales
    - Rulers, tape-measures
    - Stethoscope (light weight)
    - Auriscope (otoscope) and ophthalmoscope (lightweight)
    - Rigid endoscope (battery-operated)
    - Syringes and needles (disposable)
    - At least one boilable, re-usable syringe and needle
    - Empty drinks cans, labelled 'sharps boxes', for used needles, scalpel blades, etc.
    - Disinfectant(s), including alcohols
    - Camping (gas cylinder-operated) stove – for sterilising, lighting (and cooking!)
    - Pressure cooker for sterilising
    - Selected medicines, including local analgesics, sedatives and agents for euthanasia (plus gun if large animals may need to be killed)
    - Cotton wool
    - Dressings
    - Basic surgical ('cut-down') set and other instruments as necessary, including suture materials
    - Dental equipment, for the investigation of lesions in the oral cavity
    - Disposable skin-biopsy punch
    - Cautery (battery-operated)
    - Clippers for claws, talons, beaks
    - Ring (band) remover
    - Cloth bags and other devices for restraining small animals
    - Gloves – surgical and for handling
    - Towels
    - Oesophageal and other tubes
    - Mouth gag/wooden spatulae
    - Aluminium foil
    - Sampling and other equipment for laboratory work (see later – list c).

    Specific items for use when blood-sampling live birds and other animals for DNA and other investigative purposes:
    - Syringes (1, 2, 5, 10 mL) as appropriate
    - Needles (20, 25, 22, 28 g) as appropriate
    - Anti-coagulant serum tubes
    - Cotton wool
    - Frosted glass slides and box or tray
    - Cover slips
    - Pencils for marking glass slides, plus sharpener
    - Methanol

- Cards or solutions for blood collection
- Magnifying glass/hand-lens
- Plastic bags and labels
- Electronic/spring balance
- Ruler, tape-measure
- Clipboard, black pen and record sheets
- Basic clinical equipment
- Restraining equipment – nets, towels, gloves, hood, bags

2. *Post-mortem* equipment – when dead animals have to be examined in the field:
   - Standard necropsy items (see also Chapter 9) – portable/folding, lightweight/plastic where appropriate:
   - *Post-mortem* record sheet (see Appendix D) and pens and pencils
   - A table or equivalent for each large *post-mortem* specimen
   - A tray for each small *post-mortem* specimen – useful, not essential
   - Cork board and pins for each small *post-mortem* specimen – useful, not essential
   - Absorbent paper or equivalent for surface of tray or table
   - Spring balance(s) or battery-operated scales
   - Rulers, tape-measures
   - Saw(s)
   - Scalpels and blades
   - Knives
   - Forceps and artery forceps (haemostats)
   - Probes – solid and flexible (rubber)
   - Thread/suture material
   - Scoops for brain, etc.
   - Pen torch (flashlight)
   - Spare bulbs and batteries
   - Syringes and needles (disposable)
   - Empty drinks cans, labelled 'sharps boxes', for used needles, scalpel blades, etc.
   - Disinfectant(s), including ethanol/methanol/methylated spirits
   - Camping (gas cylinder-operated) stove – for sterilising, lighting (and cooking!)
   - Pressure cooker for sterilising
   - Cotton wool
   - Spring balance(s) or battery-operated scales
   - Scalpel handle and disposable sterile blades of several sizes and shapes. Dissecting scissors, curved haemostatic forceps, toothed and smooth-jawed fine-pattern thumb forceps and bone forceps
   - Sampling and other equipment for laboratory work (see later – list c)

3. Laboratory equipment for when samples have to be examined in the field:
   - Microscope (solar or battery-operated)
   - Immersion oil (or methyl salicylate) with swabs and xylene for cleaning
   - Pre-cleaned, frosted, ground-ended microscope slides and slide box or tray
   - Pencils for marking glass slides
   - Diamond-tipped pen for marking glass slides (if frosted not available)
   - Worm-egg counting slide
   - Coverslips of different sizes
   - Lens tissues

- Saline, saturated NaCl solution and other reagents for parasitology
- Transparent polythene strips and methylene blue/malachite green for the KATO method of cleaning faecal films for parasites, ova and cysts
- Fixatives – alcohol, formalin (see Appendix G)
- Selected stains for cytology (see Chapters 10 and 12)
- Lightweight (plastic) staining jar or staining rack
- Urine and blood chemistry test strips
- Portable centrifuge
- Polypropylene capillary tubes, some coated with heparin or EDTA, plus commercial haemoglobin and PCV reader
- Hand-held refractometer
- Transport medium for bacteria, viruses, mycoplasmas and *Trichomonas*
- Vacuum flask
- Buffer tablets, for use with local water
- Scalpel, scissors, forceps, artery forceps (haemostats)
- Wash bottles for alcohol, stains, etc.
- Lightweight pots for specimens
- Disinfectant(s), including ethanol/methanol/methylated spirits
- Camping (gas cylinder-operated) stove – for sterilising, lighting (and cooking!)
- Pressure cooker for sterilising.

Equipment that may be essential for more specialised laboratory investigations in the field can include:

- Vacuum flasks and portable, lightweight, cool-box
- Normal (isotonic) saline
- Hypertonic NaCl or sugar (sucrose) solution for flotation/sedimentation examination
- Tincture of merthiolate for staining faecal protozoa
- Fixatives for blood and other body fluids, bone marrow and endo- and ectoparasites
- Rapid-acting stains for blood and other body fluids (sputum, urine, synovial and coelomic, cerebrospinal, bone marrow, etc.) and touch/impression smear cytology (see Chapters 10 and 12)
- Gram, acid-fast and other special stains
- Lactol-phenol cotton-blue for demonstrating fungi
- Plastic pipettes
- Slotted stain jar. Lightweight, unbreakable plastic staining jars are preferable to heavy, fragile, glass Coplin jars
- Mounting media for permanent preparations of blood and bone marrow films
- Clearing and mounting media for small ectoparasites
- Light weight, slide-drying rack
- Transport media for bacteria, viruses and protozoa (see earlier)
- Microbiological test strips
- Urine and blood chemistry test strips
- Rapid diagnostic test strips
- Cardboard strips, which can be labelled in pencil or waterproof ink, and placed inside specimen containers
- Safety matches, a small Bunsen burner or disposable butane cigarette lighter.

Appendix B

- Squeeze bottles for methanol, etc.
- Specimen containers, filled with concentrated formaldehyde, for dilution with river or seawater (see Chapter 7)
- Tongue depressors, wooden applicator sticks and sterile cotton-tipped applicators. Plastic coffee spoons for use as spatulae (see text)
- Non-lubricated condoms as finger covers
- Plastic film canisters (pots) with labels attached for faecal collection, parasites, etc. With the advent of digital photography now less easy to oftain
- Sterile disposable venous and urethral catheters; latex or plastic tubing
- Plastic slide boxes, each prefilled with polished, frosted, glass microscope slides

Recommended additional items and precautions when working overseas in the field:

- Emergency pack containing business cards, photocopies of passport, visa and driving licence, letters of authorisation (see Chapter 7), protocols for snakebite (etc.), medicines and antidotes
- Multipurpose pocket knife
- Sewing kit with assorted needles, thread, scissors
- Screwdrivers, pliers and an adjustable spanner
- Elastic bands, string, dental floss, suture material, adhesive tape, insulating tape, duct tape, electrician's tape
- Spare nylon cable ties for securing lid hasp of case during travel
- Standard veterinary and other textbooks: where space is limited, the 'Merck Veterinary Manual' is recommended (but caution must be exercised in quoting from this in court as it does not include references)
- Phrase books of appropriate languages
- The 'SAS Survival Guide' (Wiseman, 1993) – contains much useful information that can be applied to difficult situations in the field
- Appropriate clothing – for instance, the shoulders should be covered when working in a Muslim community, a tie is considered a courtesy in many countries when one is meeting dignitaries.

Sharp and other possibly dangerous items should not be placed in hand luggage when travelling by air, or through land or sea security checkpoints.

It must be remembered that obtaining and possessing some items can prove difficult in certain countries. For example, alcohol is not readily available in certain Muslim areas and special arrangements may have to be made to obtain it. Alternatives, acceptable to the authorities may be the best option.

A similar situation can apply to certain drugs and medicinal compounds. The basic rule for the forensic investigator who is travelling overseas is to plan ahead and under no circumstances to take possibly sensitive items unless express written permission to import them has been granted from the host country.

Appendix C: Standard Witness Statement (United Kingdom)

JOHN E. COOPER
MARGARET E. COOPER

Statement of Witness

(C.J. Act 1967, Sec.9: M.C. Act 1980, ss 5A (3) (A) and 5 (B), M.C. Rules 1981, Rule 70.)
This report consisting of XXX (XXX) pages, each signed by me, is true to the best of my knowledge and belief and I make it knowing that, if it is tendered in evidence, I shall be liable to prosecution if I have wilfully stated in it anything which I know to be false or do not believe to be true.

I make this report in respect of the case of R v YYYY.

I am aware of Rule 33.2 of the Criminal Procedure Rules in relation to an expert's duty to the court, namely:

1. An expert must help the court to achieve the overriding objective by giving objective, unbiased opinion on matters within his experience.
2. This duty overrides any obligation to the person from whom he receives instructions or by whom he is paid.
3. This duty includes an obligation to inform all parties and the court if the expert's opinion changes from that contained in a report as evidence or given in a statement.

I have complied with my above duty as an expert and will continue to do so.

Dated the 10th day of February 2011.

Statement of

Age   Over 18

Occupation   Wildlife Biologist

This statement consisting of XX page(s) each signed by me is true to the best of my knowledge and belief and I make it knowing that, if it is tendered in evidence, I shall be liable to prosecution if I have wilfully stated in it anything which I know to be false or do not believe to be true.

Dated the   day of   20..

States:

WITNESS REPORT   page   of

Date: Signature: Continued   yes/no

WITNESS REPORT

This is a recommended format for detailed reports or opinions by a witness or expert. The information cited is fictitious but is intended to give some indication of the type and scope required in a legal case.

WILLIAM BEAN, MSc, PhD, FSB

Prepared 10th February 2012

Address of witness, including telephone and fax numbers
(international code) and e-mail/website details.

WITNESS REPORT   page   of

Date: Signature: Continued   yes/no

Appendix C

Contents of Report

Statement of witness

    Introduction

    Background

    Request

Material available prior to visit

Discussions held

The technical investigation

Findings

Records

Statements

The facts on which the opinion is based

Appendices

Appendix 1 – Experience

Appendix 2 – Findings and interpretation

Appendix 3 – Photographic evidence

References or relevant supporting literature

WITNESS REPORT   page   of

Date: Signature: Continued   yes/no

REPORT

1.0 Introduction

    1.1 Formal details

    1.2 I am a wildlife biologist living at the above address. I hold an MSc (Master of Science) degree from ……………………..University and a PhD (Doctor of Philosophy) from………………………….. University. I was elected a Fellow of the Society of Biology (FSB) in …………………………………….. I became a Chartered Biologist in …………………………..in ………………………………………..

    1.3

    1.4 I have provided Expert Witness services for both prosecution and defence for 20 years. I have acted for and advised various bodies including the European Union, Customs and Excise and numerous Police Forces. I am listed as an Expert Witness within my field on various registers. I am accredited / have applied for accreditation by the *(as appropriate)*.

    1.5 I was asked (a) to study relevant document (b) to meet ……………………. (c) to visit the premises of ………………………………………..

WITNESS REPORT   page   of

Date: Signature: Continued   yes/no

Synopsis

Appendix C

1.6 Instructions

1.7 I was provided with a statement by……………………………..and was to give my opinion in writing.

1.8 Following study of this statement I was of the opinion that ……………………….
………………………………………………………………………………………………………
…………………………………………………….

1.9 The visit and sources of information

1.10 At the time of the visit, I was in the company of ……………………………….
My statement is a compilation from the following sources:

    a. My contemporaneous notes and drawings, photographs and digital images (see attached)

    b. Video of ………

    c. Statement of ……………..

    d. Bundle of documents, reference …………………

    e. Statement of …………………..

    f. Statement of ……………………………..

    g. Photographic album exhibit reference (see Appendix 3)………..

WITNESS REPORT   page   of

Date: Signature: Continued   yes/no

2.0 THE BACKGROUND

2.1 The relevant parties

2.2 Listed below are the persons and their alleged role in the relevant events

2.3 PC (Police Constable) ………

2.4 The RSPCA, represented by …………..

2.5 The Expert Witness ……………. myself

2.6 Supported by a colleague………………

2.7 Others………………..

3.0 TECHNICAL INVESTIGATION

3.1 Background Information

The relevant legislation

3.2 Section ………….. of the …………..provides that

4. Material available to the witness prior to the visit

WITNESS REPORT   page   of

Date: Signature: Continued   yes/no

Appendix C

5. Interview records of …..

   The facts on which the opinion is based

6. CONCLUSIONS

   The facts on which my opinion is based have been detailed earlier.

7. My opinion

8. Apportionment of responsibility

   8.1

   8.2

WITNESS REPORT   page   of

Date: Signature: Continued   yes/no

9. I confirm in so far as the facts stated in my report are within my knowledge. I have made clear which they are and I believe them to be true and that the opinions I have expressed represent my true and complete professional opinions

I confirm that this statement has been compiled in accordance with (*relevant guidelines such as professional cooles of practice.*

DATE:

TIME:

NAME AND SIGNATURE:

WITNESS REPORT   page   of

Date: Signature: Continued   yes/no

Appendix C

Appendix 1

Experience
(Details of Expert)

Appendix 2

Findings and Interpretation
(Table of Findings)

Appendix 3

Photographic Evidence
(List of illustrations)

Reference	Location	Description

WITNESS REPORT   page   of

Date: Signature: Continued   yes/no

Appendix D: Specimen Forms – Wildlife Forensic Cases

JOHN E. COOPER

Form I

SUBMISSION OF WILDLIFE FORENSIC EXHIBITS
CHAIN OF CUSTODY RECORD

Laboratory case No: ....

Submitter's reference No: ....

Other relevant information ....

Special precautions necessary (e.g. hazards) ....

Exhibits submitted:

Item No.	Lab Exhibit No.	Brief Description

Chain of custody: received in sealed condition

Item No.	Received From	Received By	Date

Laboratory stamp received: ....

Laboratory stamp delivered: ....
....

The form given next is for the submission of a live animal, dead animal or sample for forensic investigation. The forms that follow are specific to certain types of examination and serve as report forms. All can be adapted, as needed, to suit requirements.

Form II

SUBMISSION FOR FORENSIC CLINICAL/ POST-MORTEM/LABORATORY EXAMINATION

Submitting Client/Agency .... Reference No ....

Lab Ref .... Date (in full) ....

Address ....

.... Post (Zip) code ....

Tel .... Fax .... E-mail...

Veterinary surgeon (veterinarian), where applicable..

Address and contact details..

Other relevant persons, e.g. police officer, wildlife inspector.........................

Species of animal (English and scientific name)...

Breed/variety .... Local name ....

Pet name (if appropriate)................................. Age .... Sex ....

Colour/markings ....

Ownership of animal or sample(s)...

Signature: ....

Date: ....

Time: ....

Sheet 1/5

Number/Ring (Band)/Tattoo/Microchip/Other methods of identification

....

Appendix D

Other comments, e.g. cruelty case, civil action, insurance claim, professional malpractice hearing, etc.

....

....

....

Background to the case/history..

....

....

....

Sample submitted..

Sent by hand/courier/post/other ....

Method and description of packing...

Tag no (see chain of custody)....................... Prior storage condition ....

Signature of courier .... Date ....

Live animal .... Dead animal ....

Organs (state)..

Other tissues..

Parasites...

Blood (and details of how presented)...

Swabs...

Signature: ....

Date: ....

Time: ....

Other..

Comment on chain of custody (attach relevant paperwork) ....
....
....
....

Questions being asked (expand as necessary)

- What is this material?
  ....
- Species/sub-species/breed?
  ....
- Age?
  ....
- What is its provenance/parentage, etc?
  ....
- Why did this animal die/how long has it been dead??
  ....
- When did this animal die?
  ....
- How did this animal die?
  ....
- Did this animal suffer pain or distress?
  ....
- Sex/entire/neutered?
  ....
- Sexually mature/fertile/in oestrus?
  ....
- Pregnant/gravid?
  ....
- Lactating/incubating?
  ....
- Origin – geographical?
  ....
- Migrant or resident?
  ....

Appendix D

- Origin – captive or free-living?
  ....
- Captive or free-living – for how long?
  ....

Signature: ....
Date: ....
Time: ....

Sheet 3/5

- Health status
  ....
- Age of skin wound, fracture or other lesions
  ....
- Other questions
  ....

Investigations requested ....

Clinical examination ....

Post-mortem examination ....

Laboratory tests:

  Microscopy ....

  Cytology ....

  Microbiology ....

  Parasitology ....

  Histology ....

  Toxicology ....

  DNA ....

  Other (specify e.g. serology) ....

Other tests as necessary to answer the questions given earlier ....

....

....

Results to be sent to ....

....

....

Signature: ....

Date: ....

Time: ....   *Sheet 4/5*

Special instructions regarding follow-up action to be taken regarding a live animal or the storage/transfer/disposal of samples/wrappings/carcase or tissues

....

....

....

....

....

Signed by recipient at (location) ....

Date .... Time ....

Signature: ....

Date: ....

Time: ....   *Sheet 4/4*

Form III

LABORATORY EXAMINATION FINDINGS (GENERAL) IN FORENSIC CASES

Reference Nos (delete if necessary):
Clinical: .... Pathology .... Case No: ....
Species (English name): .... (Scientific name): ....
Age: .... Sex: .... Number/Ring (Band)/Tattoo/Microchip No: ....
Other methods of identification: ....
Relevant history/circumstances (including time) of death:

Accompanying dockets or other relevant papers: ....

Appendix D

Any special requirements regarding techniques to be followed, fate of body/samples/special precautions needed

Submitted by: .... Date: ....

Received (date): .... by: ....

Laboratory Report

Gross examination: ....

....

Radiography, if performed..

Microscopical examination:..

....

Further Test(s)

Test	Submitted (date)	Results (Received date)	Comments
Microscopy			
Cytology			
Microbiology			
Parasitology			
Histology			
Toxicology			
DNA			
Other (Specify e.g. serology)			

Summary of findings:

....

....

Results/Comments/Interpretation:

....

....

Examination performed by: .... Date: .... Time: ....

Reported by: .... Date: .... Time: ....

Method of reporting: electronic/fax/post/by hand

Form IV

GROUP OR FLOCK EXAMINATION FINDINGS IN FORENSIC CASES

Reference Nos (delete if necessary): Clinical: .... Pathology: .... Case No: ....
Species (English name): .... (Scientific name): .... Age: .... Sex: ....
Number/Ring (Band)/Tattoo/microchip No: .... Other methods of identification: ....
Relevant history/circumstances (including time) of death:

Accompanying dockets or other relevant papers: ....
Any special requirements regarding techniques to be followed, fate of body/samples/special precautions needed
Submitted by: .... Date: .... Received (date): .... by: ....

No.	Sex, Age, Size	External Findings	Internal Findings	Other Observations	Comments

Photographs taken (tick box) ....
Summary:
....
....

Examination performed by: .... Date: .... Time: ....
Reported by: .... Date: .... Time: ....
Method of reporting: electronic/fax/post/by hand

Appendix D

Form V

INITIAL RECORDING OF HISTOPATHOLOGICAL FINDINGS IN FORENSIC CASES

Slide No.	Comments on Preparation, Including Stain	Tissues on Slide	Observations	Scoring of Changes (Where Appropriate)	Comments	Interpretation

Some suggested abbreviations: AICI = acute inflammatory cell infiltration. CICI = chronic inflammatory cell infiltration;

wnl = within normal limits. Scoring system: 1 = minimal; 2 = moderate; 3 = marked

Summary: ....

Further work needed: ....

Examination performed by: .... Date: .... Time: ....

Reported by: .... Date: .... Time: ....

Method of reporting: electronic/fax/post/by hand

Form VI

FORENSIC EXAMINATION OF LIVE ANIMALS

Reference Nos (delete if necessary):

Clinical: .... Pathology .... Case No: ....

Species (English name): .... (Scientific name): ....

Age: .... Sex: .... Number/Ring (Band)/Tattoo/Microchip No: ....

Other methods of identification: ....

Relevant history/circumstances (including time) of death:

Accompanying dockets or other relevant papers: ....
Any special requirements regarding techniques to be followed, fate of body/samples/special precautions needed

Submitted by: .... Date: ....

Received (date): .... by: ....

HISTORY AND RELEVANT DATA OBSERVATION – COMMENTS RESULTS OF CLINICAL EXAMINATION FURTHER TESTS (e.g. RADIOGRAPHY) SAMPLES TAKEN/LABORATORY RESULTS (use other form for details) SUBSEQUENT INVESTIGATIONS FOLLOW-UP	Sequential weights/measurements:-

COMMENTS (continue overleaf or attach additional papers as necessary)

Photographs taken (tick box) ....

Examination performed by: .... Date: .... Time: ....

Reported by: .... Date: .... Time: ....

Method of reporting: electronic/fax/post/by hand

Form VII

FORENSIC *POST-MORTEM* EXAMINATION FORM

Reference Nos (delete if necessary):

Clinical: .... Pathology .... Case No: ....

Species (English name): .... (Scientific name): No: ....

Age: .... Sex: .... Number/Ring (Band)/Tattoo/Microchip No: ....

Appendix D

Other methods of identification: ....

Relevant history/circumstances (including time) of death:

Accompanying dockets or other relevant papers: ....

Any special requirements regarding techniques to be followed, fate of body/samples/special precautions needed

Submitted by: .... Date: ....

Received (date): .... by: ....

Measurements: carpus tarsus other bodyweight (mass)

Condition: obese or fat/good/fair or thin/poor/numerical score if appropriate

Autolysis: fresh/minimal/moderate/marked autolysis/numerical score if appropriate

Storage since death: ambient temperature/refrigerator frozen/fixed/other

Radiography/other imaging, if performed: (findings)

EXTERNAL OBSERVATIONS, including mammary/preen gland, state of moult, ectoparasites, skin condition, lesions etc. (expand as necessary)

MACROSCOPIC EVALUATION ON OPENING THE BODY, including position and appearance of organs, lesions etc. (expand as necessary)

ORGAN SYSTEMS (expand each as necessary):
ALIMENTARY
MUSCULOSKELETAL
CARDIOVASCULAR
RESPIRATORY
URINARY
REPRODUCTIVE
LYMPHOID (including bursa of Fabricius)
NERVOUS
OTHER
SAMPLES TAKEN (expand as necessary)

.... Cyto Micro Paras Hist DNA Tox Other (e.g. serology)

.... Cyto Micro Paras Hist DNA Tox Other (e.g. serology)

.... Cyto Micro Paras Hist DNA Tox Other (e.g. serology)

.... Cyto Micro Paras Hist DNA Tox Other (e.g. serology)

.... Cyto Micro Paras Hist DNA Tox Other (e.g. serology)
.... Cyto Micro Paras Hist DNA Tox Other (e.g. serology)
.... Cyto Micro Paras Hist DNA Tox Other (e.g. serology)

LABORATORY FINDINGS (expand as necessary)

Date: .... Initials: ...... Reported by: .... To whom: ....

PRELIMINARY REPORT (based on gross findings and immediate laboratory results e.g. cytology)

Reported to: .... By whom: .... Date: .... Time: ....

FINAL REPORT (based on all available information)

FATE OF CARCASE/TISSUES
Destroyed/frozen/fixed in formalin (or other fixative)/retained as evidence/deposited in Reference Collection/sent elsewhere
Photographs taken (tick box) ....

Examination performed by: .... Date: .... Time: ....
Corroborating pathologist (if appropriate) .... Date: .... Time: ....
Reported by: .... Date: .... Time: ....
Method of reporting: electronic/fax/post/by hand   *Sheet 1/2*

Form VIII

FORENSIC CYTOLOGY REPORT

Reference Nos (delete if necessary):
Clinical: .... Pathology .... Case No: ....
Species (English name): .... (Scientific name): No: ....
Age: .... Sex: .... Number/Ring (Band)/Tattoo/microchip No: ....
Other methods of identification: ....
Relevant history/circumstances (including time) of death:

Accompanying dockets or other relevant papers: ....
Any special requirements regarding techniques to be followed, fate of body/samples/special precautions needed

Submitted by: .... Date: ....
Received (date): .... (by): ....

Sample submitted: ....

Received (date): .... (by): ....

Stain(s) and number of preparation(s): ....
....

General comments on preparation(s): ....
....
....

| Cell types | Numbers (+/++/+++) | Features | Comments |

Photographs taken (tick box) ....
Other findings: ....
....
....

Summary or interpretation of findings: ....
....
....
Examination performed by: .... Date: .... Time: ....
Reported by: .... Date:............... Time: ....
Method of reporting: electronic/fax/post/by hand

Form IX

FORENSIC HISTOPATHOLOGY REPORT

Reference Nos (delete if necessary):

Clinical:........................... Pathology......................... Case No:............................

Species (English name): .... (Scientific name): No: ....

Age:.......... Sex:.......... Number/Ring (Band)/Tattoo/microchip No:..........................

Other methods of identification: ....

Relevant history/circumstances (including time) of death:

Accompanying dockets or other relevant papers: ....

Any special requirements regarding techniques to be followed, fate of body/samples/special precautions needed

Submitted by:.. Date: ....

Received (date):.......................... by:..

Findings: ....
....
....
....
Comments: ....
....
....
Further work needed: ....
....
....
Summary or interpretation of findings: ....
....
Advice or action if appropriate: ....
....
....
Photographs taken (tick box) ....
Examination performed by: .... Date: .... Time: ....
Reported by: .... Date: .... Time: ....
Method of reporting: electronic/fax/post/by hand

Form X

FORENSIC EXAMINATION OF EGGS/EMBRYOS

Reference Nos (delete if necessary):
Clinical: .... Pathology: .... Case No: ....
Species (English name): .... (Scientific name): No: ....
Age: .... Sex: .... Number/Ring (Band)/Tattoo/Microchip No: ....
Other methods of identification: ....
Relevant history/circumstances (including time) of death:

Accompanying dockets or other relevant papers: ....
Any special requirements regarding techniques to be followed, fate of samples/special precautions needed

Submitted by: .... Date: ....
Received (date): .... by: ....

Appendix D

Weight of whole unopened egg: ……..……... Length: ……..………….. Width: ……..………...
External appearance (to include drawing or photograph):
Appearance (to include drawing or photograph): Embryo
                                                                  Air cell
                                                                  Blood vessels
                                                                  Fluids
Appearance when opened to (include drawing or photograph):
Contents:
Embryo: Length (crown-rump)
                Amniotic cavity
                Allantoic cavity
                Yolk sac
Other comments:
Microbiology:
Histopathology:
Other tests:
Samples sent elsewhere:
Weight of dried eggshell: ……..……….. Thickness (measurement or index): ……..…………..
Samples stored:
Summary or interpretation of findings

Photographs taken (tick box) ……..……………………………………..
Examination performed by: ……..…………………….. Date: ……..……. Time: ……..…
Reported by: ……..……………………………………. Date: ……..……. Time: ……..…
Method of reporting: electronic/fax/post/by hand

Form XI

PELLET REPORT FORM

Reference Nos (delete if necessary):
Clinical: ……..………….. Pathology ……..………….. Case No: ……..…………..
Species (English name): ……..………….. (Scientific name): No: ……..…………..
Age: ……..… Sex: ……..… Number/Ring (Band)/Tattoo/Microchip No: ……..…………..
Other methods of identification: ……..……………………………………………..

Relevant history/circumstances (including time) of death:

Accompanying dockets or other relevant papers: ....

Any special requirements regarding techniques to be followed, fate of samples/special precautions needed

Submitted by: .... Date: ....

Received (date): .... by: ....

1. External examination
   - Weight
   - Measurement (cm):
     i. Length
     ii. Width
     iii. Height
     iv. Circumference

   - Gross appearance/description, colour, consistency, apparent content, smell

     ....
     ....
     ....

Radiography, if performed

2. Internal examination – contents

   ....
   ....
   ....

3. Identification of pellet contents
   - Pellet material that floats, removed, placed on tissue paper and allowed to dry

     ....
     ....
     ....

   - Pellet material that sinks, preserved in dilute formalin to observe under low-power microscope

     ....
     ....
     ....

   - Laboratory results:
     i. Direct microscopy

     ....
     ....
     ....

Appendix D

    ii. Culture

....
....
....

15. Material retained/submitted elsewhere:

....
....
....

16. Comments

....
....
....

17. Photographs taken/drawings

....
....
....

Examination performed by: .... Date: .... Time: ....
Reported by: .... Date: .... Time: ....
Method of reporting: electronic/fax/post/by hand
....

Form XII

FORENSIC EXAMINATION OF GORILLA SKELETON/BONES

Study Ref: **SHEET 1/7**

Reference Nos (delete if necessary):

Clinical: .... Pathology .... Case No: ....
Species (English name): .... (Scientific name): ....
Age: infant/juvenile/young adult/old adult .... Sex: male/female/uncertain ....
Number/tattoo/microchip No: ....Other methods of identification e.g. noseprint.........
Relevant history/circumstances (including time) of death:..
Origin of bones...
Method of preservation/preparation/storage of bones..
State of preservation..

Accompanying dockets or other relevant papers: ....
Any special requirements regarding techniques to be followed, fate of samples/special precautions needed.
....

Submitted by: ....Location...........................Date:......................
Received by: ....Date:................

MATERIAL AVAILABLE/EXAMINED–GENERAL COMMENTS: ....

RADIOGRAPHY OR OTHER IMAGING PERFORMED...
ENDOSCOPY...............OTHER TESTS............ PHOTOS: Y/N

Cranium (Skull)	Mandible (Jaw)	Post-Cranial Skeleton	Other (e.g. Skin)

SUMMARY OF PATHOLOGICAL FINDINGS:			
Cranium (Skull)			
Mandible			
Post-Cranial Skeleton			

SEE SUBSEQUENT SHEETS (2-7)

SHEET 2/7

Study Ref.

Inventory of bones present, with comments as appropriate

Cranium

Mandible

Sternebrae

Appendix D

Vertebrae

Cervical: C1 | C2 | C3 | C4 | C5 | C6 | C7   Comment

Thoracic: T1 | T2 | T3 | T4 | T5 | T6 | T7 | T8 | T9 | T10 | T11

Thoracic/lumbar: T12 | (T13)   Comment   L1 | L2 | L3 | L4 | L5

Comment

Comment

Comment

Sacrum

	Right			Left			Comment
	Present	Weight	Length	Present	Weight	Length	
Clavicles							
Scapulae							
Ribs							
Humerus							
Ulna							
Radius							

Appendix D

SHEET 3/7

Study Ref.

right hand   *left hand*

sca	lun	tri	pis		pis	tri	lun	sca
trm	trd	cap	ham		ham	cap	trd	trm

mc1	mc2	mc3	mc4	mc5		mc5	mc4	mc3	mc2	mc1
pp1	pp2	pp3	pp4	pp5		pp5	pp4	pp3	pp2	pp1

mp2	mp3	mp4	mp5		mp5	mp4	mp3	mp2
dp1	dp2	dp3	dp4	dp5		dp5	dp4	dp3	dp2	dp1

Comments

Weights and measurements of selected bones:			
Bone	Weight	Measurements(s)	Comments

SHEET 4/7

Study Ref.

	Right			Left			Comment
	Present	Weight	Length	Present	Weight	Length	
Innominate							
Femur							
Patella							
Tibia							
Fibula							
Pelvis							

Appendix D

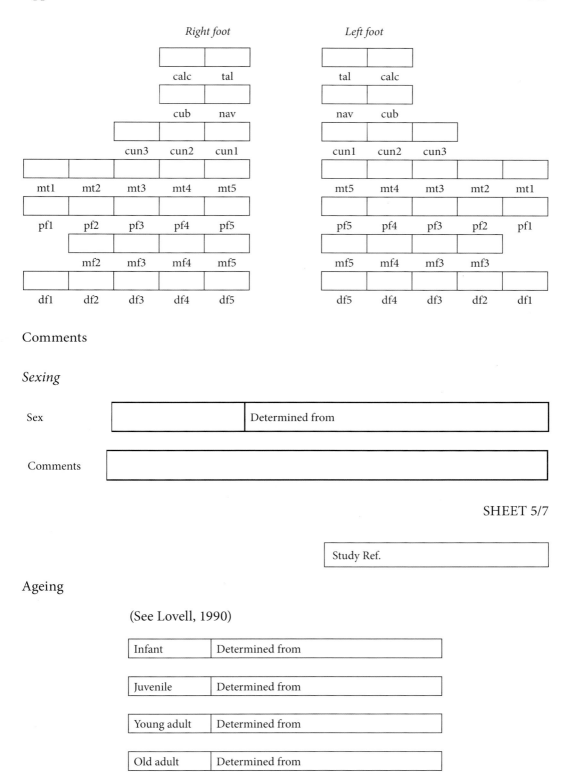

Comments

Sexing

Sex [   ] Determined from [   ]

Comments [   ]

SHEET 5/7

Study Ref. [   ]

Ageing

(See Lovell, 1990)

Infant	Determined from

Juvenile	Determined from

Young adult	Determined from

Old adult	Determined from

Appendix D

SHEET 6/7

Study Ref.

Dental record

UPPER (maxilla)

R																	L
	Other																
	Malocclusion																
	Hypoplasia																
	Abscesses																
	Caries																
	Recession																
	Calculus																
	Attrition																
	Presence																
	Permanent	8	7	6	5	4	3	2	1	2	3	4	5	5	6	7	8
R	Deciduous				e	d	c	b	a	a	B	c	d	e				L

R																	L
R	Deciduous				e	d	c	b	a	a	B	c	d	e				L
	Permanent	8	7	6	5	4	3	2	1	2	3	4	5	5	6	7	8
	Presence																
	Attrition																
	Calculus																
	Recession																
	Caries																
	Abscesses																
	Hypoplasia																
	Malocclusion																
R	Other																L

Lower deciduous

Key: N/A = not applicable   * = severe   ✓ = present   χ = absent

Appendix D

COMMENTS ON DENTAL PATHOLOGY

SHEET 7/7

Study Ref.

Weight of skull: Weight of mandible:

Selected measurements:

Symmetry:

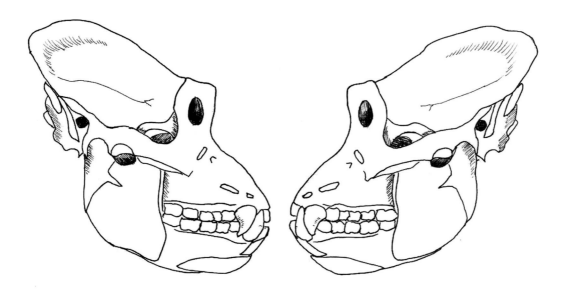

Key to lesions:
X = tooth missing ⇆ = deviation • = pathological hole in bone

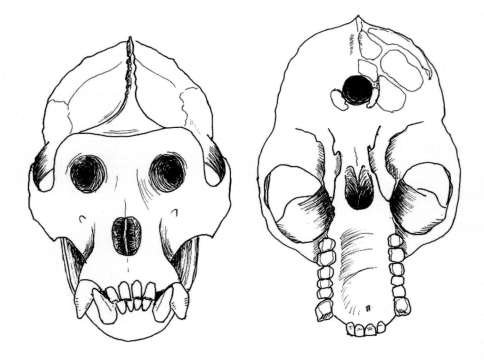

COMMENTS ON PATHOLOGY

Fate of material

Summary or interpretation of findings..
....
....
....

Examination performed by: .... Date: .... Time: ....
Reported by: .... Date: .... Time: ....
Method of reporting: electronic/fax/post/by hand

Appendix F: Health and Safety: Zoonoses and Other Hazards

JOHN E. COOPER
MARGARET E. COOPER

*'I had an aunt in Yucatan
Who bought a python from a man
And kept it for a pet.
She died, because she never knew
These simple little rules and few:
The snake is living yet'.*

Hilaire Belloc

Background

In any work with wild animals, live or dead, there are potential hazards. One of these is the risk of contracting and infectious disease.

Zoonoses are those diseases and infections that are naturally transmitted between vertebrate animals and humans (based on World Health Organization wording).

'Vertebrate animals' comprise mammals, birds, reptiles, amphibians and fish.

Clinically, zoonoses can be divided into three main groups. They can be

1. Equally hazardous to humans and most animals, e.g. rabies, anthrax
2. Only rarely or slightly impair animal health but may cause serious diseases in humans, e.g. brucellosis, hydatidosis, *Herpesvirus simiae* infection
3. Responsible for serious epizootics (epidemics) in animals but of limited significance to humans, e.g. foot-and-mouth disease, Newcastle disease

Some zoonotic infections can be serious in animals and humans but may be symptomless (produce no significant clinical signs) in both humans and animals under certain circumstances, e.g. chlamydophilosis (ornithosis or psittacosis).

Environmental Factors That Favour the Spread of Zoonoses

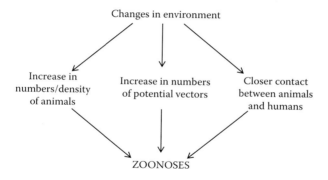

Examples of Recognised Zoonoses in Different Species of Wildlife

Ruminants	*Escherichia coli* infection/toxicosis
	Salmonellosis
	Brucellosis
	Tuberculosis
	Leptospirosis
	Campylobacteriosis
	Anthrax
	Hydatidosis
Canids and felids	Rabies
	Ringworm
	Leishmaniasis
	Toxoplasmosis
	Pasteurellosis
Rodents and lagomorphs	Salmonellosis
	Leptospirosis
	Lymphocytic choriomeningitis (LCM)
	Hanta virus infection
	Borna disease
	Tularaemia
Non-human primates	Salmonellosis
	Shigellosis
	Herpesvirus simiae ('B virus') infection
	Marburg and Ebola virus diseases
Birds	Salmonellosis
	Chlamydophilosis
	Yersiniosis
	Avian influenza
Reptiles and amphibians	Salmonellosis
	Atypical mycobacteriosis
Fish	*Erysipelothrix insidiosa* infection
	Listeriosis
	Atypical mycobacteriosis
	Edwardsiella tarda infection

Some organisms, e.g. *Salmonella* bacteria, may be acquired from a range of different animal species – and from other humans. *Homo sapiens* can transmit infectious agents to animals, as well as *vice-versa*. Some apparent zoonoses are acquired from the environment, not from animals.

Meticillin-resistant *Staphylococcus aureus* (MRSA) is an important zoonotic pathogen that has increased in prevalence over the past decade, with up to 2% of humans acting as carriers and posing a risk to in-contact animals. MRSA has been isolated from most domestic animal species, including dogs, horses and pigs and from clinically healthy wild

Appendix F

and captive marine mammals, and may have caused the death of a stranded harbour seal in Ireland (Fravel et al., 2011).

Some parasites of animals may cause reactions in humans – for example, mites from pigeons or reptiles.

The New Zoonoses

Humans who are immunosuppressed/immunocompromised (e.g. patients with AIDS, malaria or receiving chemotherapy) can be susceptible to a whole range of otherwise rarely encountered infectious agents, sometimes of animal origin. These can be considered the 'new zoonoses'. There are two categories

1. Zoonoses associated with novel exposure to unfamiliar animals (e.g. primates, rodents) or to animal products (e.g. meat, brain, transplanted heart valves).

    Examples are

    a. Ebola virus disease
    b. Marburg virus disease
    c. Spongiform encephalopathies, e.g.? bovine spongiform encephalopathy (BSE)

2. Zoonoses associated with immunodeficiency, the latter due to HIV infection, malaria or other diseases, neoplasia, irradiation, chemotherapy or splenectomy.

    Examples are

    a. Cryptococcosis
    b. Toxoplasmosis
    c. Giardiasis
    d. Listeriosis
    e. Atypical mycobacteriosis
    f. Babesiosis
    g. Aspergillosis

Zoonoses: Those at Risk

Persons who are

- Working with live/dead animals
- Working with pathogens
- Working with affected humans
- Inadvertently in contact with animals, invertebrate vectors or pathogens

Considerations When Investigating Possible Zoonoses in Humans

1. Differentiation of animal and environmental sources
2. The particular susceptibility to opportunistic infections of AIDS patients and others who may be immunocompromised
3. Possibility of animal – human – animal spread (e.g. tuberculosis)

Prevention and Control of Zoonotic Infections

1. Awareness, education and training
2. Minimising unnecessary contact with live/dead animals or their products, vectors and pathogens
3. Specific protection and monitoring of those at risk by, e.g. immunisation, health checks
4. Closer liaison between medical and veterinary professions and other disciplines
5. Correct handling and restraint of animals and their products
6. High standards of hygiene and use of barriers
7. Maintenance and monitoring of health of animals

Methods of Infection

Examples

Contact	e.g. Swimmer's itch (avian schistosomiasis)
Inhalation	e.g. Psittacosis (chlamydophilosis)
Bitten by mosquito	e.g. Equine encephalitis
Bitten by tick	e.g. Lyme disease

A Risk Assessment Is Essential

The five steps are as follows:

- Identify hazards
- Assess who might be harmed and how
- Evaluate risks and decide precautions
- Record findings
- Review assessment and update

Appendix G: Preparation and Investigation of Material

MARTYN COOKE
ANDREW C. KITCHENER
JOHN E. COOPER
JILL WEBB

This Appendix consists of five documents relating respectively, to the preparation and fixation of museum specimens, cosmetic *post-mortem* examinations, investigation of alleged contamination, forensic investigation of the skeleton and preparation of skeletal material, and the *post-mortem* examination of radio-tagged birds. It should be noted that the last of these has been reproduced *in toto* (with permission) and no changes have been made to the text or construction.

(a) The Preparation and Fixation of Specimens for Research and Museum Display

Martyn Cooke

In order to prepare a good quality specimen as a museum mount, a number of basic steps need to be taken. This work is often compromised by the need to prioritise diagnostic work and tissue-sampling before a specimen can be prepared for museum use.

Ideally, the following guidelines should be followed in order to achieve the best results.

Pre-Fixation and Preparation

- The specimen should not come into contact with water as this will cause haemolysis, resulting in loss of specimen colour and continuous discoloration of the mounting fluid.
- Should the specimen require washing, this can be carried out using the chosen fixative or normal physiological saline.
- The specimen should not be allowed to dry out as irreversible darkening can occur.
- Freezing the specimen should be avoided and refrigeration at 4°C should be used for short-term storage if necessary. Placing the specimen in a sealed polythene bag will help reduce dehydration.

Fixation

Currently, the most widely used and almost certainly the best routine fixative for both histological diagnosis and museum preparations is 4% buffered formaldehyde. Formaldehyde is a gaseous substance dissolved in water to a maximum percentage of 37%–40% w/v.

Formaldehyde is highly toxic and should be used with fume extraction. If this is not possible, a full face-mask with appropriate filter is essential. As a last resort, working in the open air is preferable to an enclosed space (see Chapter 7).

Formaldehyde must never be used in undiluted form for tissue fixation and is routinely diluted to 4% for use (1 part formaldehyde, 9 parts water). It is standard practice to include buffers to maintain pH 7.0 in order to prevent a build-up of formic acid through oxidation. Historically, the inclusion of physiological saline was considered important; there is now less scientific evidence that this has any benefit to the fixation process. However, saline is still listed in many older methods.

There are numerous names applied to formaldehyde solutions, both descriptive and trade names, that can often lead to confusion over the constituent components and the dilution. Formalin, formol and formal (sometimes formol) saline are some of the most widely used names. Solutions are often referred to as being a dilution of 10%, rather than 4%, due to the 1:9 formaldehyde/water ratio used in preparation. For the purposes of this document, we will refer only to 40% formaldehyde (concentrate) and 4% buffered formaldehyde (working solution).

A typical fixative for routine use is made up as follows:

- 40% formaldehyde (concentrate) 100 mL
- Sodium dihydrogen orthophosphate 4 g
- Di-sodium hydrogen orthophosphate 6.5 g
- Distilled water 900 mL

A more convenient, time-saving and safe approach is to purchase ready-to-use 4% buffered formaldehyde. This minimises the risk of unnecessary exposure during preparation. It can be obtained from VWR International (www.vwr.com).

VWR International supply a number of different variations; the following is recommended as an excellent routine histological and museum display fixative:

- Formalin neutral buffered pH 7.0 Histological fixative 'Gurr'.
- Code 36138 7P 5 L.
- Code 36138 8Q 25 L.
- The specimen should be placed in a large container in a way that it does not contact the sides of that container and become distorted.
- The specimen should be immersed in up to 10 times its own volume, if possible, in order to ensure adequate fixation.
- Due to the hardening effect of formalin, it is important to place the specimen in the required orientation for display purposes. Failure to achieve this may cause distortion or tearing of tissue in an attempt to force the specimen into the appropriate display position once it has hardened.
- The specimen should be placed display-side-up to avoid distortion. Cotton wool or cloth (such as cheese cloth) can be used to support parts in the correct anatomical position.
- Where possible, whole organs such as kidney and brain can be suspended with the thread attached to arterial vessels and hung into a bucket of fixative to avoid distortion of the natural shape.
- Some organs, such as lungs, cystic structures and specimens with large accumulations of fatty tissue, will float in the fixative. These specimens can be conveniently placed display-side-down in the fixative. A layer or two of disposable paper towel placed over the surface of the fluid will prevent dehydration of any parts that are protruding above the surface of the fluid.

Post-Fixation

- It is important to change the fixative after the first 24 h to ensure that fresh, active fixative is present. If the fluid continues to discolour, further changes of fresh fixative can be beneficial.
- If a cut surface is required, it is advantageous to allow the tissue to harden in the fixative before slicing. Only a thin slice should be removed as formaldehyde penetrates the tissue very slowly and may leave peripheral zones of discoloration.
- Optimal fixation is achieved after approximately 7 days, although smaller structures will be fixed more rapidly. It is advantageous to leave the specimen in fixative for between 2 weeks and 1 month. If red blood still leaches from the tissue, fixation is *not* complete.
- Following complete fixation, it is advisable to transfer the specimen to a suitable long-term preservative. Long-term storage in formaldehyde is undesirable, as formaldehyde solutions can become acidic on storage and should be monitored carefully to avoid any detrimental effect to the tissue.

Preservatives

There are several tried and tested long-term preservatives. 70% ethanol (IDA) is routinely used but has the disadvantage of rapid evaporation and can cause bleaching of the specimen colour. Some plastics deteriorate in the presence of ethanol.

By far the most suitable long-term preservative is Kaiserling III solution (KIII). KIII is an excellent tissue preservative. It permits histological diagnosis after many years' immersion, it is compatible with the majority of plastics and it is safe to work with on the open bench. An additional advantage is that KIII is the most widely used mounting solution for museum display. Unfortunately, there is no commercial supplier of this preservative and it will have to be prepared in the laboratory.

There are many different variations of Kaiserling III; the following is recommended as an excellent long-term preservative:

- Glycerine: 10,000 mL
- Sodium acetate: 3,500 g
- Water: 15,000 mL

The sodium acetate should be added to the water and mixed well before adding the glycerine. This solution should ideally have a pH value of 7.5–8.0.

(b) Cosmetic *Post-Mortem* Examination

Cosmetic *Post-Mortem* Examination of Mammals

Andrew C. Kitchener

Dead mammals from zoos are an important source of information concerning their anatomy, taxonomy, geographical variation and the pathology of their teeth and bones.

Many species are only poorly represented in research collections so that the scientific basis of many so-called sub-species is non-existent.

There is a high risk of irrevocable damage to animal specimens during *post-mortem* examinations. It is vitally important that these scarce specimens are made available for research following *post-mortem* examinations so that they can contribute to a wide variety of research.

A cosmetic *post-mortem* examination (necropsy) allows full access to all tissues required by the pathologist, but does not take significantly more time to carry out and allows for the preservation of skins and skeletons for museum-based studies.

We suggest the following main steps in carrying out a cosmetic necropsy on a mammal:

1. Weigh the mammal and record its weight.
2. Make a small central incision with *a scalpel*. Make sure to cut between the hair, not through it.
3. To increase the size of the incision, insert the scalpel blade under the skin facing upwards so that the skin is cut from underneath. The incision can be increased to extend from the anus to the throat.
    a. Please do not cut through the lips.
    b. Please do not remove any pieces of skin, unless needed for histology. If so, make cuts as small and as cleanly as possible.
4. Peel the skin back from the thorax and abdomen. If possible, rub salt (NaCl) into this skin. Access can now be made to the abdomen.
5. To gain access to the thorax, cut through the cartilaginous ribs close to the sternum and lift the sternum upwards and forwards, so that the ribs can be splayed apart. If access is required to the spinal cord, make an additional incision down the mid-line of the back and peel the skin to the side. So long as cuts are clean and all the skin is there, we can easily sew the pieces together again.
6. For larger mammals, it may be necessary to remove the limbs. This can be done from inside the skin, by detaching the scapulae from the muscle that holds them to the back and the humerus from the clavicle. For the hind limb, the head of the femur can be detached from the pelvis. Peel back the skin further to allow full access to the detachment points, but please do not cut through the skin.
7. If it is absolutely necessary to get access to the brain, make an incision from the crown on the head down to the nape of the neck. Peel back the skin and if necessary cut through the bases of the ears. This will allow full access to the cranium, which should be trepanned carefully so that the skull can be reconstructed after the *post-mortem* examination. Alternatively, from the underside of the neck, sever the skull from the cervical (axis) vertebra and peel back the skin from the top of the head and cut through the bases of the ears if necessary. Again access to the cranium is now clear.
8. After the *post-mortem* examination, if possible rub salt into the paws and face, double bag in polythene and deep-freeze. Please label the outside of the bag clearly with species' name, identification name/number and/or date of death.

Your help and co-operation in making dead mammal specimens available to museums is greatly appreciated. We shall be delighted to feedback additional pathological information to you, when skinning the carcase and preparing the skeleton.

Please contact Andrew C. Kitchener at the National Museum of Scotland, Edinburgh to arrange collection of frozen carcases by courier: Tel.: +44-131-2474240, E-mail: a.kitchener@nms.ac.uk.

Thank you

Dr Andrew C. Kitchener
Principal Curator of Vertebrates
National Museums Scotland
Chambers Street
Edinburgh EH1 1JF
United Kingdom

Cosmetic *Post-Mortem* Examination of Birds

Dead birds from zoos are an important source of information concerning anatomy, taxonomy, geographical variation and pathology of bones. Many species are only poorly represented in research collections so that the scientific basis of many so-called sub-species is non-existent.

There is a high risk of irrevocable damage to animal specimens during *post-mortem* examinations. It is vitally important that these scarce specimens are made available for research following *post-mortem* examinations so that they can contribute to a wide variety of research.

A cosmetic *post-mortem* examination (necropsy) allows full access to all tissues required by the pathologist, but does not take significantly more time to carry out and allows for the preservation of skins and skeletons for museum-based studies.

We suggest the following main steps in carrying out a cosmetic *post-mortem* examination on a bird:

1. Weigh the bird and record its weight.
2. Make a small central incision with *a scalpel*. Make sure to cut between the feathers, not through them, or cut skin not covered in feathers between the feather tracts.
3. To increase the size of the incision, insert the scalpel blade under the skin facing upwards, so that the skin is cut from underneath. The incision can be increased to extend from the cloaca to the throat.
   a. Please do not cut through the bill.
   b. Please do not remove any pieces of skin, unless needed for histology. If so, make cuts as small and as cleanly as possible.
4. Peel the skin back from the thorax and abdomen. If possible, rub salt (NaCl) into this skin. Access can now be made to the abdomen.
5. To gain access to the thorax, cut through the cartilaginous ribs close to the sternum and lift the sternum upwards and forwards so that the ribs can be splayed apart. You may find it easier to peel back the skin over the shoulder, so as to be able to remove the shoulder girdle intact, giving clear access to the thorax. To get access to the trachea and oesophagus, cut around the tongue and their attachment points at the back of the mouth. It should then be possible to pull the trachea and oesophagus down inside the thorax. Please try not to cut the neck skin.

6. If it is absolutely necessary to get access to the brain, make an incision from the crown on the head down to nape of the neck or make a circular cut around the crown between the feathers to allow the crown to be lifted forwards as a flap. Again, please cut between the feathers. Peel back the skin. This will allow full access to the cranium, which should be trepanned carefully, so that the skull can be reconstructed after the *post-mortem* examination.
7. After the *post-mortem* examination, if possible rub salt into the face, double bag in polythene and deep-freeze. Please label the outside of the bag clearly with species' name, identification name/number and/or date of death.

Your help and co-operation in making dead bird specimens available to museums is greatly appreciated. We shall be delighted to feedback any additional pathological information to you, when skinning the carcase and preparing the skeleton.

Please contact Andrew C. Kitchener at the National Museum of Scotland, Edinburgh to arrange collection of frozen carcases by courier: Tel.: +44-131-2474240, E-mail: a.kitchener@nms.ac.uk

Thank you

Andrew C. Kitchener
Principal Curator of Vertebrates
National Museums Scotland
Chambers Street
Edinburgh EH1 1JF
United Kingdom

Appendix G

Standard Data Recording Sheet for Mammal used by National Museums Scotland

Mammal	Freezer No:		Collector's No:		Species	
Locality and grid ref.					Sex (by gonads)	
Collection date		Donor/ Collector				
Wing length (mm)		Tarsus length (mm)		Tail length (mm)		Bill length (mm)	
Eye colour			Bill/facial Colour		Leg/foot Colour	

Weight Fresh (g/kg)			Processing date	
Freezer (g/kg)				

Process as:	Skin ☐ Skeleton/left wing ☐	Spirit ☐
	Confirmed ☐ ☐	☐

Skull ossification	SNO ☐	SNQFO ☐	SFO ☐

Stomach contents (brief description)	Fat subcutaneous score: (1-3)	
Save in spirit ☐		Deposited

Samples to be taken (store frozen)	Whole carcase		Brain		Ectoparasites	
	Liver		Heart			
	Kidney		Muscle			

Cause of Death:	Notes:

Appendix G

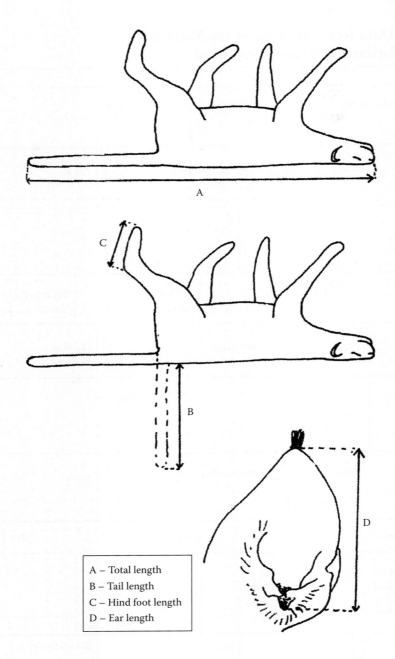

A – Total length
B – Tail length
C – Hind foot length
D – Ear length

Appendix G

Standard Data Recording Sheet for Birds used by National Museums Scotland

Bird	Freezer No:		Collector's No:		Species	
Locality and grid ref.					Sex	
Collection date			Donor/ collector			
Total length (mm)		Tail length (mm)		Hind foot length (mm)	With claws	
					Without claws	
Ear length (mm)		Body weight (g/kg)		Eye colour	Process date	
Process as:	Skin		Skeleton		Baculum		Spirit	
	Confirmed							
Stomach contents (brief description) Save in Spirit ☐								
Whole carcase		Heart		Brain		Uterus	
Liver		Tongue		Ectoparasites		Gonads	
Kidney		Muscle					
Cause of Death:			Notes:				

(c) Investigation of Small Animals/Remains in Alleged Contamination Cases

John E. Cooper
Jill Webb

Recommended Protocol

- Collate history.
- Prepare instruments and equipment.
- Receive specimen – sign for receipt, sign and date the chain of custody. Retain packing.
- Photography – assign a unique reference number before photographing (this number to be used as filename for image and subsequent images/radiographs, etc.).

- Enter – give the specimen a unique reference number.
- Record condition of packaging/sample; check seals on packaging are intact; photograph packaging, if appropriate.
- Carry out a risk assessment. Instigate appropriate health and safety precautions prior to opening package, if appropriate, and also after seeing contents.
- Open packaging (ensuring that the end opposite any security seal is cut to open and that the seal is left intact).
- Photograph specimen as presented.
- Weigh and measure the specimen.
- Carry out radiography or other scanning.
- External examination. Take samples. Photograph.
- Internal examination. Take samples. Photograph.
- Store/submit to laboratory relevant samples, e.g. tissues fixed for histology, bones and teeth retained frozen or in fixative for speciation (identification) and age determination/sexing.
- Perform gross and microscopical examination of samples.
- Pack specimen, close package with unique tamper-proof seal, ensuring all original packaging is included inside.
- Store specimen and samples (frozen and/or fixed) or return all material to sender. In latter case, sign and date the chain of custody.
- Collate and analyse findings.
- Write and issue Report (and Statement, if requested).
- Retain all remaining material for at least 5 years, or return all to client for storage.

Some Important Equipment for This Work

- *Post-mortem* and laboratory kits, including hand-lens/dissecting microscope/loupe and appropriate instrument, slides and cover slips
- Protective clothing
- Formalin (5% or 10%)
- Methanol/ethanol (70%)
- Specimen containers, including aluminium foil or paper bags for samples for toxicology
- Camera
- Tape-recorder
- Report form and pen/pencil

(d) Forensic Investigation of the Skeleton

John E. Cooper

Introduction

The skeleton can yield important information but is often neglected in routine diagnostic, health-monitoring and forensic studies.

Radiography of skeletal material should always be carried out if feasible, both before and during osteological examination. Other imaging techniques may also be appropriate

Appendix H: Scientific Names of Species and Taxa of Animals Mentioned in Text, with Notes on Taxonomy

JOHN E. COOPER
MARGARET E. COOPER

'Nomina si nesci perit et cognito rerum'.
Unless the names are known to you, the concepts will be hazy too.

Linnaeus

The Importance of Classification

Knowing the taxonomic status of an animal or animals is a prerequisite in any wildlife forensic work. It can then lead to the collation of information on the biology and natural history of the species – so often important when presenting a report or giving an opinion (see Chapter 4).

How Are Animals Classified?

As explained in Chapter 3, rules and guidance for the naming of animals are issued by the International Commission on Zoological Nomenclature (ICZN) (Vane-Wright, 2011). The binomial system (Table H.1) now used throughout the world for living and extinct animals and plants was devised by the Swedish biologist Carl von Linné (Carolus Linnaeus) in the eighteenth century.

Criteria used in classification include

- Appearance
- Behaviour
- Distribution
- Ability to cross breed
- Morphology – gross anatomy, microscopical anatomy
- Physiological parameters, e.g. blood, proteins
- Presence and identity of macro- and micro-parasites
- Chromosomes
- DNA

The Importance of Nomenclature

The binomial/scientific name is

- Internationally recognised
- Essential for science
- Now commonly used in legislation
- Invaluable in forensics/court

Table H.1 Examples of Binomial Classification of Animals

	European (Eurasian) Badger	Lanner Falcon	Honey Bee
Kingdom	Animalia	Animalia	Animalia
Phylum	Chordata	Chordata	Arthropoda
Class	Mammalia	Aves	Insecta
Order	Carnivora	Falconiformes	Hymenoptera
Family	Mustelidae	Falconidae	Apidae
Genus	*Meles*	*Falco*	*Apis*
Species	*Meles meles*	*Falco biarmicus*	*Apis mellifera*

However, names used in legislation may be one or more of the following

- English/national language
- Local/vernacular/patois
- Scientific (usually Latin/Greek)

Experts should do their best to familiarise themselves with the correct names of the species with which they are dealing. Some homework may be necessary. There is also merit in being familiar with previous, now superseded, scientific names. This is for two reasons (1) because older names may be used in the court, especially if reference is being made to books and earlier manuscripts, and (2) in cases involving material collected in the past – blown birds' eggs and taxidermy specimens, for example – any labels or descriptions of the specimen in catalogues may well refer to the animal by its former name.

List of Names of Animals in Text

The English names below are generally given as they appear in the text, with additions or modifications where this may assist the reader. A few scientific names are to be found in the relevant parts of the text. The names of parasites, such as *Sarcoptes scabiei*, are not included here.

Mammals

Aardwolf	*Proteles cristatus*
African buffalo	*Syncerus caffer*
African elephant	*Loxodonta africana*
African wild dog	*Lycaon pictus*
Alpine chamois	*Rupicapra rupicapra*
American kestrel	*Falco sparverius*
American mink	*Mustela vison*
Antelopes	Bovidae (various subfamilies)
Agouti	*Dasyprocta* sp.
Arctic fox	*Vulpes lagopus*
Asiatic black bear	*Ursus thibetanus*

Appendix H

Ass	*Equus asinus*
Baboon	*Papio* sp.
Badger	*Meles meles*
Bank vole	*Myodes* (formerly *Clethrionomys*) *glareolus*
Bears	Ursidae
Beluga whale	*Delphinapterus leucas*
Bighorn sheep	*Ovis canadensis*
Black rhinoceros	*Diceros bicornis*
Bottlenose dolphin	*Tursiops aduncus*
Brushtail possum	*Trichosurus vulpecula*
Buffalo, African	*Syncerus caffer*
California sea lion	*Zalophalus californianus*
Camel	*Camelus* sp.
Camel, Arabian	*Camelus dromedarius*
Cane rat	*Thryonomys* sp.
Caracal	*Felis caracal*
Cheetah	*Acinonyx jubatus*
Chimpanzee	*Pan troglodytes*
Common pipistrelle	*Pipistrellus pipistrellus*
Cougar (puma)	*Felis concolor*
Coyote	*Canis latrans*
Deer, red	*Cervus elaphus*
Degu	*Octodon degus*
Dolphin, common	*Delphinus delphis*
Domestic cat	*Felis catus*
Domestic dog	*Canis familiaris*
Domestic goat	*Capra hircus*
Domestic ox	*Bos taurus/indicus*
Domestic pig	*Sus scrofa*
Domestic sheep	*Ovis aries*
Donkey	*Equus asinus*
Eastern red bat	*Lasiurus borealis*
Elephants	*Loxodonta africana*
Eurasian badger	*Meles meles*
European goldfinch	*Carduelis carduelis*
European hedgehog	*Erinaceus europaeus*
Fallow deer	*Dama dama*
Ferret	*Mustela putorius furo*
Florida manatee	*Trichechus manatus latirostris*
Fox	*Vulpes vulpes*
Fruit bat/flying fox	*Pteropus* sp.
Fur seal, Antarctic	*Arctocephalus gazella*
Gazelles	Bovidae (various subfamilies)
Giant rat	*Cricetomys gambianus*
Genet	*Genetta* sp.
Giraffe	*Giraffa camelopardalis*
Golden lion tamarin	*Leontopithecus rosalia*

Gorilla, lowland	*Gorilla gorilla*
Gorilla, mountain	*Gorilla berengei*
Grey wolf	*Canis lupus*
Green (vervet) monkey	*Chlorocebus pygerythrus*
Green sea turtle	*Chelone mydas*
Guanaco	*Lama guanicoe*
Gueron	*Cercopithecus* sp.
Guinea pig	*Cavia porcellus*
Hamster, common	*Cricetus cricetus*
Hamster, golden	*Mesocricetus auratus*
Harbour porpoise	*Phocoena phocoena*
Hare	*Lepus capensis*
Hedgehog, European	*Erinaceus europaeus*
Hippopotamus	*Hippopotamus amphibius*
Howler monkey	*Alouatta* sp.
Humpback whale	*Megaptera novaeangliae*
Iberian lynx	*Lynx pardinus*
Impala	*Aepyceras melampus*
Indian mongoose	*Herpestes auropunctatus*
Indian Ocean bottlenose dolphin	*Tursiops aduncus*
Jaguar	*Panthera onca*
Japanese serow	*Capricornis crispus*
Javan rhinoceros	*Rhinoceros sondaicus annamiticus*
Kangaroos	Diprotodontia, Macropodidae
Koala	*Phascolarctos cinereus*
Leopard	*Panthera pardus*
Lion	*Panthera leo*
Little red flying-fox	*Pteropus scapulatus*
Macaque	*Macaca* sp.
Mangabey	*Cercocebus* sp.
Marmoset	*Callithrix* sp.
Miss Waldron's red colobus	*Procolobus badius walroni*
Mole	*Talpa europaea*
Mongooses	Herpestidae
Moose	*Alces alces*
Mouse, house	*Mus musculus*
Musk ox	*Ovibos moschatus*
New Zealand fur seal	*Arctocephalus forsteri*
Nile crocodile	*Crocodylus niloticus*
North Atlantic right whale	*Eubalaena glacialis*
Noctule bat	*Nyctalus noctula*
Opossum	*Didelphis marsupialis*
Orang-utan	*Pongo pygmaeus/abelii*
Peccary	*Tayassu* sp.
Polar bear	*Thalarctos maritimus*
Porcupine, North American	*Erithizon dorsatum*
Raccoon	*Procyon* sp.

Rat, black or ship	*Rattus norvegicus*
Red deer	*Cervus elaphus*
Red fox	*Vulpes vulpes*
Red grouse	*Lagopus lagopus*
Reindeer	*Rangifer tarandus platyrhynchus*
Rhesus monkey	*Macaca mulatta*
Rhinoceros	Rhinocerotidae
Rhinoceros, black	*Diceros bicornis*
Rhinoceros, Javan	*Rhinoceros sondaicus annamiticus*
Roe deer	*Capreolus capreolus*
Sea otters	*Enhydra lutris nereis*
Seal, common	*Phoca vitulina*
Seal lions	Otariidae
Serotine bat	*Eptesicus serotinus*
Short-tailed field vole	*Microtus agrestis*
Silver-haired bat	*Lasionycteris noctivagans*
Snub-nosed monkey	*Rhinopithecus* sp.
Spotted hyaena	*Crocuta crocuta*
Squirrels	Sciuridae
Stoat	*Mustela erminea*
Tibetan antelope (chiru)	*Pantholops hodgsoni*
Tiger	*Panthera tigris*
Vampire bat	*Desmodus* sp.
Wallaby	Macropodidae
Waterbuck	*Kobus defassa*
Warthog	*Phacochoerus africanus*
Whitetail deer	*Odocoileus virginianus*
Wild boar	*Sus Scrofa*
Wolf	*Canis lupus*
Wolverine	*Gulo gulo*

Birds

African marsh harrier	*Circus ranivorus*
Australian white ibis	*Threskiornis molucca*
Bald eagle	*Haliaeetus leucocephalus*
Barn owl	*Tyto alba*
Bean goose	*Anser fabalis*
Bobwhite quail	*Colinus virginianus*
Brown teal	*Anas chlorotis*
Budgerigar	*Melopsittacus undulatus*
Canada goose	*Branta canadensis*
Canary	*Serinus canaria*
Carrion crow	*Corvus corone*
Chachalacas	*Ortalis* sp.
Cockatoo, sulphur-crested	*Cacatua galerita*
Common quail	*Coturnix coturnix*

Common tern	*Sternus hirundo*
Cormorant	*Phalacrocorax carbo*
Crowned hawk-eagle	*Stephanoaetus coronatus*
Dark-eyed junco	*Junco hyemalis*
Domestic duck	*Anas platyrhynchos*
Domestic fowl	*Gallus domesticus (gallus)*
Domestic turkey	*Meleagris gallopavo*
Dunlin	*Calidris alpina*
Eagle, golden	*Aquila chrysaetos*
Egyptian goose	*Alopochen aegyptiacus*
Egyptian vulture	*Neophron percnopterus*
European eagle-owl	*Bubo bubo*
European starling	*Sturnus vulgaris*
Flamingo, lesser	*Phoeniconaias minor*
Forest owlet	*Heteroglaux blewitti*
Gannet	*Morus* sp.
Goldfinch	*Carduelis carduelis*
Goosander	*Mergus merganser*
Greater white-fronted goose	*Anser albifrons*
Great-horned owl	*Bubo virginianus*
Greylag goose	*Anser anser*
Guillemot	*Uria aalge*
Guineafowl	*Numida meleagris*
Gulls	Laridae
Gyr falcon	*Falco rusticolus*
Harrier	*Circus* sp.
Hartlaub's duck	*Cairina hartlaubi*
Herons	Ardeidae
House sparrow	*Passer domesticus*
Kereru (New Zealand pigeon)	*Hemiphaga novaeseelandiae*
Kestrel, European	*Falco tinnunculus*
Kiwi	*Apteryx* sp.
Koklass pheasant	*Pucrasia macrolopha*
Lesser kestrel	*Falco naumanni*
Lesser snow goose	*Chen caerulescens*
Little owl	*Athene noctua*
Macaw	*Ara* sp.
Mallard duck	*Anas platyrhynchos*
Marabou stork	*Leptoptilos crumeniferus*
Merlin	*Falco columbarius*
Monk parakeet	*Myiopsitta monachus*
Muscovy duck	*Cairina moschata*
Nazca booby	*Sula granti*
North Island weka	*Gallirallus australis*
Parrot and allies	Psittaciformes
Pateke (brown teal)	*Anas chlorotis*
Peafowl (peacock)	*Pavo* sp.

Pelican, brown	*Pelecanus occidentalis*
Peregrine falcon	*Falco peregrinus*
Philippine eagle	*Pithecophaga jefferyi*
Mauritius kestrel	*Falco punctatus*
Pigeon, feral	*Columba livia*
Pink pigeon	*Columba mayeri*
Red-billed quelea	*Quelea quelea*
Red grouse	*Lagopus lagopus/Lagopus scoticus*
Red kite	*Milvus milvus*
Redpoll	*Carduelis* sp.
Ring-billed gull	*Larus delawarensis*
Ring-necked pheasant	*Phasianus colchicus*
Rodrigues fody	*Foudia flavicans*
Ruffed grouse	*Bonasa umbellus*
Sacred ibis	*Threskiornis aethiopicus*
Shrikes	Laniidae
Sparrow-hawk	*Accipiter nisus*
Starling, common	*Sturnus vulgaris*
Trinidad piping-guan	*Pipile pipile*
Waterfowl	Anseriformes

Reptiles

African rock python	*Python sebae*
American alligator	*Alligator mississippiensis*
Boa constrictor	*Boa constrictor*
Burmese python	*Python molurus bivittatus*
Flap-necked chameleon	*Chamaeleo dilepis*
Freshwater crocodile	*Crocodylus johnstoni*
Garter snake	*Thamnophis* sp.
Green iguana	*Iguana iguana*
Green turtle	*Chelonia mydas*
Hawksbill turtle	*Eretmochelys imbricata*
Nile crocodile	*Crocodylus niloticus*
Rattlesnake	*Crotalus* sp.
Round island skink	*Leiolopisma telfairii*
St Lucia whiptail lizard	*Cnemidophorus vanzoi*
Tortoise, leopard	*Geochelone pardalis*
Tortoises, Mediterranean	*Testudo* sp.

Amphibians

Common toad	*Bufo bufo*
Mantella frog	*Mantella sp.*
Marine (cane) toad	*Bufo marinus*
Toad, African clawed	*Xenopus* sp.

Fish

Anchovy	Engraulidae
Brotulids	*Bassogigas* sp.
Carp	*Cyprinus carpio*
Carpetshark, cobbler	*Sutorectus tentaculatus*
Koi	*Cyprinus carpio*
Halibut	*Hippoglossus* sp./*Paralichthys* sp.
Puffer	*Diodin* sp. (and other genera)
Rock fish	Sebastidae and other species
Salmon	*Salmo salar*
Seahorse	*Hippocampus* sp.
Shark	Selachii
Sturgeon	Acipenseridae
Tench	*Tinca tinca*
Trout	*Salmo trutta*
Whiting	*Merlangius* sp. and other species

Invertebrates (for other species, see text)

Bee, honey	*Apis mellifera*
Birdwing butterflies	*Ornithoptera* sp.
Blowflies	*Calliphora* sp.
Carrion beetles	*Silpha* sp.
Cinnabar moth	*Callimorpha jacobaeae*
Freshwater pearl mussel	*Margaritifera margaritifera*
Leeches	Hirudinea
Mole cricket	Gryllotalpidae
Scorpions	Scorpione, Arachnida
Silver-spotted skipper butterfly	*Hesperia comma*
Stick insects	Phasmida
Ticks	Acarina
Tsetse fly	*Glossina* sp.

'What's the use of their having names,' the Gnat said, 'if they won't answer to them?'
'No use to them,' said Alice, 'but it's useful to the people that name them, I suppose'.

Lewis Carroll, Alice Through the Looking-Glass

Appendix I: Legal Aspects of Sample Movement in Wildlife Forensic Cases

MARGARET E. COOPER

Introduction

The collection and examination of samples is an important part of forensic work involving free-living wildlife and captive (non-domesticated species). Such samples range from blood smears and microbiological swabs to bones, skins and stomach contents. They are likely to be taken for the following purposes:

- Diagnosis or investigation of disease
- Health-monitoring
- DNA studies
- Forensic investigations

Samples for examination may need to be moved within the country of origin or sent abroad. Laws may apply to both of these circumstances and fall into the following three categories:

1. Animal health legislation
2. Conservation legislation
3. Carriage of samples

The necessary documentation must be obtained before shipment and should accompany the samples. Personal importations should be declared at the border and documents and samples should be available for inspection. Usually, a copy of the documents is retained by the border authorities.

Animal Health Legislation

Animal health legislation provides powers to manage outbreaks of disease in animals and to prevent the importation of animals and animal products that may introduce infection. Most countries in the world have legal provisions or administrative procedures that protect animal health because major outbreaks put humans or livestock at risk, attract international attention and demand for action, as in the case of avian influenza. The OIE (World Organisation for Animal Health http://www.oie.int/about-us/) (see Appendix E) manages the world animal health information system, based on reports of the occurrence of 'notifiable' diseases (see also Chapter 9) and the harmonisation of animal health measures and standards. The European Union has extensive legislation on animal health that is implemented by its member countries. See the DEFRA Animal Health website: http://animalhealth.defra.gov.uk/imports-exports/index.htm

Samples may be covered by legislation relating to pathological material that applies comparable restrictions to those for live animals. This may depend upon the nature of the sample, the species from which it was taken and the disease status of the countries concerned. Conditions may be imposed on the final destination of the samples (usually a laboratory of given standards, depending upon the risk of infection), the packaging, transportation and storage, and final disposal of the sample and its packing material.

Conservation Legislation

International

As explained in Chapter 3, the CITES Convention (http://www.cites.org/) lays down rules for the international movement (whether for trade or not) of the species listed in Annexes I, II and III thereto. Permits are issued by the Management Authority of each country. There is a special facility for the movement of biological samples in CITES Resolution Conf. 12.3 (Rev. CoP15), Section XII and Annex 4 (http://www.cites.org/eng/res/all/12/E12-03R15.pdf) and Article 18 and Annex VII (see Table I.1) of the EU Commission Regulation (EC) No 865/2006, http://eur-lex.europa.eu/LexUriServ/LexUriServ.do?uri=OJ:L:2006:166:0001:0069:EN:PDF).

The special procedure laid down in Article 18 is intended for samples that require urgent attention. It is particularly useful when there is a large quantity of samples or the exact numbers cannot be ascertained at the time of applying for the permit or there are regular movements of samples over a long period. Management Authorities may provide pre-issued and partially completed permits to people and institutions that are approved by, and registered with, the Authority. The Scientific Authority must be satisfied that such multiple shipments of samples are not detrimental to conservation status of the species concerned.

National

Conservation laws vary considerably regarding in-country movement of samples. In federal countries, there may be movement controls between states. One would not expect there to be conservation law constraints on the internal movement of samples. However, in the first place samples must be obtained legally. Authorisation is likely to be required for the capture, to obtain samples, of a live animal of a protected species and the possession of a carcase or the derivatives of a protected species may require a permit. For example, in England these are usually issued by Natural England (http://www.naturalengland.org.uk/ourwork/regulation/wildlife/species/default.aspx). Access to the land where the samples are to be taken should be authorised by the owner or occupier.

If the animal concerned is alive, it may be necessary for a veterinary surgeon to take the sample and to attend to any clinical or welfare needs. If the animal is captive and owned by someone, the veterinarian may be required to verify the condition of the animal before and after the sample (see Chapter 8). If a forensic investigation constitutes wildlife research, a permit (with conditions and fees) may be required (see Chapter 2). In the United Kingdom, if the unlikely situation were to arise in which it was necessary to evaluate

Appendix I

Table I.1 Types of Biological Samples Referred to in Article 18 and Their Use

Type of Sample	Typical Size of Sample	Use of Sample
Blood, liquid	Drops or 5 mL of whole blood in a tube with anticoagulant; may deteriorate in 36 h	Haematology and standard biochemical tests to diagnose disease; taxonomic research; biomedical research
Blood, dry (smear)	A drop of blood spread on a microscope slide, usually fixed with chemical fixative	Blood counts and screening for disease parasites
Blood, clotted (serum)	5 mL of blood in tube with or without a blood clot	Serology and detection of antibodies for evidence of disease; biomedical research
Tissues, fixed	5 mm^3 pieces of tissues in a fixative	Histology and electron-microscopy to detect signs of disease; taxonomic research; biomedical research
Tissues, fresh (excluding ova, sperm and embryos)	5 mm^3 pieces of tissues, sometimes frozen	Microbiology and toxicology to detect organisms and poisons; taxonomic research; biomedical research
Swabs	Tiny pieces of tissue in a tube on a swab	Growing bacteria, fungi, etc. to diagnose disease
Hair, skin, feathers, scales	Small, sometimes tiny pieces of skin surface in a tube (up to 10 mL in volume) with or without fixative	Genetic and forensic tests and detection of parasites and pathogens, and other test
Cell lines and tissue cultures	No limitation of sample size	Cell lines are artificial products cultured either as primary or continuous cell lines that are used extensively in testing the production of vaccines or other medical products and taxonomic research (e.g. chromosome studies and extraction of DNA)
DNA	Small amounts of blood (up to 5 mL), hair, feather follicle, muscle and organ tissue (liver, heart, etc.), purified DNA, etc.	Sex determination; identification; forensic investigations; taxonomic research; biomedical research
Secretions (saliva, venom, milk)	1–5 mL in vials	Phylogenetic research, production of anti-venom, biomedical research

Source: Annex VII EU Regulation (EC) No. 865/2006. © European Union, http://eur-lex.europa.eu/
Note: Only European Union legislation printed in the paper edition of the *Official Journal of the European Union* is deemed authentic.

forensic evidence by carrying out research in live animals that was likely to cause pain or suffering in those animals, there is provision for such 'forensic enquiries' to be carried out within the framework of the Animals (Scientific Procedures) Act 1986 (see also Chapter 2).

Carriage of Samples

Samples that are consigned to a land, sea or air carrier should be packed according to the rules of such companies. These are based on the United Nations classification of infectious substances and international packing standards for the transport of dangerous goods.

The rules are designed to prevent breakage, leakage or harm to handlers Most samples taken during wildlife forensic investigations are likely to be designated as 'Biological Substance, Category B – UN3373' and packed according to the specifications in Packing Instruction 650 (see for example, http://www.ed.ac.uk/schools-departments/health-safety/biosafety/policy/guidance-rules/transport).

Substances that may carry pathogens likely to cause severe infections in humans, emerging diseases and some animal diseases are classed as Category A biological substances and are assigned to UN 2814 and UN 2900, respectively. They must be packed according to Packing Instruction 620 (http://www3.imperial.ac.uk/safety/subjects/dangerousgoods/infectioustransport). A thorough overview of the requirements is provided by Imperial College, London on http://www3.imperial.ac.uk/safety/subjects/dangerousgoods/diagnosticspecimens.

Conclusion

As with the shipment of live animals, the documentation and procedures for moving the parts and derivatives of protected species from one country to another are complicated and time-consuming. It requires an assessment of the legal aspects, planning the necessary procedures, meticulous attention to detail, patience with bureaucracy and above all plenty of time.

Appendix J: Information and Intelligence Gathering in Wildlife Crime Investigation

NEVIN HUNTER

Introduction

Over the past decade law enforcement agencies across the United Kingdom have professionalised their use of intelligence, producing a significant impact upon crime and criminal behaviour. Intelligence has been used to make risk-based decisions focussing effort upon areas of most need to achieve results. This professional approach has been adopted to tackle wildlife crime, with not only law enforcement agencies but also non-statutory agencies now adopting a similar approach. The outcome has been significant, showing that co-ordination, co-operation and communication has been effective.

At the heart of the UK approach to tackling wildlife crime has been the development of the National Wildlife Crime Unit. It has adopted a professional intelligence approach, utilising the National Intelligence Model (NIM), to dovetail its work with law enforcement agencies. As a result, in addition to being the driving force against wildlife crime, the Unit has helped to have an impact upon other areas of criminality.

In this chapter the background to the approach taken in the United Kingdom is outlined. An overview of the NIM is provided and then its application to wildlife crime is considered.

Those involved in the fight against wildlife crime know that it is never-ending. Many see wildlife as a commodity to be acquired and to be used for profit. This approach means that, when law enforcement tackles one commodity issue, the criminal is likely to move into another. Therefore an effort is required to identify these changes in behaviour and to take early steps to address it. At the core of the UK intelligence-led approach (using the NIM) are four key aspects – prevention, intelligence, enforcement and reassurance. The last is particularly important as it provides the bedrock upon which communities (the general public) will provide additional information when criminal patterns change, thereby enabling law enforcement and other interested agencies to address this proactively.

Background

The phrase 'wildlife crime' is loosely defined (see also Chapter 2). In the United Kingdom a partnership exists between statutory and non-statutory agencies with an interest in addressing wildlife crime under the umbrella of PAW – the Partnership for Action Against Wildlife Crime (see Chapter 3 and Appendix E). PAW defines wildlife crime as follows:

> People buying, selling, harming or disturbing wild animals or plants that are protected by law or being cruel to animals

PAW cites examples, including

- Smuggling protected species and their parts (such as tortoises, ramin, ivory and caviar)
- Illegally trading in endangered species
- Poisoning of animals, such as birds of prey
- Disturbing or killing wild birds; or taking their eggs
- Poaching of game, venison and fish
- Disturbing, injuring or killing bats and damaging or obstructing their roosts
- Taking protected plants from the countryside
- The illegal use of poisons, snares or explosives to kill or injure animals
- Violence towards badgers, which includes their being buried alive or being ripped apart by dogs

The partnership has been in place since 1995 and has helped to galvanise a coordinated approach to tackling these issues across the whole of the United Kingdom.

It is all too easy to get hung-up on the term 'wildlife crime'. Remove the word wildlife and we are left with crime. Tackling crime is the primary responsibility of law enforcement agencies, but they need the support of the communities they serve in order to do effectively. The public are the eyes and ears that provide information that can be assessed as intelligence which is the lifeblood of modern enforcement. Why is this so? Simply put, it is essential that the limited resources available in any society to address criminality are used effectively by identifying priorities and focussing efforts. Intelligence provides this.

Wildlife crime investigation has a patchy history in the United Kingdom. It was not until the mid-1990s that issues relating to domestic wildlife legislation were acknowledged to be the responsibility of police forces. Import and export offences had been dealt with by customs authorities before this, but there was little, if any, co-ordination between any law enforcement agencies nor acknowledgement that wildlife crime was something that needed to be addressed. With the formation of PAW, a Police Wildlife Liaison Officer Network followed, whereby police officers across the United Kingdom took on voluntary enforcement roles over and above their normal day-to-day duties. This led to the first effort at a coordinated approach to wildlife crime.

In 2000 the UK Police Service first piloted an intelligence-led approach to deal with all crime and this led to the creation of the National Crime Intelligence Service (NCIS). Later the use of the National Intelligence Model was enshrined in legislation via the Police Reform Act 2002. This required all police forces across England and Wales to adopt the model by April 2004, implementing it to common minimum standards. Its application has been backed up by detailed comprehensive guidance documents. Applying the model has been shown to have a significant and positive impact upon every level of criminality, ranging from low-level non-compliance through to serious and organised crime.

In 2005 the National Wildlife Crime Intelligence Unit (NWCIU) was formed as part of the NCIS. In 2006 a stand-alone National Wildlife Crime Unit (NWCU) was created when NCIS was disbanded. The Unit is now based at a police station near Edinburgh, Scotland.

The NWCU has a primary role to assist in

- The prevention and detection of wildlife crime
- Obtaining and disseminating intelligence from a wide range of organisations

Appendix K: Examination of a Dead Javan Rhinoceros, Cat Tien National Park, Vietnam

The authors, editors, and WWF take no responsibility for any misrepresentation of material that may results form the translation of this document into any other language.

AUTHORS

Dr Ulrike Streicher, Danang, Vietnam
Special Agent Ed Newcomer, US Fish and Wildlife Service
Douglas McCarty, Freeland, Thailand
Sarah Brook, WWF Vietnam Programme
Bach Thanh Hai, Cat Tien National Park

ACKNOWLEDGEMENTS

The authors would like to thank Professor John Cooper - Department of Veterinary Medicine, University of Cambridge, UK, for leading the examination and providing veterinary forensic pathology advice during construction of the report. We would like to express our gratitude to veterinary and other colleagues in the UK, USA, Europe, Thailand and Vietnam for their support, including but not limited to: US Fish and Wildlife Service, WWF US, WWF Switzerland, WWF Austria, WWF France, WWF International, Critical Ecosystem Partnership Fund and BirdLife International in Indochina, William Schaedla - TRAFFIC SE Asia, Leanne Clark - WCS, Daniela Schrudde – Cat Ba Langur Conservation Project, Milly Farrell and Martyn Cooke - Royal College of Surgeons of England, and Richard Sabine and Roberto Portela Miguez - Natural History Museum, London.

The authors would like to commend CTNP for their diligence in locating and recovering the dead Javan Rhinoceros, and for their support and assistance with this examination. We are thankful to Lam Dong Police, Tan Phu Police and CTNP, for the opportunity to perform this examination and hope that the information obtained will contribute to the police investigation.

Appendix K

CONTENTS

1. INTRODUCTION	5*
1.1 Javan Rhinoceros status	5
1.2 Dead Javan Rhinoceros recovered	5
1.3 Police Investigation	6
2. METHODS	7
2.1 Preparation	7
2.2 Pathological examination	7
2.3 Site visit	8
2.4 Presentation of preliminary findings	8
2.5 Follow-up tasks	8
3. RESULTS	9
3.1 Pathological examination	9
3.1.1 Summary	9
3.1.2 Radius/ulna, metacarpals, carpals and hooves of left forelimb	10
3.1.3 Skull and horn	12
3.1.4 Right rib	12
3.1.5 Discoloration and damage	13
3.1.6 CT results	13
3.2 Site analysis	14
3.3 Ballistics evaluation	17
4. DISCUSSION	19
4.1 Pathology	19
4.1.1 Time of death (PMI)	19
4.1.2 Javan Rhinoceros survey	20
4.2 Likely effect of pathological changes	20
4.3 Cause of death	21
4.3.1 Pathology	21
4.3.2 Site analysis	22
4.4 Ballistics evaluation	23
5. CONCLUSIONS	23
6. REFERENCES AND FURTHER READING	25
7. APPENDICES	27
Appendix 1. Letter from WWF to Lam Dong Police investigation team	27
Appendix 2. Letter from CTNP requesting assistance from WWF	29
Appendix 3. Reference notes – forensic investigation of the skeleton	30

* These page numbers reflect the pagination of the original World Wildlife Fund publication and not the pagination of this appendix.

Appendix 4. List of Javan rhinoceros bones present and missing, CTNP   32
Appendix 5. Lam Dong Police ballistics report, Vietnamese   33
Appendix 6. Lam Dong Police ballistics report, unofficial English translation   35
Appendix 7. Advice on Preserving Bones of Javan Rhinoceros – John E Cooper   37

EXECUTIVE SUMMARY

In April 2010, a Javan rhinoceros was found dead in Cat Tien National Park (CTNP), Vietnam. WWF and Cat Tien National Park found a bullet embedded in the left forelimb (front left leg) and the horn had been removed.

Lam Dong Police asked Cat Tien National Park to assist with the police investigation by providing information related to the death of the rhinoceros. Cat Tien National Park requested WWF's assistance, who put together a team of experts to come to CTNP and conduct an examination of the skeleton and visit the site where the rhinoceros was found.

Pathological examination could not determine exactly when the rhinoceros died, but it is thought that the rhinoceros died in late January/early February 2010. WWF and CTNP conducted a survey of Javan rhinoceros from October 2009 to April 2010 in CTNP, collecting faecal samples for DNA analysis to determine population status. Rhinoceros footprints and dung were regularly encountered preceding Tet (Lunar New Year), but no fresh footprints or dung were located after Tet, from February to April. It is therefore suspected that the dead rhinoceros might have been the last Javan rhinoceros in Vietnam, but the DNA analysis of faecal and tissue samples that are currently being conducted will provide final confirmation on the population status.

Pathological examination of the bones revealed that the rhinoceros did not die of old age; it was estimated to be 15-25 years old, which is adult, but not elderly. Javan rhinoceros are estimated to live to up to 40 years in the wild (van Strien et al 2008).

The examination revealed that the horn was removed after the rhinoceros died. The bullet found lodged in the front left forelimb caused considerable injury and inflammation in the leg which would have resulted in impaired locomotion for the rhinoceros. The alteration of the bones suggests that at least 2 months passed between the time the rhinoceros was shot and the time of death.

Appendix K

A direct cause of death could not be found due to the absence of soft tissues. However, the injuries the bullet caused would have predisposed the rhinoceros to other hazardous factors.

It is strongly suspected that the bullet wound to the rhinoceros's leg contributed to its death. The authors of this report suggest that the final cause of death was most likely to be either: i) bacterial infection, probably chronic, arising as a result of the bullet wound, or ii) traumatic event, such as a fall, caused by the impaired locomotion of the rhinoceros due to the bullet wound and highly likely associated infection.

It was not possible to determine whether the person who shot the rhinoceros was the same person who removed the horn. It is highly likely that the horn has been traded and utilized for 'traditional medicine' purposes. The shooting of a Javan rhinoceros constitutes an illegal poaching event; poaching of a Javan rhinoceros, trade and consumption of rhinoceros horn are serious contraventions of Vietnamese law.

1. INTRODUCTION

1.1 Javan Rhinoceros status

The Javan Rhinoceros (*Rhinoceros sondaicus*) is Critically Endangered (van Strien et al, 2008) with only two small populations remaining, representing two distinct subspecies, one in Vietnam (*Rhinoceros sondaicus annamiticus*) and one in Indonesia (*Rhinoceros sondaicus sondaicus*). Cat Tien National Park (CTNP) was discovered to hold a small population of *R. sondaicus annamiticus* in 1988, after a rhinoceros was poached and body parts were found for sale at a local market place (Schaller et al, 1990).

WWF Vietnam has been working with Cat Tien National Park to conserve this globally important population since the 1990's. Although several surveys have been undertaken previously, the exact status of the population is currently not known, estimated at a maximum of 10-15 individuals in 1990 (Schaller et al, 1990), and 5-8 individuals by later surveys (Polet et al, 1999). The first comprehensive survey of Javan rhinoceros in Cat Tien National Park has recently been completed by WWF and CTNP. Faecal samples have been sent for DNA analysis to Queen's University, Canada, which should determine the true status of the population (number and sex of individuals). This information is crucial to develop conservation plans. Results of the DNA analysis are expected in December 2010.

The recent death of a Javan Rhinoceros in Cat Tien National Park is of great concern to WWF and to CTNP. It represents a considerable loss to the global Javan rhinoceros population. Furthermore, this rhinoceros could have been the last remaining individual in Vietnam, representing the global extinction of the Vietnamese subspecies.

1.2 Dead Javan Rhinoceros recovered

On April 29th 2010, local people informed Gia Vien ranger station of the whereabouts of the skeleton of a large animal. Cat Tien National Park dispatched a team of staff from the Science and International Division, District Forest Protection Department and Gia Vien ranger station, to the location to secure the skeleton and to take photographs of the skeleton and the site. The rangers remained at the site overnight. A second patrol team from CTNP headquarters, including one WWF staff member (Luong Viet Hung) and the Vice Director of CTNP Technical Department, Bach Thanh Hai, went to the site on April

Appendix K

30th. The team excavated most of the skeleton which was removed from the site to Gia Vien ranger station and then on to CTNP headquarters. Cat Tien National Park staff confirmed that the skeleton was that of a Javan Rhinoceros.

Sarah Brook (Species Coordinator) and Ruth Mathews (Programme Manager) from WWF Vietnam went to CTNP from 4th-7th May 2010, following receipt of the news of the dead rhinoceros. During their time there, WWF and CTNP laid out all of the bones from the rhinoceros skeleton to take photographs and to remove samples of tissue and teeth. The teeth and tissue samples have been sent by WWF, to the institution that is conducting the Javan Rhinoceros faecal DNA analysis (Queen's University, Canada). On 7th May, Sarah and Hung discovered a bullet embedded in the left forelimb bone of the rhinoceros. Sarah immediately informed Tran Van Thanh, Director of CTNP. Thanh notified the Police and requested their presence at CTNP that same afternoon. The bullet was removed by the police on 7th May 2010 in order to conduct ballistics analysis.

1.3 Police investigation

Cat Tien National Park was asked by Lam Dong Police to assist with the investigation into the cause of death of the Javan Rhinoceros. Cat Tien National Park requested WWF's assistance in providing the following information to support the police investigation (Appendix 2):

i) Scientific name
ii) Sex (male or female)
iii) Age
iv) Time of death
v) Cause of death
vi) Injuries caused by the bullet found in the leg bone

Consequently, WWF identified a team of international experts to come to CTNP to attempt to provide this information. This team consisted of:

- Veterinary Pathologist and forensic expert, *Professor John E Cooper*, University of Cambridge, UK
- Wildlife Veterinarian, *Dr Med. Vet. Ulrike Streicher*, Vietnam

- US Fish and Wildlife Service *Special Agent Ed Newcomer*, detailed to ASEAN-WEN, Thailand
- ASEAN-WEN Support Programme, FREELAND Operations Officer, *Douglas McCarty*, Thailand

The trip to CTNP was funded by WWF, organised and coordinated by *Sarah Brook, Species Coordinator*, WWF Vietnam. *Bach Thanh Hai, Vice Director of the Technical Department*, Cat Tien National Park, provided logistical and technical support.

2. METHODS

2.1 Preparation

Prior to the trip to CTNP, John Cooper conducted a literature search on the biology of Javan rhinoceros (*Rhinoceros sondaicus*), obtained a number of relevant scientific papers on Javan rhinoceros; and visited two museums in London, England, that hold specimens of *R. sondaicus*: these are the Royal College of Surgeons of England and the Natural History Museum. The prime purpose of these activities was to familiarise him with the anatomical features of the Javan rhinoceros.

2.2 Pathological examination

The pathological examination was led by John Cooper, who followed standard procedures for this sort of investigation (Cooper and Cooper, 2007), (Appendix 3). He was assisted by Ulrike Streicher. On 7th September, the team arranged the rhinoceros skeleton and associated tissues on the floor of the Museum in CTNP, to identify which bones and tissues were present and to try to determine which were missing; all bones/tissues were photographed and checked for abnormalities.

Any bones and other tissues that either showed clear pathological changes or warranted more detailed examination for other reasons, were identified, marked and weighed.

Following this, on 9th September, the team conducted an in-depth investigation of bones and hooves that were considered of particular relevance, together with microscopical examination of material from bones, and tests on soil and stones collected from the site where the rhinoceros had been found in April 2010.

Appendix K

2.3 Site visit

On 8th September, the team went to the location where the rhinoceros skeleton was found (0754889, 1289524), accompanied by rangers from the Gia Vien ranger station and Bach Thanh Hai, who had been present for the excavation of the bones in April 2010.

Ed Newcomer led the analysis of the site, with assistance from Doug McCarty. The team looked for any signs in the immediate environment that could help to determine the circumstances in which the rhinoceros died and how people might have gained access to the site. The rangers and Bach Thanh Hai provided more information on how the skeleton was found and excavated in April 2010. During this visit on 8th September a minor amount of excavation was conducted in the immediate location where the skeleton was found, to look for any remaining bones and intestinal contents.

2.4 Presentation of preliminary findings

On 9th September, CTNP invited representatives from Lam Dong and Tan Phu Police, along with CTNP staff and two journalists from Ho Chi Minh City, to attend a presentation of preliminary findings by the investigation team. Ho Thi Thanh Thuy of WWF Vietnam provided translation.

Lam Dong Police brought the bullet that had been extracted from the front limb bone of the rhinoceros and a copy of a police report on the ballistics analysis. The ballistics report has since been translated into English (see Appendix 5 for Vietnamese version, Appendix 6 for English version).

2.5 Follow-up tasks

The in-situ investigation was concluded on 9th September; the team returned to their respective countries on the following day and subsequently pursued queries that had arisen and other matters that required further attention.

Ulrike Streicher took the two front limbs (ulna/radius) and right rib to HCM city on 22nd September for CT scanning. This yielded more information on the internal bone structure and allowed better comparison between the healthy leg and the leg that had been damaged by the bullet.

Ed Newcomer analysed photographs of the bullet before and after it had been removed from the leg bone and enquired with colleagues regarding the calibre of the bullet.

John conferred with colleagues regarding bone-healing time in large mammals and pachyderms in particular, in order to try to pinpoint more accurately when the rhinoceros was shot relative to its time of death.

This report represents the final conclusions of the investigation team regarding the answers to the information required by Lam Dong Police, above, with the exception of the sex of the individual. The results of the DNA analysis conducted by Queen's University, Canada, (of the deceased rhinoceros teeth and tissue) will confirm the sex of the individual and the significance of the loss of this individual to the subspecies. This information is expected in December 2010.

3. RESULTS

3.1 Pathological examination

3.1.1 Summary

The remains examined consisted of bones and keratinous material (11 hooves). Virtually no soft tissues were present: a few pieces of skin, reported to have been obtained originally, had been destroyed allegedly due to smell. Later size comparisons with a skeleton kept at the Museum of the Forestry and Inventory Planning Institute (previously obtained from the Cat Tien area) confirmed that this was a Javan rhinoceros (*Rhinoceros sondaicus*).

The age of the animal was estimated on the basis of standard tests (closure of epiphyses, tooth type and wear, presence or otherwise of exostoses on the cranium (Cave, 1985)) to be adult, but not elderly – probably 15-25 years old. It may be possible to define this more accurately by a) retrospective comparison of the animal's dentition with those of specimens of *R. sondaicus* in London and Hanoi, and b) study of cementum lines and other features of the two teeth that are destined for DNA studies in Canada.

Appendix K

- Photographs of the bullet in the leg bone were shown to a WWF expert in wildlife crimes, with many years of experience managing rhino protected areas, who confirmed that the photo showed a high caliber bullet in the front left leg of the rhino. (Please read full statement below)

- Rhino horn, is a highly valued commodity in the illegal wildlife trade, the skin and faeces are also used for alleged medicinal purposes.

- Vietnamese law protects the rhino. Decree 32/2006/ND-CP prohibits hunting, catching, keeping, slaughtering, trading, or transporting rhinos, parts of rhino, or any items containing rhino.

An analysis of the photographic evidence and information regarding the recent rhino death in Vietnam was conducted by Mr. Craig Bruce, Wildlife Crimes Specialist, WWF. Based on the photographs, Mr Bruce offered the following conclusions.

1. *"The enlarged photograph and statements from those on the scene is evidence that this rhino was seriously wounded. A fairly heavy caliber bullet seems to be lodged in the bone, this could certainly not be considered a flesh wound.*

2. *I have managed large rhino populations for over a decade and been at the scene of a variety of natural and unnatural rhino deaths. Despite the age of the discovered carcass, the bullet lodged in the bone and the harvested horn are clear indicators that this rhino was poached for its horn.*

3. *In my experience opportunistically harvested horn is very rare and 98% of the time a rhino found with a horn removed has been killed by poachers for that horn. Cut marks on the skull indicate that someone was there with the right tool prepared and ready to remove a rhino horn.*

4. *It is not atypical for a rhino to be shot many times in a poaching incident, the deteriorated carcass unfortunately would not have allowed the rangers on site to confirm this but it is certainly common. The evidence we have of one wound is in all probability one of many.*

5. *It is finally therefore highly likely that this rhino was poached for its horn."*

WWF respectfully urges the Police Investigation Team to conduct a full investigation into both the shooting of the rhino and the likely illegal trading of its horn. WWF would like to express our willingness to support the investigation in any way we can which could include providing an international specialist in wildlife pathology and forensic science to assist the investigation. Determining the cause of death of this rhino and ensuring a full investigation is carried out is essential to national conservation efforts.

We hope the results of this investigation will be available to the public soon.

Yours sincerely,

Huynh Tien Dung
Chairman of Management Committee
Vietnam Country Programme - WWF Greater Mekong,

Appendix 2. Letter from CTNP requesting assistance from WWF

Date: 1st July 2010

WWF Viet Nam

On behalf of Cat Tien National Park, I would like to express our great appreciation for the support and cooperation of WWF to conservation efforts in Cat Tien National Park in general and for the Javan Rhino in particular. Regarding to the investigation of the dead rhino found in April 2010, Cat Tien National Park has been requested by Lam Dong Police to provide the following information:

1. Cause of death
2. Injuries sustained by being shot with a high caliber bullet
3. Time of death
4. Age and overall health of the individual

Since this request is beyond our ability at the moment to answer, we would like to ask for the assistance of WWF Vietnam in providing more information relating to the death of the rhino. Specifically, we would like to ask WWF to identify and support a team of international experts in wildlife crime and veterinary pathology, to determine the factors above where possible.

We would like the team of experts to come to Cat Tien National Park to conduct a full examination of the skeleton, to provide the above information and identify any other factors which may be important to this case.

Following the mission, the team should submit a full report on their methodology, findings and conclusions, to Cat Tien National Park, Lam Dong PPC and Provincial Police, MARD, Environmental Police and other relevant ministries and agencies.

Signed

Trân Văn Thành

Appendix K

Appendix 3. Reference notes – forensic investigation of the skeleton

John E Cooper, DTVM, FRCPath, DTVM, FRCPath, FSB, CBiol, FRCVS,
Diplomate, European College of Veterinary Pathologists,
RCVS Specialist in Veterinary Pathology
Department of Veterinary Medicine, University of Cambridge, UK

The skeleton is often neglected in routine diagnostic, health-monitoring and forensic studies. It can yield important information.

Radiography is valuable, before and during osteological examination.

Basic rules

1. Prepare equipment before starting to examine the bones. Check agreed protocol and explain it carefully to any assistants or new colleagues.

2. Have protective clothing available, even if not immediately needed. Remind all present of the importance of health and safety and check the Risk Assessments.

3. Handle bones carefully. They can be damaged easily. Hold them over a table top or bench, preferably with two hands. The wearing of gloves helps to minimise trauma and may reduce the risk of spread of pathogens, but they reduce dexterity. If forceps are to be used, pad them or employ those made of plastic. Always use plastic/padded forceps when handling metallic objects, such as bullets.

4. Be particularly careful when moving or transporting bones, even over a short distance. Carry them on a tray or in a box. Wrap bones individually in paper to reduce damage. If soft tissues are present, place them in jars or plastic bags to prevent or delay desiccation.

5. If the bones are dirty, clean them with care, using a damp cloth or a soft brush. Do not apply hot water and do not scrub bone surfaces, especially of joints. Record on the record sheet if the bones have been cleaned. Remember that dirty bones are likely to present more of a health hazard than do dry bones.

6. When material is not properly labelled, work on one skeleton at a time, so as to avoid mixing and possible transposition of bones. Ideally, arrange the bones in a skeletal arrangement on a table.

7. Use standard record sheets for your work, to avoid inter-observer bias and to facilitate later analysis. Use clear, correct, unambiguous language – and be consistent in what you say. Write dates in full – for example, 11th September 2009, not "11/9/09" or "9/11/09".

8. When possible, supplement written records with photographs, drawings and tape-recordings. Ensure that the photographs and drawings are labelled with the animal's reference number and the date of the study (see above).

9. Make one or more photocopies or electronic scans of completed record sheets as soon as possible and keep in a safe place separate from the original.

10. Do not remove bones for study overseas unless specifically authorised to do so, in which case both CITES permits and animal health documentation are likely to be needed (see Cooper, ME, Abstract of paper presented at PASA meeting, Kenya, 2009).

Equipment needed

Paper, pens (black), pencils including marking pen/soft pencil for labelling bones
Record sheets and clipboard
Protective clothing, including goggles
Scales/balance
Tape measure (for circumferential measurements)
Rules (for linear measurements and as a scale on photographs)
Self-sealing plastic bags
Sticky labels and dry labels
Hand lens or magnifier
Brushes
Probes
Reference texts (photocopies, printouts or books)
Camera
Tape-recorder

Appendix 4. List of Javan rhinoceros bones present and missing, Cat Tien National Park

Scientific name	Number in skeleton	Number found
Skull	1	1
Mandible	1	1
Cervical vertebrae	7	5
Thoracal vertebrae	17	12
Lumbar vertebrae	2	-
Os sacrum	1	1
Caudal vertebrae	?	-
Costa	34	23
Os sternum	1	1
Pelvis	1	1
Scapula	2	2
Humerus	2	2
Radius/Ulna	2	2
Carpals /tarsals	32	28
Metacarpals /Metatarsals	12	10
Phalanges	36	10
Hooves	12	11
Sesamoids		2 (1 patella)

TỔNG CỤC CẢNH SÁT PCTP　　　CỘNG HOÀ XÃ HỘI CHỦ NGHĨA VIỆT NAM
PHÂN VIỆN KHHS TẠI TP.HỒ CHÍ MINH　　Độc lập - Tự do - Hạnh - Phúc
　　　Số: 940/C54B

Thành phố Hồ Chí Minh, ngày 0 7 tháng 6 năm 2010

KẾT LUẬN GIÁM ĐỊNH

Kính gửi: **Cơ quan Cảnh sát điều tra - Công an tỉnh Lâm Đồng**

Ngày 19 tháng 5 năm 2010 Phân viện Khoa học hình sự tại thành phố Hồ Chí Minh đã nhận được quyết định trưng cầu giám định số 72/PC14 đề ngày 18 tháng 5 năm 2010 của Cơ quan Cảnh sát điều tra Công an tỉnh Lâm Đồng yêu cầu giám định dấu vết súng đạn trong vụ phát hiện bộ xương Tê giác ở dưới suối tại tiểu khu 513, khu Cát Lộc Vườn Quốc gia Cát tiên, thuộc địa phận huyện Cát Tiên, tỉnh Lâm Đồng, vào ngày 29 tháng 4 năm 2010

Theo sự phân công của lãnh đạo Phân viện khoa học hình sự tại thành phố Hồ Chí Minh.

Chúng tôi gồm:

Ông: Lê Đức Sơn - Giám định viên tư pháp kỹ thuật hình sự về dấu vết súng đạn, số thẻ 1703/TP - GĐTP đã tiến hành giám định tại Phân viện Khoa học hình sự tại thành phố Hồ Chí Minh từ ngày 20 tháng 5 năm 2010 đến ngày 31 tháng 5 năm 2010.

1- Nội dung vụ việc

Như ghi trong Quyết định trưng cầu giám định số: 72/PC14 đề ngày 18 tháng 5 năm 2010 của Cơ quan Cảnh sát điều tra Công an tỉnh Lâm Đồng.

2. đối tượng cần giám định

01 (một) mẫu kim loại

3. Yêu cầu giám định

3.1- "Đầu đạn thu được trên xương ở chân trước bên phải con Tê giác là loại đạn gì, được sử dụng cho loại súng nào có phải được sử dụng cho súng quân dụng hay không? nếu phải thì loại súng gì ?"

3.1- "Dấu vết trên đầu đạn đã được thu giữ gửi giám định có đủ yếu tố truy nguyên khẩu súng bắn ra hay không ?"

4 - Phương pháp và kết quả giám định

4.1-Phương pháp giám định

+Nghiên cứu hình dạng, cấu tạo, kích thước, đặc điểm và dấu vết để lại trên mẫu vật gửi giám định ghi ở mục 2.

+Nghiên cứu các tài liệu liên quan đến đạn

+So sánh đối chứng giữa hình dạng, cấu tạo, kích thước, đặc điểm và dấu vết để lại trên mẫu vật gửi giám định ghi ở mục 2 với hình dạng, cấu tạo, kích thước, đặc điểm của các loại đầu đạn của các loại đạn tiêu chuẩn được lưu tại Phân viện Khoa học hình sự.

4.2- Phương tiện giám định: sử dụng các tiện chuyên dùng như kính lúp, kính hiển vi soi nổi MZ8, kính hiển vi so sánh với ánh sáng chiếu xiên và các công cụ hỗ trợ cần thiết.

4.3- Kết quả giám định: đã ghi nhận đầy đủ các thông tin, đặc điểm liên quan đến mẫu cần giám định. Quá trình giám định bảo đảm chính xác, khách quan đủ cơ sở để kết luận giám định.

5 - Kết luận giám định

5.1- Mẫu vật bằng kim loại đã bị han rỉ thu ở vụ phát hiện bộ xương Tê giác ở dưới suối tại tiểu khu 513, khu Cát Lộc thuộc Vườn Quốc gia Cát tiên, tỉnh Lâm Đồng, vào ngày 29 tháng 4 năm 2010 gửi giám định là phần ống chứa thuốc cháy nằm trong lõi đầu đạn của loại đạn vạch đường cỡ 7,62mm x 39mm (đạn K56).

5.2- Đạn cỡ 7,62mm x 39mm (đạn K56) là loại đạn sử dụng cho các loại súng quân dụng có cỡ nòng 7,62mm, như AK47, CKC...

5.3- Dấu vết để lại trên mẫu vật bằng kim loại đã bị han rỉ ghi ở mục 5.1 không có cơ sở để giám định truy nguyên khẩu súng bắn.

Hoàn lại toàn bộ đối tượng gửi giám định như ghi trong QĐTC.

GIÁM ĐỊNH VIÊN

Lê Đức Sơn

PHÓ VIỆN TRƯỞNG VIỆN KHHS KIÊM PHÂN VIỆN TRƯỞNG

Đại tá Phạm Ngọc Hiền

Appendix 6. Lam Dong Police ballistics report, unofficial English translation

Assessment result

Report to: Police agency – Police Department, Lam Dong province

19/5/2010, Ho Chi Minh Criminal Science Institute has been informed on announcing an assessment referendum number 72/PC14 on 18/5/2010 from the Policy agency, Police Department; Lam Dong province which had required an investigation on bullets when a dead rhino had been found near the spring in sub-division 513, Cat Loc sector, Cat tien national park, Cat Tien district, Lam Dong province on 29/4/2010.

Follow the assignment from the management of Ho Chi Minh City Criminal Science Institute.
We are:
Le Duc Son – Ballistics Expert, card number 1703/TP – GDTP did make an examination in Ho Chi Minh Criminal Science Institute from 20/5/2010 till 31/5/2010.

1. Job description:
As mentioned in the examination referendum number 72/PC14 on 18/5/2010 from the Policy agency, Police Department; Lam Dong province.

2. Assessment object:
01 sample of metal

3. Assessment requirements:
3.1 "What kind of bullet tip obtained on bone in the right front leg of the Rhino, for what kind of gun – specifies more on whether it is one of military guns."
3.2 "Is this enough evidences to know the exact kind of gun if there is only a trace on the bullet tip?"

4. Assessment methodology and result:
4.1 Assessment methodology
- Study of the shape, structure, size, characters and traces on the specimen mentioned on part 2.
- Study of gun documents
- Compare and contrast between the specimen with its shape, structure, size, characters and traces and the shape, structure, size and characters of bullet tips that kept and stored at Criminal Science Institute.

4.2 Assessment tools: Specialized tools such as magnifier, microscope MZ8 (floating check), microscope which is used to compare with oblique light and other supporting tools.
4.3 Assessment result: All outcomes have been collected. Assessment process has been completely qualified and transparent to give out the final conclusion.

5. Assessment conclusion:
5.1 The rusty metal specimen that people found it in the dead Rhino's bone near the spring in sub-division 513, Cat Loc sector, Cat tien national park, Cat Tien district, Lam Dong province on 29/4/2010 is a part of a fire tube in the bullet tip which is the bullet of size of 7.62mm*39mm (bullet K56).
5.2 Bullet 7.62mm*39mm (K56) is used for military guns with caliber of 7.62mm such as AK47, CKC...

Appendix K

5.3 The remaining trace on the rusty metal specimen which mentioned on part 5.1 is not enough to tell exactly what gun it belongs to.

Resend all assessment objects as mentioned.

Appendix 7. Advice on Preserving Bones of Javan Rhinoceros – John E Cooper

The following notes apply to the preservation of non-articulated bones, as a collection. If the rhinoceros bones are articulated (put together) for mounting as a skeleton, some different instructions will be needed:

- Have protective clothing available, even if not immediately needed. Remind all present of the importance of health and safety. Bones can transmit pathogens.
- Handle bones carefully. They can be damaged easily. Hold them over a table top or bench, preferably with two hands. The wearing of gloves helps to minimise damage and minimise the risk of spread of pathogens, but they reduce dexterity.
- Be particularly careful when moving or transporting bones, even over a short distance. Whenever possible, carry them on a tray or in a box.
- Give a unique reference number to each bone, linked to the reference number of the rhinoceros, write this in soft pencil or ink on the side of the bone (not over a joint). Make a list of all bones, with numbers and identification, where known.
- Identify as many bones as possible. Write the identification (name) of the bone in English or Latin in pencil or pen as above. Use for guidance the book "Anatomy of the Domestic Animals" by Sisson, which was left with Mr Thanh. If there is any doubt about the identification of a bone, give it a reference number but do not add a name until an authoritative opinion has been obtained.
- If the bones are dirty, clean them with care, using a damp cloth or a soft brush. Do not apply hot water and do not scrub bone surfaces, especially of joints. Record on a record sheet if the bones have been cleaned. Remember that dirty bones are likely to present more of a health hazard than do dry bones.
- Do not treat the bones with chemicals or attempt to boil them until a decision is made about their long-term future.
- If the bones are to be stored at Cat Tien, they should be wrapped in dry paper and stored in secure, labelled, boxes (preferably cardboard). They should be checked regularly - at least once every three months.
- Do not use insecticides on the bones or wrappings. Instead, seal them in a secure, dry, cupboard or display cabinet.
- Check the bones every three months for signs of mould, insect damage or other changes. If any changes are visible, seek further advice.

Why are we here
To stop the degredation of the planet's natural environment and to build a future in which humans live in harmony with nature

wwf.panda.org/vietnam

© Text 2010 WWF

WWF is one of the world's largest and most experienced independent conservation organizations, with over 5 million supporters and a global Network active in more than 100 countries.

WWF's mission is to stop the degradation of the planet's natural environment and to build a future in which human live in harmony with nature, by: conserving the world's biological diversity, ensuring that the use of renewable natural resources is sustainable, and promoting the reduction of pollution and wasteful consumption.

WWF Greater Mekong
Vietnam Country Programme
D13, Thang Long
International Village
Cau Giay, Hanoi, Vietnam
Tel: 844 3719 3049
Fax: 844 37193048

Cooper, J.E. (2004b). Searching the literature. *Veterinary Record* 155, 375.
Cooper, J.E. (2007). Pathology and disease. In: Bird, D.M. and Bildstein, K., Eds. *Raptor Research and Management Techniques*. Hancock House, British Columbia, Canada, pp. 293–310.
Cooper, J.E. (2008b). Health hazards to animals associated with aluminium smelting. In: Khare, M., Sankat, C.K., Shrivastava, G., and Venkobachar, C., Eds. *Aluminium Smelting. Health, Environment and Engineering Perspectives*. Ian Randle Publishers, Kingston, Jamaica, pp. 161–168.
Cooper, J.E. (2008c). Methods in herpetological forensic work – Post-mortem techniques. In: Cooper, J.E., Baker, B.W., and Cooper, M.E., Eds. Special issue: Forensic Science and Herpetology. *Applied Herpetology* 5(4), 351–370.
Cooper, J.E. (2009a). Amyloid, apples and analogies. *Bulletin of the Royal College of Pathologists* 146, 176–177.
Cooper, J.E. (2009b). Conjoined (Siamese) twins of the leopard tortoise (*Geochelone pardalis*), with a plea for documentation of such abnormalities in reptiles. *Journal of Herpetological Medicine and Surgery* 19, 69–71.
Cooper, J.E. (2010a). Doctor! The monkeys are dying… *Bulletin of the Royal College of Pathologists* 149, 13–15.
Cooper, J.E. (2010b). Heat stress, climate change and animal welfare. *Veterinary Record* 166, 729.
Cooper, J.E. (2010c). References, reliability and readership. Editorial. *Journal of Small Animal Practice* 51, 359–360.
Cooper, J.E. (2010d). Sarcocystosis (sarcosporidiosis). In: Warrell, D.A., Cox, T.M., Firth, J.D., Benz, E.J., Jr., Eds. *Oxford Textbook of Medicine*. Oxford University Press, Oxford, UK, pp. 75–77.
Cooper, J.E. (2011a). Anesthesia, analgesia, and euthanasia of invertebrates. In: Spineless Wonders: Welfare and Use of Invertebrates in the Laboratory and Classroom. *ILAR Journal* 52(5), 196–197.
Cooper, J.E. (2011b). Organ and tissue responses in animals. Paper presented at *Symposium on The Pathology of Abuse – In Humans and Animals*, 13 January 2011 The Royal College of Pathologists, London, UK.
Cooper, J.E. (2012). The estimation of post-mortem interval (PMI) in reptiles and amphibians: Current knowledge and needs. *The Herpetological Journal* 22(2), 91–96.
Cooper, J.E. Guest Editor. (2013). Field techniques in exotic animal medicine. *Journal of Exotic Pet Medicine* 23(1).
Cooper, J.E. and Cooper, M.E. (1986). Is this eagle legal? A veterinary approach to litigation involving birds. *Proceedings of Forensic Zoology Discussion Group*. Zoo-Technology, London, UK.
Cooper, J.E. and Cooper, M.E. (1991). Legal cases involving birds: The role of the veterinary surgeon. *Veterinary Record* 129, 505–507.
Cooper, J.E. and Cooper, M.E. (1996). Veterinary and legal implications of the use of snakes in traditional dancing in East Africa. *British Herpetological Society Bulletin* 55, 29–34.
Cooper, J.E. and Cooper, M.E. (2001). Legal and ethical aspects of working with wildlife, with particular reference to Africa. *ANZCCART News* 14(4), 4–7.
Cooper, J.E. and Cooper, M.E. (2006). Ethical and legal implications of treating casualty wild animals. *In Practice* 28(1), 2–6.
Cooper, J.E. and Cooper, M.E. (2007). *Introduction to Veterinary and Comparative Forensic Medicine*. Blackwell, Oxford, UK.
Cooper, J.E. and Cooper, M.E. (2008). Skeletal pathology of primates and other wildlife. *Veterinary Record* 162, 63–64.
Cooper, J.E. and Cooper, M.E. (2011). Veterinary forensic workshop in Kenya. *Bulletin of the Royal College of Pathologists* 156, 256–257.
Cooper, J.E. and Cooper, M.E., Eds. (1998). Forensic veterinary medicine. Special issue. *Seminars in Avian and Exotic Pet Medicine* 7(4), 159–230.
Cooper, J.E. and Eley, J. (1979). *First Aid and Care of Wild Birds*. David and Charles, Newton Abbot, UK.
Cooper, J.E. and Jackson, O.F., Eds. (1981). *Diseases of the Reptilia*. Academic Press, London, UK.

Cooper, J.E. and Knowler, C. (1991). Snails and snail farming: An introduction for the veterinary profession. *Veterinary Record* 129, 541–549.

Cooper, J.E. and Seebaransingh, R. (2008). Fibropapillomatosis in a green turtle (*Chelonia mydas*) from Trinidad. *West Indian Veterinary Journal* 8, 77–80.

Cooper, J.E. and Silvanose, C.E. (2008). Cytology. In: Samour, J., Ed. *Avian Medicine*, 2nd edn. Mosby Elsevier, Edinburgh, UK, pp. 72–78.

Cooper, J.E. and Simmons, V. (2010). Toad-poisoning and raptors. *Journal of Raptor Research* 44, 75–77.

Cooper, J.E. and Williams, D.L. (1995). Veterinary perspectives and techniques in husbandry and research. In: Warwick, C., Frye, F.L., and Murphy, J.B., Eds. *Health and Welfare of Captive Reptiles*. Chapman and Hall, London, UK, pp. 98–111.

Cooper, J.E., Baker, B.W., and Cooper, M.E., Eds. (2008). Special issue: Forensic Science and Herpetology. *Applied Herpetology* 5(4), 303–402.

Cooper, J.E., Cooper, M.E., and Budgen, P. (2009). Wildlife crime scene investigation: Techniques, tools and technology. *Endangered Species Research* 9, 229–238.

Cooper, J.E., Dutton, C.J., and Allchurch, A.F. (1998). Reference collections: Their importance and relevance to modern zoo management and conservation biology. *Dodo* 34, 159–166.

Cooper, J.E., Ed. (1989c). *Disease and Threatened Birds*. ICBP Technical Publication No. 10. International Council for Bird Preservation (now BirdLife), Cambridge, UK.

Cooper, J.E., Ewbank, R., and Warwick, C. (1989). *Euthanasia of Amphibians and Reptiles*. Universities Federation for Animal Welfare, Potters Bar, UK.

Cooper, J.E., Gschmeissner, S., Parsons, A.J., and Coles, B.H. (1987). Psittacine beak and feather disease. *Veterinary Record* 120, 287.

Cooper, J.E., Needham, J.R., Applebee, K., and Jones, C.G. (1988). Clinical and pathological studies on the Mauritian Pink Pigeon *Columba mayeri*. *Ibis* 130, 57–64.

Cooper, M.E. (1987a). *An Introduction to Animal Law*. Academic Press, London, UK.

Cooper, M.E. (1989a). Legal considerations in the movement and submission of avian samples. In: Cooper, J.E., Ed. *Disease and Threatened Birds*. International Council for Bird Preservation (now Bird Life), Cambridge, UK.

Cooper, M.E. (1995a). Legal and ethical aspects of new wildlife food sources. In: Cooper, J.E., Ed. Wildlife Species for Sustainable Food Production. *Biodiversity and Conservation* 4(3), 322–335.

Cooper, M.E. (2000). Ethical considerations in wildlife health. *Proceedings of the Seminar: Wildlife Health in Conservation*. Massey University, Publication No. 207, Palmerston, New Zealand.

Cooper, M.E. (2008a). Forensics in herpetology – Legal aspects. In: Cooper, J.E., Baker, B.W., and Cooper, M.E., Eds. Special issue: Forensic Science and Herpetology. *Applied Herpetology* 5(4), 319–338.

Cooper, M.E. and Cooper, J.E. (1981). The use of animals in films: A veterinary and legal viewpoint. *British Kinematograph Sound and Television Society* 63, 544–546.

Cooper, M.E. and Rosser, A.M. (2002). International regulation of wildlife trade: Relevant legislation and organisations. *Review of the Office International des Epizooties* 21(1), 103–123.

Cooper, M.E. and Sinclair, D.A. (1990). Wildlife rehabilitation and the law: An introduction. In: Thomas, T., Ed. *The Proceedings of the Third Symposium of the British Wildlife Rehabilitation Council*, Stoneleigh (1990). British Wildlife Rehabilitation Council, London, UK, pp. 63–65.

Cooper, N. and Carling, B., Guest editors. (1995). Ecologists and ethical judgements. *Biodiversity and Conservation* 4(8), 783–785.

Corbet, G.B and Harris, S. (1991). *The Handbook of British Mammals*, 3rd edn. The Mammal Society, Blackwell, Oxford, UK.

Corbett, J. (1944). *The Man-Eaters of Kumaon*. Oxford University Press, London, UK.

Cornaby, B.W. (1974). Carrion reduction by animals in contrasting tropical habitats. *Biotropica* 6, 51–63.

Cornelius, E.H. and Harding Rains, A.J. (1976). *Letters from the Past from John Hunter to Edward Jenner*. The Royal College of Surgeons of England, London, UK.

Cornelius, S.F. (1975). Marine turtle mortalities along the Pacific coast of Costa Rica. *Copeia* 1, 186–187.

Cornell Lab of Ornithology. (2011). *All about Birds*. Ithaca, New York, USA. http://www.allaboutbirds.org/guide/Canada_Goose/lifehistory

Costa, D.P. and Kooyman, G.L. (1982). Oxygen consumption, thermoregulation, and the effect of fur oiling and washing on the sea otter, *Enhydra lutris*. *Canadian Journal of Zoology* 60(11), 2761–2767.

Cousquer, G.O., Cooper, J.E., and Cobb, M.A. (2010). Conjunctival flora in tawny owls (*Strix aluco*). *Veterinary Record* 166, 652–654.

Cranford, T.W. et al. (2008). Anatomic geometry of sound transmission and reception in Cuvier's beaked whale (*Ziphius cavirostris*). *The Anatomical Record: Advances in Integrative Anatomy and Evolutionary Biology* 291(4), 353–378.

Črček, M., and Drobnič, K. (2011). Forensic identification of 12 mammals species based on size variation of mitochondrial cytochrome b gene using multiplex PCR assay. *Forensic Science International: Genetics* 3(1), e218–e219.

Crispino, F. (2008). Nature and place of crime scene management within forensic sciences. *Science & Justice* 48, 24–28.

Cronin, M.A., Palmisciano, D.A., Vyse, E.R., and Cameron, D.G. (1991). Mitochondrial DNA in wildlife forensic science: Species identification of tissues. *Wildlife Society Bulletin* 19, 94–105.

Crowl, T.A., Crist, T.O., Parmenter, R.R., Belovsky, G., and Lugo, A.E. (2008). The spread of invasive species and infectious disease as drivers of ecosystem change. *Frontiers in Ecology and the Environment* 6(5), 238–246.

Croxall, J.P., Prince, P.A., Baird, A and Ward, P. (1985). The diets of the Southern rockhopper penguin *Eudyptes chrysocome chrysocome* at Beauchêne Island, Falkland Islands. *Journal of Zoology, London* 206, 485–496.

Cryan, P.M. (2008). Mating behavior as a possible cause of bat fatalities at wind turbines. *Journal of Wildlife Management* 72, 845–849.

Cummings, P.M., Trelka, D.P., and Springer, K., Eds. (2010). *Atlas of Forensic Histopathology*. Cambridge University Press, Cambridge, UK.

Currey, J.D. (2002). *Bones: Structure and Mechanics*. Princeton University Press, Princeton, NJ, USA.

Cuthbert, R.J., Dave, R., Chakraborty, S.S., Kumar, S., Prakash, S., Ranade, S.P., and Prakash, V. (2011). Assessing the ongoing threat from veterinary non-steroidal anti-inflammatory drugs to critically endangered *Gyps* vultures in India. *Oryx* 45(3), 420–426.

Dabas, Y.P.S. and Saxena, O.P. (1994). *Veterinary Jurisprudence and Post-mortem*. International Book Distributing, Lucknow, India.

Dabin, W., Cesanini, C., Clemerceau, I., Dhermain, F., Jauniaux, T., Van Canneyt, O., and Ridoux, V. (2004). Double-faced monster in the bottlenosed dolphin (*Tursiops truncatus*) found in the Mediterranean Sea. *Veterinary Record* 154, 306–308.

Dagnall, J.L. (1995). A simple negative staining technique for the identification of mammal hairs. *Journal of Zoology, London* 237, 670–675.

Dale, W.M. and Nagy, R. (2006). Crime scene and crime lab: Joined by technology; validated by quality. *Forensic Magazine* 3(5), 16–21.

Dalebout, M.L., Van Helden, A., Van Waerebeek, K., and Baker, C.S. (1998). Molecular genetic identification of southern hemisphere beaked whales (Cetacea: Ziphiidae). *Molecular Ecology* 7, 687–694.

Dalén, L., Götherström, A., and Angerbjörn, A. (2004). Identifying species from pieces of faeces. *Conservation Genetic* 5, 109–111.

Dalton, D.L., and Kotze, A. (2011). DNA barcoding as a tool for species identification in three forensic wildlife cases in South Africa. *Forensic Science International* 207(1), e51–e54.

Daniels, G.D. and Kirkpatrick, J.B. (2011). Attitude and action syndromes of exurban landowners have little effect on native mammals in exurbia. *Biodiversity and Conservation* 20, 3517–3535.

Daoust, P.-Y. (1991). Porcupine quill in the brain of a dog. *Veterinary Record* 128, 436–435.

Davies, L. (1999). Seasonal and spatial changes in blowfly production from small and large carcasses at Durham in lowland northeast England. *Medical and Veterinary Entomology* 13, 245–251.

Davis, B.T. (1974). George Edward Male, MD – The father of English medical jurisprudence. *Proceedings of the Royal Society of Medicine* 67, 117–120.

Davis, J.B. and Goff, M.L. (2000). Decomposition patterns in terrestrial and intertidal habitats on Oahu Island and Coconut Island, Hawaii. *Journal of Forensic Sciences* 45(4), 836–842.

Davis, P.G. (1992). The taphonomy of birds. *Fifth North American Paleontological Convention*. The Paleontological Society Special Publication No. 6, p. 264.

Davis, P.G. (1996). The taphonomy of Archaeopteryx. *Bulletin of the National Science Museum Tokyo Series* 22C, 91–106.

Davis, P.G. (1997). The bioerosion of bird bones. *International Journal of Osteoarchaeology* 7, 388–401.

Davis, P.G. and Briggs, D.E.G. (1998). The impact of decay and disarticulation on the preservation of fossil birds. *Palaios* 13, 3–13.

Davis, S.J.M. (1987). *The Archaeology of Animals*. B.T. Batsford, London, UK.

Davison, A.M. (2004). The incised wound. In: Rutty, G.N., Ed. *Essentials of Autopsy Practice*. Springer-Verlag, London, UK.

Dawnay, N., Ogden, R., McEwing, R., Carvalho, G.R., and Thorpe, R.S. (2007). Validation of the barcoding gene COI for use in forensic genetic species identification. *Forensic Science International* 173, 1–6.

Dawnay, N., Ogden, R., Thorpe, R.S., Pope, L.C., Dawson, D.A., and McEwing, R. (2008). A forensic STR profiling system for the Eurasian badger: A framework for developing profiling systems for wildlife species. *Forensic Science International: Genetics* 2(1), 47–53.

Dawnay, N., Ogden, R., Wetton, J.H., Thorpe, R.S., and McEwing, R. (2009). Genetic data from 28 STR loci for forensic individual identification and parentage analyses in 6 bird of prey species. *Forensic Science International: Genetics* 3, e63–e69.

Day, M.G. (1966). Identification of hair and feather remains in the gut and faeces of stoats and weasels. *Journal of Zoology, London* 148, 201–217.

Day, R.H. (1980). The occurrence and characteristics of plastic pollution in Alaska's marine birds. MS thesis. University of Alaska, Fairbanks, AK, USA.

De Batist, P. (2003). Kweekexperimenten met *Megaselia scalaris* (Loew, 1866) (Diptera, Phoridae), Een recent 'intruder' in de Belgische Fauna. *Entomo-Info* 14, 77–84.

De Buffrenil, V. and Schoevaert, D. (1988). On how the periosteal bone of the delphinid humerus becomes cancellous: Ontogeny of a histological specialization. *Journal of Morphology* 198, 149–164.

De Cupetre, B., Thys, S., Van Neer, W., Ervynck, A., Corremans, M., and Waelkens, M. (2009). Eagle Owl (*Bubo bubo*) pellets from Roman Sagalassos (SW Turkey): Distinguishing the prey remains from nest and roost sites. *International Journal of Osteoarchaeology* 19, 1–22.

De Gange, A.R. and Vacca, M.M. (1989). Sea otter mortality at Kodiak Island, Alaska, during summer 1987. *Journal of Mammology* 70(4), 836–838.

De Guise, S., Lagacé, A., Béland, P., Girard, C., and Higgins, R. (1994a). Non-neoplastic lesions in beluga whales (*Delphinapterus leucas*) and other marine mammals from the St Lawrence Estuary. *Journal of Comparative Pathology* 112(3), 257–271.

De Guise, S., Lagacé, A., and Béland, P. (1994b). True hermaphroditism in a St. Lawrence beluga whale (*Delphinapterus leucas*). *Journal of Wildlife Diseases* 30(2), 287–290.

De Guise, S., Lagacé, A., and Béland, P. (1994c). Tumors in St. Lawrence beluga whales (*Delphinapterus leucas*) *Veterinary Pathology* 31(4), 444–449.

De Guise, S., Martineau, D., Béland, P., and Fournier, M. (1995). Possible mechanisms of action of environmental contaminants on St. Lawrence beluga whales (*Delphinapterus leucas*). *Environ Health Perspect* 103(Suppl 4), 73–77.

De Guise, S., Flipo, D., Boehm, J.R., Martineau, D., Béland, P., and Fournier, M. (1997). Immune functions in Beluga whales (*Delphinapterus leucas*): Evaluation of natural killer cell activity. *Veterinary Immunology and Immunopathology* 58(3–4), 345–354.

De Guise, S., Martineau, D., Béland, P., and Fournier, M. (1998). Effects of in vitro exposure of Beluga whale leukocytes to selected organochlorines. *Journal of Toxicology and Environmental Health, Part A* 55(7), 479–493.

De Guise, S. (2003). Contaminants and marine mammal immunotoxicology and pathology. In: Vos, J.G., Fournier, M., and O'Shea, T.J., Eds. *Toxicology of Marine Mammals*. Taylor & Francis Group, New York, USA.

DeHaro, L. and Pommier, P. (2003). Envenomation: A real risk of keeping exotic house pets. *Veterinary and Human Toxicology* 45, 214–216.

De Jong, G.D. and Chadwick, J.W. (1999). Decomposition and arthropod succession on exposed rabbit carrion during summer at high altitudes in Colorado, USA. *Journal of Medical Entomology* 36(6), 833–845.

Dedek, J. and Steineck, T. (1994). *Wildhygiene*. Gustav Fischer Verlag, Jena, Stuttgart, Germany.

DEFRA. (2005). *Wildlife Crime: A Guide to the Use of Forensic and Specialist Techniques in the Investigation of Wildlife Crime*. DEFRA, London, UK.

DEFRA. (2012). *Secretary of State's Standards of Modern Zoo Practice*. Department for Environment, Food and Rural Affairs, London, UK.

Dekeirsschieter, J., Verheggen, F.J., Gohy, M., Hubrecht, F., Bourguignon, L., Lognay, G., and Haubruge, E. (2009). Cadaveric volatile organic compounds released by decaying pig carcasses (*Sus domesticus* L.) in different biotopes. *Forensic Science International* 189, 46–53.

Dekeirsschieter, J., Verheggen, F.J., Haubruge, E., and Brostaux, Y. (2011). Carrion beetles visiting pig carcasses during early spring in urban, forest and agricultural biotopes of Western Europe. *Journal of Insect Science* 11, 73, available online: insectscience.org/11.73.

Desser, S.S., Stuht, J., and Fallis, A.M. (1978). Leucocytozoonosis in Canada geese in upper Michigan. 1. Strain differences among geese from different localities. *Journal of Wildlife Diseases* 14, 124–131.

Dessler, A.E. (2011). *Introduction to Modern Climate Change*. Cambridge University Press, Cambridge, UK.

Devuyet, D., Hens, L., and de Lannoy, W., Eds. (2001). *How Green is the City? Sustainability Assessment and the Management for Urban Environments*. Columbia University Press, New York, USA.

Dewaele, P., Le Clercq, M., and Disney, R.H.L. (2000). Entomologie et médicine légale: Les Phorides (Diptères) sur cadavres humains. Observation inédite. *Journal de Médecine Légale Droit Medical* 43, 569–572.

Dias, T.L.P., Neto, N.A.L., and Alves, R.N. (2011). Molluscs in the marine curio and souvenir trade in NE Brazil: Species composition and implications for their conservation and management. *Biodiversity and Conservation* 20, 2393–2405.

Diaz, D.E., Ed. (2005). *The Mycotoxin Blue Book*. Nottingham University Press, Nottingham, UK.

Dickersin, K. and Chalmers, I. (2011). Recognizing, investigating and dealing with incomplete and biased reporting of clinical research: from Francis Bacon to the WHO. *Journal of the Royal Society of Medicine* 104, 532–538.

Dierauf, L.A. (1994). In: Wilkinson, D. and Driscoll, C., Eds. *Pinniped Forensic Necropsy and Tissue Collection Guide*. U.S. Department of Commerce National Oceanic and Atmospheric Administration, Washington, DC, National Marine Fisheries Service 83.

Dierauf, L.A. and Gulland, F.M.D., Eds. (2001). *CRC Handbook of Marine Mammal Medicine*. CRC Press LLC, Boca Raton, FL, USA.

DiMaio, V.J.M. (1998). *Gunshot Wounds*. CRC Press, Boca Raton, FL, USA.

DiMaio, V.J.M. and DiMaio, D.J. (2001). *Forensic Pathology*, 2nd edn. CRC Press, Boca Raton, FL, USA.

Dimond, R.L. and Montagna, W. (1976). The skin of the giraffe. *Anatomical Record* 185, 63–75.

Dinet, V. (2010). On terrestrial hunting by crocodilians. *Herpetological Bulletin* 114, 15–18.

Disney, H. (2006). The expert witness system. *Biologist* 53(1), 9–10.

Disney, R.H.L. (1980). An exotic scuttle fly, *Chonocephalus heymonsi* Stobbe (Dipt., Phoridae) from Middlesex. *Entomologist's Monthly Magazine* 116, 207–212.

Disney, R.H.L. (1994). *Scuttle flies: The Phoridae*. Chapman and Hall, London, UK.

Dix, J. (1999). *Color Atlas of Forensic Pathology*. CRC Press, Boca Raton, FL, USA.

Dix, J. and Graham, M. (2000). *Time of Death, Decomposition and Identification. An Atlas*. CRC Press, Boca Raton, FL, USA.

Dobney, K., Ervynck, A., Albarella, U., and Rowley-Conwy, P. (2004). The chronology and frequency of a stress marker (linear enamel hypoplasia) in recent and archaeological populations of *Sus scrofa* in north-west Europe, and the effects of early domestication. *Journal of Zoology* 264, 197–208.

Dobson, J. (2011). Role of the Expert witness and report-writing considerations. *Practical Law in Veterinary Times* 11 April 2011, 29–31.

Dockrell, M. (2012). *A Guide to Expert Witness Evidence*. Bloomsbury Professional, London, UK.

Dodd, C.K. (1995). Disarticulation of turtle shells in North-central Florida: How long does a shell remain in the woods? *American Midland Naturalist* 134, 378–387.

Dolbeer, R., Wright, S., Weller, J., and Begier, M. (2009). Wildlife strikes to civil aircraft in the United States 1990–2008. U.S. Department of Transportation, Federal Aviation Administration, Serial Report No. 15 DOT/FAA/AS/00-6(AAS-310). Washington, DC, USA.

Dominguez, J., Vidal, M., and Tapia, L. (2010). Morphological changes in European goldfinches (*Carduelis carduelis*) released by bird trappers. *Animal Welfare* 19(4), 385–390.

Domning, D.P. and Hayek, L.A.C. (1984). Horizontal tooth replacement in the Amazonian manatee (*Trichechus inunguis*). *Mammalia* 48(1), 105–128.

Donovan, S.E., Hall, M.J.R., Turner, B.D., and Moncrieff, C.B. (2006). Larval growth rates of the blowfly, *Calliphora vicina*, over a range of temperatures. *Medical and Veterinary Entomology* 20(1), 106–114.

Dorcas, E. and Willson, J.D. (2011). *Invasive Pythons in the United States: Ecology of an Introduced Predator*. The University of Georgia Press, Athens, GA, USA.

Dorion, R.B.J., Ed. (2004). *Bitemark Evidence*. CRC Press, Boca Raton, FL, USA.

Dove, C.J. (2000). *A Descriptive and Phylogenetic Analysis of Plumulaceous Feather Characters in Charadriiformes*. American Ornithologists' Union, Washington, DC, USA.

Dove, C.J., Rotzel, N.C., Heacker, M., and Weigt, L.A. (2008). Using DNA barcodes to identify bird species involved in birdstrikes. *Journal of Wildlife Management* 72, 1231–1236.

Draulans, D. and Van Krunkelsven, E. (2002). The impact of war on forest areas in the Democratic Republic of the Congo. *Oryx* 36(1), 35–40.

Du Toit, J.T. (2002). Wildlife harvesting guidelines for community-based wildlife management; a southern African perspective. *Biodiversity and Conservation* 11, 1403–1416.

Dudley Furniss-Roe, T. (2008). Incremental structures and wear patterns of teeth for age assessment of red deer. *BAR International Series* 1835, 131.

Duff, J.P. (2007). Suspected wild bird mortality in Britain due to stormy weather and hailstones. *Veterinary Record* 160, 884.

Duff, J.P., Colville, K., Foster, J., and Dumphreys, N. (2011). Mass mortality of great crested newts (*Triturus cristatus*) on ground treated with road salt. *Veterinary Record* 168, 282.

Duff, J.P. and Hewitt, S. (1999). Predation as a suspected cause of common toad (*Bufo bufo*) incidents in Scotland. *Veterinary Record* 144, 27.

Duff, J.P., Holliman, A., Hateley, G., Irvine, R.M., Wight, A., and Williamson, S. (2011). Water bird mortality associated with freezing winter weather and disease. *Veterinary Record* 168, 593.

Duff, J.P. and Hunt, B.W. (1995). Courtship and mortality in foxes (*Vulpes vulpes*). *Veterinary Record* 136, 367.

Duke, G.E., Jegers, A.A., Loff, G., and Evanson, O.A. (1975). Gastric digestion in some raptors. *Comparative Biochemistry and Physiology* 50A, 649–656.

Duncan, C.C. (2010). *Advanced Crime Scene Photography*. CRC Press, Boca Raton, FL, USA.

Dunham, K.M. (1994). The effect of drought on the large mammal populations of Zambezi riverine woodlands. *Journal of Zoology, London* 234, 489–526.

Dunning, K. (2007). From the McArthur to the Millennium Health Microscope: Future developments in microscope miniaturization. *Newsletter, The Royal Society of Tropical Medicine and Hygiene* 4–5.

Dupras, T.L., Schultz, J.J., Wheeler, S.M., and Williams, L.J., Eds. (2005). *Forensic Recovery of Human Remains. Archaeological Approaches*. CRC Press, Boca Raton, FL, USA.

Dürr, T. and Bach, L. (2004). Bat deaths and wind turbines – A review of current knowledge, and of the information available in the database for Germany. *Bremer Beiträge für Naturkunde und Naturschutz* 7, 253–264.

Durrell, G. (1982). *The Amateur Naturalist*. Hamish Hamilton, London, UK.

Dwyer, D. (2008). *The Judicial Assessment of Expert Evidence*. Cambridge University Press, Cambridge, UK.

Early, M. and Goff, M.L. (1986). Arthropod succession patterns in exposed carrion on the island of O'ahu, Hawaiian Islands, USA. *Journal of Medical Entomology* 23(5), 520–531.

Eaton, M.J., Meyers, G.L., Kolokotronis, S.-O., Leslie, M.S., Martin, A.P., and Amato, G. (2009). *Conservation Genetics* 11(4), 1389–1404.

Ebenhard, T. (1987). Introduced birds and mammals and their ecological effects. *Swedish Wildlife Research* 13(4), 1–107.

Eberhardt, T.L. and Elliot, D.A. (2008). A preliminary investigation of insect colonisation and succession on remains in New Zealand. *Forensic Science International* 176, 217–223.

Eck, S., Fiebig, J., Fiedler, W., Heynen, I., Nicolai, B., Töpfer, T., v. d. Elzen, R., Winkler, R., and Woog, F. (2011). *Measuring Birds – Vögel vermessen*. Deutsche Ornithologen-Gesellschaft, Wilhelmshaven.

Eden, C.S., Whiteman, H.H., Duobinis-Gray, L., and Wissinger, S.A. (2007). Accuracy assessment of skeletochronology in the Arizona tiger salamander (*Ambystoma tigrinum nebulosum*). *Copeia* 2007, 471–477.

Edmundson, P. (2002). Single joint expert training. *Veterinary Times* 23 December 2002, 22.

Edmundson, P. (2003). Learning to be an expert witness. *Veterinary Times* 18 August 2003, 14.

Eiken, H.G., Andreassen, R.J., Kopatz, A., Bjervamoen, S.G., Wartiainen, I., Tobiassen, C., Knappskog, P.M., Aspholm, P.E., Smith, M., and Aspi, J. (2009). Population data for 12 STR loci in Northern European brown bear (*Ursus arctos*) and application of DNA profiles for forensic casework. *Forensic Science International: Genetics* 2(1), 273–274.

Elkins, N. (1983). *Weather and Bird Behaviour*. T. & A.D. Poyser, Calton, UK.

Elliott, J.E., Bishop, C.A., and Morrissey, C.A. (2011). *Wildlife Ecotoxicology: Forensic Approaches*. Springer, New York, USA.

Ellis, R.A. and Montagna, W. (1962). The skin of the primates VI. The skin of the gorilla (*Gorilla gorilla*). *American Journal of Physiological Anthropology* 20, 79–94.

Ellison, G.T.H. (1990). The effect of scavenger mutilation on insect succession at impala carcasses in southern Africa. *Journal of the Zoological Society of London* 220, 679–688.

Elsheikha, H.M. and Khan, N. (2011). *Essentials of Veterinary Parasitology*, 1st edn. Caister Academic Press, Norfolk, UK.

Eltringham, S.K. (1994). Can wildlife pay its way? *Oryx* 28, 163–168.

Energy Information Administration. (2009). International energy statistics [www document]. http://www.eia.gov (accessed 16 September 2011).

England, S. (1999). The plight of mammals, reptiles and amphibians when falling into storm drains. *The Rehiberary* 260, 13–15.

Ennion, E.A.R. and Tinberger, N. (1967). *Tracks*. Oxford University Press, Oxford, UK.

Erasmus, M.A., Turner, P.V., Nykamp, S.G., and Widowski, T.M. (2010). Brain and skull lesions resulting from use of percussive bolt, cervical dislocation by stretching, cervical dislocation by crushing and blunt trauma in turkeys. *Veterinary Record* 167, 850–858.

Erickson, W.P., Kronner, J.K., and Bay, K. (2003). Stateline Wind Project Wildlife Monitoring Annual Report, Results for the Period July 2001 – December 2002 [www document]. Report prepared for FPL Energy, the Oregon Office of Energy, and the Stateline Technical Advisory Committee. http://www.oregon.gov/ENERGY/SITING/docs/SWP2002Monitoring.pdf?ga=t, pp. 5–71 (accessed 22 September 2011).

Erlandsson, M. and Munro, R. (2007). Estimation of the post-mortem interval in beagle dogs. *Science & Justice* 47, 150–154.

Erwin, J.M. and Hof, P.R., Eds. (2002). *Aging in Nonhuman Primates*. Karger, Basel, Switzerland.

Erzinclioglu, Z. (1996). *Blowflies*. The Richmond Publishing Co., Ltd, Slough, UK.

Espinoza, E.O., Baker, B.W., Moores, T.D., and Voin, D. (2008). Forensic identification of elephant and giraffe hair artifacts using HATR FTIR spectroscopy and discriminant analysis. *Endangered Species Research* 9, 239–246.

Espinoza, E.O., Espinoza, J.L., Trail, P.W., and Baker, B.W. (2012). The future of wildlife forensic science. In: Huffman, J.E. and Wallace, J.R., Eds. *Wildlife Forensics Methods and Applications*. Wiley-Blackwell, Oxford, UK, pp. 343–356.

Estes, J.A. et al. (1989). The ecology of extinctions in Kelp forest communities. *Conservation Biology* 3(3), 252–264.

Evans, E.P. (1906). *The Criminal Prosecution and Capital Punishment of Animals*. Faber & Faber, London, UK.

Eveleigh, G. (2009). Development of latent fingerprints on reptile skin. *Journal of Forensic Identification* 59(3), 285–296.

Ewins, P.J. (1985). Variation of Black Guillemot wing lengths post-mortem and between measurers. *Ringing & Migration* 6, 115–117.

Faber, H. (1997). Der Einsatz von passiven integrierten Transpondern zur individuellen Markierung von Bergmolchen (*Triturus alpestris*) im Freiland. Naturschutzrelevante Methoden der Feldherpetologie. *Mertensiella* 7, 121–132.

Fabre, J.-H. (1879). *Souvenirs entomologiques: Études sur l'instinct et les moeurs des insectes*. [Série 1]. Ch. Delagrave, Paris, France.

Federal Bureau of Investigation. (2008). *F.B.I. Handbook of Crime Scene Forensics*. Skyhorse Publishing, Washington, DC, USA.

Federal Evidence Review. (2012). Federal rules of evidence. Federal Evidence Review. http://federal-evidence.com/downloads/rules.of.evidence.pdf

Feldman, E. (2004). Legal reform and medical malpractice litigation in Japan: Justice, policy, and the expert witness system. 76 Horitsu Jiho 16, 2004. [in Japanese]

Ferguson, R. (2004). Preliminary analysis of data in the African human – crocodile conflict database. *Crocodile Specialist Group (CSG) Newsletter* 23(4), 20–21.

Ferreira, J.M., Martins, F.D.M., Ditchfield, A., and Morgante, J.S. (2005). The use of PCR-RFLP as an identification tool for two closely related species of bats of genus *Platyrrhinus*. *Genetics and Molecular Biology* 28(1), 120–122.

Fish, J.T., Miller, L.S., and Braswell, M.C., Eds. (2007). *Crime Scene Investigation*. Lexis Nexis, Newark, NJ, USA.

Fisher, B.A.J. (2003). *Techniques of Crime Scene Investigation*, 7th edn. CRC Press, Boca Raton, FL.

Fisher, J., Francis, J., and Jones, G. (2005). *The Bats of Britain Virtual Key* [www document]. http://www.bio.bris.ac.uk/research/bats/britishbats/ (accessed 01 October 2011).

FitzRoy, F.R. and Papyrakis, E. (2009). *An Introduction to Climate Change Economics and Policy*. Earthscan, London, UK.

Flageole, S. and Leclair, R., Jr. (1992). Étude démographique d'une population de salamanders (*Ambystoma maculatum*) à l'aide de la méhode squeletto-chronologique. *Canadian Journal of Zoology* 70, 740–749.

Flint, F.O. (1994). *Food Microscopy: A Manual of Practical Methods, Using Optical Microscopy*. BIOS, Oxford, UK, p. 27.

Flint, F.O. and Firth, B.M. (1988a). A toluidine blue stain mountant for the microscopy of comminuted meat products. *Analyst* 106, 1242–1243.

Flint, F.O. and Firth, B.M. (1988b). Improved toluidine blue stain mountant for the microscopy of food products. *Analyst* 113, 365–366.

Forbes, N. (2004). An exacting science: The veterinary surgeon as expert witness. *In Practice* 26, 503–506.

Forbes, N.A. (1998). Clinical examination of the avian forensic case. *Seminars in Avian and Exotic Pet Medicine* 7(4), 193–200.

Forbes, T. (1981). Early forensic medicine in England: The Angus murder trial. *Journal of the History of Medicine and Allied Sciences* 36(3), 296–309.

Forbes, T.R. (1985). *Surgeons at the Bailey*. Yale University Press, New Heaven, CT, USA.

Fossey, D. (1983). *Gorillas in the Mist*. Hodder and Stoughton, London, UK.

Foster, J.B. (1966). The giraffe of Nairobi National Park: home range, sex ratios, the herd, and food. *East African Wildlife Journal* 4, 139–148.

Fournier, A. (2011). Consequences of wooded shrine rituals on vegetation conservation in West Africa: A case study from the Bwaba cultural area (West Burkina Faso). *Biodiversity and Conservation* 20, 1895–1910.

Fox, N., Blay, N., Greenwood, A.G., Wise, D., and Potapov, E. (2005). Wounding rates in shooting foxes (*Vulpes vulpes*). *Animal Welfare* 14, 93–102.

France, D.L. (2008). *Human and Nonhuman Bone Identification: A Color Atlas*. CRC Press, Taylor & Francis Group, Boca Raton, FL, USA.

France, D.L., Griffin, T.J., Swanburg, J.G., Lindemann, J.W., Davenport, G.C., Trammell, V., Armbrust, C.T., Kondratieff, B., Nelson, A., Castellano, K., and Hopkins, D. (1992). A multidisciplinary approach to the detection of clandestine graves. *Journal of Forensic Sciences* 37(6), 1459–1465.

Franke, J. and Telecky, T. (2001). *Reptiles as Pets: An Examination of the Trade in Live Reptiles in the United States*. Humane Society of the United States, Washington, DC, USA.

Franz-Odendaal, T.A. (2004). Enamel hypoplasia provides insights into early systemic stress in wild and captive giraffes (*Giraffa camelopardalis*). *Journal of Zoology* 263, 197–206.

Fraser, D. (2008). *Understanding Animal Welfare: The Science in Its Cultural Context*. UFAW Animal Welfare Series. Wiley-Blackwell, Oxford, UK.

Fraser Darling, F. (1960). *Wildlife in an African Territory*. Oxford University Press, Oxford, UK.

Fravel, V., Van Bonn, W., Rios, C., and Gulland, F. (2011). Meticillin-resistant *Staphylococcus aureus* in a harbour seal (*Phoca vitulina*). *Veterinary Record* 169, 155.

Freckelton, I.R. and Selby, H. (2009). *Expert Evidence: Law, Practice, Procedure and Advocacy*. Thomson Reuters, Sydney, New South Wales, Australia.

Freeman, R.A. (2004). *The Famous Cases of Dr. Thorndyke: Thirty-Seven of His Criminal Investigations, Part 2*. Kessinger Publishing, Whitefish, MT, USA.

French, M.C., Haines, C.W., and Cooper, J. (1987). Investigation into the effects of ingestion of zinc shot by mallard ducks (*Anas platyrhynchos*). *Environmental Pollution* 47, 305–314.

Fretey, J. and Babin, R. (1998). Arthropod succession in leatherback turtle carrion and implications for determination of the postmortem interval. *Marine Turtle Newsletter* 79, 4–7.

Froecle, R.C. (1997). *The Scientific Expert in Court. Principles and Guidelines*. The American Association for Clinical Chemistry Press, Washington, DC, USA.

Frosch, C., Dutsov, A., Georgiev, G., and Nowak, C. (2011). Case report of a fatal bear attack documented by forensic wildlife genetics. *Forensic Science International: Genetics* 5(4), 342–344.

Frost, D. (2011). *Sporting Shooting and the Law*. The National Gamekeepers' Organisation, Darlington, UK.

Frye, F.L. (1991). *Biomedical and Surgical Aspects of Captive Reptile Husbandry*, 2nd edn. Krieger, Malabar, FL, USA.

Frye, F.L. (1999). Establishing the time of death in reptiles and amphibians. *Proceedings of the Association of Reptilian and Amphibian Vetererinarians*, Orlando FL, USA, pp. 23–25.

Frye, F.L. (2008). Methods in herpetological forensic work – Sampling and standard laboratory techniques. In: Cooper, J.E., Baker, B.W., and Cooper, M.E., Eds. Special issue: Forensic Science and Herpetology. *Applied Herpetology* 5(4), 339–350.

Fumagalli, L., Cabrita, C.J., and Castella, V. (2009). Simultaneous identification of multiple mammalian species from mixed forensic samples based on mtDNA control region length polymorphism. *Forensic Science International: Genetics* 2(1), 302–303.

Furness, R.W. (1988). Predation on ground-nesting seabirds by island populations of red deer *Cervus elaphus* and sheep *Ovis*. *Journal of Zoology, London* 216, 565–573.

Galambos, R. (1942). The avoidance of obstacles by flying bats: Spallanzani's ideas (1794) and later theories. *Isis* 34, 132–140.

García-Moreno, J. (2004). Is there a universal mtDNA clock for birds? *Journal of Avian Biology* 35, 465–468.

Garcia-Rojo, A.M. (2004). A study of the insect succession in carcasses in Alcala de Henares (Madrid administrative region, Spain) using pigs as animal models. *Boletin de la S.E.A* 34, 263–269.

Gardner, R.M. (2011). *Practical Crime Scene Processing and Investigation*. CRC Press, Boca Raton, FL, USA.

Garrett, R.J. (2003). A primer on the tools of crime scene analysis. *Journal of Forensic Identification* 53, 656–665.

Gaskin, D.E., Smith, G.J.D., Arnold, P.W., and Louisy, M.V. (1974). Mercury, DDT, dieldrin, and PCB in two species of odontoceti (Cetacea) from St. Lucia, Lesser Antilles. *Journal of the Fisheries Research Board of Canada* 31, 1235–1239.

Gatica, M.C., Monti, G., Gallo, C., Knowles, T.G., and Warriss, P.D. (2008). Effects of well-boat transportation on the muscle pH and onset of rigor mortis in Atlantic salmon. *Veterinary Record* 163, 111–116.

Gauthier-Pilters, H., and Dagg, I.A. (1981). *The Camel: Its Evolution, Ecology. Behavior & Relationship to Man*. University of Chicago Press, Chicago, IL, USA.

Gee, K.L., Holman, J.H., Causey, M.K., Rossie, A.N., and Armstrong, J.R. (2002). Aging white-tailed deer by tooth replacement and wear: A critical evaluation of a time-honored technique. *Wildlife Society Bulletin* 30, 387–393.

Gennard, D.E. (2007). *Forensic Entomology: An Introduction*. John Wiley & Sons Ltd, Chichester, UK.

Gerber, L.R. and Hilborn, R. (2001). Catastrophic events and recovery from low densities in populations of otariids: Implications for risk of extinction. *Mammal Review* 31(2), 131–150.

Gerlach, H. and Leipold, R. (1986). Über das Federverlust-Syndrom bei Kakadus. *Deutsche Tierärztliche Wochenschrift* 93, 24–26.

Ghazoul, J. and Sheil, D. (2010). *Tropical Rain Forest Ecology, Diversity and Conservation*. Oxford University Press, Oxford, UK.

Giannasi, N., Thorpe, R.S., and Malhotra, A. (2001). The use of amplified fragment length polymorphism in determining species trees at fine taxonomic levels: Analysis of the medically important snake *Trimeresurus albolabris*. *Molecular Ecology* 10, 419–426.

Gilbert, B.M., Martin, L.D., and Savage, H.G. (1990). *Mammalian Osteology*. Missouri Archaeological Society, Columbia, MO, USA.

Gilbert, J.D., Kemper, C.M., Hill, M., and Byard, R.W. (2000). Forensic studies of a stabbed infant bottlenose dolphin. *Journal of Forensic Medicine* 7, 150–152.

Gill-Frerking, H., Rosendahl, W., and Zink, A.R., Eds. (2011). *Yearbook of Mummy Studies*. Verlag Dr. Friedrich Pfell, Munich, Germany.

Gillham, M.E. (1967). Note on the ejection of crop pellets by small passeriens. *Bulletin of the International Bird Pellet Study Group* 2, 2.

Ginn, H.B. and Melville, D.S. (1983). *Moult in Birds. BTO Guide 19*. British Trust for Ornithology, Tring, Hertfordshire, UK.

Gittleman, J.L. (1997). Sexual dimorphism in the canines and skulls of carnivores: Effects of size, phylogeny, and behavioural ecology. *Journal of Zoology, London* 242, 97–117.

Glaser, C.A., Anguio, F.J., and Roosey, J.A. (1994). Animal-associated opportunistic infections among persons infected with the human immunodeficiency virus. *Clinical Infection and Disease* 18, 14–24.

Goddard, J. (1966). Mating and courtship of the black rhinoceros (*Diceros bicornis* L). *East African Wildlife Journal* 4, 69–75.

Godley, B.J. (2009). Introduction: Forensic methods in conservation research. *Endangered Species Research* 9, 167–168.

Goede, A.D. and De Bruin, M. (1984). The use of bird feather parts as a monitor for metal pollution. *Environmental Pollution (B)* 8, 281–298.

Goff, M.L. (2000). *A Fly for the Prosecution: How Insect Evidence Helps Solves Crimes*. Harvard University Press, Cambridge, MA, USA.

Goios, A., Amorim, A., and Pereira, L. (2006). Mitochondrial DNA pseudogenes in the nuclear genome as possible sources of contamination. *International Congress Series* 1288, 697–699.

Goldstein, T., Mazet, J.A.K., Zabka, T.S., Langlois, G., Colegrove, K.M., Silver, M., Bargu Ates, S., Van Dolah, F., Leighfield, T., Conrad, P.A., Barakos, J., Williams, D.C., Dennison, S., Haulena, M.A., and Gulland, F.M.D. (2008). Novel symptomatology and changing epidemiology of domoic acid toxicosis in California sea lions (*Zalophus californianus*): An increasing risk to marine mammal health. *Proceedings of the Royal Society of London, Series B: Biological Sciences* 275, 267–276.

Goldstein, T., Zabka, T.S., De Long, R.L., Wheeler, E.A., Ylitalo, G., Bargu, S., Silver, M., Leighfield, T., Van Dolah, F., Langlois, G., Sidor, I., Dunn, J.L., and Gulland, F.M.D. (2009). The role of domoic acid in abortion and premature parturition of California sea lions (*Zapholus californianus*) on San Miguel Island, California. *Journal of Wildlife Diseases* 45(1), 91–108.

Gomes, L., Gomes, G., and Desuó, I.C. (2009). A preliminary study of insect fauna on pig carcasses located in sugarcane in winter in southeastern Brazil. *Medical and Veterinary Entomology* 23, 155–159.

Gomes, V., Ribeiro, R., and Carretero, M.A. (2011). Effects of urban habitat fragmentation on common small mammals: Species versus communities. *Biodiversity and Conservation* 20, 3577–3590.

Gómez, A. and Aguirre, A.A.A. (2008). Infectious diseases and the illegal wildlife trade. *Annals of the New York Academy of Sciences* 1149, 16–19.

Gonzalez, S. and Brum-Zorrilla, N. (1995). Karyological studies of the South American rodent *Myocastor coypus* Molina 1782 (Rodentia: Myocastoridae). *Revista Chilena de Historia Natural* 86, 215–226.

Gorea, R.K. (2005). Crime scene investigation – A holistic view. *Journal of the Indian Academy of Forensic Medicine* 27, 3–4.

Gosler, A. (2004). Birds in the hand. In: W.J. Sutherland, I. Newton, and R.E. Green, Eds. *Bird Ecology and Conservation*. Oxford University Press, Oxford, UK.

Goss, R.J. (1969). *Principles of Regeneration*. Academic Press, London, UK.

Gosselin, M., Wille, S.M.R., del Mar Ramírez Fernandez, M., Samyn, N., De Boeck, G., and Bourel, B. (2011a). Entomotoxicology, experimental set-up and interpretation for forensic toxicologists. *Forensic Science International* 208(1–3), 1–9.

Gosselin, M., Di Fazio, V., Wille, S.M.R., del Mar Ramírez Fernandez, M., Samyn, N., Bourel, B., and Rasmont, P. (2011b). Methadone determination in puparia and its effect on the development of *Lucilia sericata* (Diptera: Calliphoridae). *Forensic Science International* 209(1–3), 154–159.

Gould, F.D. (1998). A natural history of North American box turtles – With information on the triage of injured and ill turtles. *Journal of Wildlife Rehabilitation* 21(1), 3–10.

Graham, S. (1996). Issues of surplus animals. In: Kleiman, D.E., Allen, M.E., Thompson, K.V., and Lumpkin, S., Eds. *Wild Mammals in Captivity*. The University of Chicago Press, Chicago, IL, USA.

Graham-Hall, J. and Smith, G. (2006). *The Expert Witness*. XPL Publishing, St. Albans, UK.

Grassberger, M. and Frank, C. (2004). Initial study of arthropod succession on pig carrion in a central European urban habitat. *Journal of Medical Entomology* 41(3), 511–523.

Grassberger, M. and Reiter, C. (2001). Effect of temperature on *Lucilia sericata* (Diptera: Calliphoridae) with special reference to the isomegalen- and isomorphen-diagram. *Forensic Science International* 120, 32–36.

Grassberger, M. and Reiter, C. (2002). Effect of temperature on development of the forensically important holarctic blowfly *Protophormia terraenovae* (Robineau-Desvoidy) (Diptera: Calliphoridae). *Forensic Science International* 128, 177–182.

Green, G.H. (1980a). Decrease in wing length of skins of ringed Plover and Dunlin. *Ringing and Migration* 3, 27–28.

Green, P.D. (1979). Identification of Cases and Preparation for Court. *Canadian Veterinary Journal* 20, 8–12.

Green, P.D. (1980b). Protocols in medicolegal veterinary medicine II. Cases involving death due to gunshot and arrow wounds. *Canadian Veterinary Journal* 21, 343–346.

Greenberg, B. (1990). Nocturnal oviposition behaviour of blow flies (Diptera: Calliphoridae). *Journal of Medical Entomology* 27(5), 807–810.

Greenberg, B. (1991). Flies as forensic indicators. *Journal of Medical Entomology* 28(5), 565–577.

Greenberg, B. and Kunich, J.C. (2002). *Entomology and the Law: Flies as Forensic Indicators*. Cambridge University Press, Cambridge, UK.

Greenwood, A.G. (1996). Veterinary support for in situ avian conservation programmes. *Bird Conservation International* 6, 285–292.

Greenwood, J.G. (1979). Post-mortem shrinkage of Dunlin *Calidris Alpina* skins. *Bulletin of the British Ornithologists' Club* 99, 143–145.

Gregory, M.R. (1978). Accumulation and distribution of virgin plastic granules on New Zealand beaches. *New Zealand Journal of Marine & Freshwater Research* 12(4), 399–414.

Gregory, N. (2011). Relationship between pathology and pain severity in humans and animals. *Animal Welfare* 19(4), 437–448.

Griffin, D.R. (1944). Echolocation by blind men, bats and radar. *Science* 100, 589–590.

Griffith, B., Scott, J.M., Carpenter, J.W., and Reed, C. (1989). Translocation as a species conservation tool: Status and strategy. *Science* 245, 477–480.

Griffiths, R.A., Roberts, J.M., and Sims, A. (1987). A natural hybrid newt, *Triturus helveticus* × *T. vulgaris*, from a pond in mid-Wales. *Journal of Zoology, London* 213, 133–140.

Grifo, F. and Lovejoy, T.E., Eds. (1997). *Biodiversity and Human Health*. Island Press, Washington, DC, USA.

Grogan, A., Wilson, R., and Wandenabeele, S. (2011). Implications of fitting monitoring devices to wild animals. *Veterinary Record* 169, 613.

Groves, C.P. and Humphrey, N.K. (1973). Asymmetry in gorilla crania: Evidence of lateralized brain function. *Nature* 244, 53–54.

Grubb, T.C. (1989). Ptilochronology: Feather growth bars as indicators of nutritional status. *Auk* 106, 314–3290.

Grubb, T.C. (2006). *Ptilochronology. Feather Time and Biology of Birds*. Oxford University Press, Oxford, UK.

Grubb, T.G. and Bowerman, W.W. (1997). Variations in breeding bald eagle responses to nets, light planes and helicopters. *Journal of Raptor Research* 31, 213–222.

Grue, H. and Jensen, B. (1979). Review of the formation of incremental lines in tooth cementum of terrestrial mammals. *Danish Review of Game Biology* 11(3), 48.

Gruner, S.V., Slone, D.H., and Capinera, J.L. (2007). Forensically important Calliphoridae (Diptera) associated with pig carrion in rural north-central Florida. *Journal of Medical Entomology* 44(3), 509–515.

Guarino, F.M., Angelini, F., and Cammarota, M. (1995). A skeletochronological analysis of the syntopic amphibian species from southern Italy. *Amphibia-Reptilia* 16, 297–302.

Guatelli-Steinberg, D. and Skinner, M. (2000). Prevalence and etiology of linear hypoplasia in monkeys and apes from Asia and Africa. *Folia Primatologica* 71, 115–132.

Gudmundson, F. (1972). Grit as an indicator of the overseas origin of certain birds occurring in Iceland. *Ibis* 114, 582.

Gulland, F. (2000). Domoic acid toxicity in the California sea lions (*Zalophus californianus*) stranded along the central California coast, May–October, 1998. U.S. Department of Commerce, NOAA Technical Memo NMFS-OPR-17.

Gulland, F., Haulera, M., Fauquier, D., Lander, M.E., Zabka, T., Duer, R., and Langlois, G. (2002). Domoic acid toxicity in the California sea lion (*Zalophus californianus*): Clinical signs, treatment and survival. *Veterinary Record* 150, 475–480.

Gunn, A. (2009). *Essential Forensic Biology*, 2nd edn. Wiley-Blackwell, Oxford, UK.

Gunnarsson, G., Waldenström, J., and Fransson, T. (2012). Direct and indirect effects of winter harshness on the survival of Mallards *Anas platyrhynchos* in northwest Europe. *Ibis* 154, 307–317.

Gupta, S.K., Bhagavatula, J., Thangaraj, K., and Singh, L. (2010). Establishing the identity of the massacred tigress in a case of wildlife crime. *Forensic Science International: Genetics* 5(1), 74–75.

Gupta, S.K., Kumar, A., and Nigam, P. (2011). Multilocus genotyping of Asian elephant ivory: A case study in suspected wildlife crime. *Journal of Forensic Research* (Special issue), 2(1), 1–3.

Gupta, S.K., Thangaraj, K., and Singh, L. (2006). A simple and inexpensive molecular method for sexing and identification of the forensic samples of elephant origin. *Journal of Forensic Sciences* 51(4), 805–807.

Gupta, S.K., Verma, S.K., and Singh, L. (2005). Molecular insight into a wildlife crime: The case of the peafowl slaughter. *Forensic Science International* 154, 214–217.

Gutowsky, S., Janssen, M.H., Arcese, P., Kyser, T.K., Ethier, D., Wunder, M.B., Bertram, D.F., McFarlane Tranquilla, L., Lougheed, C., and Norris, D.R. (2009). Concurrent declines in nestling diet quality and reproductive success of a threatened seabird over 150 years. *Endangered Species Research* 9, 247–254.

Haag, D. (1987). Brütende straßentauben als ursache einer invasion von *Dermanyssus gallinae* (De Geer, 1778). *Derpraktische Schädlingsbekämpfer* 8/88, 180.

HCEAC. (2012a). *Wildlife Crime*. Third Report of Session 2012-13. Vol 1. House of Commons Environmental Audit Committee. The Stationery Office, London, UK.

HCEAC. (2012b). *Wildlife Crime*. Third Report of Session 2012-13. Vol 2. House of Commons Environmental Audit Committee. The Stationery Office, London, UK.

Haftorn, S. (1970). Variation in measurements of the Willow Tit *Parus montanus* together with a method for sexing live birds and data on the degree of shrinkage in size after skinning. *Fauna Norvegica Series Cinclus* 5, 16–26.

Hagen, R. (2012). Cross-sectional imaging: The key to anatomy. *Veterinary Record* 170, 17–18.

Haglund, W.D., Reay, D.T., and Swindler, D.R. (1989). Canid scavenging/disarticulation sequence of human remains in the Pacific Northwest. *Journal of Forensic Sciences* 34(3), 587–606.

Haglund, W.D. and Sorg, M.H., Eds. (2002). *Forensic Taphonomy. The Postmortem Fate of Human Remains*. CRC Press, Boca Raton, FL, USA.

Hailey, A. (2000). Assessing body mass condition in the tortoise *Testudo hermanni*. *Herpetological Journal* 10, 57–61.

Haitao, S., Parham, J.F., Lau, M., and Tien-Hsi, C. (2007). Farming endangered turtles to extinction in China. *Conservation Biology* 21(1), 5–6.

Hall, R.D. (2010). *The Forensic Entomologist as Expert Witness*. Taylor & Francis Group/CRC Press, Boca Raton, FL, USA.

Hall, S.C.B. and Parmenter, C.J. (2006). Larvae of two signal fly species (Diptera: Platystomatidae), *Duomyia foliata* McAlpine and *Plagiostenopterina enderleini* Hendel, are scavengers of sea turtle eggs. *Australian Journal of Zoology* 54, 245–252.

Halliday, T.R. and Verrell, P.A. (1988). Body size and age in amphibians and reptiles. *Journal of Herpetology* 22, 253–265.

Hamilton, M.D. and Erhart, E.M. (2012). Forensic evidence collection and cultural motives for animal harvesting. In: Huffman, J.E. and Wallace, J.R., Eds. *Wildlife Forensics Methods and Applications*. Wiley-Blackwell, Oxford, UK, pp. 65–76.

Hammerton, D. (1997). *An Introduction of Water Quality in Rivers, Coastal Waters and Estuaries*. CIWEM, London, UK.

Hammond, E.C. and Jones, C. (1992). An examination of a strand of 30,000 and 70,000 year old mammoth hair. In: Bailey, G.W., Bentley, J., and Small, J.A., Eds. *Proceedings of the 50th Annual Meeting of the Electron Microscopy Society of America*, San Francisco, CA, USA. Electron Microscopy Society, Vol. 50(2), pp. 1100–1101.

Handy, M.K., May, K.N., and Powers, J.J. (1961). Some physical and physiological factors affecting poultry bruises. *Poultry Science* 40, 790–795.

Hangay, G. and Dingley, M. (1985). *Biological Museum Methods*, Vol. 1. Academic Press, London, UK.
Harcourt, R.A. (1971). The palaeopathology of animal skeletal remains. *Veterinary Record* 89, 267–272.
Hardey, C.R. and Martin, D.S. (2012). Plants and wildlife forensics. In: Huffman, J.E. and Wallace, J.R., Eds. *Wildlife Forensics Methods and Applications*. Wiley-Blackwell, Oxford, UK, pp. 145–157.
Hargitai, R., Hegyi, G., and Török, J. (2012). Winter body condition in relation to age, sex and plumage ornamentation in a migratory songbird. *Ibis* 154, 410–413.
Härkönen, T.J. (1986). *Guide to the Otoliths of the Bony Fishes of the Northeast Atlantic*. Danbiu, Hellerup, Denmark.
Harper, R.G., Kenigsberg, K., Sia, C.G., Horn, D., Sterne, D., and Bongiovi, V. (1980). Xiphopagus conjoined twins: A 300-year review of the obstetric, morphopathologic, neonatal, and surgical parameters. *American Journal of Obstetrics and Gynecology* 137, 617–629.
Harris, M.P. (1980). Post-mortem shrinkage of wing and bill of Puffins. *Ringing & Migration* 3, 27–28.
Harris, S. and Yalden, D.W., Eds. (2008). *Mammals of the British Isles; Handbook*, 4th edn. The Mammal Society, Southampton, UK.
Harrison, M.E., Page, S.E., and Limin, S.H. (2009). The global impact of Indonesian forest fires. *Biologist* 56(3), 156–163.
Hart, M. and Budgen, P. (2008). Forensic record-keeping and documentation of samples. In: Cooper, J.E., Baker, B.W., and Cooper, M.E., Eds. Special issue: Forensic Science and Herpetology. *Applied Herpetology* 5(4), 386–401.
Hartley, M., and Lysons, R. (2011). Development of the England Wildlife Health Strategy – A framework for decision makers. *Veterinary Record* 168, 158.
Harty, F.M., Ver Steeg, J.M., Heidorn, R.R., and Harty, L. (1991). Direct mortality and reappearance of small mammals in an Illinois grassland after a prescribed burn. *Natural Areas Journal* 11(2), 114–118.
Harwood, L.A., Norton, P., Day, B., and Hall, P.A. (2002). The harvest of beluga whales in Canada's western Arctic: Hunter-based monitoring of the size and composition of the catch. *Arctic* 55, 10–20.
Hastings, B.E., Gibbons, L.M., and Williams, J.E. (1992). Parasites of free-ranging mountain gorillas: Survey and epidemiological factors. *Proceedings of the American Association of Zoo Veterinarians and American Association of Wildlife Veterinarians*, Oakland, CA, USA, pp. 301–302.
Hausman, L.A. (1920). Structural characteristics of the hair of mammals. *The American Naturalist* 154, 496–523.
Hawk, D. (2011). Society for Wildlife Society Science. In: Huffman, J.E. and Wallace, J.R., Eds. *Wildlife Forensics Methods and Applications*. Wiley-Blackwell, Oxford, UK.
Hawksworth, D.L. (2011). Biodiversity and conservation of insects and other invertebrates. *Biodiversity and Conservation* 20, 2863–2866.
Hayden, T.J., Lynch, J.M., and O'Corry-Crowe, G. (1994). Antler growth and morphology in a feral sika deer (*Cervus nippon*) population in Killarney, Ireland. *Journal of Zoology, London* 232(1), 21–35.
Haynes, G. (1982). Utilization and skeletal disturbances of North American prey carcasses. *Arctic* 35(2), 266–281.
He, L., Wang, S., Miao, X., Wu, H., and Huang, Y. (2007). Identification of necrophagous fly species using ISSR and SCAR markers. *Forensic Science International* 168(2–3), 148–153.
Hebert, P.D.N., Cywinska, A., Ball, S.L., and DeWaard, J.R. (2003a). Biological identifications through DNA barcodes. *Proceedings of the Royal Society of London, Series B* 270, 313–321.
Hebert, P.D.N., Ratnasingham, S., and DeWaard, J.R. (2003b). Barcoding animal life: Cytochrome *c* oxidase subunit 1 divergences among closely related species *Proceedings of the Royal Society of London, Series B* 270, S96–S99.
Hebert, P.D.N., Stoeckle, M.Y., Zemlak, T.S., and Francis, C.M. (2004). Identification of birds through DNA barcodes. *PLoS Biology* 2, e312.
Heindenreich, M. (1997). *Birds of Prey: Medicine and Management*. Blackwell, Oxford, UK.

Heldmaier, G. and Rug, T. (1992). Body temperature and metabolic rate during natural hypothermia in endotherms. *Journal of Comparative Physiology B Biochemistry System Environmental Physiology* 162(8), 696–706.

Hemelaar, A. (1985). An improved method to estimate the number of year rings resorbed in phalanges of *Bufo bufo* (L.) and its application to populations from different latitudes. *Amphibia-Reptilia* 6, 323–341.

Hernández, M., Robinson, I., Aguilar, A., González, L.M., López-Jurado, L.F., Reyero, M.I., Cacho, E., Franco, J., López-Rodas, V., and Costas, E. (1998). Did algal toxins cause monk seal mortality? *Nature* 393(6680), 28–29.

Herremans, M. (1985). Post-mortem changes in morphology and its relevance to biometrical studies. *Bulletin of the British Ornithology Club* 105(3), 89–91.

Hersteinsson, P. and Macdonald, D.W. (1996). Diet of arctic foxes (*Alopex lagopus*) in Iceland. *Journal of Zoology, London* 240, 457–474.

Hewer, H.R. (1974). *British Seals*. New Naturalist Series 57. Collins, London, UK.

Hewison, A.J.M., Vincent, J.P., Angibault, J.M., Delorme, D., Van Laere, G., and Gaillard, J.M. (1999). Test of estimation of age from tooth wear on roe deer of known age: Variation within and among populations. *Canadian Journal of Zoology* 77, 58–67.

Hewson, R. (1968). Weights and growth rates in the mountain hare *Lepus timidus scoticus*. *Journal of Zoology, London* 154, 249–262.

Heyer, W.R., Donnelly, M.A., McDiarmid, R.W., Hayek, L.-A.C., and Foster, M.S., Eds. (1994). *Measuring and Monitoring Biological Diversity. Standard Methods for Amphibians*. Smithsonian Institution Press, Washington, DC, USA.

Hildebrand, S.F. (1930). Duplicity and other abnormalities in diamond-back terrapins. *Journal of the Elisha Mitchell Science Society* 46, 41–53.

Hildebrand, S.F. (1938). Twinning in turtles. *Journal of Heredity* 29, 243–253.

Hill, A. (2011). The illegal ivory road to China – And vanishing elephants. *SWARA Journal of the East African Wildlife Society* October–December, 17.

Hillier, M.L. and Bell, L. (2007). Differentiating human bone and animal bone: A review of histological methods. *Journal of Forensic Science* 52(2), 249–263.

Hilton, G.M. and Cuthbert, R.J. (2010). The catastrophic impact of invasive mammalian predators on birds of the UK Overseas Territories; a review and synthesis. *Ibis* 152, 443–458.

Hinton, P. and Jennings, D. (2007). Quality management of archaeology in Great Britain: Present practice and future. In: Willems, W.J.H. and van der Dries, M.H., Eds. *Quality Management in Archaeology*. Oxbow, Oxford, UK.

Hiraga, T. and Dennis, S.M. (1993). Congenital duplication. *Veterinary Clinic of North America: Food Animal Practice* 9, 145–161.

Hocking, K. and Humle, T. (2009a). *Best Practice Guidelines for the Prevention and Mitigation of Conflict between Humans and Great Apes*. IUCN/SSC Primate Specialist Group, Gland, Switzerland.

Hocking, K. and Humle, T. (2009b). Lignes directrices pour de meilleures pratiques en matière de prévention et d'atténénuation des conflits entre humains et grandes signes. IUCN, Gland, Switzerland.

Hodgkinson, T. and James, M. (2009). *Expert Evidence: Law and Practice*, 3rd edn. Sweet and Maxwell, London, UK.

Hoefs, M. (2001). Prevalence of mandibular oligodonty in wild sheep: Possible evolutionary relevance. *Journal of Mammalogy* 82, 401–406.

Hoffman, B.D. (2011). Eradication of populations of an invasive ant in northern Australia: Successes, failures and lessons for management. *Biodiversity and Conservation* 20, 3267–3278.

Hohn, A.A., Read, A.J., Fernandez, S., Vidal, O., and Findley, L.T. (1996). Life history of the vaquita, *Phocoena sinus* (Phocoenidae, Cetacea). *Journal of Zoology, London* 239(2), 235–251.

Holdaway, R.N. and Worthy, T.H. (1996). Diet and biology of the laughing owl *Sceloglaux albifacies* (Aves: Strigidae) on Takaka Hill, Nelson, New Zealand. *Journal of Zoology, London* 239, 545–572.

Holden, A.L. and Marsden, K. (1967). Organochlorine pesticides in seals and porpoises. *Nature* 216, 1274–1276.

Home Office. (2005). Code of practice. National Intelligence Model National Centre for Policing Excellence, Wyboston, Bedford, UK. http://www.npia.police.uk/en/docs/National_Intelligence_Model_C_of_P.pdf

Hone, J. (2007). *Wildlife Damage Control Principles for the Management of Damage by Vertebrate Pests*. CABI, Oxfordshire, UK.

Horenstein, M.B., Linhares, A.X., Rosso, B., and García, M.D. (2007). Species composition and seasonal succession of saprophagous calliphorids in a rural area of Córdoba, Argentina. *Biology Research* 40, 163–171.

Horenstein, M.B., Linhares, A.X., Rosso de Ferradas, B., and García, D. (2010). Decomposition and dipteran succession in pig carrion in central Argentin: Ecological aspects and their importance in forensic science. *Medical and Veterinary Entomology* 24, 16–25.

Horn, J.W., Arnett, E.B., and Kunz, T.H. (2008). Behavioural responses of bats to operating wind turbines. *Journal of Wildlife Management* 72, 123–132.

Hornocker, M. and Negri, S. (2009). *Cougar, Ecology and Conservation*. The University of Chicago Press, Chicago, IL, USA.

House of Commons Science and Technology Committee. (2005). Forensic science on trial. House of Commons Science and Technology Committee 7th Report, Session 2004/2005. http://www.publications.parliament.uk/pa/cm200405/cmselect/cmsctech/96/96i.pdf

Howell, A.B. (1925). Asymmetry in the crania of mammals. *Proceedings of the U.S. National Museum* 67(art. 27), 1–18.

Høye, T.T. (2006). Age determination in roe deer – A new approach to tooth wear evaluated on known age individuals. *Acta Theriologoca* 51, 205–214.

Hoyle, F. and Wickramasinghe, C. (1986). *Archaeopteryx, the Primordial Bird: A case of fossil forgery*. Christopher Davies, Swansea, UK.

Hoyt, J.A. (1994). *Animals in Peril: How Sustainable Use Is Wiping Out the World's Wildlife*. Avery Publishing Group, New York, USA.

Hsieh, H.-M., Chiang, H.-L., Tsai, L.-C., Lai, S.-Y., Huang, N.-E., Linacre, A., and Lee, J.-I. (2001). Cytochrome *b* gene for species identification of the conservation animals. *Forensic Science International* 122(1), 7–18.

Hsieh, H.-M., Huang, L.-H., Tsai, L.-C., Kuo, Y.-C., Meng, H.-H., Linacre, A., and Lee, J.C.-I. (2003). Species identification of rhinoceros horns using the cytochrome b gene. *Forensic Science International* 136, 1–11.

Hsieh, H.-M., Huang, L.-H., Tsai, L.-C., Liu, C.-L., Kuo, Y.-C., Hsiao, C.-T. et al. (2006). Species identification of *Kachuga tecta* using the cytochrome b gene. *Journal of Forensic Science* 51(1), 52–56.

Hudson, P.R., Rizzoli, A., Grenfel, B.T., Heesterbeek, H., and Dobson, A.P., Eds. (2002). *The Ecology of Wildlife Diseases*. Oxford University Press, Oxford, UK.

Huffman, J.E. and Wallace, J.R., Eds. (2012). *Wildlife Forensics Methods and Applications*. Wiley-Blackwell, Oxford, UK.

Hunter, J. (1794). *A Treatise on the Blood, Inflammation and Gun Shot Wounds*. G. Nicholl, London, UK.

Huss, H.H. (1995). Quality and quality changes in fresh fish. FAO Fisheries technical paper 348. Food and Agriculture Organization of the United Nations, Rome, Italy.

Hwang, C. and Turner, B.D. (2005). Spatial and temporal variability of necrophagous Diptera from urban to rural areas. *Medical and Veterinary Entomology* 19, 379–391.

Hyeroba, D., Apell, P., and Otali, E. (2011). Managing a speared alpha male chimpanzee (*Pan troglodytes*) in Kibale National Park, Uganda. *Veterinary Record* 169, 658.

Ibáñez-Álamo, J.D., Sanllorente, O., and Soler, M. (2012). The impact of researcher disturbance on nest predation rates: A meta-analysis. *Ibis* 154, 5–14.

ILAC Guide 19. (2002). *Guidelines for Forensic Science Laboratories*. International Laboratory Accreditation Commission, Rhodes, Australia.

ILAR. (2011). Spineless wonders: Welfare and use of invertebrates in the laboratory and classroom. *ILAR Journal* 52(5), 121–220.

Iller, M. (2006). *Civil Evidence the Essential Guide*. Sweet & Maxwell, London, UK.

Imam, E., Yahya, H.S., and Malik, I. (2002). A successful mass translocation of commensal rhesus monkeys *Macaca mulatta* in Vrindaban, India. *Oryx* 36(1), 87–93.

Imrie, F. and Lord, J. (1998). Biologist in the witness box. *Biologist* 452, 62–66.

Ippen, R., Nickel, S., and Schröder, H.-D. (1995). *Krankheiten des Jagdbaren Wildes*. Deutcher Landwiertschaftsverlag, Berlin, Germany.

Irvin, A.D., Cooper, J.E., and Hedges, S.R. (1972). Possible health hazards associated with the collection and handling of post-mortem zoological material. *Mammal Review* 2, 43–54.

Irwin, D., Kocher, T., and Wilson, A. (1991). Evolution of the cytochrome *b* gene of mammals. *Journal of Molecular Evolution* 32(2), 128–144.

Isiche, J., Hillerton, J.E., and Nowell, F. (1992). Colonization of the mouse cadaver by flies in southern England. *Medical and Veterinary Entomology* 6(2), 168–170.

ISO/IEC17025. (2005). General requirements for the competence of testing and calibration laboratories. International Organisation for Standardization (ISO) and International Electrotechnical Commission (IEC).

IUCN. (1998). IUCN guidelines for reintroductions. IUCN/SSC Reintroduction Specialist Group. IUCN, Gland, Switzerland.

IUCN. (2011). IUCN Standards and Petitions Subcommittee. Guidelines for Using the IUCN Red List Categories and Criteria. Version 9.0. Standards and Petitions Subcommittee. IUCN, Gland, Switzerland and Cambridge. http://www.iucnredlist.org/documents/RedListGuidelines.pdf

IVS. (2009). Instant view secure digital imaging, transmission, administration and asset management. Best Practices White Paper 2009. Instant View Secure Ltd. www.IVSecure.net

IWRC and NWRA. (1989). Wildlife Rehabilitation Minimum Standards and Accreditation Program. International Wildlife Rehabilitation Council, Suisan, CA, USA.

Jackson, O.F. (1980). Weight and measurement data on tortoises (*Testudo graeca* and *Testudo hermanni*) and their relationship to health. *Journal of Small Animal Practice* 21, 409–416.

Jacobson, E.R. (1992). The desert tortoise and upper respiratory trace disease. Prepared for the Desert Tortoise Preserve Committee, Inc., and U.S. Bureau of Land Management, Riverside, CA, USA.

Jaffe, F.A. (1991). *A Guide to Pathological Evidence for Lawyers and Police Officers*, 3rd edn. Thomson Professional Publishing, Scarborough, Ontario, Canada.

Jakob-Hoff, R. (2011). IUCN DRA tool needs survey. New Zealand Centre for Conservation Medicine, Auckland Zoological Park, New Zealand.

James, M.T. (1947). The flies that cause myiasis in man. Miscellaneous Publication 631. U.S. Department of Agriculture, Washington, DC, USA, pp. 1–175.

Jamieson, A. (2000). Forensic biology: Its past, present and future. *Biologist* 47(2), 69–73.

Jamieson, A. (2004). A rational approach to the principles and practice of crime scene investigation: I. Principles. *Science & Justice* 44, 3–7.

Jarofke, D. and Klös, H.-G. (1975). Uber einige Hauterkrankungen bei Zootierei des Zoologischen Garten Berlin. Sonderdruck aus Verhandlungsbericht des 17. Internationalen Symposiums über die Erkrankungen der Zootiere. Tunis, 1975. Akademie-Verlag, Berlin, Germany.

Jefferies, D.J. (1969). Causes of badger mortality in eastern counties of England. *Journal of Zoology, London* 157, 429–436.

Jeffrey, A., Morgan, B., and Rutty, G. (2007). The role of CT in forensic investigations. *Bulletin of the Royal College of Pathologists* 139, 22–23.

Jekl, V., Gumpenberger, M., Jeklova, E., Hauptman, K., Stehlik, L., and Knotek, Z. (2011). Impact of pelleted diets with different mineral composition on the crown size of mandibule cheek teeth and mandibular-relative density in degus (*Octodon degus*). *Veterinary Record* 168, 641. Doi: 10.1136/vr.d2012

Jenkins, P. and Mooney, H. (2006). The United States, China, and invasive species: Present status and future prospects. *Biological Invasion* 8(7), 1589–1593.

Jirón, L.F. and Cartín, V.M. (1981). Insect succession in the decomposition of a mammal in Costa Rica. *New York Entomological Society* LXXXIX(3), 158–165.

Johansson, M.P., McMahon, B.J., Höglund, J., and Segelbacher, G. (2012). Amplification success of multilocus genotypes from feathers found in the field compared with feathers obtained from shot birds. *Ibis* 154, 15–20.

Johns, B.J. (1995). Responses of chimpanzees to habituation and tourism in the Kibale forest, Uganda. *Journal of Biological Conservation* 78, 257–262.

Johnson, G.D., Krueger, H.O., and Balcomb, R.T. (1993). Effects on wildlife of Brace 10G applications to corn field in south-central Iowa. *Environmental Toxicology and Chemistry* 12, 1733–1739.

Johnson, M.B., Clifford, S.L., Goossens, B., Nyakanaa, S., Curran, B., White, L.J.T., Wickings, E.J., and Bruford, M.W. (2007). Complex phylogeographic history of central African forest elephants and its implications for taxonomy. *BMC Evolutionary Biology* 7, 244.

Johnson, M.K., Wofford, H., and Pearson, H.A. (1983a). Microhistological techniques for food habits analysis. Research Paper SO-199. Department of Agriculture, Forest Service, Southern Forest Experiment Station, New Orleans, LA, USA.

Johnson, M.K., Wofford, H., and Pearson, H.A. (1983b). Microhistological techniques for food habits analysis. Research Paper SO-199. U.S. Department of Agriculture, Forest Service, Southern Forest Experiment Station. 40 p. RP-SO-199, New Orleans, LA, USA.

Johnson, R.N. (2010). The use of DNA identification in prosecuting wildlife-traffickers in Australia: Do the penalties fit the crimes? In: Combating Wildlife Crime. Ed. Special issue. *Forensic Science, Medicine, and Pathology* 6(3), 211–216.

Johnson, R.N. (2012). Conservation genetics and wildlife forensics of birds. In: Huffman, J.E. and Wallace, J.R., Eds. *Wildlife Forensics Methods and Applications*. Wiley-Blackwell, Oxford, UK, pp. 293–317.

Johnston, J.P., Anthony, D., and Bloxham, Q.M.C. (1994). Rat eradication from Praslin Island: The St Lucia whiptail lizard conservation programme. *Dodo, Journal of Wildlife Preservation Trust* 30, 114–118.

Jones, M.L. and Swartz, S.L. (2002). Gray whale (*Eschrichtius robustus*). In: Perrin, W.F., Wursig, B., and Thewissen, J.G.M., Eds. *Encyclopedia of Marine Mammals*. Academic Press, New York, USA, pp. 524–536.

Jones, P. and Williams, R.E. (2009). *Crime Scene Processing and Laboratory Work*. CRC Press, Boca Raton, FL, USA.

Jori, F. (2001). La Criá de Roedores Tropicales (*Thryonomys swinderianus y Atherurus africanus*) como Fuente de Alimento en Gabón, Africa Central. Doctoral Thesis, Facultat de Veterinaria, Universitat Autonoma de Barcelona, Barcelona, Spain.

Kaartinen, S., Luoto, M., and Kojola, I. (2009). Carnivore-livestock conflicts: Determinants of wolf (*Canis lupus*) depredation on sheep farms in Finland. *Biodiversity and Conservation* 18, 3503–3517.

Kakkar, M., Ramani, S., Menon, G., Sankhe, L., Gaid, A., and Krishnan, S. (2011). 'Zoonoses? Not sure what that is …' An assessment of knowledge of zoonoses among medical students in India. *Transactions of the Royal Society of Tropical Medicine and Hygiene* 105, 254–261.

Kalema, G. (1995). Epidemiology of the intestinal parasite burden of mountain gorillas, *Gorilla gorilla beringei*, in Bwindi Impenetrable National Park, Uganda. *British Veterinary Zoological Society* 14, 4–6.

Kalema, G. (1999). Mountain gorilla veterinary interventions – Conservation vs welfare? *Gorilla Conservation News* 13, 10–12.

Kalema-Zikusoka, G. (2005). Protected areas, human livelihoods and healthy animals: Ideas for improvements in conservation and development interventions. In: Osofsky, S.A., Cleaveland, S., Karesh, W.B., Kock, M.D., Nyhus, P.J., Starr, L., and Yang, A., Eds. *Conservation and Development Interventions at the Wildlife/Livestock Interface: Implications for Wildlife, Livestock and Human Health*. IUCN, Gland, Switzerland and Cambridge, UK, pp. 111–120.

References and Further Reading

Kalema-Zikusoka, G., Kock, R.A., and Macfie, E.J. (2002). Scabies in free-ranging mountain gorillas (*Gorilla beringei beringei*) in Bwindi Impenetrable National Park, Uganda. *Veterinary Record* 150, 12–15.

Kalema-Zikusoka, G. and Lowenstine, L. (2001). Rectal prolapse in a free-ranging mountain gorilla (*Gorilla beringei beringei*): Clinical presentation and surgical management. *Journal of Zoo and Wildlife Medicine* 32(4), 509–513.

Kalema-Zikusoka, G., Rothman, J.M., and Fox, M.T. (2005). Intestinal parasites and bacteria of mountain gorillas (*Gorilla beringei beringei*) in Bwindi Impenetrable National Park, Uganda. *Primates* 46, 59–63.

Kaltenborn, N.P., Bjerke, T., Nyahongo, J.W., and Williams, D.R. (2006). Animal preferences and acceptability of wildlife management actions around Serengeti National Park, Tanzania. *Biodiversity and Conservation* 15, 4633–4649.

Karanuaweere, N.D., Ihalamulla, R.I., and Kumarasinghe, S.P.W. (2002). *Megaselia scalaris* (Diptera: Phoridae). Can live on ripe bananas – A potential health hazard? *Ceylon Medical Journal* 47, 9–10.

Karesh, W., Cook, R.A., Gilbert, M., and Newcombe, J. (2007). Implications of wildlife trade on the movement of avian influenza and other infectious diseases. *Journal of Wildlife Diseases* 43(3 Suppl), 55–59.

Karesh, W.B. (1995). Wildlife rehabilitation: Additional considerations for developing countries. *Journal of Zoo and Wildlife Medicine* 26(1), 2–9.

Karesh, W.B., Cook, R.A., Bennett, E.L., and Newcomb, J. (2005). Wildlife trade and global disease emergence. *Emerging Infectious Diseases* 11, 1000–1002.

Karlsson, A.O. and Holmlund, G. (2007). Identification of mammal species using species-specific DNA pyrosequencing. *Forensic Science International* 173(1), 16–20.

Keane, A. (2008). *The Modern Law of Evidence*. Oxford University Press, Oxford, UK.

Kear, K. (1962). Food selection in finches with special reference to inter-specific differences. *Proceedings of the Zoological Society of London* 138, 163–204.

Kelley, T.B., Nute, H.D., Zinszer, M.A. et al., Eds. (2008). *Underwater Crime Scene Investigation: A Guide for Law Enforcement*. Best Publishing Company, Flagstaff, AZ, USA.

Kelly, J.A., van der Linde, T.C., and Anderson, G.S. (2009). The influence of clothing and wrapping on carcass decomposition and arthropod succession during the warmer seasons in Central South Africa. *Journal of Forensic Sciences* 54(5), 1105–1112.

Kelsall, J.P. (1974). Snow goose primaries as indicators of age and sex. *Canadian Journal of Zoology* 52, 791–794.

Kelsall, J.P. (1984). The use of chemical profiles from feathers to determine the origins of birds. *Proceedings of Fifth Pan-Africa Ornithological Congress*, Lilongwe, Malawi, pp. 501–516.

Kelsall, J.P. and Burton, R. (1977). Identification of origins of lesser snow geese by X-ray spectrometry. *Canadian Journal of Zoology* 55, 718–732.

Kelsall, J.P. and Burton, R. (1979). Some problems in identification of origins of lesser snow geese by chemical profiles. *Canadian Journal of Zoology* 57, 2292–2302.

Kelsall, J.P. and Pannekoek, W.J. (1976). The mineral profile of plumage in captive lesser snow geese. *Canadian Journal of Zoology* 54, 301–305.

Kelsall, J.P., Pannekoek, W.J., and Burton, R. (1975a). Chemical variability in plumage of wild lesser snow geese. *Canadian Journal of Zoology* 53, 1369–1375.

Kelsall, J.P., Pannekoek, W.J., and Burton, R. (1975b). Variability in the chemical content of waterfowl plumage. *Canadian Journal of Zoology* 53, 1379–1386.

Kentner, E. and Streit, B. (1990). Temporal distribution and habitat preference of congeneric insect species found at rat carrion. *Pedobiologia* 34(6), 347–359.

Kenyon, K.W. and Kridler, E. (1969). Laysan albatrosses swallow ingestible matter. *Auk* 86, 339–341.

Kerbis Peterhans, J.C. (1990). The role of porcupines, leopards and hyenas in ungulate carcass disposal. Implications for paleoanthropology. PhD thesis, University of Chicago, Chicago, IL, USA.

Kerbis Peterhans, J.C.K. and Groske, T.P. (2001). The science of "man-eating" among lions *Panthera leo* with a reconstruction of the natural history of the "man-eaters of Tsavo". *Journal of East African Natural History* 96, 1–40.

Kerns, J. and Kerlinger, P. (2004). A study of bird and bat collision fatalities at the Mountaineer Wind Energy Center, Tucker County, West Virginia: Annual Report for 2003 [www document]. Report prepared for FPL Energy and the Mountaineer Energy Center Technical Review Committee. http://www.wvhighlands.org/Birds/MountaineerFinalAvianRpt-%203-15-04PKJK.pdf, pp. 1–39 (accessed 22 September 2011).

Kerr, K.C.R., Stoeckle, M.Y., Dove, C.J., Weigt, L.A., Francis, C.M., and Hebert, P.D.N. (2007). Comprehensive DNA barcode coverage of North American birds. *Molecular Ecology Notes* 7, 535–543.

Kienzler, J.M., Dahm, P.F., Fuller, W.A., and Ritter, A.F. (1984). Temperature-based estimation for the time of death in white-tailed deer. *Biometrics* 40, 849–854

Kierdorf, H. and Kierdorf, U. (1989a). Studies on the development of dentition in the European roe deer. *Zoologische Jahrbücher. Abteilung für Anatomie und Ontogenie der Tiere.* 119, 37–75. (In German, English abstract).

Kierdorf, H. and Kierdorf, U. (2000b). Roe deer antlers as monitoring units for assessing temporal changes in environmental pollution by fluoride and lead in a German forest area over 67-year period. *Archives of Environmental Contamination and Toxicology* 39, 1–6.

Kierdorf, H. and Kierdorf, U. (2003). Abnormal lower tusks in a male wild boar (*Sus scrofa* L.). *Zeitschrift fürJagdwissenschaft* 49, 150–155.

Kierdorf, H., Kierdorf, U., and Hommelsheim, S. (1991). Scanning electron microscopic observations on the development and structure of tooth enamel in Cervidae (*Mammalia: Ruminantia*). *Anatomia, Histologia, Embryologia* 20, 237–252.

Kierdorf, H., Kierdorf, U., Richards, A., and Sedlacek, F. (2000). Disturbed enamel formation in wild boars (*sus scofa* l.) from fluoride polluted areas in Central Europe. *The Anatomical Record* 259, 12–24.

Kierdorf, H., Kierdorf, U., Sedlacek, F., and Erdelen, M. (1996a). Mandibular bone fluoride levels and occurrence of fluoride induced dental lesions in populations of wild red deer (*Cervus elaphus*) from central Europe. *Environmental Pollution* 93, 75–81.

Kierdorf, H., Kierdorf, U., Sedlacek, F., and Fejerskov, O. (1996b). Structural changes in fluorosed dental enamel of red deer (*Cervus elaphus* L.) from a region with severe environmental pollution by fluorides. *Journal of Anatomy* 188, 183–195.

Kierdorf, U., and Becher, J. (1997). Mineralization and wear of mandibular first molars in red deer (*Cervus elaphus*) of known age. *Journal of Zoology, London* 241, 135–143.

Kierdorf, U. and Kierdorf, H. (1989b). A scanning electron microscopic study on surface lesions in fluorosed enamel of roe deer (*Capreolus capreolus* L.). *Veterinary Pathology* 26, 209–215.

Kierdorf, U. and Kierdorf, H. (1999). Dental fluorosis in wild deer: Its use as a biomarker of increased fluoride exposure. *Environmental Monitoring and Assessment* 57, 265–275.

Kierdorf, U. and Kierdorf, H. (2000a). Comparative analysis of dental fluorosis in roe deer (*Capreolus capreolus*) and red deer (*Cervus elaphus*): Interdental variation and species differences. *Journal of Zoology, London* 250, 87–93.

Kimmerle, E.H. and Baraybar, J.P. (2008). Eds. *Skeletal Trauma. Identification of Injuries Resulting from Human Rights Abuse and Armed Conflict.* CRC Press, Boca Raton, FL, USA.

Kintz, P., Tracqui, A., and Mangin, P. (1990). Toxicology and fly larvae on a putrified cadaver. *Journal of Forensic Sciences* 30, 243–246.

Kipps, E.K. et al. (2002). Skin density and its influence on buoyancy in the manatee (*Trichechus manatus latirostris*), harbor porpoise (*Phocoena phocoena*) and bottlenose dolphin (*Tursiops truncatus*). *Marine Mammal Science* 18(3), 765–778.

Kirk, P. (1953). *Crime Investigation.* InterScience, New York, USA.

Kirkwood, J. and Best, R. (1998). Treatment and rehabilitation of wildlife casualties: Legal and ethical aspects. *In Practice* 20, 214–216.

Kirkwood, J.K., Bennett, P.M., Jepson, P.D., Kuiken, T., Simpson, V.R., and Baker, J.R. (1997). Entanglement in fishing gear and other causes of death in cetaceans stranded on the coasts of England and Wales. *Veterinary Record* 141, 94–98.

Kirkwood, J.K. and Macgregor, S.K. (1998). Salmonellosis in provisioned free-living greenfinches (*Carduelis chloris*) and other garden birds. *Proceedings of the European Association of Zoo- and Wildlife Veterinarians Second Scientific Meeting* 21–24 May 1998, Chester, UK, pp. 229–233.

Kirkwood, J.K. and Sainsbury, A.W. (1996). Ethics of intervention for the welfare of free-living wild animals. *Animal Welfare* 5, 235–243.

Kirkwood, J.K., Sainsbury, A.W., and Bennett, P.W. (1994). The welfare of free-living wild animals: Methods of assessment. *Animal Welfare* 3, 257–273.

Kiska, J., Simon-Bouhet, B., Charlier, F., and Ridoux, V. (2010). Individual and group behavioural reactions of small delphinids to remote biopsy sampling. *Animal Welfare* 19(4), 411–418.

Kitano, T., Umetsu, K., Tian, W., and Osawa, M. (2007). Two universal primer sets for species identification among vertebrates. *International Journal of Legal Medicine* 121(5), 423–427.

Kitpipit, T., Linacre, A., and Tobe, S.S. (2009). Tiger species identification based on molecular approach. *Forensic Science International: Genetics* 2(1), 310–312.

Kjorlien, Y.P., Beattie, O.B., and Peterson, A.E. (2009). Scavenging activity can produce predictable patterns in surface skeletal remains scattering: Observations and comments from two experiments. *Forensic Science International* 188(1–3), 103–106.

Klein, H. and Marquard, R. (2005). *Feed Microscopy: Atlas for the Microscopic Examination of Feed containing Vegetable and Animal Products*. Agrimedia, Niedersachsen, Germany.

Kleinman, D.M., Dunne, E.F., and Taravella, M.J. (1998). Boa constrictor bite to the eye. *Archives of Ophthalmology* 116, 949–950.

Knecht, E. (2012). The use of hair morphology in the identification of mammals. In: Huffman, J.E. and Wallace, J.R., Eds. *Wildlife Forensics Methods and Applications*. Wiley-Blackwell, Oxford, UK, pp. 129–142.

Kneidel, K.A. (1984). Influence of carcass taxon and size on species composition of carrion-breeding Diptera. *American Midland Naturalist* 111, 57–63.

Knight, B., Ed. (2002). *The Estimation of the Time since Death in the Early Postmortem Period*. Arnold, London, UK.

Knight, M. (1948). The study of bird pellets as a subject of interest and instruction to amateur naturalists. *South-Eastern Naturalist and Antiquary* 1, 53–60.

Knight, M. (1968). *Be a Nature Detective*. Frederick Warne & Co, London, UK.

Knight, M. (2011). South Africa's rhino success story gets said new poaching chapter. *SWARA Journal of the East African Wildlife Society* October–December, 28–32.

Knox, A. (1993a). A statement from the Council of the British Ornithologists' Union. *Ibis* 135, 320–325.

Knox, A.G. (1993b). Richard Meinertzhagen – A case of fraud examined. *Ibis* 135, 320–325.

Knutson, R.M. (1987). *Flattened Fauna. A field Guide to Common Animals of Roads, Streets, and Highways*. Ten Speed Press, Berkeley, CA, USA.

Kocarek, P. (2002). Diel activity patterns of carrion-visiting Coleoptera studied by time-sorting pitfall traps. *Biologica, Bratislava* 57(2), 199–211.

Kocarek, P. (2003). Decomposition and Coleoptera succession on exposed carrion of small mammal in Opava, the Czech Republic. *European Journal of Soil Biology* 39(1), 31–45.

Kocher, T.D., Thomas, W.K., Meyer, A., Edwards, S.V., Pääbo, S., and Villablanca, F.X. (1989). Dynamics of mitochondrial DNA evolution in animals: Amplification and sequencing with conserved primers. *Proceedings of the National Academy of Sciences of the United States of America* 86, 6196–6200.

Kock, R.A. (1995). Wildlife utilization: Use it or lose it – A Kenyan perspective. In: Cooper, J.E., Ed. Special Issue on Wildlife Species for Sustainable Food Production. *Biodiversity and Conservation* 4(3), 241–256.

Köhnemann, S. and Pfeiffer, H. (2010). Application of mtDNA SNP analysis in forensic casework. *Forensic Science International: Genetics* 5(3), 216–221.

Koopman, H.N., Iverson, S.J., and Read, A.J. (2003). High concentrations of isovaleric acid in the fats of odontocetes: Variation and patterns of accumulation in blubber vs. stability in the melon. *Journal of Comparative Physiology B: Biochemical, Systemic, and Environmental Physiology* 173(3), 247–261.

Koopman, W.J.M., Kuiper, I., Klein-Geltink, D.J.A., Sabatino, G.J.H., and Smulders, M.J.M. (2011). Botanical DNA evidence in criminal cases: Knotgrass (*Polygonum aviculare* L.) as a model species. *Forensic Science International: Genetics* 6(3): 366–374.

Koppert, G., Dounais, E., Froment, A., and Pasquet, P. (1998). Consommation alimentaire dans trois populations forestières de la région côtière du Cameroun: Yass, Mvae et Bakola. In: Hladik, C.M., Hladik, A., and Pagezy, H., Eds. *L'alimentation en Forêt Tropicale. Interactions Bioculturelles et Perspectives de Developpement*. Prstom, Paris, France, pp. 477–496.

Kramarova, L.I., Kolaeva, S.G., Bronnikov, G.E., Ignatiev, A.D., and Krasts, I.V. (1993). Hypometabolic and hypothermic factors from small intestine of hibernating ground squirrels (*Citellus undulates*). *Canadian Journal of Physiology and Pharmacology* 71(3–4), 293–296.

Kranz, K.R. (1982). A note on the structure of tail hairs from a pygmy hippopotamus (*Choeropsis liberiensis*). *Zoo Biology* 1, 237–241.

Kraus, S.D., Brown, M.W., Caswell, N., Clark, C.W., Fujiwara, M., Kenney, R.D., Knowlton, A.R., Landry, S., Mayo, C.A., McLellan, W.A., Moore, M.J., Nowacek, D.P., Pabst, D.A., Read, A.J., and Rolland, R.M. (2005). North Atlantic right whales in crisis. *Science* 309(5734), 561–562.

Krautwald-Junghanns, M.E., Pees, M., Reese, S., and Tully, T. (2011). *Diagnostic Imaging of Exotic Pets*. Schutersche Verlagsgellschaft mbH and Co., Hannover, Germany.

Kruuk, H. and Sands, W.A. (1972). The aardwolf (*Proteles cristatus* Sparrman 1783) as predator of termites. *East Africa Wildlife Journal* 10, 211–227.

Kuiken, T. and Hartman, M.G. (1992). Cetacean pathology: Dissection techniques and tissue sampling. *Proceedings of the European Cetacean Society Workshop*, Leiden, the Netherlands.

Kunz, T.H., Arnett, E.B., Erickson, W.P., Hoar, A.R., Johnson, G.D., Larkin, R.P., Strickland, M.D., Thresher, R.W., and Tuttle, M.D. (2007). Ecological impacts of wind energy development on bats: Questions, research needs, and hypotheses. *Frontiers in Ecology and the Environment* 5, 315–324.

Lair, S., De Guise, S., and Martineau, D. (1998). Uterine adenocarcinoma with abdominal carcinomatosis in a beluga whale. *Journal of Wildlife Diseases* 34(2), 373–376.

Laist, D.W., Knowton, A.R., Mead, J.G., Collet, A.S., and Podesta, M. (2001). Collisions between ships and whales. *Marine Mammal Science* 17(1), 35–75.

Lajbner, Z. and Kotlík, P. (2011). PCR-RFLP assays to distinguish the Western and Eastern phylogroups in wild and cultured tench *Tinca tinca*. *Molecular Ecology Resources* 11, 374–377.

Lamb, C.R. and Lord, P.F. (1996). Reproduction of radiographs in scientific articles. Enhancement is not a dirty word. *Journal of Small Animal Practice* 37(4), 151–153.

Lance, V.A. (2003). Alligator physiology and life history: The importance of temperature. *Experimental Gerontology* 38, 801–805.

Landete-Castillejos, T., Currey, J.D., Estevez, J.A., Gaspar-López, E., García, A.J., and Gallego, L. (2007a). Influence of physiological effort of growth and chemical composition on antler bone mechanical properties. *Bone* 41, 794–803.

Landete-Castillejos, T., Estevez, J.A., Martinez, A., Ceacero, F., García, A.J., and Gallego, L. (2007b). Does chemical composition of antler bone reflect the physiological effort made to grow it? *Bone* 40, 1095–1102.

Lane, J. (1990). Eighteenth-century medical practice: A case study of Bradford Wilmer, Surgeon of Coventry, 1737–1813. *Social History of Medicine* 3, 369–86.

Lane, J. (2011). Re-defining the 3Rs concept; refinement in wildlife research and beyond. Poster presentation at *The Institute of Animal Technology Annual Congress 2011 Great Yarmouth, UK, and LASA Winter Meeting 2009* (*Animal Technology and Welfare*), Birmingham, UK.

Lang, M.D., Allen, G.R., and Horton, B.J. (2006). Blowfly succession from possum (*Trichosurus vulpecula*) carrion in a sheep-farming zone. *Medical and Veterinary Entomology* 20, 445–452.

Langlois, C. and Langis, R. (1995). Presence of airborne contaminants in the wildlife of northern Quebec. *Science of the Total Environment* 160–161, 391–402.

Langlois, N.E.I. (2007). The science behind the quest to determine the age of bruises – A review of the English language literature. *Forensic Science Medicine and Pathology* 3, 241–252.

Laurent, S. (1977). Precautions to be taken by the first investigators to arrive at the scene of a crime involving firearms. *International Criminal Police Review* 32, 274–277.

Law Commission. (2011). *Expert Evidence in Criminal Proceedings in England and Wales* (Law Com No. 325). The Law Commission, London, UK.

Lawrence, C. and Newman, R. (2000). Acting as a material witness. *In Practice* 22, 491–494.

Lawrence, M.J. and Brown, R.W. (1974). *Mammals of Britain. Their Tracks, Trails and Signs.* Blandford Press, London, UK.

Laws, R.M. (1966). Age criteria for the African elephant *Loxodonta africana*. *East Africa Wildlife Journal* 4, 1–37.

Lawson, B., Hughes, L.A., Peters, T., de Pinna, E., John, S.K., Macgregor, S.K., and Cunningham, A.A. (2001). Pulsed-field gel electrophoresis supports the presence of host-adapted *Salmonella typhimurium* strains in the British garden bird populations. *Applied and Environmental Microbiology* 77, 8139–8144.

Lawton, M.P.C. and Cooper, J.E. (2009). Wildlife crime scene visits. *Applied Herpetology* 6, 29–45.

Leakey, R. and Morrel, V. (2001). *Wildlife Wars, My Fight to Save Africa's Natural Treasures.* St Martin's Press, New York, USA.

Leclercq, M. (1996). A propos de l'entomofaune d'un cadaver de sanglier. *Bulletin et Annales de la Société Belge d'Entomologie* 132, 417–422.

Leclercq, M. and Verstraeten, C. (1992). Eboueurs entomologiques bénévoles dans les écosystémes terrestres: Observaion inédita. *Note Faunique de Gembloux* 25, 17–22.

Lee, G., Yoo, J., Leslie, B.J., and Ha, T. (2011). Single molecule analysis reveals three phases of DNA degradation by an exonuclease. *Nature Chemical Biology* 7(6), 367–374.

Lee, H.C. and Ladd, C. (2001). Preservation and collection of biological evidence. *Croatian Medical Journal* 42, 225–228.

Lee, J.C.-I. and Chang, J.-G. (1994). Random amplified polymorphic DNA polymerase chain reaction (RAPD PCR) fingerprints in forensic species identification. *Forensic Science International* 67(2), 103–107.

Lee, J.C.-I., Tsai, L.-C., Liao, S.-P., Linacre, A., and Hsieh, H.-M. (2009). Species identification using the cytochrome *b* gene of commercial turtle shells. *Forensic Science International: Genetics* 3(2), 67–73.

Leeuwenhoek, M.R. (1674, 1665, 1678). Microscopical observations from Mr. Leeuwenhoek, about blood, milk, bones, the brain, spittle, cuticula, sweat, fatt, teares; communicated in two letters to the publisher. *Philosophical Transactions* 9, 121–131.

Lefebvre, K.A., Powell, C.L., Busman, M., Doucette, G.J., Moeller, P.D.R., Silver, J.B., Miller, P.E., Hughes, M.P., Singaram, S., Silver, M.W., and Tjeerdema, R.S. (1999). Detection of domoic acid in northern anchovies and California sea lions associated with an unusual mortality event. *Natural Toxins* 7, 85–92.

Leighton, F.A. (1995). *The Toxicity of Petroleum Oils to Birds: An Overview. Wildlife and Oil Spills, Response, Research, and Contingency Planning.* Tri-State Bird Rescue & Research, Inc., Newark, DE, USA, pp. 10–22.

Leirs, H., Verhagen, R., Verheyen, W., Mwanjabe, P., and Mbise, T. (1996). Forecasting rodent outbreaks in Africa: An ecological basis for *Mastomys* control in Tanzania. *Journal of Applied Ecology* 33, 937–943.

Leonard, J.A. (2008). Ancient DNA applications for wildlife conservation. *Molecular Ecology* 17, 4186–4196.

Leopold, A. (1933). *Game Management.* University of Wisconsin Press, Madison, WI, USA.

Leslie, D., Ed. (2012). *Responsible Tourism – Concepts, Theory and Practice.* CABI, Wallingford, UK.

Letnic, M., Webb, J.K., and Shine, R. (2008). Invasive cane toads (*Bufo marinus*) cause mass mortality of freshwater crocodiles (*Crocodylus johnstoni*) in tropical Australia. *Biological Conservation* 141, 1773–1782.

Lever, C. (1977). *The Naturalised Animals of the British Isles*. Hutchinson, London, UK.

Levin, P.S., Ellis, J., Petrik, R., and Hay, M.E. (2002). Indirect effects of feral horses on estuarine communities. *Conservation Biology* 16, 1364–1371.

Li, H.-Y., Hua, T., and Yeh, W.-B. (2010). Amplification of single bulb mites by nested PCR: Species-specific primers to detect *Rhizoglyphus robini* and *R. setosus* (Acari: Acaridae). *Journal of Asia-Pacific Entomology* 13(4), 267–271.

Li, R. (2008). *Forensic Biology. Identification and DNA Analysis of Biological Evidence*. CRC Press, Boca Raton, FL, USA.

Li, Y.M. and Li, D.M. (1998). The dynamics of trade in live wildlife across the Guangxi border between China and Vietnam during 1993–1996 and its control strategies. *Biodiversity and Conservation* 7, 895–914.

Liao, C.-Y., Wang, S.-Y., Fei, C.-Y., Du, S.-J., Hsu, T.-S., and King, Y.-T. (2009). Effects of antemortem struggling behaviour on the quality of duck carcasses. *Veterinary Record* 164, 557–558.

Liddle, M. (1997). *Recreation Ecology. The Ecological Impact of Outdoor Recreation and Ecotourism*. Chapman & Hall, London, UK.

Liebermann, D.E. and Meadow, R.H. (1992). The biology of cementum increments (with an archaeological application). *Mammal Review* 22, 57–77.

Lightsey, J.D., Rommel, S.A., Costidis, A.M., and Pitchford, T.D. (2006). Methods used during gross necropsy to determine watercraft-related mortality in the Florida manatee (*Tricheus manatus latirostris*). *Journal of Zoo and Wildlife Medicine* 37(3), 262–275.

Linacre, A. (2008). The use of DNA from non-human sources. *Forensic Science International: Genetics* 1, 605–606.

Linacre, A., Ed. (2009). *Forensic Science in Wildlife Investigations*. CRC Press, London, UK.

Linacre, A., Gusmão, L., Hecht, W., Hellmann, A.P., Mayr, W.R., Parson, W., Prinz, M., Schneider, P.M., and Morling, N. (2011a). ISFG: Recommendations regarding the use of non-human (animal) DNA in forensic genetic investigations. *Forensic Science International: Genetics* 5(5), 501–505.

Linacre, A. and Lee, J.C.-I. (2005). Species determination: The role and use of the cytochrome b gene. *Methods in Molecular Biology* 297, 45–52.

Linacre, A. and Tobe, S.S. (2008). On the trail of tigers-tracking tiger in Traditional East Asian Medicine. *Forensic Science International: Genetics* 1(Suppl), 603–604.

Linacre, A. and Tobe, S.S. (2011). An overview to the investigative approach to species testing in wildlife forensic science. *Investigative Genetics* 2, 2.

Linacre, A. et al. (2011). ISFG: Recommendations regarding the use of non-human (animal) DNA in forensic genetic investigations. *Forensic Science International: Genetics* 5(5): 501–505. Doi: 10.1016/j.fsigen.2010.10.017

Link, J. (2010). *Ecosystem-Based Fisheries Management*. Cambridge University Press, Cambridge, UK.

Linnell, J.D.C., Odden, J., Smith, M.E., Aanes, R., and Swenson, J.E. (1991). Large carnivores that kill livestock: Do 'problem individuals' really exist? *Wildlife Society Bulletin* 27, 698–705.

Liping, W., Yutang, L., Zianghong, X., Hongfel, Z., and Yonglin, W. (1989). Study of the histology of digestive system of the ring-necked pheasant. Paper presented at *World Pheasant Association Meeting*, Beijing, China.

Liu, X., Zhang, L., and Hong, S. (2011). Global biodiversity research during 1900–2009: A bibliometric analysis. *Biodiversity and Conservation* 20, 807–826.

Liu, Z., Ren, B., Hao, Y., Zhang, H., Wei, F., and Li, M. (2008). Identification of 13 human microsatellite markers via cross-species amplification of fecal samples from *Rhinopithecus bieti*. *International Journal of Primatology* 29(1), 265–272.

Lochte, T. (1938). *Atlas der Menschlichen und Tierischen Haare*. Leipzig, Germany.

Lochte, T. (1952). Das mikroskopische bild des giraffen hares. *Der Zoologischer Garten* 9, 204–206.

Loe, L.E., Meisingset, E.L., Mysterud, A., Langvatn, R., and Stenseth, N.C. (2004). Phenotypic and environmental correlates of tooth eruption in red deer (*Cervus elaphus*). *Journal of Zoology, London* 262, 83–89.

Loeb, W. (1986). Clinical pathology, primates. In: Fowler, M.E., Ed. *Zoo and Wild Animal Medicine*, 2nd edn. W.B. Saunders Company, Philadelphia, PA, USA, pp. 705–709.

Long, C.V., Flint, J.A., and Lepper, P.A. (2010). Wind turbines and bat mortality: Doppler shift profiles and ultrasonic bat-like pulse reflection from moving turbine blades. *Journal of the Acoustical Society of America* 128, 2238–2245.

Long, C.V., Flint, J.A., and Lepper, P.A. (2011a). Insect attraction to wind turbines: Does colour play a role? *European Journal of Wildlife Research* 57, 323–331.

Long, C.V., Lepper, P.A., and Flint, J.A. (2011b). Ultrasonic noise emissions from wind turbines: Potential effects on bat species. *Proceedings of the 10th International Congress on Noise as a Public Health Problem*, London, UK, pp. 907–913.

Lorenz, K. (1961). *King Solomon's Ring*. Translated by Marjorie Kerr Wilson. Methuen, London, UK.

Lorenzini, R., Cabras, P., Fanelli, R., and Carboni, G.L. (2011). Wildlife molecular forensics: Identification of the Sardinian mouflon using STR profiling and the Bayesian assignment test. *Forensic Science International: Genetics* 5, 345–349.

Lovell, C.D. and Dolbeer, R.A. (1999). Validation of the United States Air Force bird avoidance model. *Wildlife Society Bulletin* 27, 167–171.

Lovell, N.C. (1990a). *Patterns of Injury and Illness in Great Apes. A Skeletal Analysis*. Smithsonian Institution Press, Washington, DC, USA.

Lovell, N.C. (1990b). Skeletal and dental pathology of free-ranging mountain gorillas. *American Journal of Physiological Anthropology* 81, 399–412.

Lubetkin, S.C. et al. (2008). Age estimation for young bowhead whales (*Balaena mysticetus*) using annual baleen growth increments. *Canadian Journal of Zoology/Revue Canadienne de Zoologie* 86(6), 525–538.

Lucas, A.M. and Stettenheim, P.R. (1972). *Avian Anatomy: Integument. Agricultural Handbook 362*. U.S. Department of Agriculture, Washington, DC, USA.

Luiz, M., Gonçalves, C., Araujo, A., Duarte, R., Pereira da Silva, J., Reinhard, K., Boucher, F., and Ferreira, L.F. (2002). Detection of *Giardia duodenalis* antigen in coprolites using a commercially available enzyme-linked immunosorbent assay. *Transactions of the Royal Society of Tropical Medicine and Hygiene* 96, 640–643.

Lundrigan, B. (1996). Standard methods for measuring mammals. In: Kleiman, D.E., Allen, M.E., Thompson, K.V., and Lumpkin, S., Eds. *Wild Mammals in Captivity*. The University of Chicago Press, Chicago, IL, USA.

Lundrigan, N. (2001). Crime scene priorities. *Law and Order Magazine* 49(5), 38–42.

Lüps, P. and Roper, J.J. (1990). Cannibalism in a female badger (*Meles meles*): Infanticide or predation? *Journal of Zoology, London* 221, 314–315.

Lusimbo, W.S. and Leighton, F.A. (1996). Effects of Prudhoe Bay crude oil on hatching success and associated changes in pipping muscles in embryos of domestic chickens (*Gallus gallus*). *Journal of Wildlife Diseases* 32(2), 209–215.

Lynn, W.G. and Ullrich, M.C. (1950). Experimental production of shell abnormalities in turtles. *Copeia* 4, 253–262.

Maas, A.K. (2006). The mobile exotic animal practice. *Journal of Exotic Pet Medicine* 15, 122–131.

Maas, S. (1998). Feral goats in Australia: Impacts and cost of control. In: Baker, R.O. and Crabb, A.C., Eds. *Proceedings of the Eighteenth Vertebrate Pest Conference*, University of California, Davis, CA, USA, pp. 100–106.

Macaldowie, C. (2011). Working on the wildside! Guest editor. *Animal Technology and Welfare* December 2011, ix.

Macdonald, D.W. (1987). *Running with the Fox*. Unwin Hayman, London, UK.

Macdonald, D.W. and Johnson, P.J. (1996). The impact of sport hunting: A case study. In: Dunstone, N. and Taylor, V.L., Eds. *The Exploitation of Mammal Populations*. Chapman and Hall, London, UK, pp. 260–205.

Mackness, B. and Sutton, R. (2000). Possible evidence for intraspecific aggression in a *Pliocene crocodile* from North Queensland. *Alcheringa* 24, 55–62.

Mackness, B.S., Cooper, J.E., Wilkinson, C. and Wilkinson, D. (2010). Palaeopathology of a crocodile femur from the Pliocene of eastern Australia. *Alcheringa: An Australasian Journal of Palaeontology* 34(4), 515–521.

Mader, D.R. (2005). *Reptile Medicine and Surgery*, 2nd edn. Elsevier, St. Louis, MO, USA.

Mahoney, R. (1966). *Laboratory Techniques in Zoology*. Butterworths, London, UK.

Makowski, J., Vary, N., McCutcheon, M., and Veys, P. (2010). *Microscopic Analysis of Agricultural Products*, 4th edn. American Oil Chemists Society, Urbana, IL, USA.

Male, G.E. (1816). *Epitome of Juridical or Forensic Medicine for the Use of Medical Men, Coroners and Barristers*. Printed for T. and G. Underwood, Fleet Street, London, UK.

Malisa, A.L., Gwakisa, P., Balthazary, S., Wasser, S.K., and Mutayoba, B.M. (2006). The potential of mitochondrial DNA markers and polymerase chain reaction-restriction fragment length polymorphism for domestic and wild species identification. *African Journal of Biotechnology* 5(18), 1588–1593.

Malo, J.E., Acebes, P., and Traba, J. (2011). Measuring ungulate tolerance to human with flight distance: A reliable visitor management tool? *Biodiversity and Conservation* 20, 3477–3488.

Malongs, R. (1996). Dynamique socio-economique du circuit commercial de viande de chasse a brazzaville. Wildlife Conservation Society, Bronx, New York, USA.

Manchester, S.J. and Bullock, J.M. (2000). The impacts of non-native species on UK biodiversity and the effectiveness of control. *Journal of Applied Ecology* 37(5), 845–864.

Manning, J.T. and Chamberlain, A.T. (1993). Fluctuating asymmetry, sexual selection and canine teeth in primates. *Proceedings of the Royal Society of London Series B* 251, 83–87.

Manning, T.H. (1978). Measurements and weights of eggs of the Canada Goose, *Branta canadensis*, analyzed and compared with those of other species. *Canadian Journal of Zoology* 56, 676–687.

Mañosa, S. (2001). Strategies to identify dangerous electricity pylons for birds. *Biodiversity and Conservation* 10, 1997–2012.

Mao, S., Dong, X., Fu, F., Seese, R.R., and Wang, Z. (2011). Estimation of post-mortem interval using an electric impedance spectroscopy technique: A preliminary study. *Science & Justice* 51, 135–138.

Mar, K.U., Lal, M., and Williams, D.L. (2007a). Ocular abnormalities in captive Indian elephants. Research article. www. asianelephantresearch.com http://www.bioone.org/servlet/linkout?suffix=i1042-7260-39-2-148-b27&dbid=16&doi=10.1638/2007-0008R1.1&key=10.1002/zoo.20099

Mar, K.U., Madhulal, V., and Williams, D.L. (2007b). Ocular abnormalities in captive Indian elephants, *EU-Asia Link Project Symposium*. Managing the health and reproduction of elephant populations in Asia, 8–10 October 2007, Faculty of Veterinary Medicine, Kasetsart University, Bangkok, Thailand, pp. 38–40.

Marchesi, L., Pedrini, P., and Sergio, F. (2002). Biases associated with diet study methods in the Eurasian eagle owl. *Journal of Raptor Research* 36(1), 11–16.

Markandya, A.T., Taylor, A., Longo, M., Murty, S., and Dhavala, K. (2008). Counting the cost of vulture decline – An appraisal of the human health and other benefits of vultures in India. *Ecological Economics* 67(2), 194–204.

Marquiss, M., Newton, I., Hobson, K.A., and Kolbeinsson, Y. (2012). Origins of irruptive migrations by Common Crossbills *Loxia curvirostra* into northwestern Europe revealed by stable isotope analysis. *Ibis* 154, 400–409.

Marra, P.P., Dove, C.J., Dolbeer, R., Dahlan, N.F., Heacker, M., Whatton, J.F., Diggs, N.E., France, C., and Henkes, G.A. (2009). Migratory Canada geese cause crash of US Airways Flight 1549. *Frontiers in Ecology and the Environment* 7, 297–301.

Marti, C.D., Bechard, M., and Jaksic, F.M. (2008). Food habits. In: Bird, D.M. and Bildstein, K.L., Eds. *Raptor Research and Management Techniques*. Hancock House, Dubuque, IA, USA.

Martin, A.R. and Reeves, R.R. (2000). Status of the monodontid whales. Report of the International Whaling Commission Scientific Committee. Annex I. J. *Cetacean Research Management* 2, 243–251.

Martin, D.L. (2012). Identification of reptile skin products using scale morphology. In: Huffman, J.E. and Wallace, J.R., Eds. *Wildlife Forensics Methods and Applications*. Wiley-Blackwell, Oxford, UK, pp. 161–195.

Martineau, D., Béland, P., Desgardins, C., and Lagacé, A. (1987). Levels of organochlorine chemicals in tissues of beluga whales (*Delphinapterus leucas*) from the St. Lawrence Estuary, Québec, Canada. *Archives of Environmental Contamination and Toxicology* 16(2), 137–147.

Martineau, D., Lagacé, A., Béland, P., Higgins, R., Armstrong, D., and Shurgart, L.R. (1988). Pathology of stranded beluga whales (*Delphinapterus leucas*) from the St. Lawrence Estuary, Québec, Canada. *Journal of Comparative Pathology* 98(3), 287–311.

Martineau, D., Lair, S., De Guise, S., and Béland, P. (1995). Intestinal adenocarcinoma in two beluga whales (*Delphinapterus leucas*) from the estuary of the St. Lawrence River. *Canadian Veterinary Journal* 36(9), 563–565.

Martineau, D., Lemberger, K., Dallaire, A., Labelle, P., Lipscombe, T.P., Michel, P., and Mikaeliam, I. (2002). Cancer in wildlife, a case study: Beluga from the St. Lawrence estuary, Québec, Canada. *Environmental Health Perspectives* 110(3), 285–292.

Martineau, D., Mikaelian, I., LaPointe, J.M., Labelle, P., and Higgins, R. (2003). Pathology of cetaceans: A case study: Beluga of the St. Lawrence Estuary. In: Vos, J.G., Bossart, G.D., Fournier, M., and O'Shea, T.J., Eds. *Toxicology of Marine Mammals, New Perspectives: Toxicology and the Environment*. Taylor & Francis Group, New York, USA, pp. 333–380.

Martínez, E., Duque, P., and Wolff, M. (2007). Succession patterns of carrion-feeding insects in Paramo, Colombia. *Forensic Science International* 166, 182–189.

Martínez, J.E., Calvo, J.F., Martínez, J.A., Zuberogoitia, I., Cerezo, E., Mantique, J., Gómez, G.J. et al. (2010). Potential impact of wind farms on territories of large eagles in southeastern Spain. *Biodiversity and Conservation* 19, 3757–3767.

Martini, J., Buris, L.F., and Buris, L. (2003). The role of forensic experts in civil and criminal procedures in Hungary. *Medicine and Law* 22(2), 329–331.

Mason, I.L. (1984). *Evolution of Domesticated Animals*. Longman, London, UK.

Matson, J.V. (2012). *Effective Expert Witnessing. Practices for the 21st Century*, 5th edn. CRC Press, Boca Raton, FL, USA.

Matuszewski, S., Bajerlein, D., Konwerski, S., and Szpila, K. (2008). An initial study of insect succession and carrion decomposition in various forest habitats of Central Europe. *Forensic Science International* 180, 61–69.

Matuszewski, S., Bajerlein, D., Konwerski, S., and Szpila, K. (2010a). Insect succession and carrion decomposition in selected forests of Central Europe. Part 1: Pattern and rate of decomposition. *Forensic Science International* 194, 85–93.

Matuszewski, S., Bajerlein, D., Konwerski, S., and Szpila, K. (2010b). Insect succession and carrion decomposition in selected forests of Central Europe. Part 2: Composition and residency patterns of carrion fauna. *Forensic Science International* 195, 42–51.

Matuszewski, S., Bajerlein, D., Konwerski, S., and Szpila, K. (2011). Insect succession and carrion decomposition in selected forests of Central Europe. Part 3: Succession of carrion fauna. *Forensic Science International* 207, 150–163.

Maylon, C. and Healy, S. (1994). Fluctuating asymmetry in antlers of fallow deer, *Dama dama*, indicates dominance. *Animal Behaviour* 48, 248–250.

Mazayad, S.A. and Rifaat, M.M. (2005). *Megaselia scalaris* causing human intestinal myiasis in Egypt. *Journal of the Egyptian Society of Parasitology* 35, 331–340.

Mazet, J.A.K., Hunt, T.D., Ziccardi, M.H. et al. (2004). Assessment of the risk of zoonotic disease transmission to marine mammal workers and the public. Survey of Occupational Risks. Report prepared for the United States Marine Mammal Commission, Wildlife Health Center, School of Veterinary Medicine, University of California, Davis, CA, USA.

McCarthy, V.O., Cox, R.A., and Haglund, B. (1989). Death caused by a constricting snake – An infant death. *Journal of Forensic Science* 34(1), 239–243.

McCausland, I.P. (1978). Histological ageing of bruises in lambs and calves. *Australian Veterinary Journal* 54, 525–528.

McClelland, R.S., Murphy, D.M., and Cone, D.K. (1997). Report of spores of *Henneguya salminicola* (Myxozoa) in human stool specimens: Possible source of confusion with human spermatozoa. *Clinical Microbiology* 35(11), 2815–2818.

McClintock, J.T. (2008). *Forensic DNA Analysis*. CRC Press, Boca Raton, FL, USA.

McCue, M.D. (2008). Fatty acid analyses may provide insight into the progression of starvation among squamate reptiles. *Comparative Biochemistry and Physiology Part A: Molecular and Integrative Physiology* 151, 239–246.

McCullough, D.R. (1996). Failure of the tooth cementum aging technique with reduced population density of deer. *Wildlife Society Bulletin* 24, 722–724.

McDermott, C.L., Cashore, B., and Kanowski, P. (2010). *Environmental Forest Policies: An International Comparison*. Earthscam, London, UK.

McDowall, I.L. (2008). DNA technology and its applications in herpetological research and forensic investigations involving reptiles and amphibians. In: Cooper, J.E., Baker, B.W., and Cooper, M.E., Eds. Special issue: Forensic Science and Herpetology. *Applied Herpetology* 5(4), 371–385.

McDowell, D. (1997). *Wildlife Crime Policy and the Law*. Australian Government Publishing Service, Canberra, Australia.

McEvoy, F. and Svlastoga, E. (2009). Certifying digital images. *Veterinary Record* 164, 312.

McGraw, S.N., Keeler, S.P., and Huffman, J.E. (2012). Forensic DNA analysis of wildlife evidence. In: Huffman, J.E. and Wallace, J.R., Eds. *Wildlife Forensics Methods and Applications*. Wiley-Blackwell, Oxford, UK, pp. 253–267.

McKay, H.V., Bishop, J.D., Feare, C.J., and Stevens, M.C. (1993). Feeding by Brent geese can reduce yield of oilseed rape. *Crop Protection* 12, 101–105.

McKee, J.K. (2001). The Taung raptor hypothesis: Caveats and new evidence. *American Journal of Physical Anthropology* 32(Suppl), 107.

McKinney, M.A., De Guise, S., Martineau, D., Béland, P., Arukwe, A., and Letcher, R. (2006). Biotransformation of polybrominated diphenyl ethers and polychlorinated biphenyls in beluga whale (*Delphinapterus leucas*) and rat mammalian model using an in vitro hepatic microsomal assay. *Aquatic Toxicology* 77(1), 87–97

McKinney, M.A., Letcher, R.J., Aars, J., Born, E.W., Branigan, M., Dietz, R., Evans, T.J., Gabrielsen, G.W., Peacock, E., and Sonne, C. (2011). Flame retardants and legacy contaminants in polar bears from Alaska, Canada, East Greenland and Svalbard, 2005–2008. *Environment International* 37(2), 365–374

McKinney, M.L. (2002). Why larger nations have disproportionate threat rates: Area increases endemism and human population size. *Biodiversity and Conservation* 11, 1317–1325.

McKnight, B.E. (1981). *The Washing Away of Wrongs Sung Tz'u*. Center for Chinese Studies, The University of Michigan, Ann Arbor, MI.

McNaught, A.D. and Wilkinson, A. (1997). *IUPAC. Compendium of Chemical Terminology*, 2nd edn. Blackwell, Oxford, UK.

McNeilage, A., Plumptre, A., Brock-Doyle, A., and Vedder, A. (2001). Bwindi impenetrable National Park, Uganda Gorilla Census 1997. *Oryx* 35(1), 39–47.

Mégnin, P. (1894). *La Faune des Cadavres: Application de L'entomologie a la Médicine Légale*. G. Masson, Paris, France.

Mellanby, K. (1967). *Pesticides and Pollution*. New Naturalist Series. Collins, London, UK.

Mellor, D. and Monamy, V., Eds. (1999). The use of wildlife for research. *Proceedings of the Conference held at the Western Plains Zoo*, Dubbo, New South Wales, Australia, ANZCCART, Adelaide.

Melton, T. and Holland, C. (2007). Routine forensic use of the mitochondrial 12S ribosomal RNA gene for species identification. *Journal of Forensic Science* 52(6), 1305–1307.

Merceron, G., Viriot, L., and Blondel, C. (2004). Tooth microwear pattern in roe deer (*Capreolus capreolus* L.) from Chizé (Western France) and relation to food composition. *Small Ruminant Research* 53, 125–132.

Merck M. (2007). *Veterinary Forensics. Animal Cruelty Investigations*. Blackwell, Ames, IA, USA.

Meriggi, A. and Lovari, S. (1996). A review of wolf predation in southern Europe: Does the wolf prefer wild prey to livestock? *Journal of Applied Ecology* 33, 1561–1571.

Mermin, J., Hutwagner, L., Vugia, D., Shallow, S. et al. (2004). Reptiles, amphibians, and human *Salmonella* infection: A population-based, case-control study. *Clinical Infectious Diseases* 38, 253–261.

Merryman, J.H. and Pérez-Perdomo, R. (2007). *The Civil Law Tradition. An Introduction to the Legal Systems of Europe and Latin America*. Stanford University Press, Stanford, CA, USA.

Metcalfe, D. and Powell, J. (2011). Should doctors spurn Wikipedia? *Journal of the Royal Society of Medicine* 104, 488–489.

Miaud, C., Guyetant, R., and Faber, H. (2000). Age, size and growth of the Alpine newt, *Triturus alpestris* (Urodela: Salamandridae), at high altitude and a review of life-history trait variation throughout its range. *Herpetologica* 56, 135–144.

Michaud, J.-P. and Moreau, G. (2009). Predicting the visitation of carcasses by carrion-related insects under different rates of degree-day accumulation. *Forensic Science International* 185, 78–83.

Mickleburgh, S.P., Hutson, A.M. and Racey, P.A. (2002). A review of the global conservation status of bats. *Oryx* 36(1), 18–34.

Mikaelian, I. de Lafontaine, Y., Monard, C., Tellier, P., Harshbarger, J., and Martineau, D. (1999). Metastatic mammary adenocarcinomas in two beluga whales (*Delphinapterus leucas*) from the St. Lawrence estuary, Quebec, Canada. *Veterinary Record* 145, 738–739.

Mikhailov, K.E. (1997). *Avian Eggshells: An Atlas of Scanning Electron Micrographs*. British Ornithologists' Club Occasional Publications No. 3. BOU, Tring, UK.

Miles, A.E. and Grigson, C. (1990). *Colyer's Variations and Diseases of the Teeth of Animals*, revised edn. Cambridge University Press, Cambridge, UK.

Miles, A.E.W. (1963). The dentition in the assessment of individual age in skeletal material. In: Brothwell, D.R., Ed. *Dental Anthropology*. Pergamon Press, London, UK.

Miles, A.E.W. and White, J.W. (1960). Ivory. *Proceedings of the Royal Society of Medicine* 53(9), 775–780.

Miles, P.M. (1952). *The Entomology of Bird Pellets*. AES Leaflet No. 24. The Amateur Entomologists Society, London, UK.

Miller, C. (1971). *The Lunatic Express*. MacMillan Company, New York, USA.

Miller, E.A., Ed. (2000). *Minimum Standards for Wildlife Rehabilitation*, 3rd edn. National Wildlife Rehabilitation Association, St. Cloud, MN, USA.

Miller, K.K. and Jones, D.N. (2005). Wildlife management in Australasia: Perceptions of objectives and priorities. *Wildlife Research* 32, 265–272.

Miller, K.K. and Jones, D.N. (2006). Gender differences in the perceptions of wildlife management objectives and priorities in Australasia. *Wildlife Research* 33, 155–159.

Miller, M.L., Lord, W.D., Goff, M.L., Donnelly, B., McDonough, E.T., and Alexis, J.C. (1994). Isolation of amitriptyline and nortriptyline from fly puparia (Phoridae) and beetle exuviae (Dermestidae) associated with mummified human remains. *Journal of Forensic Sciences* 39(5), 1305–1313.

Mills, J.R.E. (1976). A comparison of lateral jaw movement in some mammals from wear facets on the teeth. *Archives of Oral Biology* 12, 645–661.

Milner-Gulland, E.J. (2002). Is bushmeat just another conservation bandwagon? *Oryx* 36(1), 1–2.

Mindell, D.P., Sorenson, M.D., and Dimcheff, D.E. (1998). Multiple independent origins of mitochondrial gene order in birds. *Proceedings of the National Academy of Sciences of the United States of America 95*, 10693–10697.

Mishra, C. (1997). Livestock depredation by large carnivores in the Indian trans-Himalaya: Conflict perceptions and conservation prospects. *Environmental Conservation* 24, 338–343.

Mitani, T., Akane, A., Tokiyasu, T., Yoshimura, S., Okii, Y., and Yoshida, M. (2009). Identification of animal species using the partial sequences in the mitochondrial 16S rRNA gene. *Legal Medicine* 11(Suppl 1) S449–S450.

Miura, S. and Yasui, K. (1985). Validity of tooth eruption-wear patterns as age criteria in the Japanese serow, *Capricornos crispus. Journal of the Mammalogical Society of Japan* 10, 169–178.

Moggi-Cecchi, J. and Crovella, S. (1991). Occurrence of enamel hypoplasia in the dentitions of Simian primates. *Folia Primatology* 57, 106–110.

Molina-Lopez, R.A., Valverdú, N., Martin, M., Mateu, E., Obon, E., Cerdà-Cuéllar, M., and Darwich, L. (2011). Wild raptors as carriers of antimicrobial-resistant *Salmonella* and *Campylobacter* strains. *Veterinary Record* 168, 565.

Molsher, R. (1999). The ecology of feral cats, *Felis catus*, in open forest in New South Wales: Interactions with food resources and foxes. PhD thesis, University of Sydney, Sydney, New South Wales, Australia.

Montague, T. (2000). *The Brushtail Possum. Biology, Impact and Management of an Introduced Marsupial*. Manaaki Whenua Press, Lincoln, New Zealand.

Montie, E.W., Schneider, G.E., Ketten, D.R., Marino, L., Touhey, K.E., and Hahn, M.E. (2007). Neuroanatomy of the subadult and fetal brain of the Atlantic white-sided dolphin (*Lagenorhynchus acutus*) from in situ magnetic resonance images. *The Anatomical Record: Advances in Integrative Anatomy and Evolutionary Biology* 290(12), 1459–1479.

Montie, E.W., Wheeler, E., Pussini, N., Battey, T.W.K., Barakos, J., Dennison, S., Colegrove, K., and Gulland, F. (2010). Magnetic resonance imaging quality and volumes of brain structures from live and postmortem imaging of California sea lions with clinical signs of domoic acid toxicosis. *Diseases of Aquatic Organisms* 91(3), 243–256.

Moore, D.P., Williams, E.H. Jr., Mignucci-Giannoni, A.A., Bunkley-Williams, L., and Dyer, W.G. (2008). Successful surgical treatment of spear wounds in Hawksbill Turtle, *Eremochelys imbricata*, in Puerto Rico. *Journal of Tropical Biology* 56S, 271–276.

Moore, M.J., Knowlton, S., Knaus, W., McLellan, W., and Bonde, R. (2004). Morphometry, gross morphology and available histopathology in North Atlantic right whale (*Eubalaena glacialis*) mortalities (1970 to 2002). *Journal of Cetacean Research and Management* 6(3), 199–214.

Moore, M.J., McLellan, W.A., Daous, P.Y.T., Bonde, R.K., and Knowlton, A.R. (2007). Right whale mortality: A message from the dead to the living. In: Kraus, S.D. and Rolland, R.M., Eds. *The Urban Whale: North Atlantic Right Whales at the Crossroads*, Harvard University Press, Cambridge, MA, pp. 338–379.

Moore, M.K., Bemiss, J.A., Rice, S.M., Quattro, J.M., and Woodley, C.M. (2003). Use of restriction fragment length polymorphisms to identify sea turtle eggs and cooked meats to species. *Conservation Genetics* 4, 95–103.

Moore, M.K. and Kornfield, I.L. (2012). Best practices in wildlife forensic DNA. In: Huffman, J.E. and Wallace, J.R., Eds. *Wildlife Forensics Methods and Applications*. Wiley-Blackwell, Oxford, UK, pp. 201–267.

Moore, N., Whiterow, A., Kelly, P., Garthwaite, D., Bishop, J., Langton, S., and Cheeseman, C. (1999). Survey of badger *Meles meles* damage to agriculture in England and Wales. *Journal of Applied Ecology* 36, 974–988.

Moore, N.P., Cahill, J.P., Kelly, P.F., and Hayden, T.J. (1995). An assessment of five methods of age determination in an enclosed population of fallow deer (*Dama dama*). *Biology and Environment: Proceedings of the Royal Irish Academy* 95B, 27–34.

Moore, S.J. and Battley, P.F. (2003). The use of wing remains to determine condition before death in brown teal (*Anas chlorotis*). *Notornis* 50(3), 133–140.

Moorehead, W. (2011). *Introduction to Forensic Photography*. CRC Press, Boca Raton, FL, USA.

Morad, Y., Kim, Y.M., Armstrong, D.C., Huyer, D., Mian, M., and Levin, A.V. (2002). Correlation between retinal abnormalities and intracranial abnormalities in the shaken baby syndrome. *American Journal of Ophthalmology* 134, 354–359.

Moraitis, K. and Spiliopoulou, C. (2010). Forensic implications of carnivore scavenging on human remain recovered from outdoor locations in Greece. *Journal of Forensic and Legal Medicine* 17, 298–303.

Morgan, D.R., Milne, L., O'Connor, C., and Ruscoe, W.A. (2001). Bait shyness in possums induced by sublethal doses of cyanide paste bait. *International Journal of Pest Management* 47, 277–284.

Morgan, J.A.T., Welch, D.J., Harry, A.V., Street, R., Broderick, D., and Ovenden, J.R. (2011). A mitochondrial species identification assay for Australian blacktip sharks (*Carcharhinus tilstoni*, *C. limbatus* and *C. amblyrhynchoides*) using real-time PCR and high-resolution melt analysis. *Molecular Ecology Resources* 11(5), 813–819.

Moritz, A.R. (1956). Classical mistakes in forensic pathology. *American Journal of Clinical Pathology* 26, 1383–1397.

Morris, D.O. and Dunstan, R.W. (1996). A histomorphological study of sarcoptic acariasis in the dog: 19 cases. *Journal of the American Animal Hospital Association* 32, 119–124.

Morris, P.A. (1970). A method for determining absolute age in the hedgehog. *Journal of Zoology, London* 161, 273–286.

Morris, P.A. (1972). A review of mammalian age determination methods. *Mammal Review* 2, 69–104.

Morris, P.A. (1997). Animal medicine or human therapy? *Biologist* 44, 288.

Morris, P.A. (2010). *A History of Taxidermy: Art, Science and Bad Taste*. MPM Publishing, Ascot, Berkshire, UK.

Morris, P.A. and Harper, J.F. (1965). The occurrence of small mammals in discarded bottles. *Proceedings of the Zoological Society of London* 145(1), 148–153.

Morris, T. (2000). Anaesthesia in the fourth dimension. Is biological scaling relevant to veterinary anaesthesia? *Veterinary Anaesthesia and Analgesia* 27, 2–5.

Morris, Z., White, C., and Longstaffe, F. (2011). Life-stages, landscapes and human-deer interactions during the Ontario Late Woodland period: The isotopic evidence. *Deer and People: Past, Present and Future Conference*, 8–11 September 2011, University of Lincoln, England, UK, Abstracts, p. 34.

Morrison, M. (2002). Searcher bias and scavenging rates in bird/wind energy studies [www document]. Report prepared for the National Renewable Energy Laboratory, http://www.nrel.gov/wind/pdfs/30876.pdf, pp. 1–9 (accessed 22 September 2011).

Morrison, R.D. (1999). *Environmental Forensics: Principles and Applications*. CRC Press, Boca Raton, FL, USA.

Morrow, T.L. and Glover, F.A. (1970). Experimental studies on post-mortem change in mallards. Bureau of Sport Fisheries and Wildlife Special Scientific Report – Wildlife No. 134, Washington, DC, USA.

Morton, R.J. and Lord, W.D. (2006). Taphonomy of child-sized remains: A study of scattering and scavenging in Virginia, U.S.A. *Journal of Forensic Sciences* 51(3), 475–479.

Mostafa, M.B., Koma, K.L.M.K., and Acon, J. (2005). Conjoined twin piglets. *Veterinary Record* 136, 779–780.

Moulton, M.P., Cropper, W.P. Jr. and Avery, M.L. (2011). A reassessment of the role of propagule pressure in influencing fates of passerine introductions to New Zealand. *Biodiversity and Conservation* 20, 607–623.

Mudge, S.M. (2008). *Methods in Environmental Forensics*. Taylor & Francis Group/CRC, Boca Raton, FL, USA.

Muir, D.C.G. et al. (1990). Organochlorine contaminants in belugas, *Delphinapterus leucas*, from Canadian waters. *Canadian Bulletin of Fisheries and Aquatic Sciences* 224, 165–190.

Mullineaux, E., Best, D., and Cooper, J.E., Eds. (2003). *Manual of Wildlife Casualties*. British Small Animal Veterinary Association, Gloucester, UK.

Mundy, P.J. (2000). Red-billed queleas in Zimbabwe. In: Cheke, R.A., Rosenberg, L.J., and Kieser, M.E., Eds. *Workshop on Research Priorities for Migrant Pests of Agriculture in Southern Africa*, Plant Protection Research Institute, Pretoria, South Africa, 24–26 March 1999. Natural Resources Institute, Chatham, UK.

Munger, L.L. and McGavin, M.D. (1972). Sequential post-mortem changes in chicken liver at 4, 20 or 37°C. *Avian Diseases* 16, 587–605.

Muniappan, R., Reddy, G.V.P., and Raman, A., Eds. (2009). *Biological Control of Tropical Weeds Using Arthropods*. Cambridge University Press, Cambridge, UK.

Munro, H.M.C. and Thrusfield, M.V. (2001). "Battered pets": Features that raise suspicion of non-accidental injury. *Journal of Small Animal Practice* 42, 218–226.

Munro, R. (1998). Forensic necropsy. *Seminars in Avian and Exotic Pet Medicine* 7(4), 201–209.

Munro, R. and Munro, H.M.C. (2008). *Animal Abuse and Unlawful Killing. Forensic Veterinary Pathology*. Saunders Elsevier, Edinburgh, UK.

Munro, R. and Munro, H. (2011). Forensic veterinary medicine. 2. Postmortem Investigation. *In Practice* 33, 262–276.

Munson, L. (2002). The living dead: Keeping wildlife alive through scientific use of biomaterials. *Proceedings of the American Association of Zoo Veterinarians*, Milwaukee, WI, USA, pp. 269–271

Murphy, B.L. and Morrison, R.D. (2002). *Introduction to Environmental Forensics*. Academic Press, New York, USA.

Murphy, M.E. and King, J.R. (1991). Ptilochronology: A critical evaluation of assumptions and utility. *Auk* 108, 695–704.

Murray, J., Murray, E., Michael, S., Johnson, M.S., and Clarke, B. (1988). The extinction of *Partula* on Moorea. *Pacific Science* 42, 3–4.

Mzamo, K.B. and Attoquayefio, D.K. (1999). *Bufo regularis* – A potential risk to exotic dogs in Ghana: A case study. *Ghana Journal of Science* 39, 73–75.

Nabaglo, L. (1973). Participation of invertebrates in decomposition of rodent carcasses in forest ecosystems. *Ekologia Polska* 21, 251–270.

Naidoo, V., Wolter, D., Cromarty, M., Diekmann, N., Meharg, A.A., Taggart, M.A., Venter, L., and Cuthbert, R. (2009). Toxicity of non-steroidal anti-inflammatory drugs to *Gyps* vultures a new threat from ketoprofen. *Biology Letters* 6(3), 339–341. Doi: 10.1098/rsbl.2009.0818.

Naidu, A., Fitak, R.R., Munguia-Vega, A., and Culver, M. (2012). Novel primers for complete mitochondrial cytochrome *b* gene sequencing in mammals. *Molecular Ecology Resources* 12(2), 191–196.

Naidu, A., Smythe, L.A., Thompson, R.W., and Culver, M. (2011). Genetic analysis of scats reveals minimum number and sex of recently documented mountain lions. *Journal of Fish and Wildlife Management* 2(1), 106–111.

Nakaki, S.-I., Hino, D., Miyoshi, M., Nakayama, H., Moriyoshi, H., Morikawa, T., and Itohara, K. (2007). Study of animal species (human, dog and cat) identification using a multiplex single-base primer extension reaction in the cytochrome b gene. *Forensic Science International* 173(2–3), 97–102.

Nakamura, K. (1938). Studies on some specimens of double monsters of snakes and tortoises. *Memoirs of the College of Science, Kyoto University (B)* 4(2), 171.

National Institute of Justice. (1999). *Death Investigation: A Guide for the Scene Investigator*. U.S. Department of Justice, Office of Justice Programs, National Institute of Justice, Washington, DC, USA.

National Institute of Justice. (2004). *Crime Scene Investigation: A Reference for Law Enforcement Training*. U.S. Department of Justice, Office of Justice Programs, National Institute of Justice, Washington, DC, USA.

National Institute of Justice. (2008). *Electronic Crime Scene Investigation: A Guide for First Responders*, 2nd edn. U.S. Department of Justice, Office of Justice Programs, National Institute of Justice. Washington, DC, USA.

National Marine Fisheries Service. (2008). Final rule to implement speed restrictions to reduce the threat of ship collisions with north Atlantic right whales. NOAA. Administration, Department of Commerce, 50 CFR Part 224, pp. 60173–60191.

National Research Council. (1991). *Micro-Livestock: Little Known Animals with a Promising Economic Future*. National Academic Press, Washington, DC, USA.

National Research Council (US). (2009). *Strengthening Forensic Science in the United States: A Path Forward*. National Academies Press, Washington, DC, USA. Available online from: http://www.nap.edu/catalog.php?record_id=12589.

Neal, B.R. and Cock, A.G. (1969). An analysis of the selection of small African mammals by two break-back traps. *Journal of Zoology* 158(3), 335–340.

Negus, V.E. (1929). *The Mechanism of the Larynx*. Heinemann, London, UK.

Neiburger, E.J. and Patterson, B.D. (2000). The man-eaters with bad teeth. *New York State Dental Journal* 66(1), 26–29.

Nel, D.C. and Whittington, P.A., Eds. (2003). *Rehabilitation of Oiled African Penguins: A Conservation Success Story*. Birdlife South Africa and Avian Demography Unit, Cape Town, South Africa.

Nelder, M.P., McCreadie, J.W., and Major, C.S. (2009). Blow flies visiting decaying alligators: Is succession synchronous or asynchronous? *Psyche* 2009, 1–7.

Neme, L.A. (2009). *Animal Investigators*. Schribner, New York, USA.

Neme, L.A. and Leakey, R. (2010). *Animal Investigators: How the World's First Wildlife Forensics Lab is Solving Crimes and Saving Endangered Species*. Simon & Schuster, New York, USA.

Neuville, M.H. (1917). Du tégument des proboscidiens. *Bulletin de la Musée d'Histoire Naturelle de Paris* 123, 374–387.

Newbery, S. and Munro, R. (2011). Forensic veterinary medicine. I Investigation involving live animals. *In Practice* 33, 220–227.

Newell, E.A., Guatelli-Steinberg, D., Field, M., and Cooke, C. (2006). Life history, enamel formation, and linear enamel hypoplasia in the Ceboidea. *American Journal of Physical Anthropology* 131, 252–260.

Newing, H. (2001). Bushmeat hunting and management: Implications of duicker ecology and interspecific competition. *Biodiversity and Conservation* 10, 99–118.

Newton, I. (1998). *Population Limitation in Birds*. Academic Press, London, UK.

Ng, L.E., Halliwell, B., and Kim, P.W. (2006). Nephrotoxic cell death by diclofenac and meloxicam. *Biochemical and Biophysical Research Communications* 369, 873–877.

Niebauer, G.W. (1993). Rectoanal disease. In: Bojrab, M.J., Smeak, D.D., and Bloomberg, M.S., Eds. *Disease Mechanisms in Small Animal Surgery*, 2nd edn. Lea & Febiger, Philadelphia, PA, USA, pp. 273–274.

Nizeyi, J.B., Innocent, R.B., Erume, J., Kalema, G.R.N.N., Cranfield, M.R., and Graczyk, T.K. (2001). Campylobacteriosis, salmonellosis, and shigellosis in free-ranging human-habituated mountain gorillas of Uganda. *Journal of Wildlife Diseases* 37(2), 239–244.

da Nóbrega Alves, R.R., da Silva Vieira, W.L., and Santana, G.G. (2008). Reptiles used in traditional folk medicine: Conservation implications. *Biodiversity and Conservation* 17(8), 2037–2049.

Nolte, K.B., Pinder, R.D., and Lord, W.D. (1992). Insect larvae used to detect cocaine poisoning in a decomposed body. *Journal of Forensic Sciences* 37(4), 1179–1185.

Norman, D.B. (1987). Review. *Archaeopteryx*, the primordial bird: A case of fossil forgery by F. Hoyle and C. Wickramasinghe. *Ibis* 129, 408–412.

North, P. (2012). *Civil Liability for Animals*. Oxford University Press, Oxford, UK.

Norton, B.G., Hutchins, M., Stevens, E.F., and Maple, T. (1995). *Ethics on the Ark: Zoos, Animal Welfare and Wildlife Conservation*. Smithsonian Institution Press, Washington, DC, USA.

Norton, W.W. (1994). *State of the World*. Earthscan, London, UK.

Nozdryn-Potnicki, Z., Liston, P., Opuszynski, W., and Debiak, P. (2005). Section investigation of animals wounded from fire arms: Some remarks. *Medycyna Polskiego Towarzystwa Nauk Weterynaryjnych* 61(8), 887–889 [Polish].

NPIA. (2010). *Guidance in the Management of Police Information*, 2nd edn. National Policing Improvement Agency, Bedford, UK. http://www.npia.police.uk/en/docs/MoPI_refreshed_Guidance.pdf

Nuorteva, P. (1987). Empty puparia of *Phormia terraenovae* R-D (Diptera: Calliphoridae) as forensic indicators. *Annales Entomologici Fennici* 53, 53–56.

Nygård, T. (1999). Correcting eggshell indices of raptor eggs for hole size and eccentricity. *Ibis* 141, 85–90.

Oaks, J.L., Gilbert, M., Virani, M.Z., Watson, R.T., Meteyer, C.U., Rideout, B.A., Shivaprasad, H.L., Ahmeed, S., Chaudrey, M.J.I., Arshad, M., Mahmood, S., Ali, A., and Khan, A.A. (2004). Diclofenac residues as the cause of vulture population decline in Pakistan. *Nature* 10, 1038.

Oates, D. (1992). Time of death. In: Adrian, W.J., Ed. *Wildlife Forensic Field Manual*. Colorado Division of Wildlife, Fort Collins, CO, USA.

Oates, D.W., Coggin, J.L., Hartman, F.E., and Holien, E.I. (1984). A guide to the time of death in selected wildlife species. Nebraska Technical Series No. 14. Nebraska Game and Parks Commission, Lincoln, NE, USA, 72 pages.

Ober, H.K. and DeGroote, L.W. (2011). Effects of litter removal on arthropod communities in pine plantations. *Biodiversity and Conservation* 20, 1273–1286.

Odino, M., Ogada, D., and Musila, S. (2008). Furadan killing birds on a large scale in Bunyala rice fields, Western Kenya. *Kenya Birds* 12(1 and 2), 7–12.

Ogada, M. (2011). Middle East pet mentality threatens African cheetah. *SWARA Journal of the East African Wildlife Society* October–December, 52–57.

Ogden, R. (2010). Forensic Science, genetics and wildlife biology: Getting the right mix for a wildlife DNA forensics lab. *Forensic Science, Medicine, and Pathology* 6(3), 172–179.

Ogden, R. (2012). DNA applications and implementation. In: Huffman, J.E. and Wallace, J.R., Eds. *Wildlife Forensics Methods and Applications*. Wiley-Blackwell, Oxford, UK, pp. 271–289.

Ogden, R., Dawnay, N., and McEwing, R. (2008a). Development of STR profiling systems for individual identification in wildlife: A case study of the Eurasian badger, *Meles meles*. *Forensic Science International: Genetics* 1(1), 612–613.

Ogden, R., Dawnay, N., and McEwing, R. (2009). Wildlife DNA forensics – Bridging the gap between conservation genetics and law enforcement. *Endangered Species Research* 9, 179–195.

Ogden, R. and McEwing, R. (2008). Adaptation of DNA analysis techniques for the identification of illegally imported bushmeat for use on the Agilent 2100 bioanalyser. Food DNA Services, Manchester, UK.

Ogden, R., McGough, H.N., Cowan, R.S., Chua, L., Groves, M., and McEwing, R. (2008b). SNP-based method for the genetic identification of ramin *Gonystylus* spp. timber and products: Applied research meeting CITES enforcement needs. *Endangered Species Research* 9, 255–261.

Ogden, R., Mellanby, R.J., Clements, D., Gow, A.G., Powell, R., and McEwing, R. (2012). Genetic data from 15 STR loci for forensic individual identification and parentage analyses in UK domestic dogs (*Canis lupus familiaris*). *Forensic Science International: Genetics* 6(2), e63–e65.

Ogden, J.A., Lee, K.A., Gerald, J., Conlogue, G.J., and Barnett, J.S. (1981). Prenatal and postnatal development of the cervical portion of the spine in the short-finned pilot whale *Globicephala macrorhyncha*. *The Anatomical Record* 200(1), 83–94.

Ohashi, G. and Matsuzawa, T. (2010). Deactivation of snares by wild chimpanzees. *Primates* 52, 1–5.

Oishi, Y. (2011). Protective management of trees against debarking by deer negatively impacts bryophyte diversity. *Biodiversity and Conservation* 20, 2527–2536.

Ojigo, D.O. and Daborn, C.J. (2011). Integrating mobile tools for surveillance, reporting and information management by veterinary services in Kenya. Department of Veterinary Services, Kabete, Kenya.

Okella, A.L., Gibbs, E.P.J., Vandersmissen, A., and Welburn, S.C. (2011). One health and the neglected zoonoses: Turning rhetoric into reality. *Veterinary Record* 169, 281–285.

Oldfield, S., Ed. (2003). *The Trade in Wildlife: Regulation for Conservation*. Earthscan, London, UK.

Olsen, J. (1986). An unusual incident with the Bald Eagle (*Haliaeetus leucocephalus*). *Raptor Research* 20, 41.

References and Further Reading

Omar, S.V. (1992). Ageing deer from cementum bands. *Deer* 8, 569–572.

Ormrod, S. (1986a). Wildlife rescue, or wildlife roulette. *RSPCA Today* Summer, 14–15.

Ormrod, S. (1986b). How to play God right. *BBC Wildlife* March, 115–116.

Osborn, R.G., Higgins, K.F., Dieter, C.D., and Usgaard, R.E. (1996). Bat collisions with wind turbines in southwestern Minnesota. *Bat Research News* 37, 105–108.

Osborne, S. and Milsom, W.K. (1993). Ventilation is coupled to metabolic demands during progressive hypothermia in rodents. *Respiration Physiology* 92(3), 305–318.

Osofsky, S.O., Kock, R.A., Kock, M.D., Kalema-Zikusoka, G., Grahn, R., Leyland, T., and Karesh, W.B. (2005). Building support for protected areas using a One Health perspective. In: McNeily, J., Ed. *Friends for Life, New Partners in Support of Protected Areas*. IUCN Species Survival Commission. IUCN Gland, Switzerland and Cambridge, UK.

Ossent, P. and Lischer, C. (1997). Post mortem examination of the hooves of cattle, horses, pigs and small ruminants under practice conditions. *In Practice* 19, 21–29.

Osti, M., Coad, L., Fisher, J.B., and Bombard, B. (2011). Oil and gas development in the World Heritage and wider protected area network in sub-Saharan Africa. *Biodiversity and Conservation* 20, 1863–1877.

Özdemir, S. and Sert, O. (2009). Determination of Coleoptera fauna on carcasses in Ankara province, Turkey. *Forensic Science International* 183, 24–32.

O'Connor, P.A. (1983). *Advanced Taxidermy*. Saiga Publishing Co. Ltd., Hindhead, Surrey, UK.

O'Regan, H., Turner, H., and Sabin, R. (2005). Medieval big cat remains from the Royal Menagerie at the Tower of London. *International Journal of Osteoarchaeology* 16, 385–394.

O'Riordan, T. (1995). Ecologists and ethical judgements. *Biodiversity and Conservation* 4(8), 781–973.

O'Toole, L. (2011). Welcome changes to the poisons regulations. *Irish Eagle News* 4, 25–26.

Pabst, D.A., Rommel, S.A., and McLellan, W.A. (1999). The functional morphology of marine mammals. In: Reynolds, J.E., III and Rommel, S.A., Eds. *Biology of Marine Mammals*. Smithsonian Institution Press, Washington, DC, USA, pp. 15–72.

Packer, C., Ikanda, D., Kissui, B., and Kushnir, H. (2005). Lion attacks on humans in Tanzania. *Nature* 436, 927–928.

Pakanen, V.-M., Luukkonen, A., and Koivula, K. (2011). Nest predation and trampling as management risks in grazed coastal meadows. *Biodiversity and Conservation* 20, 2057–2073.

Palmer, J. (2001). *Animal Law*. Shaw & Sons, Crayford, Kent, UK.

Palmer, S.R., Soulsby, L., Torgerson, P.R., and Brown, D.W.G., Eds. (2011). *Oxford Textbook of Zoonoses: Biology, Clinical Practice, and Public Health Control*, 2nd edn. Oxford University Press, Oxford, UK.

Palombit, R.A. (1993). Lethal territorial aggression in a white-handed gibbon. *American Journal of Primatology* 31(4), 311–318.

Palsbøll, P.J., Bérubé, M., Skaug, H.J., and Raymakers, C. (2006). DNA registers of legally obtained wildlife and derived products as means to identify illegal takes. *Conservation Biology* 20(4), 1284–1293.

Pamplin, C. (2000). Expert evidence: Do you have all the facts? *Veterinary Practice* February, 12–14.

Pamplin, C., Ed. (2007). *Expert Witness Practice in the Civil Arena*. JS Publications, Newmarket, Suffolk, UK.

Pamplin, C. (2012). *Expert Witness Year Book 2011*. U.K Register of Expert Witnesses, Newmarket, Suffolk, UK.

Pamplin, C., Ed. (2011). *Expert Witness Fees*, 2nd edn. JS Publications, Newmarket, Suffolk, UK.

Pamplin, C. and White, S. (2008). *Getting Started as an Expert Witness*. JS Publications, Newmarket, Suffolk, UK.

Parik, J., Freitas, A.I., Villems, R., Brehm, A., and Gonçalves, R. (2008). Identification of endangered petrel species from poor quality feather debris using cytochrome *b* sequences. [Letter to the Editor]. *Forensic Science International* 174, 86–87.

Parkes, C. and Thornley, J. (1997). *Fair Game: The Law of Country Sports and the Protection of Wildlife*. Pelham, UK.

Parkes, C. and Thornley, J. (2000). *Deer: Law and Liabilities*. Swan Hill Press, Shrewsbury, UK.
Parliamentary Office of Science and Technology. (2005). Science in Court. Postnote, October 2005, Number 248. http://www.parliament.uk/documents/post/postpn248.pdf
Parrish, J.R. and Clayton, D.H. (1988). The physiology of "stress marks" in the feathers of North American raptors. Paper presented at the *1988 Annual Meeting of the Raptor Research Foundation*, October 1988. Minneapolis, MN, USA.
Parson, W., Pegoraro, K., Niederstätter, H., Föger, M., and Steinlechner, M. (2000). Species identification by means of the cytochrome b gene. *International Journal of Legal Medicine* 114, 23–28.
Pass, D.A. and Perry, R.A. (1984). The pathology of psittacine beak and feather disease. *Australian Veterinary Journal* 61, 69–74.
Patton, F. (2011). Understanding Chinese medicine – The scourge of the rhino. *SWARA Journal of the East African Wildlife Society* October–December, 38–40.
PAW. (2005). *Wildlife Crime: A Guide to the Use of Forensic and Specialist Techniques in the Investigation of Wildlife Crime*. DEFRA, Bristol, UK.
Payne, J.A. (1965). A summer carrion study of the baby pig *Sus scrofa* Linnaeus. *Ecology* 46(5), 592–602.
Pedersen, A., Jones, K., Nunn, C.L., and Altizer, S. (2007). Infectious diseases and extinction risk in wild mammals. *Conservation Biology* 21(5), 1269–1279.
Penfold, J. (1999). Cruelty and welfare: The vet in court. *Veterinary Times* February 1999, 14.
Pennycott, T.W., Park, A., and Mather, H.A. (2006). Isolation of different serovars of *Salmonella enterica* from wild birds in Great Britain between 1995 and 2003. *Veterinary Record* 158, 817–820.
Pennycuick, C.J. and Rudnai, J. (1970). A method of identifying individual lions *Panthera leo* with an analysis of the reliability of identification. *Journal of Zoology, London* 160, 497–508.
Penzhorn, B.L. (1997). Training veterinarians for wildlife rehabilitation. *Hooo-Hooo* 1(1), 19–20.
Peppin, L., McEwing, R., Carvalho, G.R., and Ogden, R. (2008). A DNA-based approach for the forensic identification of Asiatic black bear (*Ursus thibetanus*) in a traditional Asian medicine. *Journal of Forensic Science* 53(6), 1358–1362.
Peppin, L., McEwing, R., Webster, S., Rogers, A., Nicholls, D., and Ogden, R. (2009). Development of a field test for the detection of illegal bear products. *Endangered Species Research* 9, 263–270.
Pérez, J.M. and Palma, R.L. (2001). A new species of *Felicola* (Phthiraptera: Trichodectidae) from the endangered Iberian lynx: Another reason to ensure its survival. *Biodiversity and Conservation* 10, 929–937.
Perez, S.P., Duque, P., and Wolff, M. (2005). Successional behaviour and occurrence matrix of carrion-associated arthropods in the urban area of Medellin, Colombia. *Journal of Forensic Science* 50(2), 1–7.
Perez, T.M. and Atyeo, W.T. (1984). Site selection of the feather and quill mites of Mexican parrots. In: Griffiths, D.A. and Bowman, C.E., Eds. *Acarology VI*, Vol. 1. Ellis Horwood, Chichester, England, pp. 563–570.
Pet Food Manufacturers Association (PFMA). (2011). Pet population figures. http://www.pfma.org.uk/statistics/index.cfm?id=83&cat_id=60; http://www.pfma.org.uk/statistics/index.cfm?id=127&cat_id=60
Peters, S.O., Gunn, H.H., Imumorin, I.G., Agaviezor, B.O., and Iikeobi, C.O.N. (2011). Haematological studies on frizzled and naked neck genotypes of Nigerian native chickens. *Tropical Animal Health and Production* 43, 631–638.
Pétillon, J.-M. (2008). First evidence of a whale-bone industry in the western European Upper Paleolithic: Magdalenian artifacts from Isturitz (Pyrénées-Atlantiques, France). *Journal of Human Evolution* 54(5), 720–726.
Petterson, A. (2010). Tooth resorption in the Swedish Eurasion (*sic*) lynx (*Lynx lynx*). *Journal of Veterinary Dentistry* 27(4), 222–226.
Philibert, H.G. and Clark, R.G. (1993). Counting dead birds: Examination of methods. *Journal of Wildlife Disease* 29, 284–289.

Phillott, A.D. and Parmenter, C.J. (2007). Deterioration of green sea turtle (*Chelonia mydas*) eggs after known embryo mortality. *Chelonia Conservation Biology* 6, 262–266.

Philp, R.B. (2001). *Ecosystems and Human Health*. CRC Press, Boca Raton, FL, USA.

Pickering, R. and Bachman, D. (2009). *The Use of Forensic Anthropology*, 2nd edn. CRC Press, Boca Raton, FL, USA.

Pidancier, N., Miquel, C., and Miaud, C. (2003). Buccal swabs as a non-destructive tissue sampling method for DNA analysis in amphibians. *Herpetological Journal* 13, 175–178.

Pierce, G.J., Boyle, P.R., and Diack, J.S.W. (1996). Identification of fish otoliths and bones in faeces and digestive tracts of seals. *Journal of Zoology, London* 224, 320–328.

Pierce, G.J., Miller, A., Thompson, P.M., and Hislop, J.R.G. (1991a). Prey remains in grey seal (*Halichoerus grypus*) faeces from the Moray Firth, north-east Scotland. *Journal of Zoology, London* 224, 337–341.

Pillay, K., Dawson, D.A., Horsburgh, G.J., Perrin, M.R., Burke, T., and Taylor, T.D. (2010). Twenty-two polymorphic microsatellite loci aimed at detecting illegal trade in the Cape parrot, *Poicephalus robustus* (Psittacidae, AVES). *Molecular Ecology Resources* 10(1), 142–149.

Pimentel, D., Ed. (2011). *Biological Invasions: Economic and Environmental Costs of Alien Plant, Animal, and Microbe Species*, 2nd edn. CRC Press, Boca Raton, FL, USA.

Piper, H. (2004). The linkage of animal abuse with interpersonal violence: A sheep in wolf's clothing? *Journal of Social Work* 3(2), 161–177.

Pirsig, R.M. (1974). *Zen and the Art of Motorcycle Maintenance: An Enquiry into Values*. Morrow, New York, USA.

Pisenti, J.M., Santolo, G.M., Yamamoto, J.T., and Morzenti, A.A. (2001). Embryonic development of the American kestrel (*Falco sparverius*): External criteria for staging. *Journal of Raptor Research* 35(3), 194–206.

Pitman, C.R.S. (1931). *A Game Warden among His Charges*. Nisbet & Co., London, UK.

Pitman, C.R.S. (1942). *A Game Warden Takes Stock*. Nisbet & Co., London, UK.

Pitman, C.R.S. (1974). *A Guide to the Snakes of Uganda*. Wheldon & Wesley, Codicote, UK.

Pizzi, R. (2009). Veterinarians and taxonomic chauvinism: The dilemma of parasite conservation. *Journal of Exotic Pet Medicine* 18(4), 279–282

Pizzi, R. et al. (2011). Laparoscopic cholecystectomy under field conditions in Asiatic black bears (*Ursus thibetanus*) rescued from illegal bile farming in Vietnam. *Veterinary Record* 169, 469.

Pond, C.M. and Gilmour, I. (1997). Stable isotopes in adipose tissue fatty acids as indicators of diet in arctic foxes (*Alopex lagopus*). *Proceedings of the Nutrition Society* 56, 1067–1081.

Pond, C.M., Mattacks, C.A., Colby, R.H., and Ramsay, M.A. (1992). The anatomy, chemical composition, and metabolism of adipose tissue in wild polar bears (*Ursus maritimus*). *Canadian Journal of Zoology* 70(2), 326–341.

Poole, D.F.G. (1961). Notes on tooth replacement in the Nile crocodile *Crocodylus niloticus*. *Proceedings of the Zoological Society of London* 136(1), 131–140.

Posthaus, H., Bodmerd, T., Alvesb, L., Oevermannc, A., Schillere, I., Rhodesf, S.G., and Zimmerlid, S. (2010). Accidental infection of veterinary personnel with *Mycobacterium tuberculosis* at necropsy: A case study. *Veterinary Microbiology* 149(3–4), 374–380. Doi: 10.1016/j.vetmic.2010.11.027

Pounder, D., Jones, M., and Peschel, H. (2011). How can we reduce the number of coroner autopsies? Lessons from Scotland and the Dundee initiative. *Journal of the Royal Society of Medicine* 104, 19–24.

Prabhu, S.R., Ed. (2004). *Textbook of Oral Medicine*. Oxford University Press, Oxford, UK.

Prast, W. and Shamoun, J. (1997). BRIS Bird Remains Identification System. CD-Rom, Windows Version 1.0. World Biodiversity Database Supporting Series. ETI Expert Center for Taxonomic Identification, Amsterdam, the Netherlands.

Preativatanyou, K., Sirisup, N., Payungporn, S., Poovorawan, Y., Thavara, U., Tawatsin, A., Sungpradit, S., and Siriyasatien, P. (2010). Mitochondrial DNA-based identification of some forensically important blowflies in Thailand. *Forensic Science International* 202, 97–101.

Prescott, J.F., Poppe, C., Goltz, J., and Campbell, G.D. (1998). *Salmonella typhimurium* phage type 40 in feeder birds. *Veterinary Record* 142, 732.

Price, T.D., Burton, J.H., and Bentley, R.A. (2002). The characterization of biologically available strontium isotope ratios for the study of prehistoric migration. *Archaeometry* 44, 117–135.

Pritchard, G. (2011). Prevention and control of zoonoses on farms open to the public. *In Practice* 33, 242–251.

Prichard, J.G., Kossoris, P.D., Leibovitch, R.A., Robertson, L.D., and Lovell, F.W. (1986). Implications of trombiculid mite bites: Report of case and submission of evidence in a murder trial. *Journal of Forensic Sciences* 31, 301–306.

Pun, K.-M., Albrecht, C., Castella, V., and Fumagalli, L. (2009). Species identification in mammals from mixed biological samples based on mitochondrial DNA control region length polymorphism. *Electrophoresis* 30(6), 1008–1014.

Putman, R.J. (1995). Ethical considerations and animal welfare in ecological field studies. *Biodiversity and Conservation* 4(8), 903–915.

Quadros, J. and Monteiro-Filho, E.L.A. (1998). Effects of digestion, putrefaction and taxidermy processes on *Didelphis albiventris* hair morphology. *Journal of Zoology, London* 244, 331–334.

Quality Assurance Standards for Forensic DNA Testing Laboratories. (2009). Available from http://www.fbi.gov/about-us/lab/codis/qas_testlab.pdf

Quan, J., Ouyang, Z., Xu, W., and Miao, H. (2011). Assessment of the effectiveness of nature reserve management in China. *Biodiversity and Conservation* 20, 779–792.

Quaresme, P.F., Rêgo, F.D., Botelho, H.A., da Silva, S.R., Moura Jnr, A.J., Neto, R.G.T., Madeira, F.M., Carvalho, M.B., Paglia, A.P., Melo, M.N., and Gontijo, C.M.F. (2011). Wild, synanthropic and domestic hosts of *Leishmania* in an endemic area of cutaneous leishmaniasis in Minas Gerais State, Brazil. *Transactions of the Royal Society of Tropical Medicine and Hygiene* 105, 579–585.

Rackallo, J. (1972). Determination of the age of wounds by histochemical and biochemical methods. *Forensic Science* 1, 1–16.

Rahman, M.L., Tarrant, S., McCollin, D., and Ollerton, J. (2011). The conservation value of restored landfill sites in the East Midlands, UK for supporting bird communities. *Biodiversity and Conservation* 20, 1879–1893.

Ranson, D. (1996). *Forensic Medicine and the Law, An Introduction*. Melbourne University Press, Carlton, Victoria, Australia.

Rapport, D., Costanza, R., and Epstein, P.R., Eds. (1998). *Ecosystem Health*. Blackwell, Oxford, UK.

Rasmussen, L.E.L. and Munger, B.L. (1996). The sensorineural specializations of the trunk tip (finger) of the Asian elephant, *Elephas maximus*. *Anatomy Record* 246, 127–134.

Ratcliffe, D. (1967). Decrease ineggshell weight in certain birds of prey. *Nature* 215, 208–210.

Ratcliffe, D. (1980). *The Peregrine Falcon*. T. & A.D. Poyser, London, UK.

Ratcliffe, P.R. (1970). The occurrence of vestigial teeth in badger, roe deer and fox from the county of Argyll, Scotland. *Journal of Zoology* 162, 521–525.

Ratnasingham, S. and Hebert, P.D.N. (2007). BOLD The barcode of life data system (http://www.barcodinglife.org). *Molecular Ecology Notes* 7, 355–364.

Ravago-Gotanco, R.G., Manglicmot, M.T., and Pante, M.J.R. (2010). Multiplex PCR and RFLP approaches for identification of rabbitfish (*Siganus*) species using mitochondrial gene regions. *Molecular Ecology Resources* 10(4), 741–743.

Raverty, S.A. and Gaydos, J.K. (2004). Killer whale necropsy and disease testing protocol. 6, 63. http://www.vetmed.ucdavis.edu/whc/pdfs/orcanecropsyprotocol.pdf

Rayment, M., Asthana, P., van de Weerd, H., Gittins, J., and Talling, J. (2010). Evaluation of the EU policy on animal welfare and possible policy options for the future. Submitted by GHK and ADAS UK (Food Policy Evaluation Consortium) to DG Sanco.

Read, A.J. and Murray, K.T. (2000). Gross evidence of human-induced mortality in small cetaceans. U.S. Department of Commerce, National Oceanic and Atmospheric Administration, National Marine Fisheries Service, NOAA Technical Memorandum NMFS-OPR-15, Honolulu, HI, USA.

Read, A.J., Drinker, P., and Northridge, S. (2006). Bycatch of marine mammals in U.S. and global fisheries Captura Incidental de Mamíferos en Pesquerías de E.U.A. y Globales. *Conservation Biology* 20(1), 163–169.

Rebmann, A., David, E., and Sorg, M.H. (2000). *Cadaver Dog Handbook*. CRC Press, Boca Raton, FL, USA.

Redpath, S.M., Thirgood, S.J., and Clarke, R. (2002). Field vole abundance and hen harrier diet and breeding in Scotland. *Ibis* 144, E130–E138.

Redsicker, D.R. (2000). *The Practical Methodology of Forensic Photography*, 2nd edn. CRC Press, Boca Raton, FL, USA.

Reed, C. and Reed, B.P. (1928). The mechanisms of pellet formation in the great horned owl (*Bubo virginianus*). *Science Wash* 68, 359–360.

Reed, H.B. (1958). A study of dog carcass communities in Tennessee with special reference to the insects. *American Midland Naturalist* 59, 213–245.

Rees, P. (2002). *Urban Environments and Wildlife Law*. Blackwell, Oxford, UK.

Reibe, S. and Madea, B. (2010). How promptly do blowflies colonise fresh carcasses? A study comparing indoor with outdoor locations. *Forensic Science International* 195, 52–57.

Reichenbach-Klinke, H. and Elkan, E. (1965). *The Principal Diseases of Lower Vertebrates*. Academic Press, London, UK, pp. 531–536.

Reif, J., Marhoul, P., Čížek, O., and Konvička, M. (2011). Abandoned military training sites are an overlooked refuge for at-risk open habitat bird species. *Biodiversity and Conservation* 20, 2645–2662.

Reiners, T., Encarnação, J., and Wolters, V. (2011). An optimized hair trap for non-invasive genetic studies of small cryptic mammals. *European Journal of Wildlife Research* 57(4), 991–995.

Rendo, F., Iriondo, M., Manzano, C., and Estonba, A. (2011). Microsatellite based ovine parentage testing to identify the source responsible for the killing of an endangered species. *Forensic Science International: Genetics* 5(4), 333–335.

Rares, S. (2011). Using the "hot tub": How concurrent expert evidence aids understanding issues. *The Expert Witness Institute Newsletter* Autumn, 5–13.

Reynolds, S.J. and Schoech, S.J. (2012). A known unknown: Elaboration of the 'observer' effect on nest success? *Ibis* 154, 1–4.

Rhodin, A.G.J., Walde, A.D., Horne, B.D., van diJk, P.P., Blanck, T., and Hudson, R. Eds. (2011). *Turtle Conservation Coalition. Turtles in Trouble: The World's 25+ Most Endangered Tortoises and Freshwater Turtles – 2011*. Lunenburg, MA, USA: IUCN/SSC Tortoise and Freshwater Turtle Specialist Group, Turtle Conservation Fund, Turtle Survival Alliance, Turtle Conservancy, Chelonian Research Foundation, Conservation International, Wildlife Conservation Society, and San Diego Zoo Global.

Richards, C.S., Crous, K.L., and Villet, M.H. (2009). Models of development for blowfly sister species *Chrysomya chloropyga* and *Chrysomya putoria*. *Medical and Veterinary Entomology* 23, 56–61.

Richards, E.N. and Goff, M.L. (1997). Arthropod succession on exposed carrion in three contrasting tropical habitats on Hawaii Island, Hawaii. *Journal of Medical Entomology* 34(3), 328–339.

Richards, N. (2011). *Carbofuran and Wildlife Poisoning: Global Perspectives and Forensic Approaches*. Wiley, New York, USA.

Richter, H., Kierdorf, U., Richards, A., and Kierdorf, H. (2010). Dentin abnormalities in cheek teeth of wild red deer and roe deer from a fluoride-polluted area in Central Europe. *Annals of Anatomy* 192, 86–95.

Rickman, R. (2006). *Spindoctor*. Ford Cross Books, Bideford, Devon, UK.

de Ricqles, A. and de Buffrenil, V. (2001). Bone histology, heterochronies and the return of tetrapods to life in water: Where are we? In: Mazin, J.-M. and de Buffrenil, V., Eds. *Secondary Adaptation of Tetrapods to Life in Water*. Verlag Dr. Friedrich Pfeil, Munchen, Germany, pp. 289–310.

Rideout, B.A. (2002). Creating and maintaining a post-mortem biomaterials archive: Why you should do it and what's in it for you. *Proceedings of the American Association of Zoo Veterinarians*, Milwaukee, WI, USA, pp. 272–274.

Ridgway, S.H. (1972). *Mammals of the Sea. Biology and Medicine*. C.C. Thomas, Springfield, IL, USA.
Riffenburgh, R.S. and Sathyavagiswaran, L. (1991). The eyes of child abuse victims: Autopsy findings. *Journal of Forensic Sciences* 36, 741–747.
Riney, T. (1951). Standard terminology for deer teeth. *Journal of Wildlife Management* 15, 99–101.
Ritzman, T.B., Baker, B.J., and Schwartz, G.T. (2008). A fine line: A comparison of methods for estimating ages of linear enamel hypoplasia formation. *American Journal of Physical Anthropology* 135, 348–361.
Roberts, B.K., Aronsohn, M.G., Moses, B.L., Burk, R.L., Toll, J., and Weeren, F.R. (2000). *Bufo marinus* intoxication in dogs: 94 cases (1997–98). *Journal of the American Veterinary Medical Association* 216, 1941–1944.
Roberts, D.H. (1986). Determination of predators responsible for killing small livestock. *South African Journal of Wildlife Research* 16, 150–152.
Robertson, J. (1999). *Forensic Examination of Hair*. CRC Press, Boca Raton, FL, USA.
Robinson, E.M. (2010). *Crime Scene Photography*, 2nd edn. Academic Press, New York, USA.
Robinson, J.G. and Redford, K.H., Eds. (1991). *Neotropical Wildlife Use and Conservation*. University Press, Chicago, IL.
Robinson, P. (1982). *Bird Detective*. Elm Tree Books, London, UK.
Roca, A.L., Georgiadis, N., and O'Brien, S.J. (2007). Cyto-nuclear genomic dissociation and the African elephant species question. *Quaternary International* 70, 4–16.
Rodgers, R.D. (1985). A field technique for identifying the sex of dressed pheasants. *Wildlife Society Bulletin* 13, 528–533.
Rodríguez, A., Rodríguez, B., and Lucas, M.P. (2012). Trends in numbers of petrels attracted to artificial lights suggest population declines in Tenerife, Canary Islands. *Ibis* 154, 167–172.
Rodriguez, W.C. and Bass, W.M. (1983). Insect activity and its relationship to decay rates of human cadavers in East Tennessee. *Journal of Forensic Sciences* 28(2), 423–432.
Rollins, K.E., Meyerholz, D.K., Johnson, D.G., Capparella, A.P., and Loew, S.S. (2012). A forensic investigation into the etiology of bat mortality at a wind farm: Barotrauma or traumatic injury? *Veterinary Pathology* 49(2), 332–371, doi: 10.1177/0300985812436745.
Rommel, S. and Reynolds, J.E. (2000). Diaphragm structure and function in the Florida manatee (*Trichechus manatus latirostris*). *The Anatomical Record* 259(1), 41–51.
Rommel, S.A., Costidis, A.M., Pitchford, T.D., Lightsey, J.D., Snyder, R.H., and Haubold, E.M. (2007). Forensic methods for characterizing watercraft from watercraft-induced wounds on the Florida manatee (*Trichechus manatus latirostris*). *Marine Mammal Science* 23(1), 110–132.
Rosa, I.L., Oliveira, T.P.R., Osório, F.M., Moraes, L.E., Castro, A.L.C., Barros, G.M.L., and Alves, R.R.N. (2011). Fisheries and trade of seahorses in Brazil: Historical perspective, current trends, and future directions. *Biodiversity and Conservation* 20, 1951–1971.
Roscoe, D.E. and Stansley, W. (2012). Wildlife forensic pathology and toxicology in wound analysis and pesticide poisoning. In: Huffman, J.E. and Wallace, J.R., Eds. *Wildlife Forensics Methods and Applications*. Wiley-Blackwell, Oxford, UK, pp. 109–125.
Roscoe, E.G. and McMaster, M. (2012). Wildlife ownership. In: Huffman, J.E. and Wallace, J.R., Eds. *Wildlife Forensics Methods and Applications*. Wiley-Blackwell, Oxford, UK, pp. 1–13
Rosel, P. (1999). Identification of multiple whale species in canned meat using DNA sequence analysis. *Proceedings of the International Association of Forensic Sciences, 75th Triennial Meeting*, 22–28 August 1999, Los Angeles, CA, USA.
Ross, J.G. and Fairley, J.S. (1969). Studies of disease in the Red Fox (*Vulpes vulpes*) in Northern Ireland. *Journal of Zoology, London* 157, 375–381.
Rothberg, J.M. and Leamon, J.H. (2008). The development and impact of 454 sequencing. *Nature Biotechnology* 26(10), 1117–1124.
Rothschild, B. (2009). Scientifically rigorous reptile and amphibian osseous pathology: Lessons for forensic herpetology from comparative and paleo-pathology. *Applied Herpetology* 6, 47–79.
Rotstein, D.S. (2008). How to perform a necropsy if a toxin is suspected. *Journal of Exotic Pet Medicine* 17(1), 39–43.

Routh, A. and Sleeman, J. (1997). A preliminary survey of syndactyly in the mountain gorilla: Medical conditions and veterinary considerations of zoo animals. *Proceedings of the British Veterinary Zoological Society*, Howletts and Port Lympne Wild Animal Parks, UK, pp. 22–25.

Rowcliffe, J.M., de Merode, E., and Cowlishaw, E. (2004). Do wildlife laws work? Species protection and the application of a prey choice model to poaching decisions. *Proceedings of the Royal Society B* 271, 2631–2636.

Rowley, I. (1970). Lamb predation in Australia: Incidence, predisposing conditions, and the identification of wounds. *CSIRO Wildlife Research* 15, 79–123.

RSPCA. (2010). The welfare state: five years measuring animal welfare 2005–2009. RSPCA, Horsham, UK.

Rubel, G.A., Isenbugel, E., and Wolvekamp, P., Eds. (1991). *Atlas of Diagnostic Radiology of Exotic Pets; Small Mammals, Birds, Reptiles and Amphibians*. Schlutersche Verlagsanstalt und Druckerei GMBH & Co, Hannover and W.B. Saunders Company, Philadelphia, PA, USA.

Rubner, M. (1883). Ueber den Einfluss der Körpergrösse auf Stoffund. *Kraftweschel Z Biol* 19, 535–562.

Rudin, N. and Inman, K. (2001). *An Introduction to Forensic DNA Analysis*, 2nd edn. CRC Press, Boca Raton, FL, USA.

Rudnick, J.A., Katzner, T.E., Bragin, E.A., and DeWoody, J.A. (2007). Species identification of birds through genetic analysis of naturally shed feathers. *Molecular Ecology Notes* 7, 757–762.

Rudnick, J.A., Katzner, T.E., Bragin, E.A., Rhodes, O.E., Jr., and DeWoody, J.A. (2005). Using naturally shed feathers for individual identification, genetic parentage analyses, and population monitoring in an endangered Eastern imperial eagle (*Aquila heliaca*) population from Kazakhstan. *Molecular Ecology* 14, 2959–2967.

Rudy, C. (1987). How did this wren die? *Passenger Pigeon* 49, 142.

Ruffell, A. and Murphy, E. (2011). An apparently jawless cadaver: A case of post-mortem slippage. *Science & Justice* 51, 150–153.

Rutty, G.N. (2000). Human DNA contamination of mortuaries: Does it matter? *The Journal of Pathology* 190(4), 410–411.

Rutty, G.N. (2004). The pathology of shock versus post-mortem change. In: Rutty, G.N., Ed. *Essentials of Autopsy Practice. Recent Advances, Topics and Developments*. Springer-Verlag, London, UK.

Rychlík, I., Kubícek, O., Holcák, V., Bárta, J., and Pavlík, L. (1994). DNA fingerprinting in Falconidae. *Journal of Veterinary Medicine* 39(2–3), 111–116.

Rydell, J., Bach, L., Dubourg-Savage, M.J., Green, M., Rodrigues, L. and Hedenström, A. (2010). Bat mortality at wind turbines in northwestern Europe. *Acta Chiropterologica* 12, 261–274.

Ryder, M.L. (1974). Hair of the mammoth. *Nature* 249, 190–192.

Ryg, M. et al. (1990). Seasonal changes in body mass and body composition of ringed seals (*Phoca hispida*) on Svalbard. *Canadian Journal of Zoology* 68(3), 470–475.

Saferstein, R. (2010). *Criminalistics: An Introduction to Forensic Science*, 10th edn. Prentice Hall, Upper Saddle River, NJ, USA.

Sahajpal, V. and Goyal, S.P. (2009). Microscopic examinations in wildlife investigations. In: Linacre, A., Ed. *Forensic Science in Wildlife Investigations*. CRC Press/Taylor & Francis, Boca Raton, FL, USA.

Sahajpal, V. and Goyal, S.P. (2010). Identification of a forensic case using microscopy and forensically informative nucleotide sequencing (FINS): A case study of small Indian civet (*Viverricula indica*). *Science & Justice* 50, 94–97.

Sainsbury, A.W., Bennett, P.M., and Kirkwood, J.K. (1995). Welfare of free-living wild animals in Europe: Harm caused by human activities. *Animal Welfare* 4, 183–206.

Sampson, W.C. and Sampson, K.L. (2003). Survival of the safest: Personal protective equipment for skin at the crime scene. *Evidence Technology Magazine* 1(3), 12–15.

Sanders, J., Cribbs, J., Fienberg, H., Hulburd, G., Katz, L., and Palumbi, S. (2008). The tip of the tail: Molecular identification of seahorses for sale in apothecary shops and curio stores in California. *Conservation Genetics* 9(1), 65–71.

Sanderson, G.C., Anderson, W.L., Foley, G.L., Havera, S.P., Skowron, L.M. Brawn, J.W., Taylor, G.D., and Seets, J.W. (1998). Effects of lead, iron, and bismuth alloy shot embedded in the breast muscles of game-farm mallards. *Journal of Wildlife Diseases* 34, 688–697.

Sands, P. (2003). *Principles of International Environmental Law*, 2nd Edn. Cambridge University Press, Cambridge, UK.

Sankoff, P. and White, S., Eds. (2009). *Animal Law in Australasia: a New Dialogue*. Federation Press, Annandale, NSW, Australia.

Sanson, G.D. (1980). The morphology and occlusion of the molariform cheek teeth in some Macropodinae (Marsupiala: Macropodidae). *Australian Journal of Zoology* 28, 341–365.

Sapse, D. and Kobilinsky, L. (2011). *Forensic Science Advances and their application in the Judiciary System*. CRC Press, Boca Raton, FL, USA.

Saunders, R. (2004). Ballistic injuries of animals. *Veterinary Times* 8–10.

Schaller, G.B. (1963). *The Mountain Gorilla*. University of Chicago Press, Chicago, IL, USA.

Schaper, A., Desel, H., Ebbecke, M., De Haro, L. et al. (2009). Bites and stings by exotic pets in Europe: An 11 year analysis of 404 cases from Northeastern Germany and Southeastern France. *Clinical Toxicology (Philadelphia)* 47(1), 39–43.

Schierding, M., Vahder, S., Dau, D., and Irmler, U. (2011). Impacts on biodiversity at Baltic Sea beaches. *Biodiversity and Conservation* 20, 1973–1985.

Schmidt-Nielsen, K. (1984). *Why Is Animal Size So Important?* Cambridge University Press, Cambridge, UK.

Schoenly, K.G., Haskell, N.H., and Hall, R.D. (2007). Comparative performance and complementarity of four sampling methods and arthropod preference tests from human and porcine remains at the forensic anthropology center in Knoxville, Tennessee. *Journal of Medical Entomology* 44, 881–894.

Scholin, C.A., Gulland, F., Doucette, G.J., Berson, S., Busman, M., Chavez, F.P., Cordaro, J., Delong, R., De Vogelaere, A., Harvey, J., Haulera, M. et al. (2000). Mortality of sea lions along the central California coast linked to a toxic diatom bloom. *Nature* 403(6765), 80–84.

Schröder, T. (1997). Ultrasound measurements around wind turbine sites: A study of wind energy sites in Niedersachsen and Schleswig-Holstein. Thesis, University of Frankfurt, Frankfurt, Germany.

Schulman, L. and Lehvävirta, S. (2011). Botanic gardens in the age of climate change. *Biodiversity and Conservation* 20, 217–220.

Schulz, M., Kierdorf, U., Sedlacek, F., and Kierdorf, H. (1998). Pathological bone changes in the mandibles of wild red deer (*Cervus elaphus* L.) exposed to high environmental levels of fluoride. *Journal of Anatomy* 193, 431–442.

Schwartz, M.C., Ed. (1997). *Conservation of Highly Fragmented Landscapes*. Chapman & Hall, London, UK.

Scientific Working Group on DNA Analysis Methods (SWGDAM): Revised validation guidelines, approved July 2003. Forensic Science Communications 2004, Volume 6, Number 3. http://www2.fbi.gov/hq/lab/fsc/backissu/july2004/index.htm accessed 21 June 2011

Scott, G.B.D. (1992). *Comparative Primate Pathology*. Oxford University Press, Oxford, UK.

Scott, R.S., Ungar, P.S., Bergstrom, T.S. et al. (2005). Dental microwear texture analysis shows within-species variability in fossil hominins. *Nature* 436, 693–695.

Seaman, W.J. (1987). *Postmortem Change in the Rat: A Histologic Characterization*. Iowa State University Press, Ames, IA, USA.

Segura, N.A., Usaquén, W., Sánchez, M.C., Chuaire, L., and Bello, F. (2009). Succession patterns of cadaverous entomofauna in a semi-rural area of Bogotá, Colombia. *Forensic Science International* 187, 66–72.

Seim, H.B., III. (1986). Diseases of the anus and rectum. In: Kirk, R.W., Ed. *Current Veterinary Therapy IX Small Animal Practice*. W.B. Saunders Company, London, UK, pp. 916–920.

Senn, D.R. and Stimson, P.G., Eds. (2010). *Forensic Dentistry*. CRC Press, Boca Raton, FL, USA.

Senning, W.C. (1940). A study of age determination and growth of *Necturus maculosus* based on the parasphenoid bone. *American Journal of Anatomy* 66, 483–494.

Serbanescu, I., Brown, S.M., Ramsay, D., and Levin, A.V. (2008). Natural animal shaking: A model for non-accidental head injury in children? *Eye (London)* 22, 715–717.

Servello, F.A. and Kirkpatrick, R.L. (1986). Sexing ruffed grouse in the Southeast using feather criteria. *Wildlife Society Bulletin* 14, 280–282.

Seutin, G., White, B.N., and Boag, P.T. (1990). Preservation of avian blood and tissue samples for DNA analyses. *Canadian Journal of Zoology* 69, 82–90.

Sewell, M.H.H. and Brocklesby, D.W. (1990). *Handbook on Animal Diseases in the Tropics*, 4th edn. Baillière Tindall, London, UK.

Shahan, R. (2010). *The Expert Witness in Islamic Courts. Medicine and Crafts in the Service of Law*. University of Chicago Press, Chicago, IL, USA.

Sharanowski, B.J., Walker, E.G., and Anderson, G.S. (2008). Insect succession and decomposition patterns on shaded and sunlit carrion in Saskatchewan in three different seasons. *Forensic Science International* 179, 219–240.

Sheffield, S.R., Sullivan, J.P., and Hill, E.F. (2005). Identifying and handling contaminant – Related wildlife mortality/morbidity incidents. In: Braun, C.E., Ed. *Techniques for Wildlife Investigations and Management*, 6th edn. The Wildlife Society, Bethesda, MD, USA, pp. 213–237.

Shepherd, R. (2003). *Simpson's Forensic Medicine*, 12th edn. Arnold, London, UK.

Sherman, H. (2005). *Illustrated Guide to Crime Scene Investigation*. CRC Press, Boca Raton, FL, USA.

Shi, Y.-W., Liu, X.-S., Wang, H.-Y., and Zhang, R.-J. (2008). Seasonality of insect succession on exposed rabbit carrion in Guangzhou, China. *Insect Science* 16(5), 425–439.

Shine, C., Kettunen, M., Genovesi, P., Essl, F., Gollasch, S., Rabitsch, W., Scalera, R., Starfinger, U., and ten Brink, P. (2010). Assessment to support continued development of the EU Strategy to combat invasive alien species. Final Report for the European Commission. Institute for European Environmental Policy (IEEP), Brussels, Belgium.

Shone, V. (2009). Investigating skeletal trauma inflicted by man-eating large carnivores: A study using lions (*Panthera leo*), tigers (*Panthera tigris*) and leopards (*Panthera pardus*). Dissertation submitted as part of the requirement for MSc Forensic and Biological Anthropology. University of Bournemouth, Dorset, UK.

Shoshani, J. and Eisenberg, J.F. (1982). On the dissection of a female Asian elephant (*Elephas maximus maximus* Linnaeus, 1758) and data from other elephants. *Elephant* 2, 3–93.

Sikarskie, J.G. (1992). The role of veterinary medicine in wildlife rehabilitation. *Journal of Zoo and Wildlife Medicine* 23, 397–400.

Sikes, S.K. (1967). The African elephant, *Loxodonta africana*: A field method for the estimation of age. *Journal of Zoology, London* 154, 235–248.

Sills, E.S. (2005). Conjoined twins, delivery and the memory of medicine. *Journal of the Royal Society of Medicine*, 98, 135–136.

Silvagni, P., Lowenstine, L.J., Spraker, T., Lipscombe, T.P., and Gulland, F.M. (2005). Pathology of domoic acid toxicity in California sea lions (*Zalophus californianus*). *Veterinary Pathology* 42, 184–191.

Silver, W.E. and Souviron, R.R. (2009). *Dental Autopsy*. CRC Press, Boca Raton, FL, USA.

Simmons, R.E., Avery, D.M., and Avery, G. (1991). Biases in diets determined from pellets and remains: Correction factors for a mammal and bird-eating raptor. *Journal of Raptor Research* 25(3), 63–67.

Simmons, V. and Sohan, L. (2001). A retrospective study of the treatment of toad poisoning in dogs in Trinidad. *Journal of the Caribbean Veterinary Medical Association* 1, 23–28.

Simpson, G.C. (1940). Possible causes of change in climate and their limitations. *Proceedings of the Linnean Society of London* 152, 190–219.

Simpson, V.R. (2006). Patterns and significance of bite wounds in Eurasian otters (*Lutra lutra*) in southern and south-west England. *Veterinary Record* 158, 113–119.

Sinclair, L., Merck, M., and Lockwood, R. (2006). *Forensic Investigation of Animal Cruelty*. Humane Society Press, Washington, DC, USA.

Singh, A., Gaur, A., Shailaja, K., Satyare Bala, B., and Singh, L. (2004). A novel microsatellite (STR) marker for forensic identification of big cats in India. *Forensic Science International* 141(2–3), 143–147.

Sinsch, U., Oromi, N., and Sanuy, D. (2007). Growth marks in natterjack toad (*Bufo calamita*) bones: Histological correlates of hibernation and aestivation periods. *Herpetological Journal* 17, 129–137.

Sis, R. and Landry, A. (1992). Postmortem changes in turtles. *Proceedings of the 23rd Annual International Association Aquatic Animal Medicine (IAAAM)*, Hong Kong. IAAAM, San Leandro, CA, USA, pp. 17–19.

Skinner, M. (1986). Enamel hypoplasia in sympatric chimpanzee and gorilla. *Human Evolution* 1, 289–312.

Sleeman, J.M. (2007). Wildlife rehabilitation centers as monitors of ecosystem health. In: Fowler, M.E. and Miller, R.E., Eds. *Zoo and Wild Animal Medicine*, Vol. 6. Elsevier Press, St. Louis, MO, USA, pp. 97–104.

Smith, A. (2011a). Macro photography on a budget. *Bulletin of the Amateur Entomologists' Society* 70, 119–121.

Smith, B.J. and Smith, S.D. (1991). Normal xeroradiographic and radiographic anatomy of the bobwhite quail (*Colinus virginianus*), with reference to other galliform species. *Veterinary Record* 32(3), 127–134.

Smith, D.A., Barker, I.K., and Allen, B.O. (1988). The effect of ambient temperature on healing cutaneous wounds in the common garter snake (*Thamnophis sirtalis*). *Canadian Journal of Veterinary Research* 52, 120–128.

Smith, F. (1890). The histology of the skin of the elephant. *Journal of Anatomical Physiology* 24, 493–503.

Smith, G., Trumbo, S.T., Sikes, D.S., Scotts, M.P., and Smith, R.L. (2007). Host shift by the burying beetle, *Nicrophorus pustulatus*, a parasitoid of snake eggs. *Journal of Evolutionary Biology* 20, 2389–2399.

Smith, J.M., Stauber, E.H., and Bechard, M.J. (1993). Identification of peregrine falcons using a computerized classification system of toe-scale pattern analysis. *Journal of Raptor Research* 27, 191–195.

Smith, K.E. and Wall, R. (1997). The use of carrion as breeding sites by the blowfly *Lucilia sericata* and other Calliphoridae. *Medical and Veterinary Entomology* 11(1), 38–44.

Smith, K.F., Acevedo-Whitehouse, K., and Pedersen, A.B. (2009a). The role of infectious diseases in biological conservation. *Animal Conservation* 12(1), 1–12.

Smith, K.F., Behrens, M., Schloegel, L.M., Marano, N., Burgiel, S., and Daszak, P. (2009b). Reducing the risks of the wildlife trade. *Science* 324(5927), 594–595.

Smith, K.G.V. (1986). *A Manual of Forensic Entomology*. The Trustees of the British Museum (Natural History), London, UK.

Smith, K.M., Anthony, S.J., Switzer, W.M., Epstein, J.H., Seimon, T., Jia, H., Sanchez, M.D., Huynh, T.T., Galland, G.G., Shapiro, S.E., Sleeman, J.M., McAloose, D., Stuchin, M., Amato, G., Kolokotronis, S.-O., Lipkin, W.I., Karesh, W.B., Daszak, P., and Marano, N. (2012). Zoonotic viruses associated with illegally imported wildlife products. *PLoS ONE* 7(1): e29505. Doi: 10.1371/journal.pone.0029505.

Smith, P.A., Baber, C., Hunter, J., and Butler, M. (2008). Measuring team skills in crime scene investigation: Exploring ad hoc teams. *Ergonomics* 51, 1463–1488.

Smith, S.A. (2011b). Invertebrate resources on the internet. In: Spineless Wonders: Welfare and Use of Invertebrates in the Laboratory and Classroom. *ILAR Journal* 52(5), 165–174.

Smith-Schalkwijk, M. (1988). Plastic eggs on the beach. *Wildlife Veterinary Report* 1(2), 5.

Snary, E.L., Ramnial, V., Breed, A.C., Stephenson, B., Field, H.E., and Fooks, A.R. (2012). Qualitative release assessment to estimate the likelihood of henipavirus entering the United Kingdom. *PLoS ONE* 7(2): e27918. Doi: 10.1371/journal.pone.0027918.

Solomon, S.E. (1991). *Egg and Eggshell Quality*. Wolfe Publications, London, UK.

Sonne, C. (2010). Health effects from long-range transported contaminants in Arctic top predators: An integrated review based on studies of polar bears and relevant model species. *Environment International* 36(5), 461–491.

Sorenson, M.D., and Quinn, T.W. (1998). Numts: A challenge for avian systematics and population biology. *The Auk* 115(1), 214–221.

References and Further Reading

Spearman, R.I.C. and Hardy, J.A. (1985). Integument. In: King, A.S. and McLelland, J., Eds. *Form and Function in Birds*, Vol. 3. Academic Press, London, UK, pp. 1–56.

Spencer, P.B.S., Schmidt, D., and Hummel, S. (2010). Identification of historical specimens and wildlife seizures originating from highly degraded sources of kangaroos and other macropods. In: Combating Wildlife Crime. Ed. Special issue. *Forensic Science, Medicine, and Pathology* 6(3), 225.

Spinage, C.A. (1986). *The Natural History of Antelopes*. Croom Helm, London, UK.

Spinage, C.A. (1994). *Elephants*. T. & A.D. Poyser Ltd., London, UK.

Spinage, C.A. and Brown, W.A.B. (1988). Age determination of the West African buffalo (*Syncerus caffer brachyceros*) and the constancy of tooth wear. *African Journal of Ecology* 26, 221–227.

Spratt, D. (1993). Gin'll fix it. *Institute of Medical Laboratory Sciences Gazette* 29(1), 26–27.

Staedler, M. and Riedman, M. (1993). Fatal mating injuries in female sea otters (*Enhydra lutris nereis*). *Mammalia* 57(1), 135–139.

Staerkeby, M. (2001). Dead larvae of *Cynomya mortuorum* (L.) (Diptera: Calliphoridae) as indicators of the post-mortem interval – A case history from Norway. *Forensic Science International* 120, 77–78.

Stanley Price, M.R. (1989). *Animal Re-introductions: The Arabian Oryx in Oman*. Cambridge University Press, Cambridge, UK.

Stauber, E.H. (1984). Footprinting of raptors for identification. *Raptor Research* 18(2), 67–71.

Stephenson, T. (1997). Ageing of bruising in children. *Journal of the Royal Society of Medicine* 90, 312–314.

Stewart, A. (2008). *Wildlife Detective: A Life Fighting Wildlife Crime*. Argyll Publishing, Argyll, UK.

Stewart, A. (2009). *The Thin Green Line. Wildlife Crime Investigation in Britain and Ireland*. Argyll Publishing, Argyll, Scotland.

Stewart, R.E.A., Campana, S.E., Jones, C.M., and Stewart, B.E. (2006). Bomb radiocarbon dating calibrates beluga (*Delphinapterus leucas*) age estimates. *Canadian Journal of Zoology* (84), 840.

Stewart, A. (2011). *A Lone Furrow - the continued fight against wildlife crime*. Argyll Publishing, Argyll, UK.

Stewart, A. (2012). *Wildlife and the Law*. Argyll Publishing, Argyll, UK.

Stiles, D. (2011a). Poverty, corruption and new Asian wealth – The perfect storm battering African elephants. *SWARA Journal of the East African Wildlife Society* October–December, 24–27.

Stiles, D. (2011b). Burning ivory stockpiles. *SWARA Journal of the East African Wildlife Society* October–December, 12–13.

Stiles, D., Martin, E., and Vigne, L. (2001). Exaggerated ivory prices can be harmful to elephants. *SWARA Journal of the East African Wildlife Society* October–December, 18–23.

Stirling, H.P. (1985). *Chemical and Biological Methods of Water Analysis for Aquaculturists*. Institute of Aquaculture, University of Stirling, Stirling, Scotland.

Stocker, L. (2000). *Practical Wildlife Care*. Blackwell Science, Oxford, UK.

Stoddart, T.J. (1973). Bite-marks in perishable substances. *British Dental Journal* 135, 285–287.

Stokes, E.J. and Byrne, R.W. (2000). Cognitive capacities for behavioural flexibility in wild chimpanzees (*Pan troglodytes*): The effect of snare injury on complex manual food processing. *Animal Cognitive* 4, 11–28.

Strauss, W.L. (1950). The microscopic anatomy of the skin of the gorilla. In: Gregory, W.K., Ed. *The Anatomy of the Gorilla. The Henry Cushier Raven Memorial Volume*. Columbia University Press, New York, USA.

Strickland, G.T. (1991). *Hunter's Tropical Medicine*, 7th edn. W.B. Saunders Company, Philadelphia, PA, USA.

Strickland, M.D., Johnson, G.D., Erickson, W.P., Sarappo, S.A., and Halet, R.M. (2000). Avian use, flight behavior, and mortality on the Buffalo Ridge, Minnesota wind resource area. *Proceedings of the National Avian-Wind Power Planning Meeting III*. National Wind Coordinating Committee, Washington, DC, USA, pp. 70–79.

Stroud, R.K. (1998). Wildlife forensics and the veterinary practitioner. *Seminars in Avian and Exotic Pet Medicine* 7(4), 182–192.

Stroud, R.K. (2010). Book review: *Introduction to Veterinary and Comparative Forensic Medicine* by John E. Cooper and Margaret E. Cooper. *Journal of Wildlife Diseases* 46(1), 330–332.

Stroud, R.K. and Adrian, W.J. (1996). Forensic investigational techniques for wildlife law enforcement investigations. In: Fairbrother, A., Locke, L.N., and Hoff, C.L., Eds. *Non-infectious Diseases of Wildlife*, Iowa State University Press, Ames, IA, USA.

Struhsaker, T.T. (1997). Ecology of an African rain forest: Logging in Kibale National Park and the conflict between conservation and exploitation. *Journal of Tropical Ecology* 14, 561–561.

Sukumar, R. (2006). A brief review of the status, distribution and biology of wild Asian Elephants *Elephas maximus*. *International Zoo Yearbook* 40(1), 1–8.

Summers, B.A. (2009). Climate change and animal disease. *Veterinary Pathology* 46, 1185–1186.

Sutherland, W.J. (2004). Diet and foraging behavior. In: Sutherland, W.J., Newton, I., and Green, R.E., Eds. *Bird Ecology and Conservation*. Oxford University Press, Oxford, UK.

Swain, D. (2004). *Asian Elephants Past, Present and Future*. International Book Distributors, Dehra Dun, Uttaranchal, India.

de Swart, R.L., Ross, P.R., Vedder, L.J., Timmerman, H.H., Heisterkamp, S., Loveren, H.V., Vos, J.G., Reijinders, P.J.H., and Osterhaus, A.D.M.E. (1994). Impairment of immune function in harbor seals (*Phoca vitulina*) feeding on fish from polluted waters. *Ambio* 23, 155–159.

Swobodzinski, K. (2004). Written crime-scene documentation: Six guidelines for putting what you see into words. *Evidence Technology Magazine* 2(3), 26–27.

Sykes, N.J., White, J., Hayes, T.E., and Palmer, M.R. (2006). Tracking animals using strontium isotopes in teeth: The role of fallow deer (*Dama dama*) in Roman Britain. *Antiquity* 80, 948–959.

Szewczak, J.M. and Arnett, E.B. (2006). Ultrasound emissions from wind turbines as a potential attractant to bats: A preliminary investigation. Report prepared for Bat Conservation International [www document]. http://www.batsandwind.org/pdf/ultrasoundem.pdf, pp. 1–11 (accessed 02 June 2011).

Tabor, K.L., Fell, R.D., and Brewster, C.C. (2005). Insect fauna visiting carrion in Southwest Virginia. *Forensic Science International* 150, 73–80.

Tapley, B., Griffiths, R.A., and Bride, I. (2011). Dynamics of the trade in reptiles and amphibians within the United Kingdom over a ten-year period. *Herpetological Journal* 21, 27–34.

Taylor, M.A., Coop, R.L., and Wall, R.L. (2007). *Veterinary Parasitology*, 3rd edn. Wiley-Blackwell, Oxford, UK.

Technical Working Group on Crime Scene Investigation. (2000). *Crime Scene Investigation: A Guide for Law Enforcement*. U.S. Department of Justice, Office of Justice Programs, National Institute of Justice, Washington, DC, USA.

Teeling, E.C., Springer, M.S., Madsen, O., Bates, P., O'Brien, J., and Murphy, W.J. (2005). A molecular phylogeny for bats illuminates biogeography and the fossil record. *Science* 307, 580–584.

Teerink, B.J. (1991). *Hair of West-European Mammals*. Cambridge University Press, Cambridge, UK.

Tejedo, M., Reques, R., and Esteban, M. (1997). Actual and osteochronological estimated age of natterjack toads (*Bufo calamita*). *Herpetological Journal* 7, 81–82.

Teskey, H.J. and Turnbull, C. (1979). Diptera puparia from pre-historic graves. *The Canadian Entomologist* 111, 527–528.

Thali, M.J., Dirnhofer, R., and Vock, P., Eds. (2009). *The Virtopsy Approach: 3D Optical and Radiological Scanning and Reconstruction in Forensic Medicine*. CRC Press, Boca Raton, FL.

Thériault, G. et al. (1981). Reducing aluminum: An occupation possibly associated with bladder cancer. *Canadian Medical Association Journal* 124(4), 419–425.

Thériault, G. et al. (1984). Bladder cancer in the aluminium industry. *Lancet* 1(8383), 947–950.

Thiollay, J.-M. (2005). Effects of hunting on Guianan forest game birds. *Biodiversity and Conservation* 14, 1121–1135.

Thomas, J.A., Thomas, C.D., Simcox, D.J., and Clarke, R.T. (1986). Ecology and declining status of the silver-spotted skipper butterfly (*Hesperia comma*) in Britain. *Journal of Applied Ecology* 23(2), 365–380.

Thomas, P.M., Pankhurst, N.W., and Bremner, H.A. (1999). The effect of stress and exercise on post-mortem biochemistry of Atlantic salmon and rainbow trout. *Journal of Fish Biology* 54, 1177–1196.

Thompson, P. (1923). Bird pellets and their evidence as to the food of birds. *Essex Naturalist* 20, 115–142.

Thompson, S.W. and Luna, L.G. (1978). *An Atlas of Artifacts Encountered in the Preparation of Microscopic Tissue Sections.* Charles Louis Davis, DVM Foundation, Gurnee, IL, USA.

Thompson, T. (2001). Legal and ethical considerations of forensic anthropological research. *Science & Justice* 41(40), 261–270.

Thys, J.-P., Jacobs, F., and Byl., B. (1994). Microbiological specimen collection in the emergency room. *European Journal of Emergency Medicine* 1, 47–53.

Tibbett, M., and Carter, D.O., Eds. (2008). *Soil Analysis in Forensic Taphonomy.* Taylor & Francis Group/CRC Press, Boca Raton, FL, USA.

Timmerbeil, S. (2003). The role of expert witnesses in German and U.S. civil litigation. *Annual Survey of International and Comparative Law* 9 Iss. 1, Article 8. Available at: http://digitalcommons.law.ggu.edu/annlsurvey/vol9/iss1/8

Tobe, S.S. (2009). Review of Pourret, O., Naïm, P., and Marcot, B., Eds. *Bayesian Networks: A Practical Guide to Applications.* John Wiley & Sons, Ltd, New York. Science & Justice 49(2), 150–151.

Tobe, S.S., Govan, J., and Welch, L.A. (2011a). Recovery of human DNA profiles from poached deer remains: A feasibility study. *Science & Justice* 51(2011), 190–195.

Tobe, S.S., Kitchener, A., and Linacre, A. (2009). Cytochrome *b* or cytochrome *c* oxidase subunit I for mammalian species identification – An answer to the debate. *Forensic Science International: Genetics* 306–307.

Tobe, S.S., Kitchener, A.C., and Linacre, A.M.T. (2010). Reconstructing mammalian phylogenies: A detailed comparison of the cytochrome *b* and cytochrome oxidase subunit I mitochondrial genes. *PLoS ONE* 5(11), e14156.

Tobe, S.S., Kitchener, A.C., and Linacre, A. (2011b). Assigning confidence to sequence comparisons for species identification: A detailed comparison of the cytochrome *b* and cytochrome oxidase subunit I mitochondrial genes. *Forensic Science International: Genetics* 3(1), e246–e247.

Tobe, S.S. and Linacre, A. (2007). A method to identify a large number of mammalian species in the UK from trace samples and mixtures without the use of sequencing. *Forensic Science International: Genetics* 1(1), 625–627.

Tobe, S.S. and Linacre, A.M.T. (2008a). A technique for the quantification of human and non-human mammalian mitochondrial DNA copy number in forensic and other mixtures. *Forensic Science International: Genetics* 2(4), 249–256.

Tobe, S.S. and Linacre, A.M.T. (2008b). A multiplex assay to identify 18 European mammal species from mixtures using the mitochondrial cytochrome *b* gene. *Electrophoresis* 29(2), 340–347.

Tobe, S.S. and Linacre, A. (2009). Identifying endangered species from degraded mixtures at low levels. *Forensic Science International: Genetics* 2(1), 304–305.

Tobe, S.S. and Linacre, A. (2010). DNA typing in wildlife crime: Recent developments in species identification. *Forensic Science, Medicine, and Pathology* 6(3), 195–206.

Tobe, S.S. and Linacre, A. (2011). A new assay for identifying endangered species in Traditional East Asian Medicine. *Forensic Science International: Genetics* 3(1), e232–e233.

Tomberlin, J.K. and Sanford, M.R. (2012). Forensic entomology & wildlife. In: Huffman, J.E. and Wallace, J.R., Eds. *Wildlife Forensics Methods and Applications.* Wiley-Blackwell, Oxford, UK, pp. 81–102.

Tovar-López, G., Finch, N.P., and Stauber, E.H. (2011). Footprinting of raptors for identification: A follow-up from 1982 to 2010. *Journal of Raptor Research* 45(2), 194–195.

Trail, P.W. (2006). Avian mortality at oil pits in the United States: A review of the problem and efforts for its solution. *Environmental Management* 38, 532–544.

Trape, J.F., Vattier-Bernard, G., and Trouillet, J. (1982). A case of intestinal myiasis caused by larvae of *Megaselia scalaris* (Diptera, Phoridae) in the Congo. *Bulletin de la Société de Pathologie Exotique et de ses Fliales* 75, 443–446.

Tremblay, C., Armstrong, B., Thériault, G., and Brodeur, J. (1995). Estimation of risk of developing bladder cancer among workers exposed to coal tar pitch volatiles in the primary aluminum industry. *American Journal of Industrial Medicine* 27, 335–348.

Treves, A and Naughton-Treves, L. (1999). Risk and opportunity for humans co-existing with large carnivores. *Journal of Human Evolution* 36, 275–282.

Tribe, A. and Brown, P.R. (2000). The role of wildlife rescue groups in the care and rehabilitation of Australian fauna. *Human Dimensions of Wildlife* 5, 69–85.

Tribe, A. and Spielman, D. (1996). Restraint and handling of captive wildlife. *ANZCCART Fact Sheet; ANZCCART News* 9(1), March 1996.

Tsai, L.-C., Huang, M.T., Hsiao, C.-T., Lin, A.C.-Y., Chen, S.-J., Lee, J.C.-I., and Hsieh, H.-M. (2007). Species identification of animal specimens by cytochrome b gene. *Journal of Forensic Science* 6(1), 63–65.

Tsukrov, I., DeCew, J.C., Baldwin, K., Campbell-Malone, R., and Moore, M.J. (2009). Mechanics of the right whale mandible: Full scale testing and finite element analysis. *Journal of Experimental Marine Biology and Ecology* 374(2), 93–103.

Tudge, C. (2000). *The Variety of Life*. Oxford University Press, Oxford, UK.

Turvey, S.T., Pitman, R.L., Taylor, B.L., Barlow, J., Tomonari, A., Barret, L.A., Zhaq, X., Reeves, P.R., Stewart, B.S., Wang, K., Wei, Z., Zhang, X., Pusser, L.T., Richlen, M.R., Brandon, M.R., and Wang, D. (2007). First human-caused extinction of a cetacean species? *Biology Letters* 3(5), 537–540.

Tuttle, M.D. (2004). Wind energy and the threat to bats. *BATS Magazine* 22, 4–5.

Tyler, N.J.C. (1987). Sexual dimorphism in the pelvis bones of Svalbard reindeer, *Rangifer tarandus platyrhynchus*. *Journal of Zoology, London* 213, 147–152.

Tyler, M.J., Leong A.S.-Y., and Godthelp, H. (1994). Tumors of the ilia of modern and tertiary Australian frogs. *Journal of Herpetology* 28, 528–529.

United Nations Environment Programme (UNEP). (2002). *Global Environment Outlook 3: Past, Present and Future Perspectives*. Earthscan Publications Ltd, London, UK.

United Nations Office on Drugs and Crime. (2009). *Crime Scene and Physical Evidence Awareness for Non-Forensic Personnel*. United Nations Publications, New York, USA.

Upex, B.R. (2009). Enamel hypoplasia in modern and archaeological caprine populations: The development and application of a new methodological approach. Doctoral thesis, Durham University. Available at Durham E-Theses Online: http://etheses.dur.ac.uk/182/.

Urquhart, K.A. and McKendrick, I.J. (2003). Survey of permanent wound tracts in the carcases of culled wild red deer in Scotland. *Veterinary Record* 152, 497–501.

U.S. Department of Justice. (1998). Forensic Laboratories: Handbook for Facility Planning, Design, Construction and Moving. National Criminal Justice Reference Service, Maryland, USA.

U.S. Fish & Wildlife Service. (2001). *Florida Manatee Recovery Plan, (Trichechus manatus latirostris)*, third Revision. U.S. Fish & Wildlife Service, Atlanta, Georgia, USA, 144 pp. + appendices.

Valdes-Perezgasga, M.T., Sanchez-Ramos, F.J., Garcia-Martinez, O. and Anderson, G.S. (2010). Arthropods of forensic importance on pig carrion in the Coahuilan Semidesert, Mexico. *Journal of Forensic Sciences* 55(4), 1098–1101.

Valente, A. (1983). Hair structure of the woolly mammoth, *Mammuthus primigenius* and the modern elephants, *Elephas maximus* and *Loxodonta africana*. *Journal of Zoology* 199, 271–274.

VanLaerhoven, S.L. and Anderson, G.S. (1999). Insect succession on buried carrion in two biogeoclimatic zones of British Columbia. *Journal of Forensic Sciences* 44(1), 32–43.

VanLaerhoven, S.L. and Hughes, C. (2008). Testing different search methods for recovering scattered and scavenged remains. *Canadian Society of Forensic Science Journal* 41(4), 209–213.

Van Pelt, T.I. and Piatt, J.F. (1995). Deposition and persistence of beachcast seabird carcasses. *Marine Pollution Bulletin* 30(12), 794–803.

Van Riper, C. and Van Riper, S.G. (1980). A necropsy procedure for sampling disease in wild birds. *Condor* 82, 85–98.

Van Valkenburgh, B. (1988). Incidence of tooth-breakage among large, predatory mammals. *American Naturalist* 131, 292–293.

Van Zuylen, J. (1981). The microscopes of Antoni van Leeuwenhoek. *Journal of Microscopy* 121, 309–328.

Vane-Wright, R.J. (2011). ICZN – An increasing concern for zoological nomenclature? *Antenna* 35(3), 115–117.

Vardy, P.H. and Bryden, M.M. (1981). The kidney of *Leptonychotes weddelli* (Pinnipedia: Phocidae) with some observations on the kidneys of two other southern phocid seals. *Journal of Morphology* 167(1), 13–34.

Velásquez, J., López-Angarita, J., and Sánchez, J.A. (2011). Evaluation of the FORAM index in a case of conservation. Benthic foraminifers as indicators of ecosystem resilience in protected and non-protected coral reefs of the Southern Caribbean. *Biodiversity and Conservation* 20, 3591–3603.

Verma, S.K., Prasad, K., Nagesh, N., Sultana, M., and Singh, L. (2003). Was elusive carnivore a panther? DNA typing of faeces reveals the mystery. *Forensic Science International* 137, 16–20.

Verma, S.K. and Singh, L. (2003). Novel universal primers establish identity of an enormous number of animal species for forensic application. *Molecular Ecology Notes* 3, 28–31.

Vermeij, G.J. (1993). Biogeography of recently extinct marine species: Implications for conservation. *Conservation Biology* 7(2), 391–397.

Vermylen, Y. (2010). The role of the forensic expert in criminal procedures according to Belgian Law. *Forensic Science International* 201(1–3), 8–13.

Vicente, F. (2010). Micro-invertebrates conservation: Forgotten biodiversity. *Biodiversity and Conservation* 19, 3629–3634.

Villa, J. (1979). Two fungi lethal to frog eggs in Central America. *Copeia* 4, 650–655.

Villa, J. and Townsend, D.S. (1983). Viable frog eggs eaten by phorid fly larvae. *Journal of Herpetology* 17, 278–281.

Virani, M., Watson, R., Risebrough, B., and Oaks, L. (2000). Should Africa brace itself for an imminent vulture epidemic? Lessons from the Asian vulture crisis. Abstract of paper presented at the *Tenth Pan-African Ornithological Congress (PAOC)*, Makerere University, Kampala, Uganda 3–8 September 2000.

Vitta, A., Pumidonming, W., Tangchaisuriya, U., Poodendean, C., and Nateeworanart, S. (2007). A preliminary study on insects associated with pig (*Sus scrofa*) carcasses in Phitsanulok, Northern Thailand. *Tropical Biomedicine* 24(2), 1–5.

Voitkevich, A.A. (1966). *Feathers and Plumage of Birds*. Sidgwick and Jackson, London, UK.

Voss, S.C., Cook, D.F., and Dadour, I.R. (2011). Decomposition and insect succession of clothed and unclothed carcasses in Western Australia. *Forensic Science International* 211, 67–75.

Voss, S.C., Spafford, H., and Dadour, I.R. (2009). Annual and seasonal patterns of insect succession on decomposing remains at two locations in Western Australia. *Forensic Science International* 193, 26–36.

Wagenknecht, E. Ed. (1979). *Altersbestimmung des erlegten Wildes*, 5. Auflage. J. Neumann-Neudamm, Melsungen, Germany.

Waggoner, K., Ed. (2007). *Handbook of Forensic Services*. Federal Bureau of Investigation, Quantico, VA, USA.

Wagner, A., Schabetsberger, R., Sztatecsny, M., and Kaiser, R. (2011). Skeletochronology of phalanges underestimates the true age of long-lived Alpine newts (*Ichthyosaura alpestris*). *Herpetological Journal* 21, 145–148.

Wagner, S.A. (2008). *Death Scene Investigation*. CRC Press, Boca Raton, FL, USA.

Waine, J.C. (1998). Anatomical abnormalities of the skull of a pink pigeon (*Columba mayeri*). *Veterinary Record* 142, 403–404.

Waldron, T. (2009). *Palaeopathology*. Cambridge University Press, Cambridge, UK.

Walker, D.N. and Adrian, W.J. (2003). *Wildlife Forensic Field Manual*, 3rd edn. Association of Midwest Fish and Game Law Enforcement Officers, Lincoln, NE, USA, 4th edn. Published in 2012 (http://www.midwestgamewarden.org/forensics.phtml).

Walker, K.A., Mellish, J.E., and Weary, D.M. (2011). Effects of hot-iron branding on heart rate, breathing rate and behaviour of anaesthetised Stellar sea lions. *Veterinary Record* 169, 363.

Walker, M. (2011). Abused baby boobies grow up to abuse other chicks. www.bbc.co.uk/nature/14418174, reporting on research by Müller and Anderson. *The Auk* 128(3), 615–619.

Walker, T.J. (1957). Ecological studies of the arthropods associated with certain decaying materials in four habitats. *Ecology* 38(2), 262–276.

Walker, W.A., Durie, P.R., Hamilton, J.R., Walker-Smith, J.A., and Watkins, J.B. (1996). Pediatric gastrointestinal disease (children's case article). In: Walker, W.A., Durie, P.R., Hamilton, J.R., Walker-Smith, J.A., Watkins, J.B., Eds. *Pathophysiology, Diagnosis, Management*, Vol. 1, 2nd edn. Mosby Co., St. Louis, MO, USA, pp. 581–582.

Wall, R., French, N.P., and Morgan, K.L. (1995). Population suppression for control of the blowfly *Lucilia sericata* and sheep blowfly strike. *Ecological Entomology* 20, 91–97.

Wall, W.P. (1983). The correlation between high limb-bone density and aquatic habits in recent mammals. *Journal of Paleontology* 57(2), 197–207.

Wallace, J.R. and Ross, J.C. (2012). The application of forensic science to wildlife evidence. In: Huffman, J.E. and Wallace, J.R., Eds. *Wildlife Forensics Methods and Applications*. Wiley-Blackwell, Oxford, UK, pp. 35–49.

Wallace, J.S. (2008). *Chemical Analysis of Firearms, Ammunition, and Gunshot Residue*. CRC Press, Boca Raton, FL, USA.

Waller, J.C. (1995). The aetiologies of major limb injuries amongst chimpanzees in the Sonso area of the Budongo forest, Uganda. PhD dissertation, Department of Biological Anthropology, Oxford University, Oxford, UK.

Walsh, A.L. and Morgan, D. (2005). Identifying hazards, assessing the risks. *Veterinary Record* 157, 684–687.

Walsh, C.J., Luer, C.A., and Noyes, D.R. (2005). Effects of environmental stressors on lymphocyte proliferation in Florida manatees, *Trichechus manatus latirostris*. *Veterinary Immunology and Immunopathology* 103(3–4), 247–256.

Walsh-Haney, H.E., Byrd, J.H., Jones, S., and Johnson, S. (2010). *Nonhuman Bone Identification*. CRC Press, Boca Raton, FL, USA.

Walter, C.B., O'Neill, E., and Kirby, R. (1986). "ELISA" as an aid in the identification of fish and molluscan prey of birds in marine ecosystems. *Journal of Experimental Marine Biology and Ecology* 96, 97–102.

Wan, Q.-H., and Fang, S.-G. (2003). Application of species-specific polymerase chain reaction in the forensic identification of tiger species. *Forensic Science International* 131, 75–78.

Wang, F.X. (1998). Enforcement of China Wildlife Protection Laws and development of wildlife conservation cause in China. *Chinese Wildlife* 19(5), 3–6. [Chinese]

Wang, H.-F., López-Pujol, J., Meyerson, L.A., Qiu, J.-X., Wang, X.-K., and Ouyang, Z.-Y. (2011). Biological invasions in rapidly urbanizing areas: A case study of Beijing, China. *Biodiversity and Conservation* 20, 2483–2509.

Wang, J., Li, Z., Chen, Y., Chen, Q., and Yin, X. (2008). The succession and development of insects on pig carcasses and their significances in estimating PMI in South China. *Forensic Science International* 179, 11–18.

Warburton, B., Gregory, N.G., and Morriss, G. (2000). Effect of jaw shape in kill-traps on time to loss of palpebral reflexes in brushtail possums. *Journal of Wildlife Diseases* 36, 92–96.

Wardle, H. (2008).Courting disaster: Ethics warning for witness vets. *Veterinary Times* 25, 1–3.

Warren, M.W. and Freas, L.E., Eds. (2008). *The Forensic Anthropology Laboratory*. CRC Press, Boca Raton, FL, USA.

Warwick, C. (2004). Gastrointestinal disorders: Are healthcare professionals missing zoonotic causes? *Journal of the Royal Society of Health* 124, 137–142.

Warwick, C. (2006). Zoonoses: Drawing the battle lines. *Veterinary Times Clinical* 36, 26–28.

Warwick, C., Frye, F.L., and Murphy, J.B., Eds. (2004). *Health and Welfare of Captive Reptiles*. Chapman & Hall/Kluwer, London, UK.

Wasser, S.K., Clark, W.J., Drori, O., Kisamo, E.S., Mailand, C., Mutayoba, B., and Stephens, M. (2008). Combating the illegal trade in African elephant ivory with DNA forensics. *Conservation Biology* 22(4), 1065–1071.

Wasser, S.K., Mailand, C., Booth, R., Mutayoba, B., Kisamo, E., Clark, B., and Stephens, M. (2007). Using DNA to track the origin of the largest ivory seizure since the 1989 trade ban. *Proceedings of the National Academy of Sciences of the United States of America* 104(10), 4228–4233.

Wasser, S.K., Shedlock, A.M., Comstock, K., Ostrander, E.A., Mutayoba, B., and Stephens, M. (2004). Assigning African elephant DNA to geographic region of origin: Applications to the ivory trade. *Proceedings of the National Academy of Sciences of the United States of America* 101(41), 14847–14852.

Watson, E.J. and Carlton, C.E. (2005). Insect succession and decomposition of wildlife carcasses during fall and winter in Louisiana. *Journal of Medical Entomology* 42(2), 193–203.

Watson, E.J.G. (2004). Faunal succession of necrophilous insects associated with high-profile wildlife carcasses in Louisiana. PhD dissertation, Louisiana State University, Baton Rouge, LA, USA.

Watson, K.D. (2011). *Forensic Medicine in Western Society*. Routledge, London, UK.

Waugh, J., Evans, M.W., Millar, C.D., and Lambert, D.M. (2011). Birdstrikes and barcoding: Can DNA methods help make the airways safer? *Molecular Ecology Resources* 11, 38–45.

Weatherley, A.H. and Gill, H.S. (1987). *The Biology of Fish Growth*. Academic Press, London, UK.

Weaver, J.L., Arvidson, C., and Wood, P. (1992). Two wolves, *Canis lupus*, killed by a moose, *Alces alces*, in Jasper National Park, Alberta. *Canadian Field-Naturalist* 106(1), 126–127.

Webb. J.P., Loomies, R.B., Madon, M.B., Bennett, S.G., and Green, G.E. (1983). The chigger species *Eutrombicula belkini* GOULD (Acri: Trombiculidae) as a forensic tool in homicide investigation in Ventura County, California. *Bulletin of the Society of Vector Entomology* 8(2), 141–146.

Webb, K.E., Barnes, D.K.A., Clark, M.S., and Bowden, D.A. (2006). DNA barcoding: A molecular tool to identify Antarctic marine larvae. *Deep-Sea Research Part II* 53, 1053–1060.

Webb, K.M. and Allard, M.W. (2009). Identification of forensically informative SNPs in the domestic dog mitochondrial control region. *Journal of Forensic Science* 54(2), 289–304.

Wecht, C.H. and Rago, J.T., Eds. (2005). *Forensic Science and Law. Investigative Applications in Criminal, Civil and Family Justice*. CRC Press, Boca Raton, FL, USA.

Weir, J.S. (1972). Mineral content of elephant dung. *East African Wildlife Journal* 10, 229–230.

Weise, M. and Harvey, J. (2005). Impact of the California sea lion (*Zalophus californianus*) on salmon fisheries in Monterey Bay, California. *Fisheries Bulletin* 103(4), 685–696.

Weiss, S.L. (2009). *Forensic Photography: The Importance of Accuracy*. Prentice Hall, Upper Saddle River, NJ, USA.

Weissenberger, M., Reichert, W., and Mattern, R. (2011). A multiplex PCR assay to differentiate between dog and red fox. *Forensic Science International: Genetics* 5, 411–414.

Wells, N. (2011). *Animal Law in New Zealand*. Brookers, Wellington, New Zealand.

Wemmer, C., Krishnamurthy, V., Srestha, S., Hayek, L.-A., Thant, M., and Nanjappa, K.A. (2006). Assessment of body condition in Asian elephants (*Elephus maximus*). *Zoo Biology* 25, 187–200.

Wen, J., Hu, C., Zhang, L., Luo, P., Zhao, Z., Fan, S., and Su, T. (2010). The application of PCR-RFLP and FINS for species identification used in sea cucumbers (Aspidochirotida: Stichopodidae) products from the market. *Food Control* 21(4), 403–407.

Wertheim, P.A. (1992). Crime scene note taking. *Journal of Forensic Identification* 42, 230–236.

Wesselink, M. and Kuiper, I. (2011). Individual identification of fox (*Vulpes vulpes*) in forensic wildlife investigations. *Forensic Science International: Genetics* 3(1), e214–e215.

Weston, D.A. (2008). Investigating the specificity of periosteal reactions in pathology museum specimens. *American Journal of Physical Anthropology* 137, 48–59.

Wetton, J.H., Braidley, G.L., Tsang, C.S.F., Roney, C.A., Powell, S.L., and Spriggs, A.C. (2002). Generation of a species-specific DNA sequence library of British Mammals. A study by the Forensic Science Service. http://jncc.defra.gov.uk/pdf/dna_report.pdf

Whitaker, T.B., Slate, A.B., and Johansson, A.S. (2005). Sampling feeds for mycotoxin analysis. In: Diaz, D.E., Ed. *The Mycotoxin Blue Book*. Nottingham University Press, Nottingham, UK, pp. 1–23.

White, J.T., McGill, B.J., and Lechowicz, M.J. (2011). Human-disturbance and caterpillars in managed forest fragments. *Biodiversity and Conservation* 20, 1745–1762.

Whittaker, D.K., Thomas, V.C., and Thomas, R.I.M. (1976). Post-mortem pigmentation of teeth. *British Dental Journal* 140, 100–102.

Wieczorek, K., Beutin, L., and Osek, J. (2011). Rare VTEC serotypes of potential zoonotic risk isolated from bovine hides and carcases. *Veterinary Record* 168, 80.

Wiig, Ø. (1985). Morphometric variation in the hooded seal (*Cystophora cristata*). *Journal of Zoology, London* 206, 497–508.

Wijnstekers, W. (2011). *The Evolution of CITES*. 9th Edition. CITES Secretariat, Lausanne. Switzerland and http://www.cites.org/common/resources/Evolution_of_CITES_9.pdf

The Wildlife Forensic Science Society. Web page for the Society for Wildlife Forensic Science. www.wildlifeforensicscience.org

Willemser, R.E. and Hailey, A. (2002). Body mass condition in Greek tortoises: Regional and interspecific variation. *Herpetological Journal* 12, 105–114.

Willey, P. and Snyder, L.M. (1989). Canid modification of human remains: Implications for time-since-death estimations, *Journal of Forensic Sciences* 34(4), 894–901.

William, O.M. and John, W. (2001). Immunological differentiation of camel meat from other mammalian meats. *Journal of Camel Practice and Research* 8(1), 1–6.

Williams, C.C. (1934). A simple method for sectioning mammalian hairs for identification. *Journal of Mammalogy* 15, 251–252.

Williams, D.J., Ansford, A.J., Priday, D.S., and Forrest, A.S. (1998). *Forensic Pathology*. Churchill Livingstone, Edinburgh, UK.

Williams, D.L. (2012). *Ophthalmology of Exotic Pets*. Wiley-Blackwell, Oxford, UK.

Williams, T.D., Allen, D.D., Groff, J.M., and Glass, R.L. (1992). An analysis of California sea otter (*Enhydra lutris*) pelage and integument. *Marine Mammal Science* 8(1), 1–18.

Wilson, A., Janaway, R., Holland, A., Dodson, H., Baran, E., Pollard, A.M., and Tobin, D. (2007). Modelling the buried human body environment in upland climes using three contrasting field sites. *Forensic Science International* 169, 6–18.

Wilson, G.H., Ed. (2004). Wound healing and management. *Veterinary Clinics of North America: Exotic Animal Practice* 7, 1.

Wilson, J.F. (1998). *Law and Ethics of the Veterinary Profession*. Priority Press, Yadley, PA, USA.

Wilson-Wilde, L. (2010). Combating wildlife crime. *Forensic Science and Medical Pathology* 6(3), 149–150.

Wilson-Wilde, L., Norman, J., Robertson, J., Sarre, S., and Georges, A. (2010). Current issues in species identification for forensic science and the validity of using the cytochrome oxidase I (COI) gene. *Forensic Science, Medicine, and Pathology* 6(3), 233–241.

Wimberger, K. and Downs, C.T. (2010). Annual intake trends of a large urban animal rehabilitation centre in South Africa: A case study. *Animal Welfare* 19(4), 501–514.

Wimberger, K., Downs, C.T., and Boyes, R.S. (2010). A survey of wildlife rehabilitation in South Africa: Is there a need for improved management? *Animal Welfare* 19(4), 481–500.

Wiseman, J. (1993). *SAS Survival Guide*. Harper Collins, Glasgow, UK.

Witzenberger, K.A. and Hochkirch, A. (2011). Ex situ conservation genetics: A review of molecular studies on the genetic consequences of captive breeding programmes for endangered animal species. *Biodiversity and Conservation* 20, 1843–1861.

Wobeser, G. (1996). Forensic (medico-legal) necropsy of wildlife. *Journal of Wildlife Diseases* 32, 240–249.

Wobeser, G.A. (1981). Necropsy and sample preservation techniques. In: Wobeser, G.A., Ed. *Diseases of Wild Waterfowl*. Plenum Press, New York, USA.

Wolf, B.O., and Hatch, K.A. (2011). Aloe nectar, birds and stable isotopes: Opportunities for quantifying trophic interaction. *Ibis* 153, 1–3.

Wolfe, A. and Wright, I.P. (2004). Parasitic nematode eggs in fur samples from dogs. *Veterinary Record* 154, 408.

Wolfes, R., Mathe, J., and Seitz, A. (1991). Forensics of birds of prey by DNA fingerprinting with 33P-labeled oligonucleotide probes. *Electrophoresis* 12, 175–180.

Wolff, M., Uribe, A., Ortiz, A., and Duque, P. (2001). A preliminary study of forensic entomology in Medellín, Colombia. *Forensic Science International* 120, 53–59.

Wolstenholme, D.R. (1992). Animal mitochondrial DNA: Structure and evolution. *International Review of Cytology* 141, 173–216.

Wong, K.-L., Wang, J., But, P.P.-H., and Shaw, P.-C. (2004). Application of cytochrome b DNA sequences for the authentication of endangered snake species. *Forensic Science International* 139(1), 49–55.

Woodford, M.H., Keet, D.F., and Bengis, R.G. (2000). Woodford, M.H., Ed. *Post-mortem Procedures for Wildlife Veterinarians and Field Biologists*. OIE, Care for the Wild International and the IUCN Veterinary Specialist Group, Paris, France.

Woodroffe, R., Frank, L.G., Lindsey, P.A., ole Ramah, S.M.K., and Romañach, S. (2007). Livestock husbandry as a tool for carnivore conservation in Africa's community rangelands: A case-control study. *Biodiversity Conservation* 16, 1245–1260.

Woolf, Lord (1996). *Access to Justice. Final Report to the Lord Chancellor on the Civil Justice System in England and Wales*. http://webarchive.nationalarchives.gov.uk/+/http://www.dca.gov.uk/civil/final/index.htm

Wootton, R. and Craig, J. (1999). *Introduction to Telemedicine*. The Royal Society of Medicine, London, UK.

Wright, J.L.C., Boyd, R.K., deFreitas, A.S.W., Falk, R.A., Jamieson, W.D., Laycock, M.V., McCulloch, A.W., McInnes, A.G., Oderse, P., Pathok, V., Quilliam, M.A., Ragan, M.A., Sim, P.G., Thibautt, P., Watter, J.A., Gilgan, M., Richard, D.J.A., and Dewar, D. (1989). Identification of domoic acid, a neuroexcitatory amino-acid, in toxic mussels from Eastern Prince-Edward-Island. *Canadian Journal of Chemistry* 67(3), 481–490.

Wright, K., Trupkiewicz, J.E., and Johnson, J.D. (2004). Learning from roadkills: Implications for conservation of native snakes. *Proceedings of Annual Conference Association of Reptilian and Amphibian Veterinarians*, Napes, FL, USA, pp. 109–115.

WWF. (2011). Impact of habitat loss on species http://wwf.panda.org/about_our_earth/species/problems/habitat_loss_degradation/

Wyllie, I. (1993). *Guide to Age and Sex in British Birds of Prey*. Institute of Terrestrial Ecology, Monkswood, England.

Wyllie, I. and Newton, I. (1999). Use of carcasses to estimate the proportions of female Sparrowhawks and Kestrels which bred in their first year of life. *Ibis* 141, 489–506.

Xu, Y.C., Li, B., Li, W.S., Bai, S.Y., Jin, Y., Li, X.P., Gu, M.B., Jing, S.Y., and Zhang, W. (2005). Individualization of tiger by using microsatellites. *Forensic Science International* 151(1), 45–51.

Yalden, D.W. (1977). *The Identification of Remains in Owl Pellets*. The Mammal Society, Berkshire, UK.

Yalden, D.W. (2009). *The Analysis of Owl Pellets*, 4th edn. Mammal Society, Southampton, UK.

Yan, P., Wu, X.-B., Shi, Y., Gu, C.-M., Wang, R.-P., and Wang, C.-L. (2005). Identification of Chinese alligators (*Alligator sinensis*) meat by diagnostic PCR of the mitochondrial cytochrome b gene. *Biological Conservation* 121, 45–51.

Yannic, G., Sermier, R., Aebischer, A., Gavrilo, M.V., Gilg, O., Miljeteig, C., Sabard, B., Strøm, H., Pouivé, E., and Broquet, T. (2011). Description of microsatellite markers and genotyping performances using feathers and buccal swabs for the Ivory gull (*Pagophila eburnea*). *Molecular Ecology Resources* 11(5), 877–889.

Yates, B.C., Espinoza, E.O., and Baker, B.W. (2010). Forensic species of elephant (Elephantidae) and giraffe (Giraffidae) tail hair using light microscopy. *Forensic Science, Medicine and Pathology* 6(3), 165–171.

Yau, F.C.F., Wong, K.L., Shaw, P.C., But, P.P.H., and Wang, J. (2002). Authentication of snakes used in Chinese medicine by sequence characterized amplified region (SCAR). *Biodiversity and Conservation* 11(9), 1653–1662.

Yntema, C.L. (1960). Effects of various temperatures on the embryonic development of *Chelydra serpentina*. *Anatomical Record* 136, 305–306.

Yoo, H.S., Eah, J.Y., Kim, J.S., Kim, Y.J., Min, M.S., Paek, W.K., Lee, H., and Kim, C.B. (2006). DNA barcoding Korean birds. *Molecules and Cells* 22, 323–327.

Young, W.G. and Robson, S.K. (1987). Jaw movement from microwear on the molar teeth of the koala *Phascolarcus cinereus*. *Journal of Zoology, London* 213, 51–61.

Zajac, A.Z. and Conboy, G.A. (2012). *Veterinary Clinical Parasitology*, 8th edn. Wiley-Blackwell, Oxford, UK.

Zavalaga, C.B., Emslie, S.D., Estela, F.A., Müller, M.S., Dell'Omo, G., and Anderseon, D.J. (2012). Overnight foraging trips by chick-rearing Nazca Boobies *Sula granti* and the risk of attack by predatory fish. *Ibis* 154, 61–73.

Zhang, L.B. (1996). Supervising key cases involved in wildlife in China in 1996. In: Maochen, S., Ed. *China Forestry Yearbook 1996*. China Forestry Press, Beijing, China, p. 168 [in Chinese].

Zinck, J.M., Duffield, D.A., and Ormsbee, P.C. (2004). Primers for identification and polymorphism assessment of Vespertilionid bats in the Pacific Northwest. *Molecular Ecology Notes* 4, 239–242.

Zink, R.M. and Barrowclough, G.F. (2008). Mitochondrial DNA under siege in avian phylogeography. *Molecular Ecology* 17, 2107–2121.

Zucca, P., Di Guardo, G., Pozzi-Mucelli, R., Scaravelli, D., and Francese, M. (2004). Use of computer tomography for imaging of *Crassicauda grampicola* in a Risso's Dolphin (*Grampus griseus*). *Journal of Zoo and Wildlife Medicine* 35(3), 391–394.

Zwart, P., De Batist, P., Mutschmann, F., and Disney, R.H.L. (2005). The phorid "scuttle fly" (*Megaselia scalaris*), a threat to zoological conditions and especially to amphibians. *Bulletin of the British Veterinary Zoological Society* 5(2), 27–30.

Index

A

Abortion, 285, 414
Abscesses, 418, 436, 438, 560
Abuse of animals, 6, 58, 149, 151, 158, 213, 222, 223, 284–285, 291, 398, 509; *see also* Welfare
    animals, 6, 58, 149, 151, 158, 213, 222, 223, 284–285, 291, 398, 509
    cases, 6, 149, 155, 158, 213
    humans, 149, 160, 398
Accidents, 8, 14, 17, 21–22, 27, 31, 36–39, 46–48, 59, 67, 79, 119, 136, 164, 210, 218, 222, 227, 228, 231, 238–239, 253, 257, 265, 266, 282, 285, 286, 289, 295, 297, 298, 300–301, 308, 310, 311, 322, 364, 439, 443, 445, 452, 473, 474, 508, 510; *see also* Disasters; Road-traffic (vehicular) accidents (RTAs)
Accreditation of laboratories, 327–328, 385
Addresses, useful, 564–567
Admissibility of evidence, *see* Evidence
African wild dog, 21, 54–55, 592
Ageing (Age determination)
    of animals and derivatives, 26, 41, 184, 192–193, 199, 271–272, 276, 277, 282–284, 290, 298, 317, 321, 399–402, 450, 502
    of bruises, 120, 198–199, 282, 284
    of fractures, 120, 282, 283, 502, 543
    of lesions, 120, 198, 281–284, 298, 543
Aggression, inter-and intra-specific, 298–299
Agriculture and food production, effects of wildlife on, 15
Aircraft and aircraft accidents, 222, 300–301
Aircraft wildlife strike, 421–425
Air guns, 302–303, 307–308
Aluminium, 39, 178, 410–411, 446, 524, 582
American English, xv, 483, 592
Ammunition, 253, 274, 303, 307; *see also* Bullets
Amphibians, 22, 26–27, 34, 47, 50, 52, 53, 60, 61, 63, 70, 72, 73, 83, 107–108, 110, 124, 185, 189, 192, 193, 195, 197, 199, 207, 210, 214, 223, 229, 240, 249, 250, 259, 260, 264, 272, 278–280, 297–299, 309, 330, 334, 351, 357, 362, 366, 470, 514, 519, 569, 570, 597; *see also* Ectotherms
    English and scientific names, 592–598
Anaesthesia of animals, 185, 203, 204, 213, 226; *see also* Immobilisation

Anatomy, 122, 124, 183, 200–202, 249, 265, 271, 273, 275, 280, 285–286, 296, 306, 308, 312, 314, 315, 351, 397, 412, 416–418, 420, 421, 442, 501–502, 505, 508, 513, 574–577, 591
Animal(s), 1, 12, 91, 115, 128, 149, 161, 181, 237, 328, 392, 397, 465, 482, 501
    captive, xii, 5, 8, 13, 14, 17, 22, 28, 48, 57, 58, 78, 96, 106, 196, 201, 205, 207, 210, 214, 216, 218, 223, 224, 227–229, 231–233, 285, 286, 296, 299–301, 343, 360, 398, 482, 509, 600
    cruelty to, *see* Welfare
    domesticated, 1–6, 9, 14, 20, 26, 28, 37–39, 52, 53, 68, 73, 74, 84, 106, 117, 121, 124–126, 155, 172, 179, 183, 192, 193, 195, 199, 204, 205, 214, 216, 223, 226, 228, 229, 235, 244–245, 266, 271, 272, 275, 278, 284–285, 287, 299, 302, 303, 311, 313, 314, 318–319, 343, 363, 446, 511
    feed (food), 13, 23, 61, 74, 87–88, 121, 155, 158, 160, 224, 226, 230, 267, 288, 289, 293, 332, 338–344, 362, 364, 469, 470, 477, 509
    feral, 2–3, 36, 68, 93, 513
    free-living, 1–5, 14, 17, 28, 37, 48, 51–53, 56–57, 61, 77, 78, 92, 96, 106, 172, 185, 188, 196, 197, 205, 216, 218, 219, 223, 224, 229, 231, 233, 244, 295, 300–301, 307–308, 357, 398, 477, 513
    health, 13, 35, 57, 83, 84, 92, 98, 100, 101, 103–105, 107, 125, 164, 165, 215, 258, 319, 321, 336, 565, 569, 572, 583, 599–600
    keeping, 52, 75, 92, 105, 164, 208, 472, 483
    law, *see* Law
    live, 6, 9, 28–29, 34, 39, 40, 52, 57, 82, 88, 107, 119–121, 124, 157, 168, 172, 181–236, 262, 268, 282, 288, 290, 301, 307, 329, 330, 362, 468, 470, 473, 524, 540, 541, 544, 548, 600–602
    models, 105, 109–110, 298
    non-domesticated, 3, 52, 93, 96, 105, 106, 124, 226, 286, 509, 520
    research, 98
    terminology, 591–598
    transportation, 92, 233
    welfare, *see* Welfare
    wild, 1, 13, 97, 116, 165, 182, 238, 336, 368, 392, 397, 482, 501, 513, 521, 569, 603
Animal diets and foodstuffs, 338–343
Animal forensics, evolution of, 115–117

Animal health, 13, 35, 57, 84, 92, 98, 100, 101, 105–107, 125, 164, 165, 215, 258, 319, 321, 458, 561, 583, 599
Annex A (CITES), 103, 105
Anoxia, 296, 309; *see also* Asphyxia
Antlers, 64, 120, 195, 264, 268, 318, 402, 418, 502
Appearing in court, 10, 126, 194, 429, 476, 481–499
Aquaria, 50, 51, 428; *see also* Fish
Arabia, 87; *see also* Middle East
Archaeology, 123, 124, 171, 192, 242, 312, 315, 316, 319, 337, 403
Archaeopteryx, 323
Archaeozoology, *see* Zooarchaeology
Article 10 certificate, 104
Asphyxia, 285, 296, 308–310, 508, 510, 518, 519
Assessment of reproductive activity, 287
Asymmetry, *see* Fluctuating asymmetry
Attacks by wild animals, 17; *see also* Predation, *ante-mortem* and *post-mortem*
Auckland Zoo, xviii, 504
Authorisation, 56, 92, 96, 99, 163, 164, 463, 482, 527, 600, 613
Autolysis, 186, 241, 256, 257, 260, 264, 276–278, 280, 290, 330, 424, 508, 511, 549

B

Baboons, xv, 19, 41, 78, 298, 345, 469, 593
Bacon, Francis, 505
Bacteriology, 193, 209, 248, 263, 274, 284, 328, 333, 334, 348, 360
Baden-Powell, Robert, 470
Baited carcases, 471
Basic local alignment search tool (BLAST), 372, 375, 509
Bats, 42, 62, 63, 79, 100, 108, 188–190, 222, 223, 300, 301, 375, 376, 399, 458–464, 604, 610
Battery-operated equipment, 171, 524, 525
Bayesian network, 476, 508
Beebe, William, 122, 389
Behaviour, *see* Ethology
Behaviour of an expert, 123, 216, 387, 444; *see also* Code of practice
Belgium, 150, 319, 485, 494, 499, 565
Bestiality, 50, 115, 117, 212, 508, 520
Bible, Holy, 1
Biodiversity, 40, 49, 50, 60, 61, 64, 69, 72, 73, 75, 76, 83, 85, 102, 103, 206, 228, 428, 479, 508
Biohazards, 128; *see also* Zoonoses
Biological control, 69–70
Biological features of animals, relevance to forensic medicine, 199, 283, 330; *see also* Ectotherms
Biological substance, 602
Biological terms, 507–520
Biopsy, 191, 210, 212, 287, 296, 329, 336, 346–348, 456, 524

Biosecurity, 92, 107, 426, 428, 429, 508
Bioterrorism, 50
Bird pellets, xiv, xviii, 337, 343, 366, 393, 470
Birds, 8, 15, 93, 116, 165, 184, 240, 334, 378, 393, 397, 465, 507, 524, 564, 569, 573, 592, 604; *see also* Birds of prey; Bird strike
  English and scientific names, 591–598
Birds of prey, 8, 21, 24, 36, 37, 43, 62, 67, 73, 77, 100, 103, 104, 185, 208, 209, 211, 212, 218, 224, 235, 252, 271, 337, 344, 359, 378, 398, 471, 513, 604, 612
  identification systems, 62, 185, 398, 429–432
Bird strike, 265, 301, 421, 425
Bites
  from animals, 25, 215, 316, 509
  courtship, 215, 224, 298, 358, 449
  marks, 26, 213–216, 285, 289, 290
Bite wounds, 25, 213, 215, 216, 218, 290, 297; *see also* Teeth; Wounds
BLAST, *see* Basic local alignment search tool (BLAST)
Blast injuries, 234; *see also* Bombs
Blood, 26, 117, 128, 157, 165, 191, 240, 329, 415, 470, 486, 502, 508, 524, 541, 575, 591, 599; *see also* Haematology
  examination, 26, 57, 195, 209, 240, 256, 333, 335, 502
  hazards from, 128, 131
Blowflies, 152–156, 158; *see also* Maggots
  eggs and larvae, 152, 153, 156, 158
  life cycle, 153, 155, 156
Boats, injury from, 218, 300; *see also* Marine mammals
  use in wildlife crime, 44, 218, 465
Boats, injury from, 218, 300
Bombs, 108
Bones, 17, 119, 166, 192, 241, 330, 377, 393, 402, 470, 502, 516, 522, 555, 575, 599, 603; *see also* Osteology
  examination, 241, 243, 250, 261, 305, 311, 312, 316, 317, 555–562, 583
Bones and skeletal lesions, 311–320
Botany and botanists, 1, 508; *see also* Medicinal plants; Plants
Breeding programmes and conservation genetics, 77–78
British Trust for Ornithology (BTO), xviii, 186
Bruising, *see* Contusion
Brushings, *see* Cytology
BTO, *see* British Trust for Ornithology (BTO)
Buffalo, African, 7, 14, 245
Bufotoxin, 26; *see also* Amphibians
Bullets and bullet wounds, 28, 135, 136, 220, 221, 254, 259, 272, 274, 302–308, 337, 343, 421, 443–445, 465, 468, 494, 509, 512, 513, 517, 523, 583; *see also* Wounds
  entry and exit hole, 303, 304

Index

Burns and scalds, 230, 285, 509, 517
Bushmeat, 33, 40–43, 76, 84, 124, 208, 221, 376, 398, 509
Bustards, 77
Butterflies, 2, 64, 74, 185, 598; see also Invertebrates
Bwindi impenetrable forest, 173

C

Cabañeros National Park, 95
Cadavers and carcases, use of in research
  specimen submission form, 539–561
Callipers, vernier, 189, 190, 248, 269, 522
Cameras and film, 52, 56, 136–140, 147, 167, 176, 179, 187, 193, 391, 462–463, 475, 496, 521, 523, 582, 584; see also Photography
Captive animals, xii, 28, 58, 106, 191, 196, 218, 224, 227–229, 231, 296, 299, 360, 477, 482, 509
Captive-breeding of animals, xii, 34, 49, 51, 78, 92, 96, 287
Captivity, unlawful, 6
Carcase retrieval, 162, 412, 460, 471
Carcases, storage prior to examination, 473
Carnivore-livestock conflicts, 20–21
Carr, Kenneth, xviii, 29
Cause and circumstances of death, 257, 286–287, 502
Cause of incidents (wildlife), 13–27
Central nervous system (CNS), 202, 211, 219, 250, 310
Cetaceans, see Marine mammals
Chain of custody (evidence) (continuity)
  initiating, 145–146
  specimen form, 251, 539–562
Chameleon, flap-necked, 104, 597
Chemical analysis, 289, 334, 340, 347, 348; see also Clinical chemistry
Chickens, use as forensic model, 278
Child abuse, see Abuse
Chilling, 167, 230, 292, 295, 498;
  see also Hyperthermia and hypothermia
Chimpanzee, 41, 68, 80, 81, 219, 221, 291, 593
China, 31, 68–69, 102, 113, 149, 150, 426, 428, 483
Chytridiomycosis, 70, 107–108; see also Amphibians
Circumstances of death, 6
Circuses, 52, 216–217
CITES, see Convention on International Trade in Endangered Species of Wild Fauna and Flora (CITES)
Civil disturbance, 39–40, 308
Civil wrong (tort), 483
Claims and compensation, 18, 96, 106, 286, 310, 482, 483, 511
Classification of animals, examples, 591–592
Climate change, 64, 72–75, 84, 228, 286, 295, 405, 409, 412

Clinical cases-specimen submission form, 539–561
Clinical checklist, 196, 207
Clinical chemistry, 209, 212, 333
Clinical signs, see Examination
Clinical work; see also Live animals, investigation of
  examination, 119, 194, 196
  history, 184
  identification of animal, 188
CNS, see Central nervous system (CNS)
Code, 44, 49, 52, 54, 55, 64, 81, 156, 164, 179, 194, 253, 328, 343, 369, 371–372, 387, 482–484, 495, 510, 522, 523, 530, 540, 574, 606, 613; see also Guidance
  ethics, 52, 387, 482–483
  evidence, 93, 156, 343, 387, 482–484, 495
  laboratory animals, 44, 54
  practice, xv, 44, 49, 52, 54, 55, 64, 81, 164, 522
    (see also Good practice)
  Wildlife Inspectors, 472
Cold, effects, 230; see also Chilling; Freezing
Collection of evidence, 9, 122, 129, 143–145, 444, 465–479; see also Evidence
Collection of specimens/samples, 30, 149, 156–160, 331–332
  from dead animals, 9, 64–65, 156, 158
  entomological, 156–160
  from live animals, 157
  from the surrounding area (environment), 156, 158
Collisions, 21–22, 32, 85, 222, 231, 234, 288, 300, 301, 397, 405–407, 409, 417, 422, 423, 425, 439, 452, 461–462; see also Road-traffic (vehicular) accidents (RTAs)
Colour codes, 194, 253, 343; see also Quantification; Scoring
Community considerations, 45
Comparative forensic medicine, 119, 167, 185, 189, 190, 192, 212, 223, 243, 244, 249, 251, 261, 264, 269, 274, 287, 300, 309, 327, 328, 334, 365, 472, 501–502, 510
  inter-disciplinary nature, 122–124
Comparative medicine, 115, 117
Competition, 15, 17, 19, 48, 61, 67, 224, 298, 506;
  see also Conflict
Composite reports, 166, 477
Computerised tomography (CT), 200, 273, 421
Computers, 9, 84, 137, 145, 167, 177, 179, 180, 195, 262, 430, 476, 521, 523, 605, 606, 608
Conciliation, 506
Condition and condition scoring, 190–192, 209, 225
Conferences, workshops, seminars and meetings, 115
Confiscation of animals or equipment, 81–82
Conflict
  human, 14, 16, 19, 27, 28, 506; see also Competition
  livestock, 20–21

Conservation, 1, 3, 5, 10, 13, 16, 17, 20, 21, 27–32, 38, 40, 41, 43, 48, 49, 51–53, 57–60, 62, 64, 66, 67, 69, 71–72, 75–78, 81, 83, 85, 91, 93–103, 109, 111, 115–117, 124, 164, 165, 201, 203, 206, 219, 221, 277, 299, 319, 323, 365, 368, 371, 374, 397–399, 405, 406, 419, 425, 427, 439, 443, 455–458, 477, 479, 482, 485, 512, 514, 564–566, 599–601, 610, 611
   biology, 13, 60, 124
   conventions, 97, 98
   legislation, 41, 77, 100, 482, 599–601
   organisations, 100–101, 477
   status, 60, 93
Conservation of wildlife, 43, 91, 109;
     see also Individual species
   holistic approach, 506
Conservation Through Public Health (CTPH), 20, 458
Containers for specimens, 169
Contaminants and foreign bodies, investigation of, 364
Contaminants and similar material-specimen submission form, 539–562
Contamination of evidence, *see* Evidence
Contamination of material or sites, *see* Crime scene
Contents of book, vii–ix, 501
Continuing Professional Development (CPD), 386
Control material, *see* Reference collections
Control region, 369, 371; *see also* Displacement loop (D-loop)
Contusion, 282–284, 290, 509, 510, 513
Convention on International Trade in Endangered Species of Wild Fauna and Flora, *see* Convention on International Trade in Endangered Species of Wild Fauna and Flora (CITES)
Convention on International Trade in Endangered Species of Wild Fauna and Flora (CITES), 4, 34–35, 40, 57, 64, 92, 97, 98, 100–105, 107, 125, 164–165, 277, 368, 425, 427–429, 510, 563, 565, 583, 600, 610, 611
   derivatives, 92, 103–105, 107, 125, 164–165, 368
Conventions, 22, 30, 34, 35, 69, 76, 92, 94, 97, 98, 100–103, 117, 125, 189, 200–202, 302, 340, 342, 368, 378, 379, 391, 425, 428, 429, 566, 600, 610
Coral, smuggling of, 233, 426
Counting of carcasses, 471
Court(s), 3, 9, 10, 13, 31, 46, 81, 88, 93, 111, 118, 122, 124, 126, 129, 135–137, 139, 140, 145–147, 149, 160, 183, 191, 194, 195, 197, 198, 206, 207, 216, 225, 240, 247, 252, 265, 274, 275, 283, 284, 295, 297, 323, 327, 331, 347, 358, 363, 368, 382, 383, 385–389, 428, 429, 441, 449, 466, 468, 472–474, 476–478, 481–499, 505, 507, 508, 510–513, 515, 516, 518–520, 527, 529, 591, 592, 611
   appearance, 9, 88, 122, 327, 492
   civil, 482, 484, 492

Civil Procedure Rules, 485, 487, 489, 491
criminal, 387, 485, 492
Daubert Rule, 486
practice directions, 484, 485, 489, 497
procedure, rules of, 484, 485, 487, 490–492, 494, 497, 529
Courtship bites, *see* Bites
Crime, 1, 27, 92, 115–116, 127–149, 161, 183, 239, 329, 367–379, 381, 397, 466, 483, 501, 522, 563, 603–613
   investigation, 10, 56, 65, 109–110, 117, 119, 130, 140, 149, 151, 160, 165, 367–379, 426, 603–613
   National Wildlife Crime Unit, 110, 603, 604
   wildlife, 1, 3, 4, 6, 8–10, 27–31, 34, 41, 45, 56, 59, 62–65, 67, 78, 99, 110, 115–119, 121–122, 124, 125, 127–148, 151, 161, 164–166, 168, 173, 218, 239, 299, 323, 351, 361, 367–379, 382–386, 397, 398, 426, 429, 468, 476, 478, 501, 503, 505, 520, 522, 563, 565, 603–613
Crime scene, 9, 31, 119, 122, 126–149, 156, 158–161, 166, 168, 174, 175, 183, 207, 248, 306, 329, 368, 381, 385–388, 442, 468, 471, 472, 510, 522–523, 563
   basic supplies, 144, 147–148
   boundary, 131, 132, 135, 146, 147
   documenting and mapping, 140
   health and safety, 130–131
   illustrating, 142–143
   insect collection techniques, 159
   investigation, 9, 128, 130, 131, 144, 146, 147, 368, 563
   investigator (CSI), 129, 133, 138, 140, 146, 563
   manager (CSM), 609
   mapping, 140, 141
   numbered placards, 134, 146
   photography (photo-documentation), 129, 130, 132, 133, 136–140, 142–143, 146, 158, 175
   processing, 130–136, 138, 140
   search, 133–136, 145, 146, 158, 159
   securing and protecting, 132–133, 146
   sketch map, 142, 143
   visits, 9, 31, 125
   walk-through, 133–135, 146
Criminal Procedure Rules
   duty to, 489, 529
   magistrates, 466, 485
Crocodiles, 5, 26, 27, 65, 289, 297, 594, 597
Crow, Indian house, 172
Cruelty to animals, 186, 474, 564, 566;
     see also Welfare
CT, *see* Computerised tomography (CT)
CTPH, *see* Conservation Through Public Health (CTPH)
Culling, *see* Euthanasia

Index

Cultural considerations, 165; *see also* Fieldwork; Religious considerations
Cuvier, Baron Georges, 505
Cytochrome b (Cytb), 368
Cytology, 209, 212, 219, 248, 273, 274, 287, 329, 333, 347, 348, 526, 543, 545, 550
  specimen report form, 539–561

D

Damage by animals, 5, 13, 92, 119, 155, 175, 183, 244, 392, 422, 472, 482, 501
  to people, 19, 25, 86, 96, 106, 133, 183, 287, 303, 482, 501
  to property, 5, 13, 19, 92, 96, 106, 217, 447, 482, 501
Damages, *see* Claims and Compensation
Damage to evidence, *see* Evidence
Dangers in wildlife forensic work, 327
Darwin, Charles, 224, 566
Dead animals, investigation of, 9, 12, 34, 37, 39, 57, 88, 119–121, 124, 128, 133, 158, 162, 167, 168, 172–176, 188, 228, 239, 252, 253, 274, 298, 300, 330, 356, 362, 471
  specimen report form, 539–562
  specimen submission form, 251, 539–562
Death, 5, 14, 108, 113, 136, 151, 162, 192, 238, 329, 375, 396, 477, 487, 502, 508, 544, 571, 576
  of animals, 6, 10, 22, 31, 39, 48, 84, 116, 118, 121, 136, 154–155, 157, 160, 174, 179, 192, 197, 214, 221, 238–242, 263, 271–272, 276, 278, 280, 284, 286–288, 291, 294, 301, 308, 407, 440, 444, 456, 487, 502
  cause of, 5, 6, 14, 23, 27, 36, 37, 39, 48, 65, 106, 118, 121, 136, 157, 213, 215, 220, 224, 228, 230, 232, 239, 257, 262–263, 284–286, 288, 290, 291, 293, 294, 296, 299–301, 306–308, 310, 405, 406, 409, 413, 440, 444, 449, 450, 456, 461, 487, 510, 579, 581
  circumstances of, 6, 31, 257, 262–263, 275, 281, 286–288, 293, 421, 440, 471, 502, 544, 546, 548–552, 554, 555
  definition of, 223, 284, 518
  of humans, 5, 14, 22, 23, 27, 31, 36, 39, 48, 78, 80, 84, 118, 152, 155, 160, 210, 213–215, 238–240, 242, 258, 263, 278, 284–286, 290, 294, 296, 298, 300, 302, 405, 410, 413, 449, 514, 518
  manner of, 6, 39, 276, 286
  mechanism of, 6, 39, 286, 296, 298, 299
  sudden, *see* Sudden death
  time of, *see* Ageing; Post-mortem interval (PMI)
  unexpected, *see* Unexpected death
Decomposition, *see* Autolysis
Dedication, xii, 116, 117, 126, 327, 430, 438, 468, 478, 607–609
Deer, 6, 15, 21–22, 35, 43, 46, 93, 151, 226, 230, 239, 242, 264, 270–271, 279, 302, 307, 312, 314, 317, 318, 334, 399–404, 442, 450–452, 470, 512, 593, 595, 610
  and road-traffic (vehicular) accidents, 21–22, 222
Definitions, 507–520; *see also* Individual entries
Dehydration, 226, 227, 229, 231, 233, 234, 294–296, 301, 311, 333, 418, 511, 516, 573, 574
Dentistry and dental surgeons, 213, 216; *see also* Odontology, forensic; Teeth
Derivatives, of animals, 3–4, 9, 28, 32, 34, 65, 92, 103, 105, 107, 119–121, 125, 164–165, 168, 233, 239, 262–264, 321, 332, 367, 470, 472–473, 502, 600, 602
  CITES, 92, 103, 105, 107, 125
Dermestes beetles, 316, 586
Description and recognition of animals, 267–268
  of clinical and pathological lesions, 281
  of lesions, 281–284
Desiccation of carcases, 242, 260, 277
Detection of evidence, *see* Evidence
"Detective work", 43, 117, 299, 468, 505
Determination of age of animal at death, 192–193, 271–272, 282; *see also* Ageing
Determination of death, 65, 192, 240, 271–272
Developing countries, challenges and difficulties, xiv, 78
Developmental abnormalities, 263, 285, 286
Diagnosis
  making a diagnosis, 183, 187, 288, 394
Diagnosis (morphological/aetiological), 263
Diclofenac, 8, 38, 228
Diet, 22, 24, 25, 151, 184, 187, 203, 205, 213, 225, 265, 270, 289, 290, 318, 323, 332, 334, 337–345, 356, 366, 377, 403, 409, 412, 417, 418, 425, 435, 449, 450; *see also* Malnutrition; Obesity; Starvation
Differentiation of human and animal material, 266–267, 314
Digital images, *see* Photography
Digital x-rays (radiography), 200, 330
Disasters, 40, 84, 446
Discipline, 10, 60, 65, 82, 84, 92, 113, 115–117, 122, 124, 126, 165, 183, 200, 204, 223, 247, 257, 267, 273, 284, 289, 297–298, 310, 315, 323, 326–328, 331, 335, 388, 389, 391, 397, 399, 456, 482, 488, 490, 493, 503, 505, 519, 572
Discomfort, 6, 51, 106, 205, 221, 235, 456–457; *see also* Welfare
Disease, infectious, 5–7, 22, 23, 25, 35, 37, 41, 45, 67, 73, 78, 79, 83, 84, 173, 182, 194, 198, 203–205, 216–218, 224, 228, 233, 239, 250, 284, 285, 290, 291, 338, 451, 458, 509, 515, 569
  treatment of, 79, 203–204
Disease, non-infectious; *see also* Trauma and other headings
  treatment of, 203–204

Dissection techniques, 259–261, 420
Distress, 6, 106, 117, 183, 205, 210, 219, 229, 542; see also Welfare
DNA and DNA technology, 10, 35, 117, 136, 165, 185, 241, 326, 368, 383, 423, 472, 482, 503, 522, 543, 584, 591, 599; see also Genetic methodologies
  analysis, 9–10, 36, 210, 212, 267, 280, 314, 328–329, 347, 360, 365, 372, 373, 375–378, 388, 423–425, 442, 482
  banks/databases, 65, 371–373, 388, 424, 479
  fingerprinting, 369, 503
  routine sampling, 245, 328–329, 423, 473
DNA Barcoding, 65, 371–372, 424, 479
Dockets, 362, 364, 473, 475, 476, 511, 544, 546, 548–552, 554, 556; see also Labelling; Tagging
Domesticated animals, 3–6, 9, 14, 28, 38, 39, 52, 53, 68, 73, 74, 84, 93, 96, 105, 106, 117, 121, 124–126, 155, 172, 179, 183, 192, 193, 195, 199, 205, 214, 216, 223, 226, 229, 235, 244–245, 266, 271, 272, 278, 284–287, 299, 303, 311, 313, 314, 318, 343, 363, 446, 509, 511, 520; see also Individual species
  definition of, 2, 106, 284
  role in wildlife forensics, 6, 155, 183
  species, 1, 2, 4, 5, 14, 28, 38, 40, 52, 53, 57, 68, 92, 93, 95, 96, 98, 105–107, 155, 192, 193, 195, 204, 205, 216, 226, 244–245, 252, 266, 272, 275, 287, 290, 314, 363
Domestic violence, links with abuse, see Abuse
Donellan case, 114
Doum palm (*Hyphaene thebaica*), 469
Drawings, use in forensic work, 259
Droppings, 211, 308, 312, 344, 346, 348, 353, 362, 440, 448, 469, 470; see also Faeces, examination and assessment; Urine, examination
Drought, 7, 32, 59, 74, 205, 295, 299
Drowning, 31, 231–232, 285, 292, 296–297, 301, 511
Drugs, trade in, 4, 34, 41, 97
Durrell, Gerald, xi, xii, 122, 584
Durrell, Lee, xi–xii, 122, 584
Durrell Wildlife Conservation Trust, xi, xii, xvii

E

East Africa, xv, 6, 7, 18, 32, 45, 68, 80, 104, 166, 231, 261, 278, 295, 503; see also Individual countries
Ectotherms, 26, 48, 50–51, 153, 195, 197, 203, 228–230, 232, 240, 278–280, 282, 283, 286, 293–295, 297, 298, 330, 333, 502
Eggs, 37, 65, 67, 74, 88, 95–96, 98–99, 120, 152, 153, 155–159, 192, 195, 201, 222, 224, 252, 260, 264, 279, 280, 287, 292, 319, 329, 333, 334, 338, 339, 355–359, 362, 366, 392, 448, 449, 456, 470, 471, 502, 514, 523, 525, 552, 553, 592, 604, 612
  collections and collectors, 37, 65, 144, 156–159, 338, 449, 592, 612
  examination of, 37, 65, 260, 264, 338, 355, 357, 552–553
Eggshells, 144, 264, 321, 357–359, 448, 449, 470, 553
Electrocution, 13, 129, 136, 218, 231–232, 263, 288, 310–311, 512; see also Lightning
Electronic data, 304; see also Computers; Internet; Telemedicine
Electron-microscopy, scanning and transmission, 187, 212, 264, 265, 274, 334, 336, 348–349, 355, 390, 502
  scanning (SEM), 187, 264, 265, 274, 334, 336, 348, 390
  transmission (TEM), 334, 336, 348, 390
Elephant ivory, 36, 266, 361, 400; see also Teeth
Elephants, xv, 2–3, 17, 19, 29, 32, 36, 66–67, 80, 188, 191, 205, 242, 275, 295, 334, 345, 356, 360, 378, 398, 400, 502, 592, 593
  Asian, 2, 191, 205, 356, 378, 398, 431–439
Elkan, Edward., xix, 275
Embryos, examination, 359, 552–553
Emergencies, see Disasters
Emergency services, see First responders
Emerging diseases, 41, 73, 82; see also Zoonoses
Emesis, 202, 212, 288
Endoscopy, clinical, 172, 522
  post-mortem, 258, 261, 263, 308
Endotherm, 26, 197, 203, 226, 232, 282, 283, 297, 309; see also Individual taxa
Enforcement
  agencies, 81, 110, 133, 387, 603–605, 607–613
  international collaboration, 109–111
  of law, 41, 44, 47, 63, 105, 109–111, 116–117, 133, 146–147, 312, 483, 505, 603–605, 607, 608, 610–613
  of legislation, 99, 105, 109–111, 425, 483, 505
English
  American, xv, 483, 592
  British (European/Commonwealth), xv, 114, 400, 483
Entertainment, use of animals in, 48, 49, 51–52, 404
Entomology, 10, 52, 64, 65, 84, 149–160, 168, 243, 244, 327, 512, 515, 566
  collection and packaging of samples, 156–160
Entomotoxicology, 155–156
Envenomation, by snakes and other animals, 26–27, 475
Environmental forensics, 3, 4, 9, 10, 60, 63–64, 117, 121, 228, 415
Environmental stressors, 234–235; see also Stress

Index

Environment and environmental
   assessment, 17, 108, 114, 179, 309, 361, 502
   change, 8, 60, 64, 84, 197, 199, 228, 422
   forensics, 3, 4, 9, 10, 60, 63–64, 117, 121, 228, 415
   sampling, 25, 361
Equipment for forensic work, 9, 118, 161, 163, 167, 171, 179, 180, 183, 194, 254, 522–523
   clinical, 118, 125, 171, 326
   entomological, 10, 327
   laboratory, 118, 125, 273, 521
   post-mortem, 171, 245, 246, 273, 524, 525, 540–544
   special, 10, 56, 123–126, 183, 199, 216, 239, 273, 290, 312, 327, 328, 383, 384, 398, 521, 523
Estimating how long an animal has been dead, 502; *see also Post-mortem* interval (PMI)
Ethical considerations in forensic work, xv, 56
Ethology, 216–217, 512
European Union (EU), 34, 37, 49, 73, 79, 86, 94, 98, 102–105, 107, 360, 367, 368, 443, 447, 483, 499, 511, 512, 517, 531, 599, 600, 612, 613
Europe, Council of (CoE), 94, 98, 367
Euthanasia, 51, 82, 207, 253
Event scene investigation, *see* Crime scene, investigation
Evidence, 4, 17, 92, 114, 129, 149, 164, 183, 243, 328, 368, 381, 412, 465, 481, 502, 510, 522, 528, 550, 563, 574, 601
   collecting and preserving, 65, 136, 143–146, 471, 479, 523
   evidence flags, 133, 138, 147
   exhibits, 17, 118, 219, 298, 323, 388
   expert, 10, 92, 122, 385, 387, 476, 482, 484–486, 491–493, 497, 498
   forensic, 47, 92, 93, 97, 116, 130, 147, 156, 213, 312, 381, 387, 412, 479, 481, 482, 493, 494, 505, 518, 584, 601
   health and safety, 50, 92, 107, 164, 246, 381, 472, 574, 584
   identification, 9
   initial assessment, 134, 461
   insect, 149, 151, 152, 156–158, 160, 168
   marking, 132, 167, 468, 473–474, 522
   packing and sealing, 145, 147, 387, 468
   photography (photo-documentation), 129, 132, 136–140; *see also* Photography
   physical, 32, 129, 130, 136, 496, 512
   searching for and locating, 133, 146
   specimen submission form, 251
   storage, 9, 10, 88, 122, 145, 345, 387, 466–468, 474–475
   trace, 122, 133, 149, 151, 156, 160, 167, 210, 212, 465, 522
   tracking, 129, 132, 133, 145, 299, 482, 523
   transfer, 145, 146, 468, 475
Examination of animals, 8, 15, 119, 136, 168, 183, 238, 336, 450, 472, 501, 516, 540, 577; *see also* Clinical work; *Post-mortem* examination
   clinical signs, 194, 197, 203, 217, 219, 230, 310, 346
   of dead animals in the field, 9, 37, 39, 88, 172–176, 239, 241, 323, 473, 525
   of the gastro-intestinal tract (GIT), 243, 288–289, 366
   indications for, 194, 207
   of live animals in the field, 9, 39, 172, 183, 193–194, 209, 236, 548
   specific, 194, 225, 345, 540
   techniques for, 260, 353
   of teeth, 213–214, 290–292; *see also* Oral cavity; Teeth
Exchange of information, 110, 117, 503
Exhibits, 6, 17, 19, 54, 118, 185, 187, 219, 230, 240, 286, 298, 314, 319, 328, 385, 388, 393, 399, 419, 467, 474, 539; *see also* Museums
Exostoses, 317, 319, 320
Experimental animals, 333; *see also* Laboratory animals; Models, animal
Experimental studies, 25, 280, 298, 504
Expert evidence, 10, 92, 122, 385, 387, 388, 476, 482, 484–486, 491–493, 497, 498
Expert witness, 10, 24, 46, 97, 123, 198, 388, 478, 481–488, 490–499, 512, 513, 516, 517, 532, 534, 566
   accreditation, 493, 531
   advice, 10, 494, 498, 499
   appearance in court, 492–493
   assessing, 494
   conferences, 490
   cross-examination of, 10, 490, 492
   duties, 97, 487–488, 492, 497, 498
   evidence, 10, 97, 484–486, 490, 492–499, 513, 516, 517
   fees, 494–495, 498
   finding, 494
   immunity, 497, 498
   impartiality, 388, 496, 497
   instructions, 485, 487, 490, 494, 498
   Jones v. Kaney, 497–498
   liability, 497, 498
   meetings, 490, 497–499
   other countries, 123, 485, 498–499
   pre-trial questions, 491
   report, 10, 388, 481–488, 490–499
   training, *see* Training
Explosions, 8, 46, 72, 231, 308, 509; *see also* Blast injuries
Explosives, 8, 59, 274, 475, 604; *see also* Ordnance, unexploded
Extermination of animals, *see* Euthanasia; Pests
Eyes and ophthalmoscopy, 211–213

F

Faeces, examination and assessment, 55, 193, 210, 343, 344; *see also* Droppings
Falconry and falconry terminology, 43, 429, 477; *see also* Birds of prey
Fangs of snakes, *see* Envenomation, by snakes and other animals
Fat and fat reserves, 191, 293; *see also* Condition
Feathers, examination and assessment, xiv, 220, 330, 346–348, 355, 471; *see also* Plumage
 effect of oil on, 292
Feeding, evidence of, 23, 469
Fetus, examination of, 260; *see also* Neonates and neonatal deaths
Field equipment and kits, 168; *see also* Fieldwork
Fieldwork, 118, 119, 161, 162, 165–168, 171, 177–179, 259, 299, 397
 basic *post-mortem* technique for, 176
 cultural considerations, 165
 equipment, 118, 171, 255
 laboratory work, 172, 177–179, 382, 385, 387, 390
 legal considerations, 164–165
 overseas, 162, 163, 165, 170, 171, 527
 personnel, 109, 131, 163, 165–166, 239, 248
 practical considerations, 162–164
 recording and collating findings, 179–180
 sample-taking, 179
 techniques, 161–180
Fingerprints, DNA, 369, 503
Firearms, 13, 32, 141, 144, 164, 302, 305, 393, 509, 513, 517, 518, 612; *see also* Bullets and bullet wounds; Shooting of animals
Firearms and unexploded ordnance, 307–308; *see also* Explosives; Ordnance, unexploded
Fires and fire damage, 72
First responders (Emergency services), 127–147, 173, 174, 291
Fish, 22, 25, 44–48, 50, 60, 69, 83, 102, 126, 147, 189, 218, 223, 249, 250, 264, 323, 356, 368, 406, 419, 478, 570, 598; *see also* Ectotherms
 English and scientific names, 598
Fishing, 32, 43–46, 48, 87, 405
Fitness, *see* Condition
Five Freedoms, The, 48, 49, 106, 204, 205
Fixation of specimens, 261, 321, 365; *see also* Formalin
Flamingos, 175, 248, 261, 262, 277, 278
Fluctuating asymmetry, 192, 317, 318
Fody, Rodrigues, 201, 597
Food, 8, 13, 93, 155, 176, 187, 262, 331, 393, 403, 470, 509, 563, 612
Food safety, *see* Bushmeat; Meat
Foodstuffs (feed) for animals, 338–343
Footprints, 122, 125, 128, 185, 469, 470, 472, 513; *see also* Scratches on trees, rocks etc.; Tracks, trails, and signs of animals and humans
Foreign bodies and contaminants, investigation of, 364
Foreign material, 213, 343, 391, 392; *see also* Contaminants
Forensic biology, 13, 117, 504
Forensic entomology, 10, 149–160, 243
Forensic evidence, 47, 92, 93, 97, 116, 130, 147, 156, 213, 312, 381, 387, 412, 481–483, 493, 494, 505, 518, 584, 601; *see also* Evidence
 specimen submission form, 539–562
Forensic investigation, 1, 13, 91, 117, 161, 182, 243, 326, 368, 382, 399, 477, 502, 522, 540, 573, 599
 wildlife, 1, 5, 9, 10, 117, 118, 122–126, 163, 167, 329, 332, 335, 368, 399, 477, 502, 503, 506, 602
Forensic medicine, definition and history, 113, 114, 261, 268, 276, 283, 284, 286, 310, 501, 502, 510, 520
Forensic method, applications of, 9, 60, 114, 368, 379, 502
Forensic science service (FSS), 502, 503
Forensics, definition of, 3–10, 13, 83, 117, 118, 171, 381, 392–394
 animal, evolution of, 115–117
Forensic terms, 114, 512, 513, 520
Formalin and fixation, 167, 176, 177, 241, 352, 365, 394, 526
Fossey, Dian, xix, 81, 197
Fossils, *see* Palaeontology
Fourier transform infrared spectroscopy (FT-IR), 280, 340, 391–393
Fractures, 120, 187, 199, 214, 220, 231, 256, 282, 283, 285, 291, 297, 299, 300, 311, 312, 319, 355, 407, 408, 418, 421, 453, 462, 482, 502, 513; *see also* Healing
Freedom of information, 10, 49, 106, 204, 205, 255, 512; *see also* Human rights
Freezing
 effects on animals, 185
 of samples, 338, 363, 474
FSS, *see* Forensic science service (FSS)
FT-IR, *see* Fourier transform infrared spectroscopy (FT-IR)
Future of wildlife forensic science, 117, 502, 504, 505

G

Gamemeat, 41, 268, 442; *see also* Bushmeat
Game-ranching, 76–77
Gas releases, 46–47
Gastric lavage, 202, 203, 329
GenBank, 372–375
Gender determination (determination of sex), 270–271, 423; *see also* Sexing
Genetic methodologies, 367–379
Geological samples, 392
"Global health", 12, 13, 40, 83–84, 412

Index

Global recession, effect on forensic science, 502
Glossary of terms, 2, 197, 412, 476, 507–520
GLP, *see* Good laboratory practice (GLP)
Good laboratory practice (GLP), 278, 328, 368
Good practice, 137, 138, 275, 290
Gorilla, mountain, xviii, xix, 20, 55, 58, 78, 169, 188, 197, 204, 206, 208, 219, 221, 319, 320, 344, 360, 399, 455–458, 594
Governmental enquiries, 114, 502
Gross (macroscopical) examination of material, 262–263; *see also* Morphological examination
Group, herd or flock examinations, 546
    specimen report form, 546
Guidance, 101, 105, 108, 164, 194, 233, 247, 251, 259, 308, 321, 365, 385, 387, 388, 483, 495, 515, 521, 591, 604, 605; *see also* Code of practice
Guidelines, *see* Code, of practice; Good practice
Guns, 142, 144, 164, 302–304, 465, 487, 524
Gunshot and gunshot wounds, 302–304, 507, 509; *see also* Wounds
Gunshot residue, 393

H

Habitat(s), 1, 3, 6, 14, 19–21, 27, 28, 34, 35, 39, 41, 44, 60, 61, 64, 66, 67, 72–73, 76–78, 81, 85, 86, 91, 93–95, 97–99, 103, 110, 154, 156, 249, 284, 318, 330, 362, 367–368, 398–400, 404, 405, 409, 422, 427, 428, 455, 464, 481–482, 520
Habitat destruction, 20, 34, 61, 398, 405
Habituation of animals, 78, 216
Haematology, 209, 211, 212, 274, 333, 335, 514, 522, 601; *see also* Blood
Haemorrhage, *see* Contusion; Trauma
Hair, examination and assessment, 55, 249, 267, 311, 350–356
Hand-lens, 178, 187, 212, 264, 274, 319, 331, 334, 335, 340, 343, 347, 349, 352, 353, 357, 400–401, 471, 523, 525, 582, 584
Hanging, 215, 308–309, 513
Hazards, health and safety, 108–109, 130–131, 352, 422, 472, 584
Healing of wounds and fractures, 14, 187, 214, 220, 282, 283, 285, 291, 297–299, 312, 319, 408, 421, 453, 482, 502, 512; *see also* Wounds
Health and safety (H&S), 13, 47–48, 50, 53, 54, 92, 98, 105, 107–109, 130–131, 164, 241, 246, 248, 252–254, 361, 381, 472, 513, 569–572, 582–584, 602, 607
Health hazards, 39, 261, 352, 472, 475, 583; *see also* Zoonoses
Health-monitoring, 55, 56, 78, 162, 207, 457, 458, 514, 582, 599
Hedgehog, European, 96, 401, 593, 594
Helicopters, effect on animals, *see* Aircraft
HEPA filters, 280
Herbal medicines, *see* Medicinal plants
Herbaria, *see* Botany and botanists; Kew gardens
Herpetology and herpetologists, *see* Amphibians; Reptiles
Hides, *see* Skins
Histology and histopathological examination, 415
    scoring of lesions, 276, 317
    specimen report form, 539–562
History and identity, 184–193, 251
History, clinical and *post-mortem*, importance, 184
Hooves, 120, 260, 266, 268, 306, 336, 442, 502
"Horizon scanning", 505, 611
Horns, 32, 33, 42–43, 64, 120, 192, 195, 243, 260, 264, 266, 270–271, 306, 355, 360, 371, 372, 402, 418, 502
Hospitalisation of animals, 207
Human forensic work, 199, 241, 268, 297, 355, 470, 522
Human medicine and health, 23, 24, 73, 76, 84, 86, 231, 285, 286, 399, 409, 410
Human rights, 94; *see also* Justice
Hungary, 123, 398, 439–452, 485, 499
Hunter, John, 27, 105, 110, 114, 263, 283, 297, 326
Hunting of animals, 13, 43–46, 53, 101, 378; *see also* Shooting
Hybrids, 68, 93, 265, 369, 430
Hygiene, 23, 52–53, 248, 279, 321, 362, 437, 458, 475, 572
Hyperthermia and hypothermia, 31, 228–230, 233, 295–296, 514
Hypostasis, 276, 280, 282, 514
Hypoxia, *see* Asphyxia

I

IATA Regulations, 233
ibis scarlet, 252, 289, 422, 595, 597
ICZN, *see* International Commission on Zoological Nomenclature (ICZN)
Identification, 25, 34, 35, 59, 62, 65, 88, 184–188
    of dead animals (carcases), 34, 65
    in the field, 188
    of individual animals, 184–187
    of live animals, 65, 185
    of skeletal and non-skeletal remains, 243, 290, 319
    of species, 65, 188
    of specimens and evidence, 65
    of whole animals or their parts, 264–268
Iguana, green, as bushmeat, 208
Ill-treatment, 6, 28, 106, 205, 222, 481–482; *see also* Welfare

Imaging, 9–10, 121, 199–201, 233, 272–273, 298, 305, 310, 312, 388, 389, 391, 421, 462–463, 582–584; *see also* Radiography and radiology
Immobilisation equipment, 170
Immunosuppression, *see* Maladaptation
Impartial opinion, *see* Expert
Impression smears, *see* Touch preparations
Imprint on window, 470
Imprints (hand-reared animals), 216
Improvisation in the field, *see* Fieldwork
Inanition, 45, 224; *see also* Starvation
India, 16, 19, 34, 38, 53, 67, 77, 102, 150, 191, 228, 322, 398, 432, 435, 438, 473, 484
Individual species, 4, 57, 58, 60, 79, 96, 206, 235, 351, 425
Individual taxa, 192, 343
Infections from bites, *see* Bites
Infectious diseases, 5, 7, 22, 23, 25, 35, 37, 41, 45, 67, 73, 78, 79, 83, 84, 173, 205, 216, 217, 224, 228, 233, 239, 250, 284, 285, 291, 338, 451, 458, 515, 569; *see also* Diseases; Zoonoses
Infertility, investigation, 232; *see also* Reproductive system
Information, 1, 12, 94, 117, 132, 149, 162, 183, 239, 331, 368, 387, 399, 467, 484, 502, 510, 521, 530, 539, 563, 575, 591, 599, 603; *see also* Surveillance
Ingesta, 120, 203, 265, 272, 276, 294, 393, 502
   neck ligatures for sampling, 203
Initial walk through (crime scene), 133–134
Injury, 5, 7, 18, 19, 28, 39, 52, 78, 92, 106, 118, 155, 183, 189, 191, 197, 199, 205, 210, 213, 215, 218, 219, 221, 222, 232, 234, 259, 290, 295, 296, 300, 302, 303, 305, 315, 319, 421, 440, 444, 445, 461–462, 507–510, 512, 515, 518–520; *see also* Trauma
   inflicted by animals, 22, 27, 45
   inflicted by humans, 13–14, 17, 31–32, 501
   inquests, 502
Insects, 37, 38, 64, 65, 83, 149–160, 168, 171, 234, 235, 248, 322, 342, 362, 392, 458, 462, 464, 508, 509, 512, 515, 519, 592; *see also* Entomology
   insect successional colonisation, 153, 154
Inspectors, wildlife, *see* Wildlife crime, personnel
Institute of Zoology (IoZ), London (London Zoo), xviii, 49, 479
Instruments, clinical and *post-mortem*, 269
Insurance claims, 23, 50, 114, 482, 502, 513
Intelligence, 110, 126, 386, 603–613
   National Intelligence Model (NIM), 603–611
Interdisciplinary approach, 10, 84, 121–124, 385
   studies in forensic work, xiv, 10, 122–124, 505
International Commission on Zoological Nomenclature (ICZN), 3, 93, 502, 591
International forensic work, *see* Overseas

International Society for Forensic Genetics (ISFG), 368, 378, 385
International Union for Conservation of Nature (IUCN), 3, 19, 27, 33, 43, 60, 67, 69, 80, 82, 93, 563; *see also* World Conservation Union
Internet, 51, 52, 108, 109, 124, 133, 194, 235, 337, 383, 477, 484, 507, 609, 612; *see also* Computers
Interventions, 16, 53–58, 66, 72, 210, 221, 455–457, 514, 609
Intestinal contents, *see* Faeces, examination and assessment
Introduced species, 67–68, 512; *see also* Pests
Introductions, *see* Releases, Introductions/Re-Introductions and Translocations
Invertebrates, 1, 2, 13–16, 22, 37, 44, 47–52, 60, 63, 65, 75, 83, 85, 87, 100, 117, 122, 124, 131, 137, 167, 185, 189, 190, 195–197, 212, 220, 222, 223, 225, 230, 240, 242–244, 248–251, 259, 260, 264, 277, 280, 281, 298, 301, 309, 311, 329, 334, 338, 341, 343, 345, 347, 355, 357, 362, 364, 392, 398, 427–428, 501, 502, 515, 598
   English and scientific names, 591–598
   identification, 362
   in wildlife forensic work, xiv, 1, 501
Investigation
   crime, 10, 56, 65, 109–110, 117, 119, 130, 140, 149, 151, 160, 165, 367–379, 426, 603–613
   wildlife, 4, 11–89, 108, 110, 113–126, 167, 228, 323, 328, 367, 378, 382, 413, 466, 612
   of wounds, 219–220; *see also* Wounds
Ionising radiation, 232, 253
ISFG, *see* International Society for Forensic Genetics (ISFG)
Islam, 483, 499, 527
Isotopes, *see* Stable isotopes
Ivory, 32, 33, 36, 40, 103, 120, 266, 359–360, 377, 378, 400, 426, 502, 604; *see also* Elephant ivory

J

Jeffreys, Sir Alec, 503
Jersey (British Channel Islands), xii, xvii, 3, 44
   pink, 3
Journals and other publications, 115, 228
Judicial processes, 129, 505; *see also* Court(s)
Judicial system(s), 93, 94, 483–485, 498, 499
Justice, xv, 64, 102, 113, 128, 204, 383–384, 386, 485, 493, 503, 510, 514, 515, 521

K

Keel bone, of birds, 430
Kenya, xv, 16, 18, 19, 32, 33, 36, 38, 41, 43, 55, 64, 74, 88, 102, 110, 115, 169, 242, 262, 277, 278, 289, 295, 319, 402, 484, 503

Index

Kenya Wildlife Service (KWS), 33, 36, 503
Kew gardens, 23, 68, 69, 74, 95, 96, 301, 432
Kew Millennium Seed Bank, 479
Killing of animals, *see* Euthanasia
Knight, Maxwell, xix, 122, 302, 337, 338
KWS, *see* Kenya Wildlife Service (KWS)

L

Labelling
   of biological substances, 602
   of lesions and specimens, 9, 10, 479
Laboratories, use of, 327, 328, 475
Laboratory animals, 226, 235, 260; *see also* Experimental animals; Models, animal; Research
Laboratory investigations, 56, 65, 119, 125, 162, 177, 241, 243, 273–275, 326–328, 338, 346, 350, 360, 381, 522, 526–527
   in the field, 177, 526–527
   interpretation and reporting of results, 125, 166, 338, 381
   methods and pitfalls, 366
   specimen submission form, 539–562
Laceration, 299, 300, 302, 303, 311, 407, 461, 507, 509, 510, 515
Languages and language differences, xv, 9, 162, 169, 176, 476, 490, 527, 583, 592; *see also* Fieldwork
Lavage, 202–203, 212, 288, 329
Law, 3, 13, 91, 114, 133, 165, 204, 312, 367, 439, 465, 482, 502, 507, 522, 563, 599, 603
   animal, 95, 102
   animal health, 107
   animal waste, disposal of, 107–108
   civil (tort), 23, 92, 93, 106, 483, 510, 511, 514, 519
   common, 483, 484, 508, 510, 520
   criminal, 93, 114, 483, 502, 510, 515
   customary, 101, 484
   employment, 109
   enforcement, 41, 44, 47, 63, 105, 109–111, 116, 117, 133, 146, 312, 505, 603–605, 607, 608, 610–613; *see also* Enforcement, of legislation
   European, 94
   European institutions and, 94
   general aspects, 93–95
   health and safety, 92, 98, 109, 513
   international, 3, 6, 27, 165, 447, 479, 520
   keeping animals, 105, 483
   kinds, 93, 483
   levels, 94, 483
   liability for animals, 106
   local, 55, 94, 101, 109
   national, 3, 6, 27, 94, 98–101, 104, 165, 367, 368, 479, 485, 511, 520
   occupational health and safety, 108–109
   regional, 3, 6, 27, 94, 97–98, 520
   religious, 483, 484
   UK, 94–95, 99
   veterinary, 109
   welfare, 4, 106
   wildlife, 66, 94–102, 110, 111, 116, 505, 611
Leakey, Louis, xix, 33, 168, 418, 421, 523
Legal controls on samples, 364
Legal system(s), 140, 423, 483–485, 505, 510
Legal terms, 507
Legislation, 9, 16, 21, 28, 34, 40, 41, 46, 48, 54, 57, 62, 64, 88, 91–111, 121, 123, 164, 165, 204, 221, 241, 250, 252, 254, 256, 267, 303, 312, 320, 356, 429, 475, 482–485, 505, 510, 511, 517, 518, 591, 592
   Acts/statutes, 102, 204
   animal health, 35, 57, 107, 599–600
   development of, xv, 98, 482
   international, 4, 35, 40, 398, 600
   national, 40, 57, 93, 94, 99, 102–103, 398, 483, 600–601
Leisure activities, 85
Lens, magnifying, *see* Hand-lens
Lesions, 19, 39, 120, 139, 176, 186, 193–195, 197–200, 209, 214, 216, 219–223, 229, 230, 232, 233, 257, 258, 260, 262, 263, 268, 269, 272, 276, 281–285, 288–290, 294, 295, 297–302, 309–311, 315, 317, 319, 347, 350, 355, 357, 363, 392, 406, 410, 418, 419, 421, 436, 443, 444, 446, 450, 453, 461, 503, 509, 512, 513, 515, 516, 520, 524, 543, 549, 561
   describing and ageing (age determination), 198, 221, 302
Light, effect on animals, 234, 291, 295
Lightning strike, 288, 311
Linnaeus, Carolus, 591
Linnean Society of London, 73, 479, 566
Literature searches, 477
Litigation, 5, 10, 13, 17, 46, 71, 76, 86, 92, 206, 336, 481, 482, 488, 490, 497, 506
Live animals, investigation of
   specimen report form, 539–562
   specimen submission form, 539–562
Live animals, investigation of, 6, 9, 39, 107, 119, 121, 182–236
Livestock, domesticated, 5, 14, 20, 37–38, 183, 217, 482
Locard, Edmond, Principle of, 130, 136, 392, 465, 472
London Zoo, *see* Institute of Zoology (IoZ), London (London Zoo)
Loupe, magnifying/dissecting, *see* Hand-lens

M

Madagascar, xi, 448
Maggots, 276, 277, 281, 291, 339, 457, 509, 515; *see also* Blowflies
Magnetic fields, effects, 235

Magnetic resonance imaging (MRI), 200, 258, 263, 273, 414, 421
Maladaptation syndrome, 197, 198
Malnutrition, 75, 191, 218, 224, 294, 338
Malpractice, 13, 50, 92, 114, 482, 502, 541; see also Professional misconduct
Malthus, Thomas, 224
Mammals
   English and scientific names, 592–598
Management of wildlife, 16, 44
Management of wildlife habitat, 72–73
Mantella sp., frozen during transportation, 448
Maori culture and beliefs, 165
Marabou stork, 23, 596
Marine mammals, 22, 25, 32, 36, 37, 48, 59, 73, 79, 122, 161, 162, 168, 171, 189, 207, 213, 218, 228, 234, 252, 263, 285, 296, 309, 397, 404–421, 523, 571
   abscesses, 418
   anorexia, 418
   anthropogenic mortality, 405
   baculum, 418
   bioaccumulation, 409, 414
   biomagnification, 409
   bio-sentinel/sentinel, 409
   blubber, 408, 415, 421
   blunt trauma, 406–409
   bone, 73, 171, 417, 418
   by-catch, 48, 405
   climate change, 48, 73, 228, 405, 409, 412
   cold stress syndrome (manatees), 404–406, 414, 417–420
   computed tomography (CT), 421
   diatom (*Pseudo-nitzscia* spp.), 296, 413, 414
   domoic acid toxicosis, 413–415
   emaciation, dehydration, 234, 296, 418
   imaging/radiography, 419, 421
   laceration, 407
   mammary gland (adenocarcinoma), 410
   manatee, 404–407, 409, 414, 417–420
   necropsy, 162, 207, 252, 405–410, 415, 417, 419, 420, 523
   neoplasia (cancer, tumour), 410
   neurotoxin (domoic acid), 414
   pneumothorax, 406–407, 417
   propeller, 406, 407, 409
   toxicosis, 414, 415
Marking of evidence, see Evidence
Marking (identification) of specimens, see Identification
Mauritius, xi, 75, 77
Meadow, Sir Roy, 476, 490
Measuring, see Morphometrics; Weighing of animals and organs
Meat, 32, 33, 40, 41, 77, 199, 229, 267, 301, 340, 393, 405, 442, 444, 571; see also Bushmeat

Medical studies, 13, 15, 83, 114, 183, 490, 502; see also Laboratory animals
Medical terms, 507–520
Medicinal plants, 1, 30; see also Traditional medicines
Meinertzhagen, Richard, 322
MES, see Mwaluganje elephant sanctuary (MES)
Metal detector, use of in forensic work, 134, 168, 305, 523
Methicillin-resistant *Staphylococcus aureus* (MRSA), 253, 570–571
Microchip and microchipping (transponders), 121, 185, 186, 268, 364, 430, 431, 437, 473, 540, 544, 546, 548, 550–553, 555
Microsatellites, 345, 375, 377–379
Microscopy, 122, 178, 187, 193, 265, 335, 336, 340, 341, 351, 382, 383, 389–394, 543, 545
   electron, 209, 255, 267, 314, 331, 334–336, 363, 390, 391, 474, 601
   in the field, 172, 178, 317
   light, 262, 264, 265, 267, 314, 335, 340–343, 347–348, 353, 356, 390–394
Middle East, 32, 53, 62, 77, 121, 185, 268, 364, 398, 429–432, 473; see also Arabia
Minimally invasive techniques, 55–57, 210, 212
Mini-necropsy, 260, 275
Miscarriages of justice, 493, 503
Misconduct, see Professional misconduct
Mitochondrial DNA (mtDNA), 265, 334, 348, 369–370, 376, 377, 424
Models, animal, 63, 77, 84, 105, 151, 226, 298, 404; see also Laboratory animals
Molecular biology, 122, 327, 329, 379, 504; see also DNA and DNA technology
Monitoring (screening) of animals prior to release, 207–208
Monitor lizards, 26
Morphological examination, 10, 188; see also Gross examination
Morphometrics, 58, 169, 189–190, 225, 239, 248, 264, 268–270, 273, 294, 506; see also Weighing
Moult, 191, 350, 549; see also Feathers, examination and assessment
Movement of animals, 57, 84
MRI, see Magnetic resonance imaging (MRI)
MRSA, see Methicillin-resistant *Staphylococcus aureus* (MRSA)
mtDNA, see Mitochondrial DNA (mtDNA)
Multiple investigations, value of, 330–331
Mummification, 241, 258, 260, 261, 275, 277, 352, 516

Index

Museums, 18, 28–30, 37, 65, 216, 247, 253, 259, 261, 314–315, 319, 321–323, 352, 365, 424, 479, 564, 573–581, 584; *see also* Reference collections
Mwaluganje elephant sanctuary (MES), xv, 2
Myiasis, 155, 227, 329, 472
Myxomatosis, 79, 83, 301

N

Names, scientific, 3, 93, 150, 151, 502, 540, 544, 546, 548, 550–553, 555, 591–598
National Center for Biotechnology Information (NCBI), 372
National Museums of Kenya (NMK), xviii, 319
National Wildlife Crime Unit (NWCU), 110, 603–605, 609–612
Natural England (NE), xviii, 100, 186, 600
Natural History Museum, 322, 323, 479; *see also* Meinertzhagen
Natural history, relevance to forensic work, xi, xii, 285
Naturalist, xi, xii, 30, 43, 47, 61, 62, 122–123, 188, 338, 470, 504–505
NCBI, *see* National Center for Biotechnology Information (NCBI)
NE, *see* Natural England (NE)
Necropsy; *see also* Post-mortem examination
specimen submission form, 539–562
Neglect, 21, 31, 48, 51, 57, 63, 81, 140, 149, 151, 155, 158, 160, 164, 204, 222, 223, 227, 294, 473, 511, 582
Neonates and neonatal deaths, 284, 285; *see also* Reproductive system
Nested PCR, 372, 379
Networking, 47, 60, 69, 110, 180, 458, 476, 508, 523, 604, 609
New Zealand, 14, 25, 55, 70, 75, 88, 95–96, 99, 101, 102, 106, 150, 165, 204, 233, 293, 299, 300, 320–321, 337, 382, 398, 425–429, 452–455, 478, 565, 594, 596
New Zealand Centre for Conservation Medicine (NZCCM), xviii, 504
Ngamba Island, 80
Night nest, of gorilla, 177, 456–457
NMK, *see* National Museums of Kenya (NMK)
Non-accidental injury (NAI), *see* Abuse
Non-governmental organisations (NGOs), 10, 44, 438, 458, 474, 609, 611; *see also* Individual bodies
Non-infectious diseases, 79, 194, 203–204, 216, 218, 290
Normal values, *see* Reference values
Nose prints, of gorilla, 188, 214, 268
Nucleotide sequencing, 368, 370; *see also* Forensically informative nucleotide sequencing (FINS)

Nutritional deficiencies and inbalances, 359; *see also* Malnutrition
NWCU, *see* National Wildlife Crime Unit (NWCU)
NZCCM, *see* New Zealand Centre for Conservation Medicine (NZCCM)

O

Obesity, 226
Observational studies, Observation, importance of, 504
Occupational health and safety, 108–109; *see also* Health and safety
Odontology, forensic, 26, 216, 290; *see also* Teeth
Oil and oil pollution, 86, 262
"One health", 83–84
Ophthalmoscopy, 124, 171, 211–213, 251, 260, 436, 477, 524; *see also* Eyes and ophthalmoscopy
Oral cavity and teeth, 196, 213–214, 216; *see also* Teeth
Orang-utan, 19, 80, 81, 92, 594
Ordnance, unexploded, 307–308
Organisations, information about, 100–101, 115, 327, 356, 477
Organ size and *post-mortem* change, 270
Ornithology, *see* Birds
Orphaned animals, 58; *see also* Rehabilitation, of wildlife; Sanctuaries, for wildlife
Osteology, 124, 267, 273, 312, 313, 315–317, 323, 503, 582
Overseas work, 163, 527; *see also* Developing countries, challenges and difficulties

P

Packaging, 122, 125, 144, 156–157, 160, 252, 331, 387, 468, 475, 523, 582, 600; *see also* Wrappings
Pain, 6, 26, 37, 44, 46, 51, 54, 57, 58, 106, 185, 204, 205, 210, 214, 219, 221, 223, 230, 231, 263, 303, 306, 501, 505, 518, 542, 601; *see also* Welfare
Palaeontology and palaeopathology, 220, 268, 277, 312, 315, 318–319, 323, 504–505, 516
Paralysis, *see* Trauma
Parasitological examination, 336, 348, 352
Partnership for Action against Wildlife Crime (PAW), 1, 27, 45, 64, 65, 88, 110, 111, 116–117, 121, 125, 210, 222, 228, 265, 299, 302, 307, 308, 326, 368, 382, 478, 563, 565, 603, 604
Pathogens, 17, 27, 41, 47, 67, 77, 84, 86, 117, 131, 208, 228, 235, 241, 251, 253, 261, 290, 292, 296, 298, 311, 315, 329, 352, 358, 362, 367, 405, 409, 418, 450, 458, 472, 475, 515–517, 519, 570–572, 583, 601, 602

PAW, *see* Partnership for Action against Wildlife Crime (PAW)
Peafowl, 23, 24, 371, 453, 597
Peer-review, 247, 285–286, 477, 504
Pellets (castings) of birds, examination and assessment, xiv, xviii, 337–339
Penalties, *see* Confiscations
Permit(s), 4, 30, 37, 52, 55, 56, 59, 67, 70, 72, 79, 99, 103–105, 107, 123, 163, 164, 174, 191, 193, 200, 209, 216, 221, 232, 239, 241–242, 256, 258, 259, 270, 273, 294, 305, 336, 340, 341, 347, 352, 354, 360, 368, 378, 389, 392, 418, 421, 428, 447, 463, 575, 583, 600, 612
Personnel, 14, 109, 116, 130–134, 144, 163, 165–166, 170, 173–176, 183, 200–201, 209, 239, 246–248, 253, 254, 258, 261, 361, 363, 366, 381, 467
Pesticides, 13, 37, 38, 59, 62, 70, 125, 131, 134, 288, 321, 330, 612; *see also* Poisons
Pests, 8, 16, 39, 44, 59, 75, 83, 95–96, 99, 117, 220, 361, 508; *see also* Introduced species
  and conservation programmes, 71–72
  species, 3, 15, 37, 70–71, 221, 223, 228, 298
Pet animals, role in wildlife forensics, 482
Pet trade, 4, 52–53
Photography (photo-documentation)
  considerations after the crime scene, 139–140
  considerations before arriving at the crime scene, 137–138
  of the scene and of evidence (Evidence photography), 136–139
Physical evidence, 32, 129, 130, 136, 496, 512
Physiology, 16, 39, 122, 183, 191, 222, 224, 229, 286, 292, 296, 297, 333, 339, 349, 397, 412, 416, 501, 508, 514, 518, 573, 574, 591
Pig, model for forensic studies, 404
Placentae and placental membranes, 120, 285, 400, 414, 470, 502
Plants
  medicinal, 1, 30
  in wildlife forensics, 1, 3, 6, 27, 28, 30, 36, 39, 42, 60, 63, 67–70, 73, 76, 77, 84, 86, 91, 96, 98, 107, 117, 121, 122, 131, 140, 165, 178, 216, 227, 228, 288, 291, 321, 323, 341–343, 392, 393, 417, 443–444, 446, 449, 470, 471, 479, 482, 508, 511, 515, 516, 518–520, 591, 603, 604
Plastic containers, for specimens, 159
Plastic forceps, use, 254
Plumage of birds, 211, 227, 345, 444; *see also* Feathers, examination and assessment; Moult
PMI, *see* Post-mortem interval (PMI)
Poaching, 2, 4, 5, 32–36, 42, 43, 62, 65, 81, 82, 116, 125, 135, 149, 169, 173, 174, 205, 295, 350, 360, 367, 377, 378, 398, 418, 426, 442, 472, 604, 610

Poisons and poisoning, 6, 8, 13, 26–27, 32, 36–39, 70, 71, 88, 119, 126, 129, 131, 136, 144, 173, 193, 202, 207, 212, 228, 248, 251, 253, 254, 256, 263, 265, 266, 281, 282, 288, 290–293, 307–309, 337, 340, 358, 361, 377, 414, 450, 471, 472, 475, 482, 487, 509, 517, 519, 601, 604, 612; *see also* Toxicology
  sampling, 32, 39, 193, 291, 361
Pollutants, 37, 47, 59, 179, 227–228, 286, 405, 413
Pollution, environmental, 59, 168, 358, 359; *see also* Oil
Poorer countries, 13, 15–16, 55, 76, 80, 82, 161, 169, 180, 221, 254, 301, 339, 466, 496; *see also* Poverty
Population growth (human), xvi, 31, 69
Portable equipment for fieldwork, 9, 168–171; *see also* Fieldwork
Post-mortem change and estimation of time of death, 275–281; *see also* Autolysis; *Post-mortem* interval (PMI)
Post-mortem examination (necropsy)
  aims of, 239–240
  common errors in, 256–257
  in DNA studies, 176, 177
  of eggs, 152, 153, 279, 280
  of embryos, fetuses and neonates, 260
  equipment, choice of, 186, 246, 250–251, 254, 525
  in the field, 173–177
  full *vs.* partial necropsy, 257–258
  history and provenance of specimens, 251
  imaging, 272–273, 312
  mistakes and omissions, 255–257
  questions that may need to be answered, 275–288
  radio-tagged birds, 573, 588
  record-keeping, 246, 252, 259
  sampling, 216, 245, 273
  some specific findings and considerations, 293
  special investigations necessitating necropsy and/or dissection, 288–311
  specimen submission, 331
  stages of, 249–251
  toxic substances, 37
Post-mortem interval (PMI), 149, 151–155, 160, 278–280, 315, 504; *see also* Autolysis
Predation, *ante-mortem* and *post-mortem*, 276, 278
Presentation of evidence, *see* Evidence
Private collections, 4, 51, 82, 323, 478; *see also* Zoos
Private laboratories, 126
Professional misconduct, 498, 513
Protected species, 5, 28, 36, 79, 93, 96, 99, 102, 105, 117, 129, 164, 220, 264, 293, 443, 481–482, 488, 600, 602, 604
Protocols, 10, 55, 98, 144, 157, 174, 176, 177, 194, 196, 208, 247, 253, 256, 259, 262, 319, 329, 338–340, 364, 381, 419, 432, 436–437, 460, 466, 468, 472, 475, 485, 527, 581–583, 606, 609; *see also* Code of practice

Index

Provenance of specimens, 251, 479
Psychological or behavioural disorders, 216–217
Ptilochronology, *see* Feathers, examination and assessment
Publishing of data (findings), 279, 283
Puncture wounds, 215, 453; *see also* Wounds
Putrefaction, *see* Autolysis

Q

QM, *see* Quality management (QM)
Quality assurance, 326, 382, 385, 386, 517
Quality control, in laboratories, 125, 328, 382, 386, 472, 493, 494; *see also* Good laboratory practice (GLP)
Quality management (QM), 125, 174, 247, 327, 382–389, 468, 476, 504, 515
Quality of evidence, 493, 504
Quantification of data, 224, 270, 280, 390; *see also* Scoring
Quelea, red-billed, 70, 71, 597
Questions about animals, 120
Questions that may need to be answered by a necropsy, 275–288

R

Radiation, ionising, 232, 253
Radiography and radiology, 9–10, 65, 88, 119, 170, 187, 199–202, 212, 216, 232, 233, 243, 253, 260, 261, 272–274, 283, 298, 303, 305, 307, 308, 310, 312, 315, 316, 319, 322, 328, 330, 339, 358, 359, 399, 421, 443–445, 462, 545, 548, 549, 554–556, 581–584; *see also* Imaging
 post mortem, 199, 272–273, 312
Radioisotope studies, *see* Stable isotopes
Radio-tagged birds, examination of, xviii, 573
Rape, effects of ingestion, 450, 451
Raptors, *see* Birds of prey
Receipt of carcases, 251–252
Recording and collating findings in the field, 179–180
Record-keeping, 10, 88, 114, 122, 125, 140, 210, 246, 252, 259, 437, 468
Reference collections, 216, 289, 321–323, 353, 388, 400, 478–479, 502, 550; *see also* Museums
References and further reading, 234, 326, 361, 507
Refrigeration, 270, 282, 347, 353, 363, 474, 573
Regurgitated pellets, *see* Pellets
Rehabilitation, of wildlife, 38, 78–81; *see also* Orphaned animals; Sanctuaries, for wildlife
Rehydration of carcases, 261
Re-introductions, *see* Releases, introductions/re-introductions and translocations

Releases, introductions/re-introductions and translocations, 66–69, 80
Religious considerations, 19, 432, 435, 436, 483, 484
Report forms, for specimens, 320, 540
Reports and reporting
 expert witness, 482–499
 writing, 10, 92, 126, 194, 384, 387, 477, 481–499, 582
Reproductive system, 65, 192, 287; *see also* Infertility, investigation
Reptiles, 22, 27, 31, 34, 43, 47, 50, 52, 53, 60, 63, 65, 66, 77–79, 83, 110, 120, 124, 185, 189–192, 195, 197, 199, 201, 207, 208, 212, 214, 220, 223–225, 229, 233, 240, 249, 250, 259, 279–280, 282, 285, 287, 289, 293, 294, 298, 301, 308, 309, 314, 330, 334, 340, 344, 345, 351, 357, 363, 366, 421, 426, 448, 514, 519, 569–571, 597; *see also* Ectotherms
 English and scientific names, 570, 597
Research, 16, 17, 22, 25, 27, 37, 42, 46, 49–51, 53–57, 60, 70–71, 75, 83, 84, 93, 98, 99, 103, 105, 109, 120–122, 124, 149, 151, 160, 162, 164, 186–188, 192, 202, 220, 229, 239–243, 258, 265, 271, 275, 278, 280, 289, 290, 298, 301, 307, 318, 320, 321, 327, 336, 337, 343, 345, 349–351, 365, 372, 373, 383, 384, 386, 388, 390, 397–399, 403–404, 410, 413, 437, 439, 455, 478, 504–505, 573–577, 601, 606
 needs, 17, 27, 288, 365, 388, 502–505, 573
 use of wildlife in, 53–56
Restriction enzymes, 375, 376
Restriction fragment length polymorphism (RFLP) analysis, 371, 373, 375–376, 379, 442
Retention of evidence and samples after a case, 478
Retention of wildlife in captivity, 207
RFLP analysis, *see* Restriction fragment length polymorphism (RFLP) analysis
Rhinoceros, Javan, xviii, 62, 306, 594
Rifles, *see* Firearms
Rigor mortis, 152, 213, 269, 276, 277, 281, 509, 517
Ringing (banding) of birds, 185
Risk assessment, 23, 47–48, 82, 107–109, 164, 246, 248, 251, 254, 328, 472, 475, 572, 582–584; *see also* Health and safety
Road-traffic (vehicular) accidents (RTAs), 21–22, 79, 222, 282, 300–301, 310
Royal College of Veterinary Surgeons (RCVS) library, xviii, 488, 498
Royal Society for the Prevention of Cruelty to Animals (RSPCA), 186, 246, 368, 474, 534
Royal Society for the Protection of Birds (RSPB), xix, 38, 62, 100, 111, 116, 337, 474
RSPB, *see* Royal Society for the Protection of Birds (RSPB)
RSPCA, *see* Royal Society for the Prevention of Cruelty to Animals (RSPCA)
RTAs, *see* Road-traffic (vehicular) accidents (RTAs)

Rubbish, 87
Ruler, evidence marker for scale, 138, 139
Rumen, lesions, 273, 450, 451
Rwanda, 18, 22, 39, 58, 81, 169, 197, 206, 245, 319, 455, 457; *see also* Genocide; Gorilla, mountain

S

12S, 370, 371, 379
16S, 370, 371, 379
Saliva, 26, 63, 120, 214, 215, 264, 451, 502, 519, 601
    examination and assessment, 502
Samples and sampling, 9, 32, 103, 118, 144, 156, 164, 184, 245, 325, 368, 381, 403, 466, 513, 522, 540; *see also* Laboratory investigations
    biological, 600, 601
    carriage, 599, 601–602
    chain of custody, 246, 364–365, 387
        specimen submission form, 251, 539–562
    data, 331, 365
    dead animals, 57, 119, 228, 237–323, 329
    in the field, 179, 361
    legal controls on samples, 364
    live animals, 107, 119, 157, 181–236, 329, 473, 600
    movement, 362–363, 387, 599–602
    packing, 336, 600–602
    Packing Instruction, 602
    retention, 363, 365, 474, 478
    special treatment, 362
    specimen submission form, 251, 539–562
    storage, 345, 363–364, 475, 478
    tracking, 362–363
    transportation and storage, 338, 361, 600
    types of, 329, 331, 337
Sanctuaries, for wildlife, 16, 31, 45, 78, 81, 452–453; *see also* Orphaned animals; Rehabilitation, of wildlife
Scales, for weighing reptiles or fish, 345–349, 357, 502
Scanning electron-microscope (SEM), 187, 231, 264, 265, 274, 311, 334–336, 340, 347–349, 353, 355, 359, 390, 392, 393, 502
Scavenging, 22–23, 133, 149, 175, 242, 244–245, 262, 276, 281, 397, 403–404, 412, 460, 522; *see also* Predation, *ante-mortem* and *post-mortem*
Scene of crime, *see* Crime scene
Scene of crime visit and preparation of evidence, 125
Scientific method, 129, 130, 505
Scientific names, 3, 93, 150–151, 502, 540, 544, 546, 548, 550–553, 555, 591–598
Scientific (Latin/Greek) names, 592
Scoring systems, 194, 205, 226, 343, 547
Scott, Robert Falcon, 180
Scouts, 470

Scrapings, *see* Cytology
Scratches on trees, rocks etc., 470; *see also* Footprints; Tracks, trails, and signs of animals and humans
Screening, 41, 200, 207–209, 226, 233, 340, 343, 361, 390, 392, 393, 514, 601; *see also* Monitoring
    of birds of prey prior to release, 208, 209
Scuttle fly (*Megaselia scalaris*), 243, 362, 506
Sealing of evidence, *see* Evidence
Sea lion, Californian, 413–415, 418, 593
Seasonality, 154–55
SEM, *see* Scanning electron-microscope (SEM)
Semen, examination and assessment, 171, 212, 334, 470
Sentinels, animals as, 5, 8, 12, 59, 84, 409
Shamba Musa, xv–xvi, 18, 19, 78
Shock, 39, 157, 194, 219, 222, 226, 229, 231, 288, 292, 297, 304, 449, 518
Shooting of animals, 43–46, 62, 72–73, 78, 79, 93, 100, 221, 257, 272, 303–305, 307, 308, 465, 471; *see also* Guns; Gunshot and gunshot wounds; Hunting of animals
Shortcomings, 126
Short tandem repeats (STRs), 35, 377, 378
Shotguns and shotgun wounds, 46, 303, 304; *see also* Wounds
Single nucleotide polymorphism (SNP), 370–371, 373–375, 378, 379
Skeletons and skeletal lesions, 276, 277, 285, 306, 312, 314, 315, 319, 321, 365, 555–562, 576–578, 582–586; *see also* Bones
Skink, Round Island, 260, 597
Skinning of carcase, 215, 258–259, 304, 576, 578
Skins, 18–19, 27, 34, 37, 64, 65, 73, 119, 120, 139, 158, 160, 196, 198, 199, 214, 215, 219, 220, 225, 229–231, 233, 244, 249, 253, 257–259, 261, 264, 272, 275, 277, 278, 281, 282, 291, 292, 295–297, 299, 300, 304, 309, 310, 321, 322, 333, 334, 348, 350–352, 355, 357, 361, 365, 392, 409, 417–421, 430, 431, 436, 453, 456, 507, 509–511, 514–516, 524, 576–578; *see also* Sloughed skin
Slaughter of animals, *see* Euthanasia
Sloughed skin, 334, 357, 502
Smuggling and smuggled animals, 62, 233, 426
Snakebite, 257, 527; *see also* Envenomation, by snakes and other animals
Snakes, 26, 27, 30, 31, 34, 52, 53, 65, 131, 189–190, 201, 211, 212, 248, 249, 282, 286, 294, 357, 371, 375, 446, 447, 470, 475, 597; *see also* Ectotherms; Reptiles
Snares, 7, 79, 99, 129, 218, 220–221, 231, 243, 253, 259, 301, 457, 604; *see also* Traps, nets, snares and trapping of animals
SNP, *see* Single nucleotide polymorphism (SNP)
Social interactions and behaviour, *see* Ethology
Societies, groupings and organisations, 116–117

Index

Society for Wildlife Forensic Science, 116, 504
Sodomy, see Bestiality
Sources of information, 1, 34, 60, 533, 563–567
Spain, 24, 48, 95, 150, 222
Spawn of amphibians, 470
Special investigations necessitating necropsy and/or dissection, 288–311
Species detection, 370, 374–377
Species identification and differentiation, 65, 158, 188, 192, 264–266, 334, 335, 360, 368–375, 378, 379, 422, 423; see also Identification
Species of animal
    CITES, see Convention on International Trade in Endangered Species of Wild Fauna and Flora (CITES)
    domesticated, 40, 68, 74, 92, 93, 106, 191, 192, 204, 205, 252, 259, 267, 269, 275, 285, 290, 314, 410
    endangered, 75, 77, 92, 97, 98, 101, 105, 107, 125, 604; see also Convention on International Trade in Endangered Species of Wild Fauna and Flora (CITES)
    migratory, 97
    non-domesticated, 4, 28, 52, 53, 92, 93, 95, 96, 98, 105–107, 191, 193, 205, 226, 286, 314, 482, 509, 512, 599
    protected, 5, 28, 36, 79, 93, 95, 96, 99, 102, 105, 117, 129, 164, 220, 264, 265, 293, 301, 443, 444, 471–472, 482, 488, 600, 602, 604
    wild/free-living, 2–4, 57, 59, 75, 92, 105
Species protection, 60–63, 66, 72, 79, 85, 97–99, 301, 443
Species specific primers, 371, 373–376, 379
Specimens, 9, 17, 92, 121, 149, 160, 167, 188, 244, 326, 373, 381, 401, 467, 496, 503, 517, 526, 539; see also Exhibits
    background history and confirming identity/recording provenance, 251
    collection, 30, 64–65, 319, 331–332, 478
Sport, use of wildlife in, 46
Stable isotopes, 9–10, 60, 119, 121, 332, 334, 356, 403, 425
Stab wounds, see Wounds
Standardisation of laboratory techniques, see Quality control, in laboratories
Starling, European, 8, 59, 596
Starvation, 31, 57, 67, 118, 191, 204, 218, 223–227, 233, 234, 289, 292–295, 299, 301, 511, 518
Statement, of witness, 489, 492, 513, 529–537
Statistics in court, use and misuse, 476
Stomach contents, examination and assessment, see Ingesta
Stomach/crop flushing (lavage), 202–203
Storage, of carcasses, evidence and samples, 9, 363, 467
Strandings, see Marine mammals
Strangulation, 301, 308–309, 508, 513, 518, 519; see also Asphyxia
Stress, 8–10, 14, 16, 27, 39, 41, 48, 54, 57, 59, 73, 78, 93, 118, 122, 123, 160, 193, 204, 205, 212, 214, 221, 222, 229, 235, 242, 257, 267, 275, 292, 294–296, 298, 301, 317, 319, 321, 323, 332, 344, 366, 387, 401, 402, 412, 416, 419, 445, 448, 450, 471, 479, 501, 506; see also Environmental stressors
    types of stressor, 54, 55, 230, 298
Stroud, Richard, xviii, 32, 37, 41, 116, 117, 243, 267, 302–305, 330, 368
STRs, see Short tandem repeats (STRs)
Submission forms, for specimens, 251, 539–562
Sudden death, 287, 288, 450, 518, 519
Suffering, 4, 13, 39, 50, 58, 106, 204, 205, 207, 226, 295, 308, 414, 436, 437, 448, 501, 505, 601; see also Welfare
Suffocation, 230, 285, 518; see also Asphyxia
Supporting investigations, laboratory tests, 209–210

T

Tagging of live animals and specimens (evidence), 2, 252, 473–474
Tape-recorder and tape-recording, 179, 193, 195, 472, 473, 475, 476, 523, 582, 583
Tattoos, see Identification
Taxidermy, 64–65, 100, 308, 351, 482, 584, 592, 612
Taxonomy, 298, 371, 482, 501, 519, 575, 591–592; see also International Commission on Zoological Nomenclature (ICZN); Scientific names
Teaching of forensic science, 109, 200, 297, 321, 323, 478, 493, 501
Technical Working Group on DNA Analysis Methods (TWGDAM), 368, 378
Teeth
    dentition of mammals, 192, 213, 214, 268, 321, 334, 359, 360, 397, 399–403
    examination of teeth, 192, 214, 289–291, 317
    including Ivory, 120, 359–360, 502, 604; see also Dentistry; Odontology
Telemedicine, 472–473
Textbooks (references and further reading), 94, 204, 234, 326, 335, 345, 354, 361, 477, 494, 507, 527
Theiler, Sir Arnold, 323
Third World countries, see Developing countries
Time of death, 152, 239, 275–281, 335, 508, 509, 512, 517, 544, 546, 548–552, 554, 555; see also Post-mortem interval (PMI)
Tissues, examination of, 26, 31, 186, 213, 241, 263, 278, 289
    retention of, 320–321
    specimen submission form, 251
Tobias, Philip, xviii, 320, 321

Tort, 106, 482, 483, 510, 511, 519
Tortoise, African leopard, 186
Tortoises, xi, 34, 103, 121, 185, 186, 188, 189, 191, 192, 218, 268, 293, 372, 597, 604, 611, 612; *see also* Ectotherms; Reptiles
Tourism, 85–86, 432, 434, 437, 458
Toxicology, 119, 151, 155–156, 212, 228, 263, 274, 275, 326, 329, 334, 358, 361, 413, 522, 543, 545, 582, 601
Toxicoses, 263; *see also* Poisons and poisoning; Toxicology
Trace evidence, 122, 133, 149, 151, 156, 160, 167, 210, 212, 465, 522; *see also* Evidence
Tracks, trails, and signs of animals and humans, 468–470
Trade, 4, 32–35, 38, 40, 41, 43, 52, 53, 65, 70, 82, 86, 92, 96, 97, 99, 101–106, 110, 125, 233, 356, 367, 368, 377, 398, 425–429, 482, 483, 510, 563, 574, 600, 610–612; *see also* Pet trade
 in animals and animal parts, 3, 10, 32–35
Traditional medicines, 33, 42–43, 103, 110, 367, 375, 377
Traffic accidents, *see* Road-traffic (vehicular) accidents
Training, xi, xii, 52, 54, 56, 78, 108, 110, 124, 130, 144, 239, 247, 313, 314, 328, 386, 388, 419, 429, 439, 440, 460–461, 468, 470, 483, 486, 493–494, 502, 503, 505, 572, 611; *see also* Witnesses, coaching
Transillumination, in examination and detection of lesions, 215, 263, 357, 376
Translation, 114, 473; *see also* Languages and language differences
Translocations, *see* Releases, introductions/re-introductions and translocations
Transponders, *see* Microchip and microchipping (transponders)
Transportation injuries, 310, 448
Transportation of animals, 35, 53, 77, 92, 107, 233, 338, 448, 563; *see also* IATA Regulations; Movement of animals
Transposition of samples or evidence, 364–365; *see also* Evidence
Traps, nets, snares and trapping of animals, 220–221, 301; *see also* Snares
Trauma, 13, 14, 19, 31, 32, 212, 213, 218, 219, 223, 233, 253, 257, 259, 277, 284, 285, 287, 289–291, 295, 297–300, 307, 310, 319, 346, 358, 404, 406–407, 417, 436, 449, 453, 461, 462, 490, 502, 503, 509–512, 515, 518, 519, 583
 blunt and sharp, 406–409, 510
 to the central nervous system, 310
Trinidad and Tobago, xix, 45
Tsavo National Park, 32, 33
Turbine, wind, 188, 222, 300, 399, 458–464, 471
Turtle, sea, 67, 88, 189, 234, 296, 594

TWGDAM, *see* Technical Working Group on DNA Analysis Methods (TWGDAM)
Types of sample, 264, 329, 331, 337, 387, 476, 601

U

UFAW, *see* Universities Federation for Animal Welfare (UFAW)
Uganda, 20, 30, 32, 56, 58, 80, 101, 173, 219, 221, 313, 319–321, 398, 399, 455, 458
Ultrasonography, *see* Imaging
Umbrella, use, 4, 164, 167, 168, 603
Unexpected death, 287–288, 518
United Kingdom, 3, 13, 21–23, 28, 30, 34, 38, 39, 41, 42, 44, 45, 49, 53, 55–57, 61, 62, 64, 65, 68, 70, 74, 81, 82, 95, 96, 99–102, 107, 109, 110, 114–119, 122, 123, 126, 183, 185, 186, 204, 221, 257, 266, 273, 284, 291, 293, 295, 299, 302–303, 312, 319, 320, 323, 326, 327, 337, 367, 368, 378, 382, 385, 398, 401, 439, 472, 474, 476, 478, 483–485, 493, 497, 503, 504, 508, 511, 515, 517, 518, 529–537, 563–565, 600–605, 609–612; *see also* Appropriate institutions
United Nations, The (UN), 73, 81, 98, 128, 331, 564, 601, 602
United States (US) Fish & Wildlife Service (USFWS), 47, 116, 126, 147, 323, 368, 420, 478, 565, 566
United States of America, *see* Appropriate institutions
Universal primers, 65, 371–372, 374, 375, 379
Universities Federation for Animal Welfare (UFAW), 70, 71, 566
University of Cambridge, xviii, 565
University of Chester, 118, 503
University of Nairobi, 503
University of the West Indies (UWI), xviii, 108, 339
University of the Witwatersrand, xviii, 319
Unnecessary suffering, 204; *see also* Suffering; Welfare of animals
Urates, 209, 210, 212, 344, 470
Urine, examination, 55, 193, 209, 212, 255, 273
USFWS, *see* United States (US) Fish & Wildlife Service (USFWS)
UWI, *see* University of the West Indies (UWI)

V

Vagal inhibition, 288, 296, 519
Validity of evidence, *see* Evidence
Variable number tandem repeats (VNTRs), 369
Vehicular accidents, 21–22, 222, 300–301; *see also* Road-traffic (vehicular) accidents
Veterinary forensic medicine, history, 115–116, 439–441, 520
Veterinary medicine, 9, 163, 200, 204, 240, 297, 398, 399, 439, 487, 503, 505, 513, 520

Index

Veterinary surgeon (veterinarian), 13, 15, 32, 44, 50, 52–54, 56, 58, 65, 67, 72, 78, 79, 84, 107, 109, 115, 116, 119, 124, 147, 151, 164, 183, 195, 196, 202, 216, 217, 221, 223, 247, 248, 259, 284, 313, 314, 323, 331, 345, 357, 398, 432, 438, 439, 444, 446, 447, 454, 455, 482, 486–488, 498, 513, 520, 540, 564, 566, 600
Victims, of human action (wildlife), 31–32, 48, 210, 235
Vietnam, 34, 62, 219, 306
Violence, 6, 58, 173, 222, 223, 604; *see also* Abuse
Virchow, Rudolph, 321
Virunga Mountains, 170
VNTRs, *see* Variable number tandem repeats (VNTRs)
Voucher specimens, *see* Reference collections
Vultures, 8, 23, 38, 56, 77, 228, 377, 596;
    see also Birds of prey

W

Wallace, Alfred Russel, 116, 119, 224, 274
War, 32, 39–40, 73, 308, 321; *see also* Civil disturbance
Washings, *see* Cytology
Waste disposal, including litter, 86, 87
Water-testing, 275; *see also* Fish
Websites, 34, 64, 70, 94, 97, 100, 101, 109, 111, 126, 160, 164, 174, 337, 477, 485, 494, 507, 530, 564, 600; *see also* Internet
Weighing of animals and organs, 36, 169, 189–193, 225, 268–270, 420, 422, 522; *see also* Measuring; *see also* Morphometrics
Weka, 25, 299, 398, 452–455, 596
Welfare of animals, xiv, 35, 49, 52, 96, 194, 501
    in aquaria, 50, 51, 428
    assessment of, 48, 58
    attitudes towards, 43, 66, 165, 206
    of casualties, 40, 72, 79–81
    in circuses, 52, 217
    in entertainment, 48, 49, 51–52
    of exotic animal pets, 52–53, 194, 220, 361, 445, 447
    extremists, 247, 418, 513, 516
    laboratory, 20, 51, 209, 215, 326–329, 332–335, 363, 366, 381–394
    legal aspects, 98, 115, 599–602
    practical aspects, 51
    sanctuaries, 16, 31, 45, 68, 78, 81, 173, 453
    of zoo animals, 49
Wikipedia, 484
Wild animals, 1, 13, 98, 116, 165, 182, 238, 336, 368, 392, 397, 482, 501; *see also* Wildlife
    definition of, 1
    free-living (free-ranging), 1, 4, 5, 14, 17, 28, 48, 51, 53, 56, 58, 59, 61, 77, 78, 92, 96, 105, 118, 172, 185, 195–197, 205, 214, 224, 231, 233, 244, 295, 308, 357, 398, 513, 520, 599
    habituation, 78

Wildlife, 1, 11, 91, 113, 127, 151, 161, 182, 239, 326, 367, 381, 397, 465, 481, 501
    captive, 48, 51, 59, 96, 98, 205, 233, 268, 293, 336, 338, 343, 482
    casualties, 72, 79–81; *see also* Rehabilitation
    confiscations, 81–82, 211
    conservation, *see* Conservation
    definition of, 1
    law, 66, 94–102, 110, 111, 116, 505, 611
    meat (bush/wild), *see* Bushmeat
    pests, *see* Pests
    rehabilitation, 38, 40, 78–81, 96, 99, 293, 563
    release, introduction, reintroduction, 66–69
    research, 16, 17, 22, 25, 27, 37, 42, 46, 53–56, 75, 83, 84, 105, 109, 122, 124, 186, 239, 242, 318–322, 336, 337, 343, 344, 349–351, 356, 365, 372, 373, 378, 383, 384, 386, 388, 390, 397, 399, 401, 403–404, 410, 420, 437–439, 455, 478, 502, 504, 505, 573, 576, 577, 600, 601
    resources, use of, 75
    species for sustainable production, 75–76
    welfare, *see* Welfare
    zoos, 4, 29, 48–51, 103, 105, 114, 205, 217, 223, 285, 288, 293, 295, 420, 432, 435, 563, 575, 577
Wildlife crime, xii, 1, 27, 99, 116, 128, 151, 161, 218, 239, 351, 367, 382, 397, 468, 501, 520, 522, 563, 603
    confiscation, 81–82, 170, 442
    definition of, 3–4
    evolution of, 28–31
    human-induced injury, 31–32
    investigation, 10, 56, 65, 109, 117, 119, 140, 165, 367–379, 603–613
    poisoning, 36–39, 126, 291–293, 361, 612
    record-keeping, 10, 114, 122, 125, 140, 210, 246, 252
    reference material and databases, 124, 180, 264, 321, 372, 373, 478, 584
    types of, 27–31
Wildlife forensic investigation, aspects of, 1, 78, 125, 126, 204
Wildlife forensics, xi, xii, xiii, 1, 12, 91, 115, 136, 161, 182, 239, 326, 370, 381, 397, 468, 501
    definition of, 3–4
    methods, 9–10, 60
    position of, 118–119, 152, 276
    scope of, 5–8, 114
    significance of "one health" and "global health", 83–84
    special features, 119–121
    status of, 117
    vs. wildlife crime, 3, 4, 34, 63, 121
Wildlife investigation, types of, 11–89, 121, 228
Wildlife trade, 4, 33, 34, 41, 101–105, 233, 368, 398, 425–429

Wild plants, 1, 3, 6, 27, 28, 30, 67–69, 98, 165, 515, 520, 603
Wind farms, 48, 222, 300, 459, 461, 463; *see also* Turbine
Witness report, 530–537; *see also* Report
Witness statement, 489, 492, 513, 529–537
Wolf, 1, 20, 21, 332, 334, 356, 445, 595
Working as a team, 447
Worldwide fund for nature (WWF), xviii, 31, 62, 567
Wounds, 14, 22, 25–26, 46, 117, 118, 120, 139, 149, 151, 152, 155–158, 160, 166, 171, 193, 199, 211, 213–221, 223, 240, 254, 257, 259, 272, 276, 278, 282, 283, 285–308, 310, 316, 362, 364, 406, 407, 420, 430, 436, 438, 444, 453, 455, 502, 507–509, 511, 512, 515, 520, 543; *see also* Bite wounds; Trauma; Wounds
  assessment, 220
  investigation, 32, 215, 219–220, 304
  from weapons, 301–307, 520
Wrappings, 136, 246, 387, 475, 583
WWF, *see* Worldwide Fund for Nature (WWF)

X

X-rays, *see* Radiography and radiology

Z

Zebra, signs of, 469
Zhuang Zi, 506
Zooarchaeology, 403
Zoological nomenclature, 3, 93, 502, 591; *see also* International Commission on Zoological Nomenclature (ICZN)
Zoological Society of London, *see* Institute of Zoology (IoZ), London (London Zoo)
Zoonoses, 14, 23, 24, 52, 182, 212, 217, 253, 472, 569–572
Zoophilia, *see* Bestiality
Zoos, xi, xii, 4, 17, 23, 28–29, 48–51, 77, 82, 92, 98, 103, 105, 107, 114, 124, 179, 192, 205, 216–218, 223, 235, 266, 270, 271, 277, 280, 285, 287, 288, 290, 293, 295, 296, 311, 313–316, 328, 420, 432–435, 479, 504, 563, 575, 577; *see also* Private collections